Student Solutions Manual

Intermediate Algebra
Algebra Within Reach

SIXTH EDITION

Ron Larson
The Pennsylvania State University,
The Behrend College

Prepared by

Ron Larson
The Pennsylvania State University,
The Behrend College

Australia • Brazil • Japan • Korea • Mexico • Singapore • Spain • United Kingdom • United States

For product information and technology assistance, contact us at **Cengage Learning Customer & Sales Support, 1-800-354-9706**

For permission to use material from this text or product, submit all requests online at **www.cengage.com/permissions**
Further permissions questions can be emailed to
permissionrequest@cengage.com

ISBN-13: 978-1-285-41985-5
ISBN-10: 1-285-41985-5

Brooks/Cole
20 Channel Center Street
Boston, MA 02210
USA

Cengage Learning is a leading provider of customized learning solutions with office locations around the globe, including Singapore, the United Kingdom, Australia, Mexico, Brazil, and Japan. Locate your local office at: **www.cengage.com/global**

Cengage Learning products are represented in Canada by Nelson Education, Ltd.

To learn more about Brooks/Cole, visit **www.cengage.com/brookscole**

Purchase any of our products at your local college store or at our preferred online store **www.cengagebrain.com**

Printed in the United States of America
1 2 3 4 5 6 7 17 16 15 14 13

CONTENTS

Chapter 1 Fundamentals of Algebra **1**

Chapter 2 Linear Equations and Inequalities **17**

Chapter 3 Graphs and Functions **54**

Chapter 4 Systems of Equations and Inequalities **90**

Chapter 5 Polynomials and Factoring **145**

Chapter 6 Rational Expressions, Equations, and Functions **171**

Chapter 7 Radicals and Complex Numbers **208**

Chapter 8 Quadratic Equations, Functions, and Inequalities **243**

Chapter 9 Exponential and Logarithmic Functions **286**

Chapter 10 Conics **318**

Chapter 11 Sequences, Series, and the Binomial Theorem **350**

CHAPTER 1
Fundamentals of Algebra

Section 1.1 The Real Number System 2

Section 1.2 Operations with Real Numbers 3

Section 1.3 Properties of Real Numbers 6

Mid-Chapter Quiz 8

Section 1.4 Algebraic Expressions 9

Section 1.5 Constructing Algebraic Expressions 12

Review Exercises 14

Chapter Test 15

CHAPTER 1
Fundamentals of Algebra

Section 1.1 The Real Number System

1. $\left\{-6, -\sqrt{6}, -\frac{4}{3}, 0, \frac{5}{8}, 1, \sqrt{2}, 2, \pi, 6\right\}$

(a) Natural numbers: $\{1, 2, 6\}$

(b) Integers: $\{-6, 0, 1, 2, 6\}$

(c) Rational numbers: $\left\{-6, -\frac{4}{3}, 0, \frac{5}{8}, 1, 2, 6\right\}$

(d) Irrational numbers: $\left\{-\sqrt{6}, \sqrt{2}, \pi\right\}$

3. $\left\{-4.2, \sqrt{4}, -\frac{1}{9}, 0, \frac{3}{11}, \sqrt{11}, 5.\bar{5}, 5.543\right\}$

(a) Natural numbers: $\left\{\sqrt{4}\right\}$

(b) Integers: $\left\{\sqrt{4}, 0\right\}$

(c) Rational numbers: $\left\{4.2, \sqrt{4}, -\frac{1}{9}, 0, \frac{3}{11}, 5.\bar{5}, 5.543\right\}$

(d) Irrational numbers: $\left\{\sqrt{11}\right\}$

5. (a) The point representing the real number 3 lies between 2 and 4.

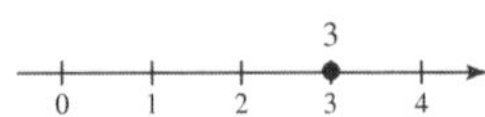

(b) The point representing the real number $\frac{5}{2}$ lies between 2 and 3.

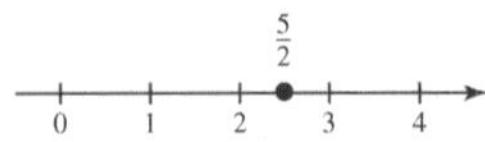

(c) The point representing the real number $-\frac{7}{2}$ lies between -4 and -3.

(d) The point representing the real number -5.2 lies between -6 and -5, but closer to -5.

−5.2

−6 −5 −4 −3 −2 −1 0

7. $\frac{4}{5} < 1$ because $\frac{4}{5}$ is to the left of 1 on the real number line.

9. $-5 < 2$ because -5 is to the left of 2 on the real number line.

11. $-5 < -2$ because -5 is to the left of -2 on the real number line.

13. $\frac{5}{8} > \frac{1}{2}$ because $\frac{5}{8}$ is to the right of $\frac{1}{2}$ on the real number line.

15. $-\frac{2}{3} > -\frac{10}{3}$ because $-\frac{2}{3}$ is to the right of $-\frac{10}{3}$ on the real number line.

17. Distance $= 10 - 4 = 6$

19. Distance $= 7 - (-12) = 7 + 12 = 19$

21. Distance $= 18 - (-32) = 18 + 32 = 50$

23. Distance $= 0 - (-8) = 0 + 8 = 8$

25. Distance $= 35 - 0 = 35$

27. Distance $= (-6) - (-9) = (-6) + 9 = 3$

29. $|10| = 10$

31. $|-225| = 225$

33. $-\left|-\frac{3}{4}\right| = -\frac{3}{4}$

35. $|-6| > |2|$ because $|-6| = 6$ and $|2| = 2$, and 6 is greater than 2.

37. $|47| > |-27|$ because $|47| = 47$ and $|-27| = 27$, and 47 is greater than 27.

39. *Label:* The weight on the elevator $= x$

Inequality: $x \le 2500$

41. *Label:* Contestant's weight $= x$

Inequality: $x > 200$

43. *Label:* Person's height $= x$

Inequality: $x \ge 52$

45. *Label:* Balance of checking account $= x$

Inequality: $200 \le x \le 700$

47. The number line shows $-2.5 < 2$ because -2.5 is to the left of -2.

49. The fractions are converted to decimals and plotted on a number line to determine the order.

51. $\{-5, -4, -3, -2, -1, 0, 1, 2, 3\}$

53. $\{5, 7, 9\}$

55. $a = -1, b = \frac{1}{2}$

$-1 < \frac{1}{2}$

57. $a = -\frac{9}{2}, b = -2,$

$-\frac{9}{2} < -2$

59. $-|-85| = -85$

61. $-|3.5| = -3.5$

63. $|-\pi| = \pi$

65. The opposite of -7 is 7.

The distance of both -7 and 7 from 0 is 7.

67. The opposite of 5 is -5 .

The distance of both -5 and 5 from 0 is 5.

69. The opposite of $-\frac{3}{5}$ is $\frac{3}{5}$.

The distance of both $-\frac{3}{5}$ and $\frac{3}{5}$ from 0 is $\frac{3}{5}$.

71. The opposite of $\frac{5}{3}$ is $-\frac{5}{3}$.

The distance of both $\frac{5}{3}$ and $-\frac{5}{3}$ from 0 is $\frac{5}{3}$.

73. The opposite of -4.25 is 4.25.

The distance of both -4.25 and 4.25 from 0 is 4.25.

75. $x < 0$

77. $u \geq 16$

79. You have more than 30 coins and fewer than 50 coins in a jar.

81. Because $|-4| = 4$ and $|4| = 4$, the two possible values of a are -4 and 4.

83. Because $|-2 - 3| = |-5| = 5$ and $|8 - 3| = |5| = 5$, the two possible values of a are -2 and 8.

85. Sample answers: $-3, -100, -\frac{4}{1}$

87. Sample answers: $\sqrt{2}, \pi, -3\sqrt{3}$

89. Sample answers: $\frac{3}{4}, 1\frac{1}{2}, 0.1\overline{6}$

91. Sample answers: $-\frac{1}{2}, \pi, -\sqrt{2}$

93. True. If a number can be written as ratio of two integers, it is rational. If not, the number is irrational.

95. $0.15 = \frac{15}{100}$ and $0.\overline{15} = 0.151515\ldots = \frac{15}{99}$

Section 1.2 Operations with Real Numbers

1. $-8 + 12 = +(12 - 8) = 4$

3. $13 + (-6) = +(13 - 6) = 7$

5. $-17 + (-6) = -(17 + 6) = -23$

7. $-8 - 12 = -8 + (-12) = -(8 + 12) = -20$

9. $13 - (-9) = 13 + 9 = 22$

11. $-15 - (-18) = -15 + 18 = +(18 - 15) = 3$

13. $\frac{3}{8} + \frac{7}{8} = \frac{3 + 7}{8} = \frac{10}{8} = \frac{5}{4}$

15. $\frac{3}{4} - \frac{1}{4} = \frac{3 - 1}{4} = \frac{2}{4} = \frac{1}{2}$

17.
$$\begin{aligned}\frac{3}{5} + \left(-\frac{1}{2}\right) &= \frac{3(2)}{5(2)} - \frac{1(5)}{2(5)}\\ &= \frac{6}{10} - \frac{5}{10}\\ &= \frac{6 - 5}{10}\\ &= \frac{1}{10}\end{aligned}$$

19. $\frac{5}{8} - \frac{1}{8} = \frac{5-1}{8} = \frac{4}{8} = \frac{4}{2 \cdot 4} = \frac{1}{2}$

21. $3\frac{1}{2} + 4\frac{3}{8} = \frac{7}{2} + \frac{35}{8}$

$= \frac{7(4)}{2(4)} + \frac{35}{8}$

$= \frac{28}{8} + \frac{35}{8}$

$= \frac{28 + 35}{8}$

$= \frac{63}{8}$

23. $10\frac{5}{8} - 6\frac{1}{4} = \frac{85}{8} - \frac{25}{4}$

$= \frac{85}{8} - \frac{25(2)}{4(2)}$

$= \frac{85}{8} - \frac{50}{8}$

$= \frac{85 - 50}{8}$

$= \frac{35}{8}$

25. $5(-6) = -30$

27. $(-8)(-6) = 48$

29. $2(4)(-5) = 8(-5) = -40$

31. $(-1)(12)(-3) = (-12)(-3) = 36$

33. $\frac{1}{2}\left(\frac{1}{6}\right) = \frac{1}{12}$

35. $-\frac{3}{2}\left(\frac{8}{5}\right) = -\frac{24}{10} = -\frac{12}{5}$

37. $\left(-\frac{5}{8}\right)\left(-\frac{4}{5}\right) = \frac{1}{2}$

39. $\frac{-18}{-3} = \frac{-6 \cdot -3}{-3} = 6$

41. $\frac{-48}{16} = \frac{-3 \cdot 16}{16} = -3$

43. $-10 \div 0$ is undefined.

Division by zero is undefined.

45. $-\frac{4}{5} \div \frac{8}{25} = -\frac{4}{5} \cdot \frac{25}{8} = \frac{(-4)(25)}{(5)(8)} = -\frac{5}{2}$

47. $\left(-\frac{1}{3}\right) \div \left(-\frac{5}{6}\right) = \left(-\frac{1}{3} \div -\frac{5}{6}\right)$

$= \left(\frac{-1}{3} \cdot \frac{-6}{5}\right) = \frac{(-1)(-6)}{(3)(5)} = \frac{2}{5}$

49. $4\frac{1}{8} \div 4\frac{1}{2} = \frac{33}{8} \div \frac{9}{2} = \frac{33}{8} \cdot \frac{2}{9} = \frac{(33)(2)}{(8)(9)} = \frac{11}{12}$

51. $-4\frac{1}{4} \div \left(-5\frac{5}{8}\right) = -\frac{17}{4} \div \left(-\frac{45}{8}\right)$

$= -\frac{17}{4} \cdot \left(-\frac{8}{45}\right) = \frac{17(8)}{4(45)} = \frac{34}{45}$

53. $(-7) \cdot (-7) \cdot (-7) = (-7)^3$

55. $\left(\frac{1}{4}\right) \cdot \left(\frac{1}{4}\right) \cdot \left(\frac{1}{4}\right) \cdot \left(\frac{1}{4}\right) = \left(\frac{1}{4}\right)^4$

57. $-(7 \cdot 7 \cdot 7) = -7^3$

59. $2^5 = (2)(2)(2)(2)(2) = 32$

61. $(-2)^4 = (-2)(-2)(-2)(-2) = 16$

63. $-4^3 = -(4)(4)(4) = -64$

65. $\left(\frac{4}{5}\right)^3 = \left(\frac{4}{5}\right)\left(\frac{4}{5}\right)\left(\frac{4}{5}\right) = \frac{64}{125}$

67. $\left(-\frac{1}{2}\right)^2 = \left(-\frac{1}{2}\right)\left(-\frac{1}{2}\right) = \frac{1}{4}$

69. $-\left(-\frac{1}{2}\right)^5 = -\left(-\frac{1}{2}\right)\left(-\frac{1}{2}\right)\left(-\frac{1}{2}\right)\left(-\frac{1}{2}\right)\left(-\frac{1}{2}\right) = -\left(-\frac{1}{32}\right) = \frac{1}{32}$

71. $(0.3)^3 = (0.3)(0.3)(0.3) = 0.027$

73. $5(-0.4)^3 = 5(-0.4)(-0.4)(-0.4) = 5(-0.064) = -0.32$

75. $16 - 6 - 10 = (16 - 6) - 10 = 10 - 10 = 0$

77. $24 - 5 \cdot 2^2 = 24 - 5 \cdot 4$

$= 24 - (5 \cdot 4) = 24 - 20 = 4$

79. $28 \div 4 + 3 \cdot 5 = (28 \div 4) + (3 \cdot 5)$

$= 7 + 15$

$= 22$

81. $14 - 2(8 - 4) = 14 - 2(4)$

$= 14 - 8$

$= 6$

83. $17 - 5(16 \div 4^2) = 17 - 5(16 \div 16)$
$$= 17 - 5(1)$$
$$= 17 - 5$$
$$= 12$$

85. $5^2 - 2[9 - (18 - 8)] = 25 - 2[9 - 10]$
$$= 25 - 2[-1]$$
$$= 25 + 2$$
$$= 27$$

87. $5^3 + |-14 + 4| = 125 + |-10|$
$$= 125 + 10$$
$$= 135$$

89. $\frac{6 + 8(3)}{7 - 12} = [6 + 8(3)] \div (7 - 12)$
$$= (6 + 24) \div (7 - 12)$$
$$= 30 \div (-5)$$
$$= -6$$

91. Apply the order of operations as follows: Parentheses, Exponents, Multiplication and Division, Addition and Subtraction.

93. To subtract the real number b from the real number a, add the opposite of b to a.

95. $85 - |-25| = 85 - 25 = 60$

97. $-(-11.325) + |34.625| = 11.325 + 34.625 = 45.95$

99. $-\left|-6\frac{7}{8}\right| - 8\frac{1}{4} = -6\frac{7}{8} - 8\frac{1}{4}$
$$= -\frac{55}{8} - \frac{33(2)}{4(2)}$$
$$= -\frac{55}{8} - \frac{66}{8}$$
$$= \frac{-55 - 66}{8}$$
$$= -\frac{121}{8}$$

101. $\frac{4^2 - 5}{11} - 7 = [(4^2 - 5) \div 11] - 7$
$$= [(16 - 5) \div 11] - 7$$
$$= (11 \div 11) - 7$$
$$= 1 - 7$$
$$= -6$$

103. $\frac{6 \cdot 2^2 - 12}{3^2 + 3} = [(6 \cdot 2^2) - 12] \div (3^2 + 3)$
$$= (24 - 12) \div (9 + 3)$$
$$= 12 \div 12$$
$$= 1$$

105. $\frac{3 + \frac{3}{4}}{\frac{1}{8}} = \left(3 + \frac{3}{4}\right) \div \frac{1}{8}$
$$= \left(\frac{12}{4} + \frac{3}{4}\right) \div \frac{1}{8}$$
$$= \frac{15}{4} \div \frac{1}{8}$$
$$= \frac{15}{4} \cdot \frac{8}{1}$$
$$= \frac{15(8)}{4} = 30$$

107. $\frac{1}{4} + \frac{2}{9} + \frac{1}{10} + x + \frac{1}{3} = 1$

So,
$$x = 1 - \left(\frac{1}{4} + \frac{2}{9} + \frac{1}{10} + \frac{1}{3}\right)$$
$$= 1 - \left(\frac{45}{180} + \frac{40}{180} + \frac{18}{180} + \frac{60}{180}\right)$$
$$= 1 - \left(\frac{45 + 40 + 18 + 60}{180}\right)$$
$$= 1 - \frac{163}{180}$$
$$= \frac{180}{180} - \frac{163}{180}$$
$$= \frac{17}{180}.$$

109. $2618.68 + $1236.45 − $25.62 − $455.00 − $125.00 − $715.95 = $2533.56

The balance at the end of the month was $2533.56.

111. $l = 14$ centimeters, $w = 8$ centimeters

$A = lw$

$A = 14 \cdot 8 = 112$ square centimeters

113. $b = 10$ feet, $h = 7$ feet

$A = \frac{1}{2}bh$

$A = \frac{1}{2} \cdot 10 \cdot 7 = 35$ square feet

115. True. A nonzero rational number is an integer divided by an integer. The reciprocal of such a number is still an integer divided by an integer, and so it is a rational number.

117. True. Any negative real number raised to an even numbered power will be a positive real number.

119. False. Division is not commutative.

121. If the numbers have like signs, the product or quotient is positive. If the numbers have unlike signs, the product or quotient is negative.

123. To add fractions with unlike denominators, you first find the least common denominator.

$$\frac{2}{3} + \frac{3}{2} = \frac{2(2)}{3(2)} + \frac{3(3)}{2(3)} = \frac{4}{6} + \frac{9}{6} = \frac{13}{6}$$

Section 1.3 Properties of Real Numbers

1. $18 - 18 = 0$

Additive Inverse Property

3. $\frac{1}{12} \cdot 12 = 1$

Multiplicative Inverse Property

5. $(8 - 5)(10) = 8 \cdot 10 - 5 \cdot 10$

Distributive Property

7. $15(-3) = (-3)15$

9. $5(6 + z) = 5 \cdot 6 + 5 \cdot z$

11. $x + 4 = 5$

$(x + 4) - 4 = 5 - 4$

Addition Property of Equality

13. $20(2 + 5) = 20 \cdot 2 + 20 \cdot 5$

15. $(x + 6)(-2) = x \cdot (-2) + 6 \cdot (-2)$ or $-2x - 12$

17. $-6(2y - 5) = -6(2y) + (-6)(-5)$ or $-12y + 30$

19. $7x + 2x = (7 + 2)x = 9x$

21. $\frac{7x}{8} - \frac{5x}{8} = (7 - 5)\left(\frac{x}{8}\right) = \frac{2x}{8} = \frac{x}{4}$

23.

$ac = bc,\ c \neq 0$	Write original equation.
$\frac{1}{c}(ac) = \frac{1}{c}(bc)$	Multiplication Property of Equality
$\frac{1}{c}(ca) = \frac{1}{c}(cb)$	Commutative Property of Multiplication
$\left(\frac{1}{c} \cdot c\right)a = \left(\frac{1}{c} \cdot c\right)b$	Associative Property of Multiplication
$1 \cdot a = 1 \cdot b$	Multiplicative Inverse Property
$a = b$	Multiplicative Identity Property

25.

$a = (a + b) + (-b)$	Write original equation.
$a = a + [b + (-b)]$	Associative Property of Addition
$a = a + 0$	Additive Inverse Property
$a = a$	Additive Identity Property

27. $13 + 12 = 12 + 13$

Commutative Property of Addition

29. $(-4 \cdot 10) \cdot 8 = -4(10 \cdot 8)$

Associative Property of Multiplication

31. $10(2x) = (10 \cdot 2)x$

Associative Property of Multiplication

33. $10x \cdot \frac{1}{10x} = 1$

Multiplicative Inverse Property

35. $2x - 2x = 0$

Additive Inverse Property

37. $3(2 + x) = 3 \cdot 2 + 3x$

Distributive Property

39. $(x + 1) - (x + 1) = 0$

Additive Inverse Property

41.

$x + 5 = 3$	Write original equation.
$(x + 5) + (-5) = 3 + (-5)$	Addition Property of Equality
$x + (5 + (-5)) = 3 - 5$	Associative Property of Addition
$x + 0 = -2$	Additive Inverse Property
$x = -2$	Additive Identity Property

43.

$2x - 5 = 6$	Write original equation.
$(2x - 5) + 5 = 6 + 5$	Addition Property of Equality
$2x + (-5 + 5) = 11$	Associative Property of Addition
$2x + 0 = 11$	Additive Inverse Property
$2x = 11$	Additive Identity Property
$\frac{1}{2}(2x) = \frac{1}{2}(11)$	Multiplication Property of Equality
$\left(\frac{1}{2} \cdot 2\right)x = \frac{11}{2}$	Associative Property of Multiplication
$1 \cdot x = \frac{11}{2}$	Multiplicative Inverse Property
$x = \frac{11}{2}$	Multiplicative Identity Property

45.

$-4x - 4 = 0$	Write original equation.
$-4x - 4 + 4 = 0 + 4$	Addition Property of Equality
$-4x + (-4 + 4) = 4$	Associative Property of Addition
$-4x + 0 = 4$	Additive Inverse Property
$-4x = 4$	Additive Identity Property
$-\frac{1}{4}(-4x) = -\frac{1}{4}(4)$	Multiplication Property of Equality
$\left[-\frac{1}{4} \cdot (-4)\right]x = -1$	Associative Property of Multiplication
$1 \cdot x = -1$	Multiplicative Inverse Property
$x = -1$	Multiplicative Identity Property

47. Every real number except zero has an additive inverse. The additive inverse (or opposite) of a number is the same distance from zero as that number. Because there is no distance from zero to zero, zero does not have an additive inverse.

49. No.

Subtraction: $8 - 2 = 6 \neq -6 = 2 - 8$

Division: $21 \div 7 = 3 \neq \frac{1}{3} = 7 \div 21$

51. $32 + (4 + y) = (32 + 4) + y$

53. $9(6M) = (9 \cdot 6)M$

55. $3(x + 5) = 3x + 15$

57. $-2(x + 8) = -2x - 16$

59. $16(1.75) = 16\left(2 - \frac{1}{4}\right) = 16(2) - 16\left(\frac{1}{4}\right) = 32 - 4 = 28$

61. $7(62) = 7(60 + 2) = 7(60) + 7(2) = 420 + 14 = 434$

63.
$$\begin{aligned} 9(6.98) &= 9(7 - 0.02) \\ &= 9(7) - 9(0.02) \\ &= 63 - 0.18 \\ &= 62.82 \end{aligned}$$

65. $a(b + c) = ab + ac$

67.
$$\begin{aligned} 4 + (x + 5) + (3x + 2) &= 4 + (5 + x) + (3x + 2) \\ &= (4 + 5) + x + (3x + 2) \\ &= 9 + (x + 3x) + 2 \\ &= 9 + 4x + 2 \\ &= 4x + 9 + 2 \\ &= 4x + 11 \end{aligned}$$

69. (a) $2(x + 6) + 2(2x) = 2x + 12 + 4x$

$= 2x + 4x + 12$

$= 6x + 12$

(b) $(x + 6)(2x) = x(2x) + 6(2x) = 2x^2 + 12x$

71. The additive inverse of a real number a is the number $-a$. The sum of a number and its additive inverse is the additive identity 0. For example, $8 + (-8) = 0$.

73. Given two real numbers a and b, the sum a plus b is the same as the sum b plus a.

75. Sample answer: $4 \odot 7 = 2 \cdot 4 + 7 = 8 + 7 = 15$

$7 \odot 4 = 2 \cdot 7 + 4 = 14 + 4 = 18$

Because $15 \neq 18$, $4 \odot 7 \neq 7 \odot 4$. So, the operation is not commutative.

$$\begin{aligned} 3 \odot (4 \odot 7) &= 3 \odot (2 \cdot 4 + 7) \\ &= 3 \odot 15 \\ &= 2 \cdot 3 + 15 \\ &= 6 + 15 \\ &= 21 \end{aligned}$$

$$\begin{aligned} (3 \odot 4) \odot 7 &= (2 \cdot 3 + 4) \odot 7 \\ &= 10 \odot 7 \\ &= 2 \cdot 10 + 7 \\ &= 20 + 7 \\ &= 27 \end{aligned}$$

Because $21 \neq 27$, $3 \odot (4 \odot 7) \neq (3 \odot 4) \odot 7$. So, the operation is not associative.

Mid-Chapter Quiz for Chapter 1

1. $-4.5 > -6$

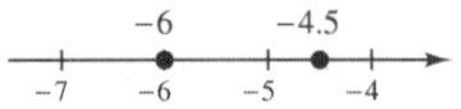

2. $\frac{3}{4} < \frac{3}{2}$

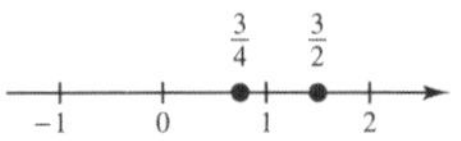

3. $|-15 - 7| = |-22| = 22$

4. $|-8.75 - (-2.25)| = |-8.75 + 2.25| = |-6.5| = 6.5$

5. $|-7.6| = 7.6$

6. $-|9.8| = -9.8$

7. $32 + (-18) = 14$

8. $-12 - (-17) = -12 + 17 = 5$

9. $\frac{3}{4} + \frac{7}{4} = \frac{3 + 7}{4} = \frac{10}{4} = \frac{5}{2}$

10. $\frac{2}{3} - \frac{1}{6} = \frac{4}{6} - \frac{1}{6} = \frac{4 - 1}{6} = \frac{3}{6} = \frac{1}{2}$

11. $(-3)(2)(-10) = (-6)(-10) = 60$

12. $\left(-\frac{4}{5}\right)\left(\frac{15}{32}\right) = \frac{(-4)(15)}{(5)(32)} = -\frac{3}{8}$

13. $\frac{7}{12} \div \frac{5}{6} = \frac{7}{12} \cdot \frac{6}{5} = \frac{(7)(6)}{(12)(5)} = \frac{7}{10}$

14. $\left(-\frac{3}{2}\right)^3 = \left(-\frac{3}{2}\right)\left(-\frac{3}{2}\right)\left(-\frac{3}{2}\right) = -\frac{27}{8}$

15. $3 - 2^2 + 25 \div 5 = 3 - 4 + 25 \div 5$

$= 3 - 4 + 5 = -1 + 5 = 4$

16. $\frac{18 - 2(3 + 4)}{6^2 - (12 \cdot 2 + 10)} = [18 - 2(3 + 4)] \div [6^2 - (12 \cdot 2 + 10)] = (18 - 14) \div (36 - 34) = 4 \div 2 = 2$

17. (a) $8(u - 5) = 8 \cdot u - 8 \cdot 5$ Distributive Property

(b) $10x - 10x = 0$ Additive Inverse Property

18. (a) $(7 + y) - z = 7 + (y - z)$ Associative Property of Addition

(b) $2x \cdot 1 = 2x$ Multiplicative Identity Property

19. $\$1406.98 - \$375.03 - \$59.20 - \$225.00 + \$320.45 = \$1068.20$

20. $\$45(2)(12)(8) = \$8640$

21.
$$\begin{aligned} 1 &= \tfrac{1}{3} + \tfrac{1}{4} + \tfrac{1}{8} + x \\ 1 - \tfrac{1}{3} - \tfrac{1}{4} - \tfrac{1}{8} &= x \\ \tfrac{24}{24} - \tfrac{8}{24} - \tfrac{6}{24} - \tfrac{3}{24} &= x \\ \tfrac{7}{24} &= x \end{aligned}$$

The sum of the parts of a circle is equal to 1.

Section 1.4 Algebraic Expressions

1. Terms: $10x$, 5

Coefficients: 10, 5

3. Terms: $-6x^2$, 12

Coefficients: -6, 12

5. Terms: $-3y^2$, $2y$, -8

Coefficients: -3, 2, -8

7. Terms: $-4a^3$, $1.2a$

Coefficients: -4, 1.2

9. Terms: $4x^2$, $-3y^2$, $-5x$, 21

Coefficients: 4, -3, -5, 21

11. Terms: $-5x^2y$, $2y^2$, xy

Coefficients: -5, 2, 1

13. Terms: $\frac{1}{4}x^2$, $-\frac{3}{8}x$, 5

Coefficients: $\frac{1}{4}$, $-\frac{3}{8}$, 5

15. $3x + 4x = (3 + 4)x = 7x$

17. $-2x^2 + 4x^2 = (-2 + 4)x^2 = 2x^2$

19. $7x - 11x = (7 - 11)x = -4x$

21. $9y - 5y + 4y = (9 - 5 + 4)y = 8y$

23.
$$\begin{aligned} 3x - 2y + 5x + 20y &= (3x + 5x) + (-2y + 20y) \\ &= (3 + 5)x + (-2 + 20)y \\ &= 8x + 18y \end{aligned}$$

25.
$$\begin{aligned} 7x^2 - 2x - x^2 &= 7x^2 - x^2 - 2x \\ &= (7 - 1)x^2 - 2x \\ &= 6x^2 - 2x \end{aligned}$$

27. $-3z^4 + 6z - z + 8 + z^4 - 4z^2 = (-3z^4 + z^4) - 4z^2 + (6z - z) + 8 = -2z^4 - 4z^2 + 5z + 8$

29. $x^2 + 2xy - 2x^2 + xy + y = x^2 - 2x^2 + 2xy + xy + y = (1 - 2)x^2 + (2 + 1)xy + y = -x^2 + 3xy + y$

31.
$$\begin{aligned} 10(x - 3) + 2x - 5 &= 10x - 30 + 2x - 5 \\ &= (10x + 2x) + (-30 - 5) \\ &= (10 + 2)x + (-30 - 5) \\ &= 12x - 35 \end{aligned}$$

33. $x - (5x + 9) = x - 5x - 9 = (1 - 5)x - 9 = -4x - 9$

35.
$$\begin{aligned} 5a - (4a - 3) &= 5a - 4a + 3 \\ &= (5 - 4)a + 3 \\ &= a + 3 \end{aligned}$$

37.
$$\begin{aligned} -3(3y - 1) + 2(y - 5) &= -9y + 3 + 2y - 10 \\ &= -9y + 2y + 3 - 10 \\ &= (-9 + 2)y - 7 \\ &= -7y - 7 \end{aligned}$$

39.
$$\begin{aligned} -3(y^2 - 2) + y^2(y + 3) &= -3y^2 + 6 + y^3 + 3y^2 \\ &= (-3 + 3)y^2 + 6 + y^3 \\ &= 6 + y^3 \end{aligned}$$

41.
$$\begin{aligned} x(x^2 + 3) - 3(x + 4) &= x^3 + 3x - 3x - 12 \\ &= x^3 + (3 - 3)x - 12 \\ &= x^3 - 12 \end{aligned}$$

43. $9a - [7 - 5(7a - 3)] = 9a - [7 - 35a + 15]$
$= 9a - [-35a + 22]$
$= 9a + 35a - 22$
$= (9 + 35)a - 22$
$= 44a - 22$

45. $3[2x - 4(x - 8)] = 3[2x - 4x + 32]$
$= 3[-2x + 32]$
$= -6x + 96$

47. $8x + 3x[10 - 4(3 - x)] = 8x + 3x[10 - 12 + 4x]$
$= 8x + 3x[-2 + 4x]$
$= 8x - 6x + 12x^2$
$= 2x + 12x^2$

49. $2[3(b - 5) - (b^2 + b + 3)] = 2[3b - 15 - b^2 - b - 3]$
$= 6b - 30 - 2b^2 - 2b - 6$
$= (-2b^2) + (6b - 2b) + (-30 - 6)$
$= -2b^2 + 4b - 36$

51. (a) When $x = \frac{2}{3}$, the expression $5 - 3x$ has a value of $5 - 3\left(\frac{2}{3}\right) = 5 - 2 = 3$.

(b) When $x = 5$, the expression $5 - 3x$ has a value of $5 - 3(5) = 5 - 15 = -10$.

53. (a) When $x = -1$, the expression $10 - 4x^2$ has a value of $10 - 4(-1)^2 = 10 - 4 = 6$.

(b) When $x = \frac{1}{2}$, the expression $10 - 4x^2$ has a value of $10 - 4\left(\frac{1}{2}\right)^2 = 10 - 1 = 9$.

55. (a) When $y = 2$, the expression $y^2 - y + 5$ has a value of $(2)^2 - 2 + 5 = 4 - 2 + 5 = 7$.

(b) When $y = -2$, the expression $y^2 - y + 5$ has a value of $(-2)^2 - (-2) + 5 = 4 + 2 + 5 = 11$.

57. (a) When $x = 0$, the expression $\frac{1}{x^2} + 3 = \frac{1}{0^2} + 3$ is undefined.

(b) When $x = 3$, the expression $\frac{1}{x^2} + 3$ has a value of $\frac{1}{3^2} + 3 = \frac{1}{9} + 3 = \frac{1}{9} + \frac{27}{9} = \frac{28}{9}$.

59. (a) When $x = 1$ and $y = 5$, the expression $3x + 2y$ has a value of $3(1) + 2(5) = 3 + 10 = 13$.

(b) When $x = -6$, and $y = -9$, the expression $3x + 2y$ has a value of $3(-6) + 2(-9) = -18 + -18 = -36$.

61. (a) When $x = 2$ and $y = -1$, the expression $x^2 - xy + y^2$ has a value of $(2)^2 - (2)(-1) + (-1)^2 = 4 + 2 + 1 = 7$.

(b) When $x = -3$ and $y = -2$, the expression $x^2 - xy + y^2$ has a value of $(-3)^2 - (-3)(-2) + (-2)^2 = 9 - 6 + 4 = 7$.

63. (a) When $x = 4$ and $y = 2$, the expression $\frac{x}{y^2 - x} = \frac{4}{2^2 - 4} = \frac{4}{4 - 4} = \frac{4}{0}$ is undefined.

(b) When $x = 3$ and $y = 3$, the expression $\frac{x}{y^2 - x}$ has a value of $\frac{3}{3^2 - 3} = \frac{3}{9 - 3} = \frac{3}{6} = \frac{1}{2}$.

65. (a) When $x = 2$ and $y = 5$, the expression $|y - x|$ has a value of $|5 - 2| = |3| = 3$.

(b) When $x = -2$ and $y = -2$, the expression $|y - x|$ has a value of $|-2 - (-2)| = |0| = 0$.

67. (a) When $r = 40$ and $t = 5\frac{1}{4}$, the expression rt has a value of $(40)\left(5\frac{1}{4}\right) = (40)\left(\frac{21}{4}\right) = 210$.

(b) When $r = 35$ and $t = 4$, the expression rt has a value of $(35)(4) = 140$.

69. The year 2005 corresponds to $t = 5$ in the model. The graph gives the sales of hunting equipment to be approximately \$3.5 billion, while the model provides $0.37(5) + 1.6 = \$3.45$ billion.

71. In an algebraic expression, terms are separated by addition whereas factors are separated by multiplication.

73. To combine like terms in an algebraic expression, first determine which terms are like terms. Then, add the coefficients of the like terms and attach the common variable factor.

For example: $5x^4 + \left(-2x^4\right) = \left[5 + (-2)\right]x^4 = 3x^4$

75. $4 - 3x = -3x + 4$ illustrates the Commutative Property of Addition.

77. $-5(2x) = (-5 \cdot 2)x$ illustrates the Associative Property of Multiplication.

79. $(5 - 2)x = 5x - 2x$ illustrates the Distributive Property.

81. $$\begin{aligned} 3(x + 2) - 5(x - 7) &= 3x + 6 - 5x + 35 \\ &= 3x - 5x + 6 + 35 \\ &= (3 - 5)x + (6 + 35) \\ &= -2x + 41 \end{aligned}$$

83. $$\begin{aligned} 2[x + 2(x + 7)] &= 2[x + 2x + 14] \\ &= 2[(1 + 2)x + 14] \\ &= 2(3x + 14) \\ &= 6x + 28 \end{aligned}$$

85. $$\begin{aligned} 2x - 3[x - (4 - x)] &= 2x - 3[x - 4 + x] \\ &= 2x - 3[x + x - 4] \\ &= 2x - 3[(1 + 1)x - 4] \\ &= 2x - 3(2x - 4) \\ &= 2x - 6x + 12 \\ &= (2 - 6)x + 12 \\ &= -4x + 12 \end{aligned}$$

87. (a) When $x = 3$, the expression $-3 + 4x$ has a value of $-3 + 4(3) = -3 + 12 = 9$.

(b) When $x = -2$, the expression $-3 + 4x$ has a value of $-3 + 4(-2) = -3 - 8 = -11$.

89. (a) When $a = 2$ and $b = -3$, the expression $b^2 - 4ab$ has a value of $(-3)^2 - 4(-3)(2) = 9 + 24 = 33$.

(b) When $a = 6$ and $b = -4$, the expression $b^2 - 4ab$ has a value of $(-4)^2 - 4(6)(-4) = 16 + 96 = 112$.

91. (a) When $x = 0$ and $y = 5$, the expression $\dfrac{-y}{x^2 + y^2}$ has a value of $\dfrac{-5}{0^2 + 5^2} = \dfrac{-5}{0 + 25} = \dfrac{-5}{25} = -\dfrac{1}{5}$.

(b) When $x = 1$ and $y = -3$, the expression $\dfrac{-y}{x^2 + y^2}$ has a value of $\dfrac{-(-3)}{1^2 + (-3)^2} = \dfrac{3}{1 + 9} = \dfrac{3}{10}$.

93. $lwh = 6(6)(7) = 252$

The volume is 252 cubic feet.

95. $lwh = 27(18)(8) = 3888$

The volume is 3888 cubic inches.

97. When $p = 11$, $n = 7$, $d = 0$, and $q = 3$, the expression $0.01p + 0.05n + 0.10d + 0.25q$ has a value of $0.01(11) + 0.05(7) + 0.10(0) + 0.25(3) = 0.11 + 0.35 + 0 + 0.75 = \$1.21$.

99. When $p = 43$, $n = 27$, $d = 17$, and $q = 15$, the expression $0.01p + 0.05n + 0.10d + 0.25q$ has a value of $0.01(43) + 0.05(27) + 0.10(17) + 0.25(15) = 0.43 + 1.35 + 1.70 + 3.75 = \$7.23$.

101. $A = \frac{1}{2}b(b - 3) = \frac{1}{2}b^2 - \frac{3}{2}b$

$$\begin{aligned} A &= \tfrac{1}{2}(15)(15 - 3) \\ &= \tfrac{1}{2}(15)(12) \\ &= 90 \end{aligned}$$

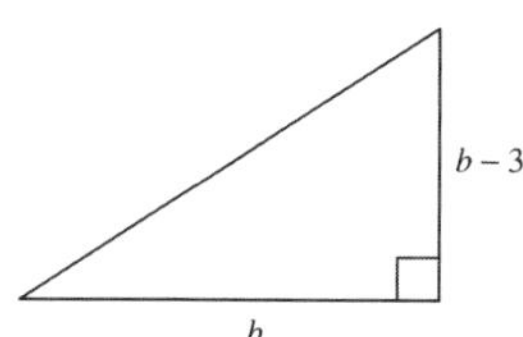

102. $A = h\left(\frac{5}{4}h + 10\right) = \frac{5}{4}h^2 + 10h$

$A = 12\left[\frac{5}{4}(12) + 10\right] = 12[15 + 10] = 12[25] = 300$

103. (a) Square $n = 4$: $\dfrac{4(4-3)}{2} = \dfrac{4(1)}{2} = 2$ diagonals

Pentagon $n = 5$: $\dfrac{5(5-3)}{2} = \dfrac{5(2)}{2} = 5$ diagonals

Hexagon $n = 6$: $\dfrac{6(6-3)}{2} = \dfrac{6(3)}{2} = 9$ diagonals

(b) For any natural number n, $n(n-3)$ is a product of an even and an odd natural number. So, the product is even and $\dfrac{n(n-3)}{2}$ is a natural number.

105. No. To evaluate the expression is to find the value of the expression for given values of the variables x, y, and z. Because z does not have a value, the expression cannot be evaluated.

107. Factors in a term are separated by multiplication. Because the coefficient and the variable(s) in a term are multiplied together, the coefficient and the variable(s) are all factors of the term. For example, the algebraic expression $3x + 2$ has the terms $3x$ and 2. The term $3x$ has the factors 3 and x.

109. Yes. To determine the value of x, set the expression equal to 100. Because $180 - 10x = 100$, the value of x is 8.

Section 1.5 Constructing Algebraic Expressions

1. The sum of 23 and a number n is translated into the algebraic expression $23 + n$.

3. The sum of 12 and twice a number n is translated into the algebraic expression $12 + 2n$.

5. Six less than a number n is translated into the algebraic expression $n - 6$.

7. Four times a number n minus 10 is translated into the algebraic expression $4n - 10$.

9. Half of a number n is translated into the algebraic expression $\frac{1}{2}n$.

11. The quotient of a number x and 6 is translated into the algebraic expression $\dfrac{x}{6}$.

13. Eight times the ratio of N and 5 is translated into the algebraic expression $8 \cdot \dfrac{N}{5}$.

15. The number c is quadrupled and the product is increased by 10 is translated into the algebraic expression $4c + 10$.

17. Thirty percent of the list price L is translated into the algebraic expression $0.30L$.

19. The sum of a number n and 5 divided by 10 is translated into the algebraic expression $\dfrac{n+5}{10}$.

21. The absolute value of the difference between a number and 8 is translated into the algebraic expression $|n - 8|$.

23. The product of 3 and the square of a number decreased by 4 is translated into the algebraic expression $3x^2 - 4$.

25. A verbal description of $t - 2$ is a number decreased by 2.

27. A verbal description of $y + 50$ is the sum of a number and 50 or a number increased by 50.

29. A verbal description of $2 - 3x$ is 2 decreased by 3 times a number.

31. A verbal description of $\dfrac{z}{2}$ is the ratio of a number and 2.

33. A verbal description of $\frac{4}{5}x$ is four-fifths of a number.

35. A verbal description of $8(x - 5)$ is 8 times the difference of a number and 5.

37. A verbal description of $\dfrac{x + 10}{3}$ is the sum of a number and 10, divided by 3.

39. A verbal description of $y^2 - 3$ is the square of a number, decreased by 3.

41. *Verbal Description:* The amount of money (in dollars) represented by n quarters

Label: n = number of nickels

Algebraic Description: $0.25n$ = amount of money (in dollars)

43. *Verbal Description:* The amount of money (in dollars) represented by m dimes

Label: m = number of dimes

Algebraic Description: $0.10m$ = amount of money (in dollars)

45. *Verbal Description:* The amount of money (in cents) represented by m nickels and n dimes

Labels: m = number of nickels

n = number of dimes

Algebraic Description: $5m + 10n$ = amount of money (in cents)

47. *Verbal Description:* The distance traveled in t hours at an average speed of 55 miles per hour

Label: t = number of hours

Algebraic Description: $55t$ = distance

49. *Verbal Description:* The time to travel 320 miles at an average speed of r miles per hour

Label: r = average speed

Algebraic Description: $\frac{320}{r}$ = time

51. *Verbal Description:* The amount of antifreeze in a cooling system containing y gallons of coolant that is 45% antifreeze

Label: y = number of gallons

Algebraic Description: $0.45y$ = amount of antifreeze

53. Perimeter $= 2(2w) + 2(w) = 4w + 2w = 6w$

Area $= 2w \cdot w = 2w^2$

55. Perimeter $= 3 + 2x + 6 + x + 3 + x = 4x + 12$

Area $= (x \cdot 3) + (3 \cdot 2x) = 3x + 6x = 9x$

57. *Verbal Description:* The sum of three consecutive integers, the first of which is n

Labels: n = first integer

$n + 1$ = second integer

$n + 2$ = third integer

Algebraic Description: $n + (n + 1) + (n + 2)$

$= 3n + 3 =$ sum

59. The phrase *reduced by* implies subtraction.

61. The known quantity is the cost per gallon. The unknown quantity is the number of gallons.

63. *Verbal Description:* The amount of wage tax due for a taxable income of I dollars that is taxed at the rate of 1.25%

Label: I = number of dollars

Algebraic Description: $0.0125I$ = amount of wage tax

65. *Verbal Description:* The sale price of a coat that has a list price of L dollars if the sale is a "20% off" sale

Label: L = number of dollars

Algebraic Description: $0.80L$ = sale price

67. *Verbal Description:* The total hourly wage for an employee when the base pay is $8.25 per hour plus 60 cents for each of q units produced per hour

Label: q = number of units produced

Algebraic Description: $8.25 + 0.60q$ = total hourly wage

69. *Verbal Description:* The sum of a number n and five times the number

Labels: n = the number

$5n$ = five times the number

Algebraic Description: $n + 5n = 6n =$ sum

71. *Verbal Description:* The sum of three consecutive odd integers, the first of which is $2n + 1$

Labels: $2n + 1$ = first odd integer

$2n + 3$ = second odd integer

$2n + 5$ = third odd integer

Algebraic Description:

$(2n + 1) + (2n + 3) + (2n + 5) = 6n + 9 =$ sum

73. *Verbal Description:* The product of two consecutive even integers, divided by 4

Labels: $2n$ = first even integer

$2n + 2$ = second even integer

Algebraic Description:

$$\frac{2n(2n+2)}{4} = \frac{4n(n+1)}{4} = n(n+1) = n^2 + n = \text{product}$$

75. Area $= \frac{1}{2}(\text{base})(\text{height}) = \frac{1}{2}(b)(0.75b) = 0.375b^2$

77. Area $=$ side $\cdot$ side $= s \cdot s = s^2$

79. Area $=$ length $\cdot$ width $= b(b - 50) = b^2 - 50b$

The unit measure for the area is square meters.

81.

n	0	1	2	3	4	5
$5n - 3$	-3	2	7	12	17	22
Differences	5	5	5	5	5	

The differences are constant.

83. The third row difference for the algebraic expression $an + b$ would be a.

85. $4x$ is the equivalent to (a) x multiplied by 4 and (c) the product of x and 4.

87. Using a specific case may make it easier to see the form of the expression for the general case.

Review Exercises for Chapter 1

1. (a) Natural numbers: $\left\{52, \sqrt{9}\right\}$

(b) Integers: $\left\{-4, 0, \sqrt{9}, 52\right\}$

(c) Rational numbers: $\left\{-4, -\frac{1}{8}, 0, \frac{3}{5}, \sqrt{9}, 52\right\}$

(d) Irrational numbers: $\left\{\sqrt{2}\right\}$

3. $\{1, 2, 3, 4, 5, 6\}$

5. (a) 4

(b) -3

(c) $\frac{3}{4}$

(d) -2.4

7. $-5 < 3$

9. $-\frac{8}{5} < -\frac{2}{5}$

11. $d = |11 - (-3)| = |11 + 3| = |14| = 14$

13. $d = |-13.5 - (-6.2)| = |-13.5 + 6.2| = |-7.3| = 7.3$

15. $|-5| = 5$

17. $-|-7.2| = -7.2$

19. $15 + (-4) = 11$

21. $-63.5 + 21.7 = -41.8$

23. $\frac{4}{21} + \frac{7}{21} = \frac{11}{21}$

25. $-\frac{5}{6} + 1 = -\frac{5}{6} + \frac{6}{6} = \frac{1}{6}$

27. $8\frac{3}{4} - 6\frac{5}{8} = \frac{35}{4} - \frac{53}{8} = \frac{70}{8} - \frac{53}{8} = \frac{17}{8}$

29. $-7 \cdot 4 = -28$

31. $120(-5)(7) = -4200$

33. $\frac{3}{8} \cdot \left(-\frac{2}{15}\right) = -\frac{6}{120} = -\frac{1}{20}$

35. $\frac{-56}{-4} = 14$

37. $-\frac{7}{15} \div -\frac{7}{30} = -\frac{7}{15} \cdot \frac{30}{-7} = 2$

39. $(-6)^4 = (-6)(-6)(-6)(-6) = 1296$

41. $-4^2 = (-1)(4)(4) = -16$

43. $120 - (5^2 \cdot 4) = 120 - (25 \cdot 4) = 120 - 100 = 20$

45.
$$\begin{aligned} 8 + 3\left[6^2 - 2(7-4)\right] &= 8 + 3[36 - 2(3)] \\ &= 8 + 3[36 - 6] \\ &= 8 + 3[30] \\ &= 8 + 90 = 98 \end{aligned}$$

47. $395 + 9(45) = 395 + 405 = 800$

You paid $800 for the entertainment system.

49. Additive Inverse Property

51. Distributive Property

53. Associative Property of Addition

55. Multiplicative Identity Property

57.

$-x + 2 = 4$	Write original equation.
$(x + 2) + (-2) = 4 + (-2)$	Addition Property of Equality
$x + [2 + (-2)] = 2$	Associative Property of Addition
$x + 0 = 2$	Additive Inverse Property
$x = 2$	Additive Identity Property

59. Terms: $4y^3, -y^2, \frac{17}{2}y$

Coefficients: $4, -1, \frac{17}{2}$

61. Terms: $-1.2x^3, \frac{1}{x}, 52$

Coefficients: $-1.2, 1, 52$

63. $6x + 3x = (6 + 3)x = 9x$

65. $3u - 2v + 7v - 3u = (3u - 3u) + (-2v + 7v) = 5v$

67. $5(x - 4) + 10 = 5x - 20 + 10 = 5x - 10$

69. $3[b + 5(b - a)] = 3[b + 5b - 5a]$
$= 3b + 15b - 15a$
$= 18b - 15a$

71. (a) When $x = 3$, the expression $x^2 - 2x - 3$ has a value of $(3)^2 - 2(3) - 3 = 9 - 6 - 3 = 0$.

(b) When $x = 0$, the expression $x^2 - 2x - 3$ has a value of $(0)^2 - 2(0) - 3 = 0 - 0 - 3 = -3$.

73. (a) When $x = 4$ and $y = -1$, the expression $y^2 - 2y + 4x$ has a value of $(-1)^2 - 2(-1) + 4(4) = 1 + 2 + 16 = 19$.

(b) When $x = -2$ and $y = 2$, the expression $y^2 - 2y + 4x$ has a value of $2^2 - 2(2) + 4(-2) = 4 - 4 - 8 = -8$.

75. Twelve decreased by twice the number n is translated into the algebraic expression $12 - 2n$.

77. The sum of the square of a number y and 49 is translated into the algebraic expression $y^2 + 49$.

79. The sum of twice a number and 7

81. The difference of a number and 5, all divided by 4

83. $0.18I$ = tax on I dollars at 18%

85. $l \cdot (l - 5) = l^2 - 5l$ = area of rectangle with length l and width $(l - 5)$

Chapter Test for Chapter 1

1. (a) $+\frac{5}{2} < |-3|$

(b) $-\frac{2}{3} > -\frac{3}{2}$

2. $d = |-4.4 - 6.9| = |-11.3| = 11.3$

3. $-14 + 9 - 15 = (-14 + 9) - 15 = -5 - 15 = -20$

4. $\frac{2}{3} + \left(-\frac{7}{6}\right) = \frac{4}{6} + \left(-\frac{7}{6}\right) = -\frac{3}{6} = -\frac{1}{2}$

5. $-2(225 - 150) = -2(75) = -150$

6. $(-3)(4)(-5) = (-12)(-5) = 60$

7. $\left(-\frac{7}{16}\right)\left(-\frac{8}{21}\right) = \frac{1}{6}$

8. $\frac{5}{18} \div \frac{15}{8} = \frac{5}{18} \cdot \frac{8}{15} = \frac{4}{27}$

9. $\left(-\frac{3}{5}\right)^3 = -\frac{27}{125}$

10. $\frac{4^2 - 6}{5} + 13 = \frac{16 - 6}{5} + 13 = \frac{10}{5} + 13 = 2 + 13 = 15$

11. (a) Associative Property of Multiplication

(b) Multiplicative Inverse Property

12. $-6(2x - 1) = -6(2x) + -6 \cdot (-1) = -12x + 6$

13. $3x^2 - 2x - 5x^2 + 7x - 1 = -2x^2 + 5x - 1$

14. $x(x + 2) - 2(x^2 + x - 13) = x^2 + 2x - 2x^2 - 2x + 26 = (x^2 - 2x^2) + (2x - 2x) + 26 = -x^2 + 26$

15. $a(5a - 4) - 2(2a^2 - 2a) = 5a^2 - 4a - 4a^2 + 4a = a^2$

16. $4t - [3t - (10t + 7)] = 4t - [3t - 10t - 7] = 4t - [-7t - 7] = 4t + 7t + 7 = 11t + 7$

17. Evaluating an expression is solving the expression when values are provided for its variables.

(a) When $x = -1$:

$$\begin{aligned} 7 + (x-3)^2 &= 7 + (-1-3)^2 \\ &= 7 + (-4)^2 \\ &= 7 + 16 \\ &= 23 \end{aligned}$$

(b) When $x = 3$:

$$\begin{aligned} 7 + (x-3)^2 &= 7 + (3-3)^2 \\ &= 7 + 0^2 \\ &= 7 + 0 \\ &= 7 \end{aligned}$$

18. *Verbal Model:* 17 · [Length of each piece] = [Total length]

Equation: $17 \cdot n = 102$

$n = 6$

Each piece should be 6 inches.

19. *Verbal Model:* [Volume of 1 cord] = [Length] · [Width] · [Height]

Equation: $V = 4 \cdot 4 \cdot 8$

$V = 128$ cubic feet

Verbal Model: [Volume of 5 cords] = [5] · [Volume of 1 cord]

Equation: $V = 5 \cdot 128 = 640$

There are 640 cubic feet in 5 cords of wood.

20. The product of a number n and 5 is decreased by 8 is translated into the algebraic expression $5n - 8$.

21. *Verbal Description:* The sum of two consecutive even integers, the first of which is $2n$

Labels: $2n$ = first even integer

$2n + 2$ = second even integer

Algebraic Description: $2n + (2n + 2) = 4n + 2$

22. Perimeter $= 2l + 2(0.6l) = 2l + 1.2l = 3.2l$

Area $= l(0.6l) = 0.6l^2$

When $l = 45$:

Perimeter $= 3.2(45) = 144$

Area $= 0.6(45)^2 = 1215$

CHAPTER 2
Linear Equations and Inequalities

Section 2.1 Linear Equations **18**

Section 2.2 Linear Equations and Problem Solving **22**

Section 2.3 Business and Scientific Problems **27**

Mid-Chapter Quiz **33**

Section 2.4 Linear Inequalities **36**

Section 2.5 Absolute Value Equations and Inequalities **41**

Review Exercises **43**

Chapter Test **51**

CHAPTER 2
Linear Equations and Inequalities

Section 2.1 Linear Equations

1. (a) $x = 0$

$$3(0) - 7 \stackrel{?}{=} 2$$
$$-7 \neq 2$$

Not a solution

(b) $x = 3$

$$3(3) - 7 \stackrel{?}{=} 2$$
$$9 - 7 = 2$$
$$2 = 2$$

Solution

3. (a) $x = 4$

$$4 + 8 \stackrel{?}{=} 3(4)$$
$$12 = 12$$

Solution

(b) $x = -4$

$$-4 + 8 \stackrel{?}{=} 3(-4)$$
$$4 \neq -12$$

Not a solution

5. (a) $x = -4$

$$\tfrac{1}{4}(-4) \stackrel{?}{=} 3$$
$$-1 \neq 3$$

Not a solution

(b) $x = 12$

$$\tfrac{1}{4}(12) \stackrel{?}{=} 3$$
$$3 = 3$$

Solution

7.

$3x + 15 = 0$	Original equation
$3x + 15 - 15 = 0 - 15$	Subtract 15 from each side.
$3x = -15$	Combine like terms.
$\frac{3x}{3} = \frac{-15}{3}$	Divide each side by 3.
$x = -5$	Simplify.

9.

$$4x = x + 10$$
$$4x - x = x - x + 10$$
$$3x = 10$$

Equivalent

11.

$$x + 5 = 12$$
$$2(x + 5) = 2(12)$$
$$2x + 10 = 24$$

Not equivalent

13.

$$3(4 - 2t) = 5$$
$$12 - 6t = 5$$

Equivalent

15.

$$2x - 7 = 3$$
$$2x - 7 + 7 = 3 + 7$$
$$2x = 10$$
$$\frac{2x}{2} = \frac{10}{2}$$
$$x = 5$$

Not equivalent

17.

$$x - 3 = 0$$
$$x - 3 + 3 = 0 + 3$$
$$x = 3$$

Check: $3 - 3 \stackrel{?}{=} 0$
$$0 = 0$$

19.

$$3x - 12 = 0$$
$$3x = 12$$
$$\frac{3x}{3} = \frac{12}{3}$$
$$x = 4$$

Check: $3(4) \stackrel{?}{=} 12$
$$12 = 12$$

21.

$$6x + 4 = 0$$
$$6x + 4 - 4 = 0 - 4$$
$$6x = -4$$
$$\frac{6x}{6} = \frac{-4}{6}$$
$$x = -\frac{4}{6}$$
$$x = -\frac{2}{3}$$

Check: $6\left(-\frac{2}{3}\right) + 4 \stackrel{?}{=} 0$
$$-4 + 4 \stackrel{?}{=} 0$$
$$0 = 0$$

23.
$$\begin{aligned} 3t + 8 &= -2 \\ 3t + 8 - 8 &= -2 - 8 \\ 3t &= -10 \\ \frac{3t}{3} &= \frac{-10}{3} \\ t &= -\frac{10}{3} \end{aligned}$$

Check:
$$\begin{aligned} 3\left(-\frac{10}{3}\right) + 8 &\stackrel{?}{=} -2 \\ -10 + 8 &\stackrel{?}{=} -2 \\ -2 &= -2 \end{aligned}$$

25.
$$\begin{aligned} 7 - 8x &= 13x \\ 7 - 8x + 8x &= 13x + 8x \\ 7 &= 21x \\ \frac{7}{21} &= \frac{21x}{21} \\ \frac{1}{3} &= x \end{aligned}$$

Check:
$$\begin{aligned} 7 - 8\left(\frac{1}{3}\right) &\stackrel{?}{=} 13\left(\frac{1}{3}\right) \\ 7 - \frac{8}{3} &\stackrel{?}{=} \frac{13}{3} \\ \frac{21}{3} - \frac{8}{3} &\stackrel{?}{=} \frac{13}{3} \\ \frac{13}{3} &= \frac{13}{3} \end{aligned}$$

27.
$$\begin{aligned} 3x - 1 &= 2x + 14 \\ 3x - 2x - 1 &= 2x + 14 - 2x \\ x - 1 &= 14 \\ x - 1 + 1 &= 14 + 1 \\ x &= 15 \end{aligned}$$

Check:
$$\begin{aligned} 3(15) - 1 &\stackrel{?}{=} 2(15) + 14 \\ 45 - 1 &\stackrel{?}{=} 30 + 14 \\ 44 &= 44 \end{aligned}$$

29.
$$\begin{aligned} 8(x - 8) &= 24 \\ 8x - 64 &= 24 \\ 8x - 64 + 64 &= 24 + 64 \\ 8x &= 88 \\ \frac{8x}{8} &= \frac{88}{8} \\ x &= 11 \end{aligned}$$

Check:
$$\begin{aligned} 8(11 - 8) &\stackrel{?}{=} 24 \\ 8(3) &\stackrel{?}{=} 24 \\ 24 &= 24 \end{aligned}$$

31.
$$\begin{aligned} 3(x - 4) &= 7x + 6 \\ 3x - 12 &= 7x + 6 \\ 3x - 7x - 12 &= 7x - 7x + 6 \\ -4x - 12 &= 6 \\ -4x - 12 + 12 &= 6 + 12 \\ -4x &= 18 \\ \frac{-4x}{-4} &= \frac{18}{-4} \\ x &= -\frac{9}{2} \end{aligned}$$

Check:
$$\begin{aligned} 3\left(-\frac{9}{2} - 4\right) &\stackrel{?}{=} 7\left(-\frac{9}{2}\right) + 6 \\ 3\left(-\frac{9}{2} - \frac{8}{2}\right) &\stackrel{?}{=} -\frac{63}{2} + \frac{12}{2} \\ 3\left(-\frac{17}{2}\right) &\stackrel{?}{=} -\frac{51}{2} \\ -\frac{51}{2} &= -\frac{51}{2} \end{aligned}$$

33.
$$\begin{aligned} t - \tfrac{2}{5} &= \tfrac{3}{2} \\ t - \tfrac{2}{5} + \tfrac{2}{5} &= \tfrac{3}{2} + \tfrac{2}{5} \\ t &= \tfrac{19}{10} \end{aligned}$$

Check:
$$\begin{aligned} \tfrac{19}{10} - \tfrac{2}{5} &\stackrel{?}{=} \tfrac{3}{2} \\ \tfrac{19}{10} - \tfrac{4}{10} &\stackrel{?}{=} \tfrac{3}{2} \\ \tfrac{15}{10} &\stackrel{?}{=} \tfrac{3}{2} \\ \tfrac{3}{2} &= \tfrac{3}{2} \end{aligned}$$

35.
$$\begin{aligned} \frac{t}{5} - \frac{t}{2} &= 1 \\ 10\left(\frac{t}{5} - \frac{t}{2}\right) &= (1)10 \\ 2t - 5t &= 10 \\ -3t &= 10 \\ \frac{-3t}{-3} &= \frac{10}{-3} \\ t &= -\frac{10}{3} \end{aligned}$$

Check:
$$\begin{aligned} \frac{-\frac{10}{3}}{5} - \frac{-\frac{10}{3}}{2} &\stackrel{?}{=} 1 \\ \frac{10}{-15} + \frac{10}{6} &\stackrel{?}{=} 1 \\ -\frac{2}{3} + \frac{5}{3} &\stackrel{?}{=} 1 \\ \frac{3}{3} &\stackrel{?}{=} 1 \\ 1 &= 1 \end{aligned}$$

37. $\frac{8x}{5} - \frac{x}{4} = -3$

$20\left(\frac{8x}{5} - \frac{x}{4}\right) = 20(-3)$

$32x - 5x = -60$

$27x = -60$

$\frac{27x}{27} = \frac{-60}{27}$

$x = -\frac{20}{9}$

Check: $\frac{8\left(-\frac{20}{9}\right)}{5} - \frac{\left(-\frac{20}{9}\right)}{4} \stackrel{?}{=} -3$

$\frac{\left(-\frac{160}{9}\right)}{5} - \frac{\left(-\frac{20}{9}\right)}{4} \stackrel{?}{=} -3$

$-\frac{32}{9} + \frac{5}{9} \stackrel{?}{=} -3$

$-\frac{27}{9} \stackrel{?}{=} -3$

$-3 = -3$

39. $0.3x + 1.5 = 8.4$

$10(0.3x + 1.5) = (8.4)10$

$3x + 15 = 84$

$3x + 15 - 15 = 84 - 15$

$3x = 69$

$\frac{3x}{3} = \frac{69}{3}$

$x = 23$

Check: $0.3(23) + 1.5 \stackrel{?}{=} 8.4$

$6.9 + 1.5 \stackrel{?}{=} 8.4$

$8.4 = 8.4$

41. $1.2(x - 3) = 10.8$

$1.2x - 3.6 = 10.8$

$10(1.2x - 3.6) = (10.8)10$

$12x - 36 = 108$

$12x - 36 + 36 = 108 + 36$

$12x = 144$

$\frac{12x}{12} = \frac{144}{12}$

$x = 12$

Check: $1.2(12 - 3) \stackrel{?}{=} 10.8$

$1.2(9) \stackrel{?}{=} 10.8$

$10.8 = 10.8$

43. $\frac{2}{3}(2x - 4) = \frac{1}{2}(x + 3) - 4$

$6\left[\frac{2}{3}(2x - 4)\right] = \left[\frac{1}{2}(x + 3) - 4\right]6$

$4(2x - 4) = 3(x + 3) - 24$

$8x - 16 = 3x + 9 - 24$

$8x - 16 = 3x - 15$

$8x - 3x - 16 = 3x - 3x - 15$

$5x - 16 = -15$

$5x - 16 + 16 = -15 + 16$

$5x = 1$

$\frac{5x}{5} = \frac{1}{5}$

$x = \frac{1}{5}$

Check: $\frac{2}{3}\left[2\left(\frac{1}{5}\right) - 4\right] \stackrel{?}{=} \frac{1}{2}\left(\frac{1}{5} + 3\right) - 4$

$\frac{2}{3}\left(\frac{2}{5} - \frac{20}{5}\right) \stackrel{?}{=} \frac{1}{2}\left(\frac{1}{5} + \frac{15}{5}\right) - 4$

$\frac{2}{3}\left(-\frac{18}{5}\right) \stackrel{?}{=} \frac{1}{2}\left(\frac{16}{5}\right) - 4$

$-\frac{12}{5} \stackrel{?}{=} \frac{8}{5} - \frac{20}{5}$

$-\frac{12}{5} = -\frac{12}{5}$

45. $4y - 3 = 4y$

$4y - 3 + 3 = 4y + 3$

$4y = 4y + 3$

$4y - 4y = 4y + 3 - 4y$

$0 = 3$

$0 \neq 3$

No solution

47. $4(2x - 3) = 8x - 12$

$8x - 12 = 8x - 12$

$8x - 12 - 8x = 8x - 12 - 8x$

$-12 = -12$

Infinitely many solutions

49. The fountain reaches its maximum height when the velocity of the stream of water is zero.

$$\begin{aligned} 0 &= 48 - 32t \\ 0 + 32t &= 48 - 32t + 32t \\ 32t &= 48 \\ \frac{32t}{32} &= \frac{48}{32} \\ t &= \frac{3}{2}\text{ seconds} = 1.5\text{ seconds} \end{aligned}$$

51. *Equivalent equations* are equations that have the same set of solutions.

53. Dividing by zero cannot be done because it is undefined.

55. $6(x + 3) = 6x + 3$

$6x + 18 \neq 6x + 3$

Contradiction

57.

$$\begin{aligned} \tfrac{2}{3}x + 4 &= \tfrac{1}{3}x + 12 \\ \tfrac{2}{3}x - \tfrac{1}{3}x + 4 &= \tfrac{1}{3}x - \tfrac{1}{3}x + 12 \\ \tfrac{1}{3}x + 4 &= 12 \\ \tfrac{1}{3}x + 4 - 4 &= 12 - 4 \\ \tfrac{1}{3}x &= 8 \\ 3\left(\tfrac{1}{3}x\right) &= 3(8) \\ x &= 24 \end{aligned}$$

Conditional equation

59.

$$\begin{aligned} 12(x + 3) &= 7(x + 3) \\ 12x + 36 &= 7x + 21 \\ 12x + 36 - 7x &= 7x + 21 - 7x \\ 5x + 36 &= 21 \\ 5x + 36 - 36 &= 21 - 36 \\ 5x &= -15 \\ \frac{5x}{5} &= \frac{-15}{5} \\ x &= -3 \end{aligned}$$

Check: $12\left[(-3) + 3\right] \stackrel{?}{=} 7\left[(-3) + 3\right]$

$$\begin{aligned} 12[0] &\stackrel{?}{=} 7[0] \\ 0 &= 0 \end{aligned}$$

61.

$$\begin{aligned} -9y - 4 &= -9y \\ -9y + 9y - 4 &= -9y + 9y \\ -4 &= 0 \\ -4 &\neq 0 \end{aligned}$$

No solution

63.

$$\begin{aligned} 7(x + 6) &= 3(2x + 14) + x \\ 7x + 42 &= 6x + 42 + x \\ 7x + 42 &= 7x + 42 \\ 7x + 42 - 7x &= 7x + 42 - 7x \\ 42 &= 42 \end{aligned}$$

Infinitely many solutions

65.

$$\begin{aligned} 38h + 162 &= 257 \\ 38h + 162 - 162 &= 257 - 162 \\ 38h &= 95 \\ \frac{38h}{38} &= \frac{95}{38} \\ h &= \frac{5}{2} = 2.5\text{ hours} \end{aligned}$$

So, the repair work took 2.5 hours.

67. (a)

t	1	1.5	2	3	4	5
Width	250	200	166.7	125	100	83.3
Length	250	300	333.4	375	400	416.5
Area	62,500	60,000	55,577.8	46,875	40,000	34,694.5

(b) Because the perimeter is fixed, as t increases the length increases and the width and area decrease. The maximum area occurs when the length and width are equal.

69. False. Multiplying both sides of an equation by zero does not yield an equivalent equation because this does not follow the Multiplication Property of Equality.

71. No. An identity in standard form would be written as $ax + b = 0$, with $a = b = 0$. But, the equation would not be in standard form.

73.
$$3x = 3x + 1$$
$$3x - 3x = 3x - 3x + 1$$
$$0 \neq 1$$

The equation is a contradiction. Let $x = 1$ pound, then 3 pounds $\neq$ 4 pounds.

75.
$$5w + 3 = 28$$
$$5w + 3 - 3 = 28 - 3$$
$$5w = 25$$
$$\frac{5w}{5} = \frac{25}{5}$$
$$w = 5$$

The equation is a conditional equation. Let w be the number of full (5-day) work weeks. So, 28 days are worked.

77. $\frac{5}{6} - \frac{2}{3} = \frac{5}{6} - \frac{4}{6} = \frac{1}{6}$

79. $-12 - (6 - 5) = -12 - 1 = -13$

81. $\frac{2(1)}{1 + 1} = \frac{2}{2} = 1$

$\frac{2(5)}{5 + 1} = \frac{10}{6} = \frac{5}{3}$

83. $|3(-1) - 7| = |-3 - 7| = |-10| = 10$

$|3(1) - 7| = |3 - 7| = |-4| = 4$

85. $\frac{n}{4}$

87. $\frac{1}{2}n - 5$

Section 2.2 Linear Equations and Problem Solving

1. *Verbal Model:* 26 · [Amount of each paycheck] + [Bonus] = [Income for year]

Labels: Amount of each paycheck $= x$

Bonus $= 2800$

Income for year $= 37{,}120$

Equation:
$$26x + 2800 = 37{,}120$$
$$26x = 34{,}320$$
$$x = 1320$$

Each paycheck will be \$1320.

3. Percent: 30%

Parts out of 100: 30

Decimal: 0.30

Fraction: $\frac{30}{100} = \frac{3}{10}$

5. Percent: 7.5%

Parts out of 100: 7.5

Decimal: 0.075

Fraction: $\frac{75}{1000} = \frac{3}{40}$

7. Percent: $66\frac{2}{3}\%$

Parts out of 100: $66\frac{2}{3}$

Decimal: 0.66. . .

Fraction: $\frac{2}{3}$

9. Percent: 100%

Parts out of 100: 100

Decimal: 1.00

Fraction: 1

11. *Verbal Model:* [Compared number] = [Percent] · [Base number]

Labels: Compared number $= a$

Percent $= p$

Base number $= b$

Equation:
$$a = p \cdot b$$
$$a = (0.35)(250)$$
$$a = 87.5$$

So, 35% of 250 is 87.5.

13. *Verbal Model:*

[Compared number] = [Percent (decimal form)] · [Base number]

Labels: Compared number $= a$

Percent $= 0.425$

Base number $= 816$

Equation:
$$a = 0.425(816)$$
$$a = 346.8$$

So, 346.8 is 42.5% of 816.

15. *Verbal Model:* $\boxed{\text{Compared number}} = \boxed{\text{Percent}} \cdot \boxed{\text{Base number}}$

Labels: Compared number $= a$

Percent $= p$

Base number $= b$

Equation:
$$a = p \cdot b$$
$$a = (0.125)(1024)$$
$$a = 128$$

So, 128 is 12.5% of 1024.

17. *Verbal Model:* $\boxed{\text{Compared number}} = \boxed{\text{Percent}} \cdot \boxed{\text{Base number}}$

Labels: Compared number $= a$

Percent $= p$

Base number $= b$

Equation:
$$a = p \cdot b$$
$$a = (0.004)(150{,}000)$$
$$a = 600$$

So, 600 is 0.4% of 150,000.

19. *Verbal Model:* $\boxed{\text{Compared number}} = \boxed{\text{Percent}} \cdot \boxed{\text{Base number}}$

Labels: Compared number $= a$

Percent $= p$

Base number $= b$

Equation:
$$a = p \cdot b$$
$$a = (2.50)(32)$$
$$a = 80$$

So, 80 is 250% of 32.

21. *Verbal Model:* $\boxed{\text{Compared number}} = \boxed{\text{Percent}} \cdot \boxed{\text{Base number}}$

Labels: Compared number $= a$

Percent $= p$

Base number $= b$

Equation:
$$a = p \cdot b$$
$$84 = (0.24)(b)$$
$$\frac{0}{0.24} = b$$
$$350 = b$$

So, 84 is 24% of 350.

23. *Verbal Model:* $\boxed{\text{Compared number}} = \boxed{\text{Percent}} \cdot \boxed{\text{Base number}}$

Labels: Compared number $= a$

Percent $= p$

Base number $= b$

Equation:
$$a = p \cdot b$$
$$42 = (1.2)(b)$$
$$\frac{42}{1.2} = b$$
$$35 = b$$

So, 42 is 120% of 35.

25. *Verbal Model:*

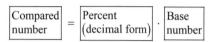

$\boxed{\text{Compared number}} = \boxed{\text{Percent (decimal form)}} \cdot \boxed{\text{Base number}}$

Labels: Compared number $= 22$

Percent $= 0.008$

Base number $= b$

Equation:
$$22 = 0.008b$$
$$\frac{22}{0.008} = b$$
$$2750 = b$$

So, 22 is 0.8% of 2750.

27. *Verbal Model:* $\boxed{\text{Compared number}} = \boxed{\text{Percent}} \cdot \boxed{\text{Base number}}$

Labels: Compared number $= a$

Percent $= p$

Base number $= b$

Equation:
$$a = p \cdot b$$
$$496 = (p)(800)$$
$$\frac{496}{800} = p$$
$$0.62 = p$$
$$p = 62\%$$

So, 496 is 62% of 800.

29. *Verbal Model:* $\boxed{\text{Compared number}} = \boxed{\text{Percent}} \cdot \boxed{\text{Base number}}$

Labels: Compared number $= a$

Percent $= p$

Base number $= b$

Equation:

$$\begin{aligned} a &= p \cdot b \\ 2.4 &= (p)(480) \\ \frac{2.4}{480} &= p \\ 0.005 &= p \\ p &= 0.5\% \end{aligned}$$

So, 2.4 is 0.5% of 480.

31. *Verbal Model:* $\boxed{\text{Compared number}} = \boxed{\text{Percent}} \cdot \boxed{\text{Base number}}$

Labels: Compared number $= a$

Percent $= p$

Base number $= b$

Equation:

$$\begin{aligned} a &= p \cdot b \\ 2100 &= (p)(1200) \\ \frac{2100}{1200} &= p \\ 175\% &= p \end{aligned}$$

So, 2100 is 175% of 1200.

33. *Verbal Model:* $\boxed{\text{Commission}} = \boxed{\text{Percent}} \cdot \boxed{\text{Price of home}}$

Labels: Commission $= a$

Percent $= p$

Price of home $= b$

Equation:

$$\begin{aligned} a &= p \cdot b \\ 12{,}250 &= p \cdot 175{,}000 \\ \frac{12{,}250}{175{,}000} &= \frac{p \cdot 175{,}000}{175{,}000} \\ 0.07 &= p \end{aligned}$$

So, it is a 7% commission.

35. $\dfrac{120 \text{ meters}}{180 \text{ meters}} = \dfrac{12}{18} = \dfrac{2}{3}$

37. $\dfrac{40 \text{ milliliters}}{1 \text{ liter}} = \dfrac{0.04 \text{ liter}}{1} = \dfrac{4}{100} = \dfrac{1}{25}$

39. (a) Unit price $= \dfrac{2.32}{14.5} = \$0.16$ per ounce

(b) Unit price $= \dfrac{0.99}{5.5} = \$0.18$ per ounce

The $14\frac{1}{2}$-ounce bag is a better buy.

41. (a) Unit price $= \dfrac{1.69}{4} = \$0.4225$ per ounce

(b) Unit price $= \dfrac{2.39}{6} = \$0.3983$ per ounce

The 6-ounce tube is a better buy.

43.

$$\begin{aligned} \frac{x}{6} &= \frac{2}{3} \\ 3 \cdot x &= 6 \cdot 2 \\ 3x &= 12 \\ x &= 4 \end{aligned}$$

45.

$$\begin{aligned} \frac{y}{36} &= \frac{6}{7} \\ 7 \cdot y &= 36 \cdot 6 \\ 7y &= 216 \\ y &= \frac{216}{7} \end{aligned}$$

47.

$$\begin{aligned} \frac{x}{7} &= \frac{4}{5.5} \\ 5.5 \cdot x &= 7 \cdot 4 \\ 5.5x &= 28 \\ x &= 5.\overline{09} \end{aligned}$$

49.

$$\begin{aligned} \frac{x}{6} &= \frac{2}{4} \\ 4 \cdot x &= 6 \cdot 2 \\ 4x &= 12 \\ x &= 3 \end{aligned}$$

51. *Verbal Model:*

$$\boxed{\frac{\text{Tax 1}}{\text{Assessed value 1}}} = \boxed{\frac{\text{Tax 2}}{\text{Assessed value 2}}}$$

Labels: Tax 1 $= x$

Assessed value 1 $= 160{,}000$

Tax 2 $= 1650$

Assessed value 2 $= 110{,}000$

Proportion:

$$\begin{aligned} \frac{x}{160{,}000} &= \frac{1650}{110{,}000} \\ 110{,}000 \cdot x &= 1650 \cdot 160{,}000 \\ 110{,}000x &= 264{,}000{,}000 \\ x &= 2400 \end{aligned}$$

So, the tax is \$2400.

53. $$\frac{5}{105} = \frac{x}{360}$$
$$360 \cdot 5 = 105 \cdot x$$
$$1800 = 105x$$
$$\frac{1800}{105} = x$$
$$17.1 \approx x$$

So, about 17.1 gallons of fuel are used.

55. The ratio of a to b is a/b if a and b have the same units.

Examples: Price earnings ratio, gear ratio

57. Not always. You need to cross-multiply to solve for an unknown quantity.

59. *Verbal Model:*

[Compared number] = [Percent] · [Base number]

Labels: Compared number = a

Percent = 0.20

Base number = 225

Equation:
$$a = p \cdot b$$
$$= (0.20)(225)$$
$$= 45$$

So, 45 is 20% of 225.

61. *Verbal Model:*

[Compared number] = [Percent] · [Base number]

Labels: Compared number = 13

Percent = 0.26

Base number = b

Equation:
$$a = p \cdot b$$
$$13 = 0.26 \cdot b$$
$$\frac{13}{0.26} = b$$
$$50 = b$$

So, 13 is 26% of 50.

63. *Verbal Model:*

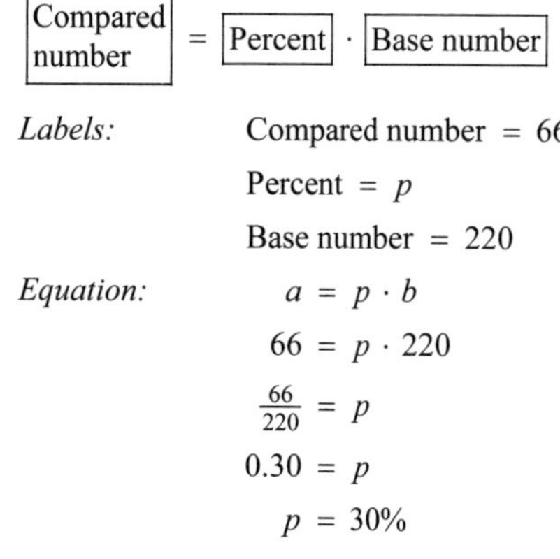

[Compared number] = [Percent] · [Base number]

Labels: Compared number = 66

Percent = p

Base number = 220

Equation:
$$a = p \cdot b$$
$$66 = p \cdot 220$$
$$\frac{66}{220} = p$$
$$0.30 = p$$
$$p = 30\%$$

So, 66 is 30% of 220.

65. $$\text{Unit price} = \frac{\text{Total price}}{\text{Total units}} = \frac{\$1.10}{20 \text{ ounces}} \approx \$0.06 \text{ per ounce}$$

67. $$\text{Unit price} = \frac{\text{Total price}}{\text{Total units}} = \frac{\$2.29}{20 \text{ ounces}} \approx \$0.11 \text{ per ounce}$$

69. $$\frac{y}{6} = \frac{y-2}{4}$$
$$4y = 6(y-2)$$
$$4y = 6y - 12$$
$$12 = 2y$$
$$6 = y$$

71. $$\frac{z-3}{3} = \frac{z+8}{12}$$
$$12(z-3) = 3(z+8)$$
$$\frac{12(z-3)}{3} = \frac{3(z+8)}{3}$$
$$4(z-3) = z+8$$
$$4z - 12 = z + 8$$
$$3z - 12 = 8$$
$$3z = 20$$
$$z = \frac{20}{3}$$

73. *Verbal Model:* [Number laid off] = [Percent] · [Number of employees]

Labels: Number laid off $= a$

Percent $= p$

Number of employees $= b$

Equation:

$$a = p \cdot b$$
$$25 = (p)(160)$$
$$\frac{25}{160} = \frac{(p)(160)}{160}$$
$$\frac{25}{160} = p$$
$$15.625 = p$$

So, 15.625% of the workforce was laid off.

75. *Verbal Model:* [Tip] = [Percent (decimal form)] · [Cost of meal]

Labels: Tip $= a$

Percent $= 0.15$

Cost of meal $= 32.60$

Equation:

$$a = 0.15(32.60)$$
$$a = 4.89$$

So, you should leave a \$4.89 tip.

77. *Verbal Model:* [Defective parts] = [Percent] · [Total parts]

Labels: Defective parts $= a$

Percent $= p$

Total parts $= b$

Equation:

$$a = p \cdot b$$
$$3 = (0.015)(b)$$
$$\frac{3}{0.015} = \frac{(0.015)(b)}{0.015}$$
$$\frac{3}{0.015} = b$$
$$200 = b$$

So, the sample contained 200 parts.

79. $\frac{\text{Tax}}{\text{Pay}} = \frac{\$12.50}{\$625} = \frac{125}{6250} = \frac{1}{50}$

81. *Verbal Model:*

$$\frac{\boxed{\text{Total defective units}}}{\boxed{\text{Total units}}} = \frac{\boxed{\text{Defective units}}}{\boxed{\text{Sample}}}$$

Labels: Total defective units $= x$

Total units $= 200{,}000$

Defective units $= 1$

Sample $= 75$

Proportion:

$$\frac{x}{200{,}000} = \frac{1}{75}$$
$$75 \cdot x = 200{,}000 \cdot 1$$
$$75x = 200{,}000$$
$$x = 2667$$

So, the expected number of defective units is 2667.

83.

$$\frac{h}{86} = \frac{6}{11}$$
$$11 \cdot h = 86 \cdot 6$$
$$11h = 516$$
$$h = \frac{516}{11}$$
$$h \approx 46.9$$

So, the height of the tree is about 46.9 feet.

85. To change percents to decimals divide by 100. To change decimals to percents multiply by 100.

Examples: $42\% = \frac{42}{100} = 0.42$

$0.38 = (0.38)(100)\% = 38\%$

87. Mathematical modeling is the use of mathematics to solve problems that occur in real-life situations. For examples review the real-life problems in the exercise set.

89. $-\frac{4}{15} \cdot \frac{15}{16} = -\frac{60}{240} = -\frac{1}{4}$

91. $(12 - 15)^3 = (-3)^3 = -27$

93. Commutative Property of Addition

95. Distributive Property

97.

$$2x - 5 = x + 9$$
$$2x = x + 14$$
$$x = 14$$

99.

$$2x + \tfrac{3}{2} = \tfrac{3}{2}$$
$$2x = 0$$
$$x = 0$$

101.

$$-0.35x = 70$$
$$x = -200$$

Section 2.3 Business and Scientific Problems

1. *Verbal Model:* [Selling price] = [Cost] + [Markup]

Labels: Selling price = 64.33
Cost = 45.97
Markup = x

Equation: $64.33 = 45.97 + x$
$x = 64.33 - 45.97$
$x = 18.36$

The markup is $18.36.

Verbal Model: [Markup] = [Markup rate] · [Cost]

Labels: Markup = 18.36
Markup rate = x
Cost = 45.97

Equation: $18.36 = x \cdot 45.97$
$$\frac{18.36}{45.97} = x$$
$0.40 \approx x$

The markup rate is 40%.

3. *Verbal Model:* [Selling price] = [Cost] + [Markup]

Labels: Selling price = 250.80
Cost = x
Markup = 98.80

Equation: $250.80 = x + 98.80$
$250.80 - 98.80 = x$
$152.00 = x$

The cost is $152.00.

Verbal Model: [Markup] = [Markup rate] · [Cost]

Labels: Markup = 98.80
Markup rate = x
Cost = 152.00

Equation: $98.80 = x \cdot 152.00$
$$\frac{98.80}{152.00} = x$$
$0.65 = x$

The markup rate is 65%.

5. *Verbal Model:* [Selling price] = [Cost] + [Markup]

Labels: Selling price = 26,922.50
Cost = x
Markup = 4672.50

Equation: $26{,}922.50 = x + 4672.50$
$26{,}922.50 - 4672.50 = x$
$22{,}250.00 = x$

The cost is $22,250.00.

Verbal Model: [Markup] = [Markup rate] · [Cost]

Labels: Markup = 4672.50
Markup rate = x
Cost = 22,250.00

Equation: $4672.50 = x \cdot 22{,}250.00$
$$\frac{4672.50}{22{,}250.00} = x$$
$0.21 = x$

The markup rate is 21%.

7. *Verbal Model:* [Markup] = [Markup rate] · [Cost]

Labels: Markup = x
Markup rate = 85.2%
Cost = 225.00

Equation: $x = 0.852 \cdot 225.00$
$x = 191.70$

The markup is $191.70.

Verbal Model: [Selling price] = [Cost] + [Markup]

Labels: Selling price = x
Cost = 225.00
Markup = 191.70

Equation: $x = 225.00 + 191.70$
$x = 416.70$

The selling price is $416.70.

9. *Verbal Model:* [Sale price] = [List price] − [Discount]

Labels: Sale price = 25.74

List price = 49.95

Discount = x

Equation:

$$25.74 = 49.95 - x$$
$$x = 49.95 - 25.74$$
$$x = 24.21$$

The discount is $24.21.

Verbal Model: [Discount] = [Discount rate] · [List price]

Labels: Discount = 24.21

Discount rate = x

List price = 49.95

Equation:

$$24.21 = x \cdot 49.95$$
$$\frac{24.21}{49.95} = x$$
$$0.485 \approx x$$

The discount rate is 48.5%.

11. *Verbal Model:* [Sale price] = [List price] − [Discount]

Labels: Sale price = x

List price = 300.00

Discount = 189.00

Equation:

$$x = 300.00 - 189.00$$
$$x = 111.00$$

The sale price is $111.00.

Verbal Model: [Discount] = [Discount rate] · [List price]

Labels: Discount = 189.00

Discount rate = x

List price = 300.00

Equation:

$$189.00 = x \cdot 300.00$$
$$\frac{189.00}{300.00} = x$$
$$0.63 = x$$

The discount rate is 63%.

13. *Verbal Model:* [Sale price] = [Percent] · [List price]

Labels: Sale price = 27.00

Percent = 0.60

List price = x

Equation:

$$27.00 = 0.60x$$
$$\frac{27.00}{0.60} = x$$
$$45.00 = x$$

The list price is $45.00.

Verbal Model: [Discount] = [Discount rate] · [List price]

Labels: Discount = x

Discount rate = 0.40

List price = 45.00

Equation:

$$x = 0.40 \cdot 45.00$$
$$x = 18.00$$

The discount is $18.00.

15. *Verbal Model:* [Sale price] = [List price] − [Discount]

Labels: Sale price = 831.96

List price = x

Discount = 323.54

Equation:

$$831.96 = x - 323.54$$
$$831.96 + 323.54 = x$$
$$1155.50 = x$$

The list price is $1155.50.

Verbal Model: [Discount] = [Discount rate] · [List price]

Labels: Discount = 323.54

Discount rate = p

List price = 1155.50

Equation:

$$323.54 = p \cdot 1155.50$$
$$\frac{323.54}{1155.50} = p$$
$$0.28 = p$$

The discount rate is 28%.

17. *Verbal Model:* $\boxed{\text{Sale price}} = \boxed{\text{List price}} - \boxed{\text{Discount}}$

Labels: Sale price $= 45$

List price $= 75$

Discount $= x$

Equation:

$$45 = 75 - x$$
$$45 - 75 = -x$$
$$30 = x$$

The discount is \$30.

19. *Verbal Model:* $\boxed{\text{Sale price}} = \boxed{\text{List price}} - \boxed{\text{Discount}}$

Labels: Sale price $= 16$

List price $= 20$

Discount $= x$

Equation:

$$16 = 20 - x$$
$$x = 20 - 16$$
$$x = 4$$

Verbal Model: $\boxed{\text{Discount}} = \boxed{\text{Discount rate}} \cdot \boxed{\text{List price}}$

Labels: Discount $= 4$

Discount rate $= x$

List price $= 20$

Equation:

$$4 = x \cdot 20$$
$$\frac{4}{20} = x$$
$$0.20 = x$$

The discount rate is 20%.

21. *Verbal Model:*

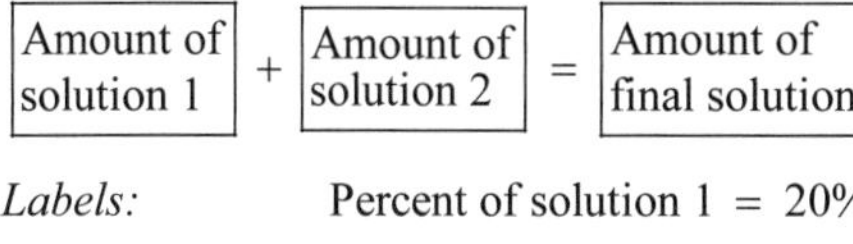

Labels: Percent of solution 1 $= 20\%$

Gallons of solution 1 $= x$

Percent of solution 2 $= 60\%$

Gallons of solution 2 $= 100 - x$

Percent of final solution $= 40\%$

Gallons of final solution $= 100$

Equation:

$$0.20x + 0.60(100 - x) = 0.40(100)$$
$$0.20x + 60 - 0.60x = 40$$
$$-0.40x = -20$$
$$x = 50$$
$$100 - x = 50$$

50 gallons of solution 1 and 50 gallons of solution 2 are needed.

23. *Verbal Model:*

$\boxed{\text{Amount of solution 1}} + \boxed{\text{Amount of solution 2}} = \boxed{\text{Amount of final solution}}$

Labels: Percent of solution 1 $= 15\%$

Quarts of solution 1 $= x$

Percent of solution 2 $= 60\%$

Quarts of solution 2 $= 24 - x$

Percent of final solution $= 45\%$

Quarts of final solution $= 24$

Equation:

$$0.15x + 0.60(24 - x) = 0.45(24)$$
$$0.15x + 14.4 - 0.60x = 10.8$$
$$-0.45x = -3.6$$
$$x = 8$$
$$24 - x = 16$$

8 quarts of solution 1 and 16 quarts of solution 2 are needed.

25. *Verbal Model:* $\boxed{\text{Cost of seed 1}} + \boxed{\text{Cost of seed 2}} = \boxed{\text{Cost of final seed mix}}$

Labels:

Number of pounds of seed 1 $= x$

Cost per pound of seed 1 $= 12$

Number of pounds of seed 2 $= 100 - x$

Cost per pound of seed 2 $= 20$

Number of pounds of final seed mix $= 100$

Cost per pound of final seed mix $= 14$

Equation:

$$12x + 20(100 - x) = 14(100)$$
$$12x + 2000 - 20x = 1400$$
$$-8x = -600$$
$$x = 75$$

75 pounds of seed 1 and 25 pounds of seed 2 are needed.

27. *Verbal Model:* $\boxed{\text{Distance}} = \boxed{\text{Rate}} \cdot \boxed{\text{Time}}$

Labels: Distance $= x$

Rates $= 480$ and 600

Time $= \frac{4}{3}$

Equation:

$$x = 480\left(\tfrac{4}{3}\right) + 600\left(\tfrac{4}{3}\right)$$
$$x = 1440$$

The planes are 1440 miles apart after $1\frac{1}{3}$ hours.

29. *Verbal Model:* $\boxed{\text{Distance}} = \boxed{\text{Rate}} \cdot \boxed{\text{Time}}$

Labels: Distance = 317

Rate for first part of trip = 58

Time for first part of trip = x

Rate for second part of trip = 52

Time for second part of trip = $5\frac{3}{4} - x$

Equation:

$$58x = 58 \cdot x \quad \text{(1st part of trip)}$$

$$52\left(5\tfrac{3}{4} - x\right) = 52 \cdot \left(5\tfrac{3}{4} - x\right) \quad \text{(2nd part of trip)}$$

$$317 = 58x + 52\left(5\tfrac{3}{4} - x\right)$$

$$317 = 58x + 299 - 52x$$

$$18 = 6x$$

$$3 = x$$

The first part of the trip took 3 hours, and the second part took $5\frac{3}{4} - 3 = 2\frac{3}{4}$ hours.

31. (a) Your rate = $\frac{1}{5}$ job per hour

Friend's rate = $\frac{1}{8}$ job per hour

(b) *Verbal Model:*

$\boxed{\text{Work done}} = \boxed{\text{Work done by first person}} + \boxed{\text{Work done by second person}}$

Labels: Work done = 1

Rate for you = $\frac{1}{5}$

Time for you = t

Rate for friend = $\frac{1}{8}$

Time for friend = t

Equation:

$$1 = \left(\tfrac{1}{5}\right)(t) + \left(\tfrac{1}{8}\right)(t)$$

$$1 = \left(\tfrac{1}{5} + \tfrac{1}{8}\right)t$$

$$1 = \left(\tfrac{13}{40}\right)t$$

$$\frac{1}{13/40} = t$$

$$3\tfrac{1}{13} = \tfrac{40}{13} = t$$

It will take $3\frac{1}{13}$ hours.

33. *Verbal Model:* $\boxed{\text{Interest}} = \boxed{\text{Principal}} \cdot \boxed{\text{Rate}} \cdot \boxed{\text{Time}}$

Labels: Interest = I

Principal = 5000

Rate = 6.5%

Time = 6

Equation:

$$I = (5000)(0.065)(6)$$

$$I = 1950$$

The interest is $1950.

35. *Verbal Model:* $\boxed{\text{Interest}} = \boxed{\text{Principal}} \cdot \boxed{\text{Rate}} \cdot \boxed{\text{Time}}$

Labels: Interest = 500

Principal = P

Rate = 7%

Time = 2

Equation:

$$500 = (P)(0.07)(2)$$

$$500 = P(0.14)$$

$$\frac{500}{0.14} = P$$

$$3571.43 \approx P$$

The principal required is $3571.43.

37. $E = IR$

$$\frac{E}{I} = R$$

39.

$$S = L - rL$$

$$S = L(1 - r)$$

$$\frac{S}{1 - r} = L$$

41.

$$h = 36t + \frac{1}{2}at^2 + 50$$

$$h - 36t - 50 = \frac{1}{2}at^2$$

$$2(h - 36t - 50) = at^2$$

$$2h - 72t - 100 = at^2$$

$$\frac{2h - 72t - 100}{t^2} = a$$

43.

$$S = 2\pi r^2 + 2\pi rh$$

$$S - 2\pi r^2 = 2\pi rh$$

$$\frac{\left(S - 2\pi r^2\right)}{2\pi r} = h$$

$$\frac{S}{2\pi r} - r = h$$

45. *Verbal Model:* Perimeter = 2 [Width] + 2 [Height]

Labels: Perimeter = 40

Height = x

Width = $x - 4$

Equation:

$$40 = 2x + 2(x - 4)$$
$$40 = 2x + 2x - 8$$
$$40 = 4x - 8$$
$$48 = 4x$$
$$12 = x$$

The height is 12 inches.

47. The formula for the volume V of a cube with side length s is $V = s^3$.

49. Perimeter is measured in linear units, such as inches, yards, and meters. Area is measured in square units, such as square feet, square centimeters, and square miles. Volume is measured in cubic units, such as cubic inches, cubic feet, and cubic meters.

51. *Verbal Model:* [Selling price] = [Cost] + [Markup]

Labels: Selling price = 157.14

Cost = 130.95

Markup = x

Equation:

$$157.14 = 130.95 + x$$
$$x = 157.14 - 130.95$$
$$x = 26.19$$

The markup is $26.19.

53. *Verbal Model:* [Markup] = [Markup rate] · [Cost]

Labels: Markup = 37.33

Markup rate = p

Cost = 46.67

Equation:

$$37.33 = p \cdot 46.67$$
$$\frac{37.33}{46.67} = p$$
$$0.80 \approx p$$

The markup rate is 80%.

55. *Verbal Model:* [Total cost] = [Cost per minute] · [Number of minutes]

Labels: Total cost = 0.70

Cost per minute = 0.02

Number of minutes = x

Equation:

$$0.70 = 0.02x$$
$$35 = x$$

The length of the call is 35 minutes.

Verbal Model: [Discount] = [Discount rate] · [List price]

Labels: Discount = x

Discount rate = 20%

List price = 0.70

Equation:

$$x = 0.20 \cdot 0.70$$
$$x = 0.14$$

Verbal Model: [Selling price] = [List price] − [Discount]

Labels: Selling price = x

List price = 0.70

Discount = 0.14

Equation:

$$x = 0.70 - 0.14$$
$$x = 0.56$$

The call would have cost $0.56.

57. *Verbal Model:* [Cost] + [Markup] = [Selling price]

Labels: Cost $= x$

Markup $= 0.10x$

Selling price $= 59.565$

Four tires cost: $3(\$79.42) = \$238.26$

Each tire costs: $\dfrac{\$238.26}{4} = \$59.565$

Equation:
$$x + 0.10x = 59.565$$
$$1.10x = 59.565$$
$$x = 54.15$$

The cost to the store for each tire is \$54.15.

59. *Verbal Model:* [Total sales] = [Adult sales] + [Children sales]

Labels: Total sales $= 2200$

Number of adult tickets $= 3x$

Price of adult tickets $= 6$

Number of children's tickets $= x$

Price of children's tickets $= 4$

Equation:
$$2200 = 6(3x) + 4x$$
$$2200 = 18x + 4x$$
$$2200 = 22x$$
$$100 = x$$

100 children's tickets are sold.

61. *Verbal Model:*

[Original antifreeze solution] − [Some antifreeze solution] + [Pure antifreeze] = [Final antifreeze solution]

Labels:

Number of gallons of original antifreeze $= 5$

Percent of antifreeze in original mix $= 40\%$

Number of gallons of antifreeze withdrawn $= x$

Number of gallons of pure antifreeze $= x$

Percent of pure antifreeze $= 100\%$

Number of gallons of final solution $= 5$

Percent of antifreeze in final solution $= 50\%$

Equation:
$$0.40(5) - 0.40x + 1.00x = 0.50(5)$$
$$2 - 0.40x + 1.00x = 2.5$$
$$0.60x = 0.5$$
$$x = \tfrac{5}{6}$$

$\frac{5}{6}$ gallon must be withdrawn and replaced.

63. *Verbal Model:* [Distance] = [Rate] · [Time]

Labels: Distance $= 5000$

Rate $= 17{,}500$

Time $= t$

Equation:
$$5000 = 17{,}500 \cdot t$$
$$\frac{5000}{17{,}500} = t$$
$$\frac{2}{7}\text{ hour} = t$$
$$17.14 \approx t$$

About 17.14 minutes are required.

65. *Verbal Model:* [Work done] = [Work done by smaller pump] + [Work done by larger pump]

Labels: Work done $= 1$

Rate of smaller pump $= \frac{1}{30}$

Rate of larger pump $= \frac{1}{15}$

Time for each pump $= t$

Equation:
$$1 = \left(\tfrac{1}{30}\right)(t) + \left(\tfrac{1}{15}\right)(t)$$
$$1 = \left(\tfrac{1}{30} + \tfrac{1}{15}\right)t$$
$$1 = \tfrac{3}{30}t$$
$$\frac{1}{3/30} = t$$
$$10 = t$$

It will take 10 minutes.

67. *Common formula:* $V = \pi r^2 h$

Equation: $V = \pi\left(3\tfrac{1}{2}\right)^2 12$

$V = 147\pi$

$V \approx 461.8$ cubic centimeters

69. The sale price of an item is the list price minus the discount rate times the list price.

71. No, it quadruples. The area of a square of side s is s^2. If the length of the sides is $2s$, the area is $(2s)^2 = 4s^2$.

73. (a) $21 + (-21) = 0$, so it is -21.

(b) $21\left(\frac{1}{21}\right) = 1$, so it is $\frac{1}{21}$.

75. (a) $-5x + 5x = 0$, so it is $5x$.

(b) $-5x\left(-\dfrac{1}{5x}\right) = 0$, so it is $-\dfrac{1}{5x}$.

77. $2x(x - 4) + 3 = 2x^2 - 8x + 3$

79. $x^2(x - 4) - 2x^2 = x^3 - 4x^2 - 2x^2 = x^3 - 6x^2$

81. $52 = 0.40x$

$130 = x$

83. $117 = p(900)$

$13\% = p$

Mid-Chapter Quiz for Chapter 2

1.
$$\begin{aligned} 4x - 8 &= 0 \\ 4x - 8 + 8 &= 0 + 8 \\ 4x &= 8 \\ \frac{4x}{4} &= \frac{8}{4} \\ x &= 2 \end{aligned}$$

Check:
$$\begin{aligned} 4(2) - 8 &\stackrel{?}{=} 0 \\ 8 - 8 &= 0 \\ 0 &= 0 \end{aligned}$$

2.
$$\begin{aligned} -3(z - 2) &= 0 \\ \frac{-3(z - 2)}{-3} &= \frac{0}{-3} \\ z - 2 &= 0 \\ z - 2 + 2 &= 0 + 2 \\ z &= 2 \end{aligned}$$

Check:
$$\begin{aligned} -3(2 - 2) &\stackrel{?}{=} 0 \\ -3(0) &\stackrel{?}{=} 0 \\ 0 &= 0 \end{aligned}$$

3.
$$\begin{aligned} 2(y + 3) &= 18 - 4y \\ 2y + 6 &= 18 - 4y \\ 2y + 4y + 6 &= 18 - 4y + 4y \\ 6y + 6 - 6 &= 18 - 6 \\ 6y &= 12 \\ \frac{6y}{6} &= \frac{12}{6} \\ y &= 2 \end{aligned}$$

Check:
$$\begin{aligned} 2(2 + 3) &\stackrel{?}{=} 18 - 4(2) \\ 2(5) &\stackrel{?}{=} 18 - 8 \\ 10 &= 10 \end{aligned}$$

4.
$$\begin{aligned} 5t + 7 &= 7(t + 1) - 2t \\ 5t + 7 &= 7t + 7 - 2t \\ 5t + 7 &= 5t + 7 \\ 7 &= 7 \end{aligned}$$

Infinitely Many

5.
$$\begin{aligned} \frac{1}{4}x + 6 &= \frac{3}{2}x - 1 \\ 4\left(\frac{1}{4}x + 6\right) &= 4\left(\frac{3}{2}x - 1\right) \\ x + 24 &= 6x - 4 \\ x - x + 24 &= 6x - 4 - x \\ 24 &= 5x - 4 \\ 24 + 4 &= 5x - 4 + 4 \\ 28 &= 5x \\ \frac{28}{5} &= \frac{5x}{5} \\ \frac{28}{5} &= x \end{aligned}$$

Check:
$$\begin{aligned} \frac{1}{4}\left(\frac{28}{5}\right) + 6 &\stackrel{?}{=} \frac{3}{2}\left(\frac{28}{5}\right) - 1 \\ \frac{7}{5} + \frac{30}{5} &\stackrel{?}{=} \frac{42}{5} - \frac{5}{5} \\ \frac{37}{5} &= \frac{37}{5} \end{aligned}$$

6.
$$\begin{aligned} \frac{2b}{5} + \frac{b}{2} &= 3 \\ \frac{4b}{10} + \frac{5b}{10} &= 3 \\ \frac{9b}{10} &= 3 \\ 9b &= 30 \\ b &= \frac{30}{9} = \frac{10}{3} \end{aligned}$$

Check:
$$\begin{aligned} \frac{2\left(\frac{10}{3}\right)}{5} + \frac{\left(\frac{10}{3}\right)}{2} &\stackrel{?}{=} 3 \\ \frac{\left(\frac{20}{3}\right)}{5} + \frac{\left(\frac{10}{3}\right)}{2} &\stackrel{?}{=} 3 \\ \frac{20}{15} + \frac{10}{6} &\stackrel{?}{=} 3 \\ \frac{4}{3} + \frac{5}{3} &\stackrel{?}{=} 3 \\ \frac{9}{3} &\stackrel{?}{=} 3 \\ 3 &= 3 \end{aligned}$$

7.
$$\frac{4-x}{5}+5=\frac{5}{2}$$
$$10\left(\frac{4-x}{5}+5\right)=10\left(\frac{5}{2}\right)$$
$$2(4-x)+50=25$$
$$8-2x+50=25$$
$$-2x+58=25$$
$$-2x+58-58=25-58$$
$$-2x=-33$$
$$\frac{-2x}{-2}=\frac{-33}{-2}$$
$$x=\frac{33}{2}$$

Check:
$$\frac{4-\frac{33}{2}}{5}+5\stackrel{?}{=}\frac{5}{2}$$
$$\frac{\frac{8}{2}-\frac{33}{2}}{5}+5\stackrel{?}{=}\frac{5}{2}$$
$$-\frac{25}{2}\cdot\frac{1}{5}+5\stackrel{?}{=}\frac{5}{2}$$
$$-\frac{5}{2}+\frac{10}{2}\stackrel{?}{=}\frac{5}{2}$$
$$\frac{5}{2}=\frac{5}{2}$$

8.
$$3x+\frac{11}{12}=\frac{5}{16}$$
$$3x+\frac{11}{12}-\frac{11}{12}=\frac{5}{16}-\frac{11}{12}$$
$$3x=\frac{15}{48}-\frac{44}{48}$$
$$3x=-\frac{29}{48}$$
$$\frac{3x}{3}=-\frac{29}{48}\div 3$$
$$x=-\frac{29}{48}\cdot\frac{1}{3}$$
$$x=-\frac{29}{144}$$

Check:
$$3\left(-\frac{29}{144}\right)+\frac{11}{12}\stackrel{?}{=}\frac{5}{16}$$
$$-\frac{29}{48}+\frac{44}{48}\stackrel{?}{=}\frac{5}{16}$$
$$\frac{15}{48}\stackrel{?}{=}\frac{5}{16}$$
$$\frac{5}{16}=\frac{5}{16}$$

9.
$$0.25x+6.2=4.45x+3.9$$
$$0.25x-0.25x+6.2=4.45x+3.9-0.25x$$
$$6.2=4.2x+3.9$$
$$6.2-3.9=4.2x-3.9$$
$$2.3=4.2x$$
$$\frac{2.3}{4.2}=\frac{4.2x}{4.2}$$
$$0.55\approx x$$

10.
$$0.42x+6=5.25x-0.80$$
$$0.42x+6-5.25x=5.25x-0.80-5.25x$$
$$-4.83x+6=-0.80$$
$$-4.83x+6-6=-0.80-6$$
$$-4.83x=-6.80$$
$$\frac{-4.83x}{-4.83}=\frac{-6.80}{-4.83}$$
$$x\approx 1.41$$

11. 0.45 is 45 hundredths, so $0.45=\frac{45}{100}$ which reduces to $\frac{9}{20}$ and because percent means hundredths, $0.45=45\%$.

12. *Verbal Model:* [Compared number] = [Percent] · [Base number]

Labels: Compared number $=a$

Percent $=p$

Base number $=b$

Equation:
$$a=p\cdot b$$
$$500=(2.50)(b)$$
$$\frac{500}{2.50}=b$$
$$200=b$$

500 is 250% of 200.

13.
$$\text{Unit price}=\frac{\text{Total price}}{\text{Total units}}$$
$$=\frac{\$4.85}{12\text{ ounces}}=\$0.40\text{ per ounce}$$

14. *Verbal Model:* $\frac{\text{Number defective}}{\text{Sample}}=\frac{\text{Total defective}}{\text{Shipment}}$

Labels: Number defective $=1$

Sample $=150$

Total defective $=x$

Shipment $=750{,}000$

Equation:
$$\frac{1}{150}=\frac{x}{750{,}000}$$
$$750{,}000=150x$$
$$5000=x$$

The expected number of defective units is 5000.

15. Store computer:

Verbal Model: $\boxed{\text{Discount}} = \boxed{\text{Discount rate}} \cdot \boxed{\text{List price}}$

Labels: Discount $= x$

Discount rate $= 0.25$

List price $= 1080$

Equation: $x = (0.25)(1080)$

$x = 270.00$

Verbal Model: $\boxed{\text{Selling price}} = \boxed{\text{List price}} - \boxed{\text{Discount}}$

Labels: Selling price $= x$

List price $= 1080$

Discount $= 270$

Equation: $x = 1080 - 270$

$x = 810$

Mail-order catalog computer:

Verbal Model: $\boxed{\text{Selling price}} = \boxed{\text{List price}} + \boxed{\text{Shipping}}$

Labels: Selling price $= x$

List price $= 799$

Shipping $= 14.95$

Equation: $x = 799 + 14.95$

$x = 813.95$

The store computer is the better buy.

16. *Verbal Model:* $\boxed{\text{Total wages}} = \boxed{\text{Regular wages}} + \boxed{\text{Overtime wages}}$

Labels: Total wages $= 616$

Regular wages $= 40(12.25)$

Overtime wages $= x(18)$

Number of hours $= x$

Equation:

$$616 = 40(12.25) + x(18)$$
$$616 = 490 + 18x$$
$$126 = 18x$$
$$7 = x$$

You worked 7 hours of overtime.

17. *Verbal Model:* $\boxed{\text{Amount of solution 1}} + \boxed{\text{Amount of solution 2}} = \boxed{\text{Amount of final solution}}$

Labels: Percent of solution 1 $= 25\%$

Gallons of solution 1 $= x$

Percent of solution 2 $= 50\%$

Gallons of solution 2 $= 50 - x$

Percent of final solution $= 30\%$

Gallons of final solution $= 50$

Equation:

$$0.25x + 0.50(50 - x) = 0.30(50)$$
$$0.25x + 25 - 0.50x = 15$$
$$25 - 0.25x = 15$$
$$-0.25x = -10$$
$$x = 40$$
$$50 - x = 10$$

40 gallons of solution 1 and 10 gallons of solution 2 are required.

18. *Verbal Model:* $\boxed{\text{Distance}} = \boxed{\text{Rate}} \cdot \boxed{\text{Time}}$

Labels: Distance $= 300$

Rate of first part $= 62$

Time for first part $= x$

Rate of second part $= 46$

Time for seond part $= 6 - x$

Equation:

$$300 = 62x + 46(6 - x)$$
$$300 = 62x + 276 - 46x$$
$$24 = 16x$$

1.5 hours $= x$ (first part of trip at 62 miles/hour)

4.5 hours $= 6 - x$ (second part of trip at 46 miles/hour)

19. *Verbal Model:* $\boxed{\text{Work done}} = \boxed{\text{Part time by you}} + \boxed{\text{Part done by friend}}$

Labels: Work done $= 1$

Time for each portion $= t$

Per hour work rate for you $= \frac{1}{3}$

Per hour work rate for friend $= \frac{1}{5}$

Equation:

$$1 = \left(\tfrac{1}{3}\right)t + \left(\tfrac{1}{5}\right)t$$
$$1 = \tfrac{8}{15}t$$
$$\tfrac{15}{8} = t$$

It will take $\frac{15}{8}$ hours, or 1.875 hours

20. Perimeter of square I $= 20$

$$4s = 20$$
$$s = 5$$

Perimeter of square II $= 32$

$$4s = 32$$
$$s = 8$$

Length of side of square III $= 5 + 8 = 13$

Area $= s^2 = 13^2 = 169$

The area of square III is 169 square inches.

Section 2.4 Linear Inequalities

1. The length of the interval $[-3, 5]$ is $|5 - (-3)| = 8$.

3. The length of the interval $(-9, 2]$ is $|2 - (-9)| = 11$.

5. The length of the interval $(-3, 0)$ is $|0 - (-3)| = 3$.

7.

9. $x > 3.5$

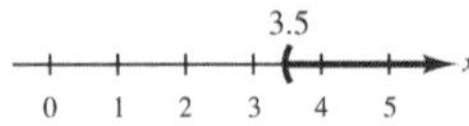

11. $x \geq \frac{1}{2}$

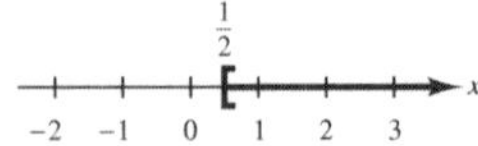

13. $-5 < x \leq 3$

15. $4 > x \geq 1$

17. $\frac{3}{2} \geq x > 0$

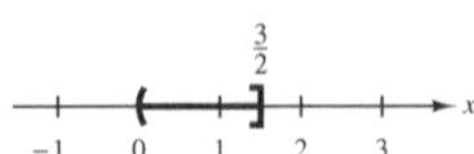

19. $3.5 < x \leq 4.5$

21.

$$3x - 2 < 12$$
$$3x - 2 + 2 < 12 + 2$$
$$3x < 14$$
$$\frac{3x}{3} < \frac{14}{3}$$
$$x < \frac{14}{3}$$

$$3x < 10$$
$$\frac{3x}{3} < \frac{10}{3}$$
$$x < \frac{10}{3}$$

The inequalities are not equivalent.

23.

$$7x - 6 \leq 3x + 12$$
$$7x - 6 + 6 \leq 3x + 12 + 6$$
$$7x \leq 3x + 18$$
$$7x - 3x \leq 3x + 18 - 3x$$
$$4x \leq 18$$
$$\frac{4x}{4} \leq \frac{18}{4}$$
$$x \leq \frac{9}{2}$$

$$4x \leq 18$$
$$\frac{4x}{4} \leq \frac{18}{4}$$
$$x \leq \frac{9}{2}$$

The inequalities are equivalent.

25.

$$x - 4 \geq 0$$
$$x - 4 + 4 \geq 0 + 4$$
$$x \geq 4$$

27.

$$x + 7 \leq 9$$
$$x + 7 - 7 \leq 9 - 7$$
$$x \leq 2$$

29.

$$2x < 8$$
$$\frac{2x}{2} < \frac{8}{2}$$
$$x < 4$$

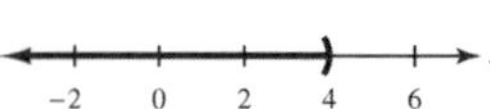

31.
$$-9x \geq 36$$
$$\frac{-9x}{-9} \leq \frac{36}{-9}$$
$$x \leq -4$$

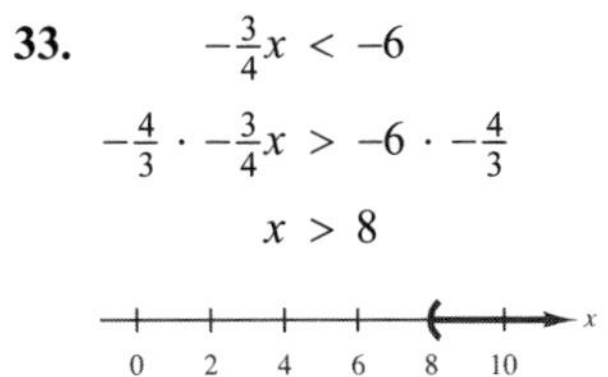

33.
$$-\tfrac{3}{4}x < -6$$
$$-\tfrac{4}{3} \cdot -\tfrac{3}{4}x > -6 \cdot -\tfrac{4}{3}$$
$$x > 8$$

35.
$$5 - x \leq -2$$
$$5 - x - 5 \leq -2 - 5$$
$$-x \leq -7$$
$$-1 \cdot x \geq -7 \cdot -1$$
$$x \geq 7$$

37.
$$2x - 5.3 > 9.8$$
$$2x - 5.3 + 5.3 > 9.8 + 5.3$$
$$2x > 15.1$$
$$\frac{2x}{2} > \frac{15.1}{2}$$
$$x > 7.55$$

39.
$$5 - 3x < 7$$
$$5 - 3x - 5 < 7 - 5$$
$$-3x < 2$$
$$\frac{-3x}{-3} > \frac{2}{-3}$$
$$x > -\frac{2}{3}$$

41.
$$3x - 11 > -x + 7$$
$$3x - 11 + x > -x + 7 + x$$
$$4x - 11 > 7$$
$$4x - 11 + 11 > 7 + 11$$
$$4x > 18$$
$$\frac{4x}{4} > \frac{18}{4}$$
$$x > \frac{9}{2}$$

43.
$$-3x + 7 < 8x - 13$$
$$-3x - 8x + 7 < 8x - 8x - 13$$
$$-11x + 7 < -13$$
$$-11x + 7 - 7 < -13 - 7$$
$$-11x < -20$$
$$\frac{-11x}{-11} > \frac{-20}{-11}$$
$$x > \frac{20}{11}$$

45.
$$-3(y + 10) \geq 4(y + 10)$$
$$-3y - 30 \geq 4y + 40$$
$$3y - 3y - 30 \geq 3y + 4y + 40$$
$$-30 \geq 7y + 40$$
$$-40 - 30 \geq 7y + 40 - 40$$
$$-70 \geq 7y$$
$$-10 \geq y$$

47.
$$0 < 2x - 5 < 9$$
$$0 + 5 < 2x - 5 + 5 < 9 + 5$$
$$5 < 2x < 14$$
$$\frac{5}{2} < \frac{2x}{2} < \frac{14}{2}$$
$$\frac{5}{2} < x < 7$$

49.

$$8 < 6 - 2x \le 12$$
$$8 - 6 < 6 - 6 - 2x \le 12 - 6$$
$$2 < -2x \le 6$$
$$\frac{2}{-2} > \frac{-2x}{-2} \ge \frac{6}{-2}$$
$$-1 > x \ge -3$$
$$-3 \le x < -1$$

51.

$$-1 < -0.2x < 1$$
$$\frac{-1}{-0.2} > \frac{-0.2x}{-0.2x} > \frac{1}{-0.2}$$
$$5 > x > -5$$
$$-5 < x < 5$$

53.

$$2x - 4 \le 4 \quad \text{and} \quad 2x + 8 > 6$$
$$2x - 4 + 4 \le 4 + 4 \quad \text{and} \quad 2x + 8 - 8 > 6 - 8$$
$$2x \le 8 \quad \text{and} \quad 2x > -2$$
$$\frac{2x}{2} \le \frac{8}{2} \quad \text{and} \quad \frac{2x}{2} > \frac{-2}{2}$$
$$x \le 4 \quad \text{and} \quad x > -1$$
$$-1 < x \le 4$$

55.

$$8 - 3x > 5 \quad \text{and} \quad x - 5 \ge 10$$
$$8 - 3x - 8 > 5 - 8 \qquad x - 5 + 5 \ge 10 + 5$$
$$-3x > -3 \qquad x \ge 15$$
$$\frac{-3x}{-3} < \frac{-3}{-3}$$
$$x < 1$$

There is no solution.

57.

$$7x + 11 < 3 + 4x \quad \text{or} \quad \frac{5}{2}x - 1 \ge 9 - \frac{3}{2}x$$
$$7x - 4x + 11 < 3 + 4x - 4x \qquad \frac{5}{2}x + \frac{3}{2}x - 1 \ge 9 - \frac{3}{2}x + \frac{3}{2}x$$
$$3x + 11 < 3 \qquad 4x - 1 \ge 9$$
$$3x + 11 - 11 < 3 - 11 \qquad 4x - 1 + 1 \ge 9 + 1$$
$$3x < -8 \qquad 4x \ge 10$$
$$\frac{3x}{3} < -\frac{8}{3} \qquad \frac{4x}{4} \ge \frac{10}{4}$$
$$x < -\frac{8}{3} \qquad x \ge \frac{5}{2}$$

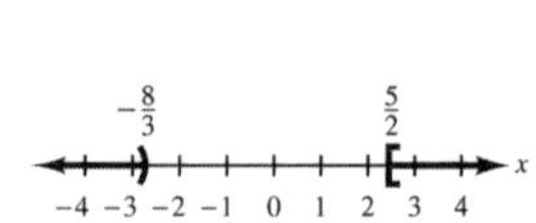

59.

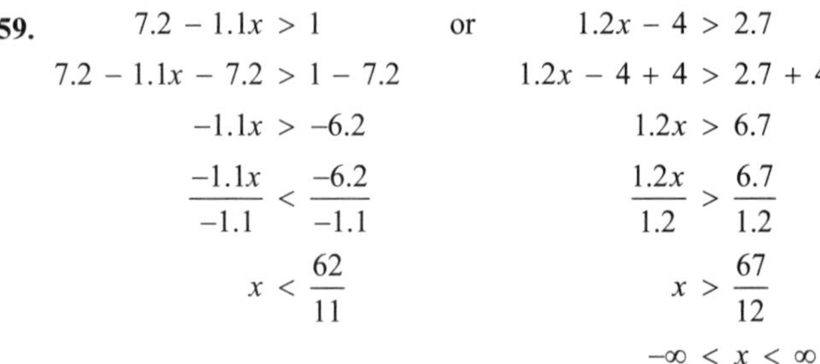

$$7.2 - 1.1x > 1 \quad \text{or} \quad 1.2x - 4 > 2.7$$
$$7.2 - 1.1x - 7.2 > 1 - 7.2 \qquad 1.2x - 4 + 4 > 2.7 + 4$$
$$-1.1x > -6.2 \qquad 1.2x > 6.7$$
$$\frac{-1.1x}{-1.1} < \frac{-6.2}{-1.1} \qquad \frac{1.2x}{1.2} > \frac{6.7}{1.2}$$
$$x < \frac{62}{11} \qquad x > \frac{67}{12}$$
$$-\infty < x < \infty$$

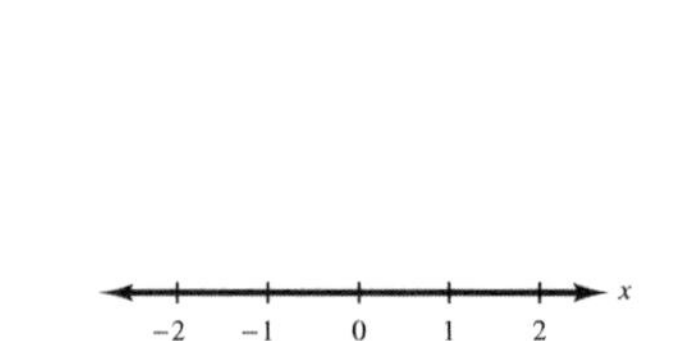

61. *Verbal Model:*

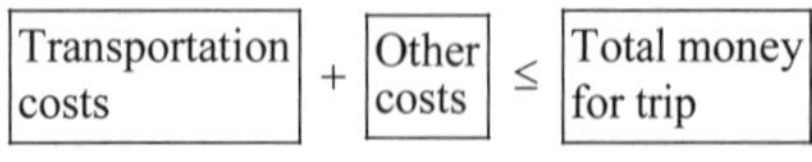

[Transportation costs] + [Other costs] ≤ [Total money for trip]

Labels: Transportation costs = 1900

Other costs = C

Total money = 4500

Inequality:

$$1900 + C \le 4500$$
$$1900 + C - 1900 \le 4500 - 1900$$
$$C \le 2600$$

C must be no more than $2600.

63. *Verbal Model:* 90 ≤ [Perimeter] ≤ [120]

Label: Perimeter = $2(x + 22)$

Inequality:

$$90 \le 2(x + 22) \le 120$$
$$\frac{90}{2} \le \frac{2(x + 22)}{2} \le \frac{120}{2}$$
$$45 \le x + 22 \le 60$$
$$45 - 22 \le x + 22 - 22 \le 60 - 22$$
$$23 \le x \le 38$$

65. The graph of $x < -2$ contains a parenthesis. A parenthesis is used when the endpoint is excluded.

67. Yes, because dividing by a number is the same as multiplying by its reciprocal.

69. Matches graph (a).

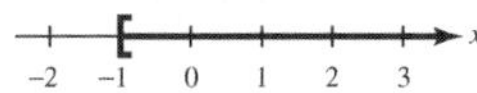

70. Matches graph (e).

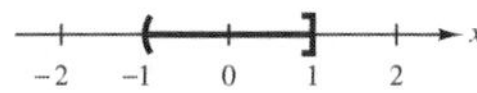

71. Matches graph (d).

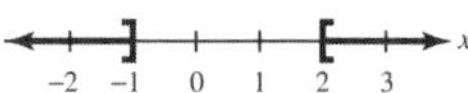

72. Matches graph (b).

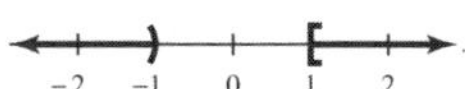

73. Matches graph (f).

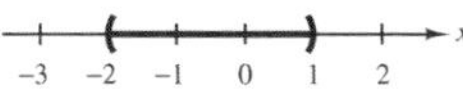

74. Matches graph (c).

75.
$$\begin{aligned}
\frac{x}{4} &> 2 - \frac{x}{2}\\
4\left(\frac{x}{4}\right) &> \left(2 - \frac{x}{2}\right)4\\
x &> 8 - 2x\\
x + 2x &> 8 - 2x + 2x\\
3x &> 8\\
\frac{3x}{3} &> \frac{8}{3}\\
x &> \frac{8}{3}
\end{aligned}$$

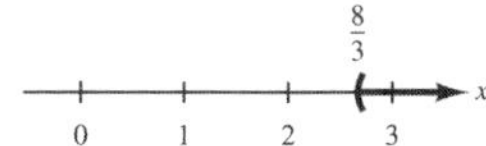

77.
$$\begin{aligned}
\frac{x-4}{3} + 3 &\le \frac{x}{8}\\
24\left(\frac{x-4}{3} + 3\right) &\le \left(\frac{x}{8}\right)24\\
8(x-4) + 72 &\le 3x\\
8x - 32 + 72 &\le 3x\\
8x + 40 &\le 3x\\
8x - 3x + 40 &\le 3x - 3x\\
5x + 40 &\le 0\\
5x + 40 - 40 &\le 0 - 40\\
5x &\le -40\\
\frac{5x}{5} &\le -\frac{40}{5}\\
x &\le -8
\end{aligned}$$

79.
$$\begin{aligned}
\frac{3x}{5} - 4 &< \frac{2x}{3} - 3\\
15\left(\frac{3x}{5} - 4\right) &< \left(\frac{2x}{3} - 3\right)15\\
9x - 60 &< 10x - 45\\
9x - 10x - 60 &< 10x - 45 - 10x\\
-x - 60 &< -45\\
-x - 60 + 60 &< -45 + 60\\
-x &< 15\\
x &> -15
\end{aligned}$$

81.
$$\begin{aligned}
-4 \le 2 - 3(x+2) &< 11\\
-4 \le 2 - 3x - 6 &< 11\\
-4 \le -4 - 3x &< 11\\
-4 + 4 \le -4 - 3x + 4 &< 11 + 4\\
0 \le -3x &< 15\\
\frac{0}{-3} \ge \frac{-3x}{-3} &> \frac{15}{-3}\\
0 \ge x &> -5
\end{aligned}$$

83. $-3 < \frac{2x-3}{2} < 3$

$-6 < 2x - 3 < 6$

$-6 + 3 < 2x - 3 + 3 < 6 + 3$

$-3 < 2x < 9$

$\frac{-3}{2} < \frac{2x}{2} < \frac{9}{2}$

$-\frac{3}{2} < x < \frac{9}{2}$

85. $1 > \frac{x-4}{-3} > -2$

$-3 < x - 4 < 6$

$-3 + 4 < x - 4 + 4 < 6 + 4$

$1 < x < 10$

87. *Verbal Model:*

[Temp in Miami] > [Temp in Washington] > [Temp in New York]

The average temperature in Miami, therefore, is greater than (>) the average temperature in New York.

89. *Verbal Model:* [Operating cost] < [$12,000]

Label: Operating cost $= 0.35m + 2900$

Inequality:

$0.35m + 2900 < 12{,}000$

$0.35m + 2900 - 2900 < 12{,}000 - 2900$

$0.35m < 9100$

$\frac{0.35m}{0.35} < \frac{9100}{0.35}$

$m < 26{,}000$

The maximum number of miles is 26,000.

91. *Verbal Model:* [Second plan] > [First plan]

Labels: First plan: $12.50 per hour

Second plan: $8 + $0.75*n* per hour where *n* represents the number of units produced.

Inequality: $8 + 0.75n > 12.5$

$0.75n > 4.5$

$n > 6$

If more than 6 units are produced per hour, the second payment plan yields the greater hourly wage.

93. The multiplication and division properties differ. The inequality symbol is reversed if both sides of the inequality are multiplied or divided by a negative real number.

95. The solution set of a linear inequality is a bounded interval if all x-values are contained by two endpoints of its graph. The solution set of a linear inequality is an unbounded interval if it is not bounded.

97. $a < x < b$; A double inequality is always bounded.

99. $x > a$ or $x < b$; The solution set includes all values on the real number line.

101. $|4| \stackrel{?}{=} |-5|$

$4 < 5$

103. $|-7| \stackrel{?}{=} |7|$

$7 = 7$

105. $3(6) \stackrel{?}{=} 27$ $\quad$ $3(9) \stackrel{?}{=} 27$

$18 \neq 27$ $\quad$ $27 = 27$

Not a solution $\quad$ Solution

107. $7(2) - 5 \stackrel{?}{=} 7 + 2$ $\quad$ $7(6) - 5 \stackrel{?}{=} 7 + 6$

$14 - 5 \stackrel{?}{=} 9$ $\quad$ $42 - 5 \stackrel{?}{=} 14$

$9 = 9$ $\quad$ $37 \neq 14$

Solution $\quad$ Not a solution

109. $2x - 17 = 0$

$2x - 17 + 17 = 0 + 17$

$2x = 17$

$\frac{2x}{2} = \frac{17}{2}$

$x = \frac{17}{2}$

111. $32x = -8$

$\frac{32x}{32} = \frac{-8}{32}$

$x = -\frac{1}{4}$

Section 2.5 Absolute Value Equations and Inequalities

1. $|4x + 5| = 10, x = -3$

$$|4(-3) + 5| \stackrel{?}{=} 10$$
$$|-12 + 5| \stackrel{?}{=} 10$$
$$|-7| \stackrel{?}{=} 10$$
$$7 \neq 10$$

Not a solution

3. $|6 - 2w| = 2, w = 4$

$$|6 - 2(4)| \stackrel{?}{=} 2$$
$$|6 - 8| \stackrel{?}{=} 2$$
$$|-2| \stackrel{?}{=} 2$$
$$2 = 2$$

Solution

5. $|x| = 4$

$x = 4$ or $x = -4$

7. $|t| = -45$

No solution

9. $|h| = 0$

$h = 0$

11. $|5x| = 15$

$5x = 15$ or $5x = -15$

$x = 3$ $\quad x = -3$

13. $|x + 1| = 5$

$x + 1 = 5$ or $x + 1 = -5$

$x = 4$ $\quad x = -6$

15. $|4 - 3x| = 0$

$$4 - 3x = 0$$
$$-3x = -4$$
$$x = \tfrac{4}{3}$$

17. $\left|\dfrac{2s + 3}{5}\right| = 5$

$\dfrac{2s + 3}{5} = 5$ or $\dfrac{2s + 3}{5} = -5$

$2s + 3 = 25$ $\quad 2s + 3 = -25$

$2s = 22$ $\quad 2s = -28$

$s = 11$ $\quad s = -14$

19. $|5 - 2x| + 10 = 6$

$$|5 - 2x| = -4$$

No solution

21. $\left|\dfrac{x - 2}{3}\right| + 6 = 6$

$$\left|\frac{x - 2}{3}\right| = 0$$
$$\frac{x - 2}{3} = 0$$
$$x - 2 = 0$$
$$x = 2$$

23. $3|2x - 5| + 4 = 7$

$$3|2x - 5| = 3$$
$$|2x - 5| = 1$$

$2x - 5 = 1$ or $2x - 5 = -1$

$2x = 6$ $\quad 2x = 4$

$x = 3$ $\quad x = 2$

25. $|2x + 1| = |x - 4|$

$2x + 1 = x - 4$ or $2x + 1 = -(x - 4)$

$x + 1 = -4$ $\quad 2x + 1 = -x + 4$

$x = -5$ $\quad 3x + 1 = 4$

$3x = 3$

$x = 1$

27. $|x + 8| = |2x + 1|$

$x + 8 = 2x + 1$ or $x + 8 = -(2x + 1)$

$8 = x + 1$ $\quad x + 8 = -2x - 1$

$7 = x$ $\quad 3x + 8 = -1$

$3x = -9$

$x = -3$

29. $|3x + 1| = |3x - 3|$

$3x + 1 = 3x - 3$ or $3x + 1 = -(3x - 3)$

$1 \neq -3$ $\quad 3x + 1 = -3x + 3$

$6x + 1 = 3$

$6x = 2$

$x = \tfrac{1}{3}$

The only solution is $x = \tfrac{1}{3}$.

31. $|4x - 10| = 2|2x + 3|$

$4x - 10 = 2(2x + 3)$ or $4x - 10 = -2(2x + 3)$

$4x - 10 = 4x + 6$ $\quad$ $4x - 10 = -4x - 6$

$-10 \neq 6$ $\quad$ $8x - 10 = -6$

$8x = 4$

$x = \frac{4}{8} = \frac{1}{2}$

The only solution is $x = \frac{1}{2}$.

33. $x = 2$

$|2| < 3$

$2 < 3$

Solution

35. $x = 9$

$|9 - 7| \geq 3$

$|2| \geq 3$

$2 \geq 3$

Not a solution

37. $|y| < 4$

$-4 < y < 4$

39. $|x| \geq 6$

$x \geq 6$ or $x \leq -6$

41. $|x + 6| > 10$

$x + 6 > 10$ or $x + 6 < -10$

$x > 4$ $\quad$ $x < -16$

43. $|2x| < 14$

$-14 < 2x < 14$

$-7 < x < 7$

45. $\left|\frac{y}{3}\right| \leq \frac{1}{3}$

$-\frac{1}{3} \leq \frac{y}{3} \leq \frac{1}{3}$

$-1 \leq y \leq 1$

47. $|2x + 3| > 9$

$2x + 3 < -9$ or $2x + 3 > 9$

$2x < -12$ $\quad$ $2x > 6$

$x < -6$ $\quad$ $x > 3$

49. $|2x - 1| \leq 7$

$-7 \leq 2x - 1 \leq 7$

$-6 \leq 2x \leq 8$

$-3 \leq x \leq 4$

51. $|3x + 10| < -1$

No solution

Absolute value is never negative.

53. $\frac{|a + 6|}{2} \geq 16$

$|a + 6| \geq 32$

$a + 6 \geq 32$ or $a + 6 \leq -32$

$a \geq 26$ or $a \leq -38$

55. $|0.2x - 3| < 4$

$-4 < 0.2x - 3 < 4$

$-1 < 0.2x < 7$

$\frac{-1}{0.2} < x < \frac{7}{0.2}$

$-5 < x < 35$

57. $|t - 42.238| \leq 0.412$

$-0.412 \leq t - 42.238 \leq 0.412$

$-0.412 + 42.238 \leq t - 42.238 + 42.238 \leq 0.412 + 42.238$

$41.826 \leq t \leq 42.65$

The fastest time is 41.826 seconds and the slowest time is 42.65 seconds.

41.826 42.65

40 41 42 43 44 45 t

59. The solutions of $|x| = a$ are $x = a$ and $x = -a$. For example, to solve $|x - 3| = 5$:

$x - 3 = 5$ or $x - 3 = -5$

$x = 8$ $\quad$ $x = -2$

61. The statement, "The distance between x and zero is a," can be represented using absolute values as $|x| = a$.

63. $|4x + 1| = \frac{1}{2}$

$4x + 1 = \frac{1}{2}$ or $4x + 1 = -\frac{1}{2}$

$4x = -\frac{1}{2}$ $4x = -\frac{3}{2}$

$x = -\frac{1}{8}$ $x = -\frac{3}{8}$

65. $|x - 4| = 9$

67. $|x| \le 2$

69. $|x - 19| > 2$

71. $|x - 4| \ge 2$

73. $\left|\frac{3x - 2}{4}\right| + 5 \ge 5$

$$\left|\frac{3x - 2}{4}\right| \ge 0$$

$$-\infty < x < \infty$$

Absolute value is always positive.

75. $|x| < 3$

77. $|2x - 3| > 5$

79. (a) $|s - x| \le \frac{3}{16}$

(b) $\left|5\frac{1}{8} - x\right| \le \frac{3}{16}$

$$-\frac{3}{16} \le 5\frac{1}{8} - x \le \frac{3}{16}$$

$$-\frac{3}{16} \le \frac{41}{8} - x \le \frac{3}{16}$$

$$-\frac{85}{16} \le -x \le -\frac{79}{16}$$

$$\frac{85}{16} \ge x \ge \frac{79}{16}$$

$$4\frac{15}{16} \le x \le 5\frac{5}{16}$$

81. $\left|\frac{h - 68.5}{2.7}\right| \le 1$

$$-1 \le \frac{h - 68.5}{2.7} \le 1$$

$$-2.7 \le h - 68.5 \le 2.7$$

$$65.8 \le h \le 71.2$$

The heights h lie on the interval $65.8 \text{ inches} \le h \le 71.2 \text{ inches}$.

83. The graph of $|x - 4| < 1$ can be described as all real numbers that are within one unit of four.

85. $|2x - 6| \le 6$ because:

$$-6 \le 2x - 6 \le 6$$

$$0 \le 2x \le 12$$

$$0 \le x \le 6$$

87. $4(n + 3)$

89. $x - 7 > 13$

$$x > 20$$

91. $4x + 11 \ge 27$

$$4x \ge 16$$

$$x \ge 4$$

Review Exercises for Chapter 2

1. (a) $45 - 7(3) = 3$

$$45 - 21 = 3$$

$$24 = 3$$

Not a solution

(b) $45 - 7(6) = 3$

$$45 - 42 = 3$$

$$3 = 3$$

Solution

3. (a) $\frac{28}{7} + \frac{28}{5} \stackrel{?}{=} 12$

$$\frac{140}{35} + \frac{196}{35} \stackrel{?}{=} 12$$

$$\frac{336}{35} \stackrel{?}{=} 12$$

$$\frac{48}{5} \ne 12$$

Not a solution

(b) $\frac{35}{7} + \frac{35}{5} \stackrel{?}{=} 12$

$$5 + 7 = 12$$

Solution

5.
$$3x + 21 = 0$$
$$3x + 21 - 21 = 0 - 21$$
$$3x = -21$$
$$\frac{3x}{3} = \frac{-21}{3}$$
$$x = -7$$

Check:
$$3(-7) + 21 \stackrel{?}{=} 0$$
$$-21 + 21 = 0$$

7.
$$5x - 120 = 0$$
$$5x - 120 + 120 = 0 + 120$$
$$5x = 120$$
$$\frac{5x}{5} = \frac{120}{5}$$
$$x = 24$$

Check:
$$5(24) - 120 \stackrel{?}{=} 0$$
$$120 - 120 = 0$$

9.
$$x + 4 = 9$$
$$x - 4 + 4 = 9 - 4$$
$$x = 5$$

Check:
$$5 + 4 \stackrel{?}{=} 9$$
$$9 = 9$$

11.
$$-3x = 36$$
$$\frac{-3x}{-3} = \frac{36}{-3}$$
$$x = -12$$

Check:
$$-3(-12) \stackrel{?}{=} 36$$
$$36 = 36$$

13.
$$-\tfrac{1}{8}x = 3$$
$$(-8)\left(-\tfrac{1}{8}x\right) = (3)(-8)$$
$$x = -24$$

Check:
$$-\tfrac{1}{8}(-24) \stackrel{?}{=} 3$$
$$3 = 3$$

15.
$$5x + 4 = 19$$
$$5x + 4 - 4 = 19 - 4$$
$$5x = 15$$
$$\frac{5x}{5} = \frac{15}{5}$$
$$x = 3$$

Check:
$$5(3) + 4 \stackrel{?}{=} 19$$
$$15 + 4 \stackrel{?}{=} 19$$
$$19 = 19$$

17.
$$17 - 7x = 3$$
$$17 - 7x - 17 = 3 - 17$$
$$-7x = -14$$
$$\frac{-7x}{-7} = \frac{-14}{-7}$$
$$x = 2$$

Check:
$$17 - 7(2) \stackrel{?}{=} 3$$
$$17 - 14 \stackrel{?}{=} 3$$
$$3 = 3$$

19.
$$7x - 5 = 3x + 11$$
$$7x - 3x - 5 = 3x - 3x + 11$$
$$4x - 5 = 11$$
$$4x - 5 + 5 = 11 + 5$$
$$4x = 16$$
$$\frac{4x}{4} = \frac{16}{4}$$
$$x = 4$$

Check:
$$7(4) - 5 \stackrel{?}{=} 3(4) + 11$$
$$28 - 5 \stackrel{?}{=} 12 + 11$$
$$23 = 23$$

21.
$$\begin{aligned} 3(2y-1) &= 9+3y \\ 6y-3 &= 9+3y \\ 6y-3y-3 &= 9+3y-3y \\ 3y-3 &= 9 \\ 3y-3+3 &= 9+3 \\ 3y &= 12 \\ \frac{3y}{3} &= \frac{12}{3} \\ y &= 4 \end{aligned}$$

Check:
$$\begin{aligned} 3(2(4)-1) &\stackrel{?}{=} 9+3(4) \\ 3(7) &\stackrel{?}{=} 9+12 \\ 21 &= 21 \end{aligned}$$

23.
$$\begin{aligned} 4y-4(y-2) &= 8 \\ 4y-4y+8 &= 8 \\ 8 &= 8 \end{aligned}$$

Infinitely many solutions

25.
$$\begin{aligned} 4(3x-5) &= 6(2x+3) \\ 12x-20 &= 12x+18 \\ 12x-12x-20 &= 12x-12x+18 \\ -20 &= 18 \end{aligned}$$

No solution

27.
$$\begin{aligned} \frac{4}{5}x-\frac{1}{10} &= \frac{3}{2} \\ 10\left[\frac{4}{5}x-\frac{1}{10}\right] &= \left[\frac{3}{2}\right]10 \\ 8x-1 &= 15 \\ 8x-1+1 &= 15+1 \\ 8x &= 16 \\ \frac{8x}{8} &= \frac{16}{8} \\ x &= 2 \end{aligned}$$

Check:
$$\begin{aligned} \frac{4}{5}(2)-\frac{1}{10} &\stackrel{?}{=} \frac{3}{2} \\ \frac{8}{5}-\frac{1}{10} &\stackrel{?}{=} \frac{3}{2} \\ \frac{16}{10}-\frac{1}{10} &\stackrel{?}{=} \frac{3}{2} \\ \frac{15}{10} &\stackrel{?}{=} \frac{3}{2} \\ \frac{3}{2} &= \frac{3}{2} \end{aligned}$$

29.
$$\begin{aligned} 1.4t+2.1 &= 0.9t \\ 1.4t+2.1-0.9t &= 0.9t-0.9t \\ 0.5t+2.1 &= 0 \\ 0.5t+2.1-2.1 &= 0-2.1 \\ 0.5t &= -2.1 \\ \frac{0.5t}{0.5} &= \frac{-2.1}{0.5} \\ t &= -4.2 \end{aligned}$$

Check:
$$\begin{aligned} 1.4(-4.2)+2.1 &\stackrel{?}{=} 0.9(-4.2) \\ -5.88+2.1 &\stackrel{?}{=} -3.78 \\ -3.78 &= -3.78 \end{aligned}$$

31. *Verbal Model:* [Total pay] = [Pay per week] $\cdot\, x$ + [Pay for training]

Labels: Total pay = 2635
Pay per week = 320
Pay for training = 75

Equation:
$$\begin{aligned} 2635 &= 320x+75 \\ 2560 &= 320x \\ 8 &= x \end{aligned}$$

The internship is 8 weeks long.

33.

Percent	Parts out of 100	Decimal	Fraction
68%	68	0.68	$\frac{17}{25}$

35.

Percent	Parts out of 100	Decimal	Fraction
60%	60	0.6	$\frac{3}{5}$

37. *Verbal Model:* [Compared number] = [Percent] $\cdot$ [Base number]

Labels: Compared number = a
Percent = 1.30
Base number = 50

Equation:
$$\begin{aligned} a &= p\cdot b \\ a &= 1.30\cdot 50 \\ a &= 65 \end{aligned}$$

So, 65 is 130% of 50.

39. *Verbal Model:* $\boxed{\text{Compared number}} = \boxed{\text{Percent}} \cdot \boxed{\text{Base number}}$

Labels: Compared number $= 645$

Percent $= 0.215$

Base number $= b$

Equation:

$$645 = 0.215 \cdot b$$
$$\frac{645}{0.215} = b$$
$$3000 = b$$

So, 645 is $21\frac{1}{2}\%$ of 3000.

41. *Verbal Model:* $\boxed{\text{Compared number}} = \boxed{\text{Percent}} \cdot \boxed{\text{Base number}}$

Labels: Compared number $= 250$

Percent $= p$

Base number $= 200$

Equation:

$$250 = p \cdot 200$$
$$\frac{250}{200} = p$$
$$1.25 = p$$

So, 250 is 125% of 200.

43. *Verbal Model:* $\boxed{\text{Commission}} = \boxed{\text{Percent rate}} \cdot \boxed{\text{Sales}}$

Labels: Commission $= 9000$

Percent rate $= x$

Sales $= 150{,}000$

Equation:

$$9000 = x \cdot 150{,}000$$
$$\frac{9000}{150{,}000} = x$$
$$0.06 = x$$

It is a 6% commission.

45. *Verbal Model:* $\boxed{\text{Defective parts}} = \boxed{\text{Percent rate}} \cdot \boxed{\text{Sample}}$

Labels: Defective parts $= 6$

Percent rate $= 1.6\%$

Sample $= x$

Equation:

$$6 = 0.016 \cdot x$$
$$\frac{6}{0.016} = x$$
$$375 = x$$

The sample contained 375 parts.

47. (a) $\text{Unit price} = \dfrac{\text{Total price}}{\text{Total units}} = \dfrac{\$9.79}{39 \text{ ounces}} = \$0.25 \text{ per ounce}$

(b) $\text{Unit price} = \dfrac{\text{Total price}}{\text{Total units}} = \dfrac{\$1.79}{8 \text{ ounces}} = \$0.22 \text{ per ounce}$

The 8-ounce can is the better buy.

49. $\dfrac{\text{Tax}}{\text{Pay}} = \dfrac{9.90}{396} = \dfrac{1.10}{44} = \dfrac{0.1}{4} = \dfrac{1}{40}$

51.

$$\frac{7}{8} = \frac{y}{4}$$
$$8y = 28$$
$$y = \frac{28}{8}$$
$$y = \frac{7}{2}$$

53.

$$\frac{b}{6} = \frac{5 + b}{15}$$
$$15b = 6(5 + b)$$
$$15b = 30 + 6b$$
$$9b = 30$$
$$b = \frac{30}{9}$$
$$b = \frac{10}{3}$$

55. *Verbal Model:* $\boxed{\dfrac{\text{Leg 1}}{\text{Leg 2}}} = \boxed{\dfrac{\text{Leg 1}}{\text{Leg 2}}}$

Proportion:

$$\frac{2}{6} = \frac{x}{9}$$
$$6x = 18$$
$$x = 3$$

57. *Verbal Model:* $\boxed{\dfrac{\text{Tax 1}}{\text{Assessed value 1}}} = \boxed{\dfrac{\text{Tax 2}}{\text{Assessed value 2}}}$

Labels: Tax 1 $= 1680$

Assessed value 1 $= 105{,}000$

Tax 2 $= x$

Assessed value 2 $= 125{,}000$

Proportion:

$$\frac{1680}{105{,}000} = \frac{x}{125{,}000}$$
$$125{,}000 \cdot 1680 = 105{,}000 \cdot x$$
$$210{,}000{,}000 = 105{,}000x$$
$$2000 = x$$

The tax is \$2000.

59. *Verbal Model:* $\boxed{\dfrac{\text{Flagpole's height}}{\text{Length of flagpole's shadow}}} = \boxed{\dfrac{\text{Lamp post's height}}{\text{Length of lamp post's shadow}}}$

Labels: Flagpole's height $= h$

Length of flagpole's shadow $= 30$

Lamp post's height $= 5$

Length of lamp post's shadow $= 3$

Proportion: $\dfrac{h}{30} = \dfrac{5}{3}$

$$3 \cdot h = 30 \cdot 5$$
$$3h = 150$$
$$h = 50$$

The flagpole's height is 50 feet.

61. *Verbal Model:* $\boxed{\text{Selling price}} = \boxed{\text{Cost}} + \boxed{\text{Markup}}$

Labels: Selling price $= 149.93$

Cost $= 99.95$

Markup $= x$

Equation: $149.93 = 99.95 + x$

$$149.93 - 99.95 = x$$
$$49.98 = x$$

The markup is \$49.98.

Verbal Model: $\boxed{\text{Markup}} = \boxed{\text{Markup rate}} \cdot \boxed{\text{Cost}}$

Labels: Markup $= 49.98$

Markup rate $= x$

Cost $= 99.95$

Equation: $49.98 = x \cdot 99.95$

$$\frac{49.98}{99.95} = x$$
$$0.50 \approx x$$

The markup rate is about 50%.

63. *Verbal Model:* $\boxed{\text{Sale price}} = \boxed{\text{List price}} - \boxed{\text{Discount}}$

Labels: Sale price $= 53.96$

List price $= 71.95$

Discount $= x$

Equation: $53.96 = 71.95 - x$

$$x = 71.95 - 53.96$$
$$x = 17.99$$

The discount is \$17.99.

Verbal Model: $\boxed{\text{Discount}} = \boxed{\text{Discount rate}} \cdot \boxed{\text{List price}}$

Labels: Discount $= 17.99$

Discount rate $= x$

List price $= 71.95$

Equation: $17.99 = x \cdot 71.95$

$$\frac{17.99}{71.95} = x$$
$$0.25 \approx x$$

The discount rate is about 25%.

65. *Verbal Model:* [Sales tax] = [Tax rate] · [Cost]

Labels: Sales tax $= x$

Tax rate $= 0.06$

Cost $= 2795$

Equation: $x = 0.06 \cdot 2795$

$x = 167.7$

The sales tax is \$167.70.

Verbal Model: [Total bill] = [Sales tax] + [Cost]

Labels: Total bill $= x$

Sales tax $= 167.70$

Cost $= 2795$

Equation: $x = 167.70 + 2795$

$x = 2962.70$

The total bill is \$2962.70.

Verbal Model: [Amount financed] = [Total Bill] − [Down payment]

Labels: Amount financed $= x$

Total bill $= 2962.70$

Downpayment $= 800$

Equation: $x = 2962.70 - 800$

$x = 2162.70$

The amount financed is \$2162.70.

67. *Verbal Model:*

[Amount of solution 1] + [Amount of solution 2] = [Amount of final solution]

Labels: Percent of solution 1 = 30%

Liters of solution 1 $= x$

Percent of solution 2 = 60%

Liters of solution 2 $= 10 - x$

Percent of final solution = 50%

Liters of final solution = 10

Equation: $0.30x + 0.60(10 - x) = 0.50(10)$

$0.30x + 6 - 0.60x = 5$

$-0.30x = -1$

$x = 3\frac{1}{3}$

$10 - x = 6\frac{2}{3}$

$3\frac{1}{3}$ liters of solution 1 and $6\frac{2}{3}$ liters of solution 2 are required.

69. *Verbal Model:* [Distance] = [Rate] · [Time]

Labels: Distance $= d$

Rate = 1500 mph

Time = $2\frac{1}{3}$ hours

Equation: $d = 1500 \cdot 2\frac{1}{3}$

$d = 3500$

The distance is 3500 miles.

71. *Verbal Model:* [Distance] = [Rate] · [Time]

Labels: Distance = 100 miles

Rates = 48 mph and 40 mph

Time $= t$

Equation: $d = rt$

$t = \dfrac{d}{r}$

$t = \dfrac{100}{48} + \dfrac{100}{40}$

$t = 4.58\overline{3}$ or $\dfrac{55}{12}$

Verbal Model: [Average speed] = [Total distance] ÷ [Total time]

Labels: Average speed $= r$

Total distance = 200 miles

Total time = $4.58\overline{3}$ hours

Equation: $r = 200 \div 4.58\overline{3}$

$r \approx 43.6$

The average speed is 43.6 miles per hour.

73. *Verbal Model:* [Work done] = [Work done by person 1] + [Work done by person 2]

Labels: Work done = 1

Rate of person 1 $= \dfrac{1}{4.5}$

Rate of person 2 $= \dfrac{1}{6}$

Time $= t$

Equation: $1 = \dfrac{t}{4.5} + \dfrac{t}{6}$

$27 = 6t + 4.5t$

$27 = 10.5t$

$\dfrac{27}{10.5} = t$

$2.57 \approx t$

The time required is about 2.57 hours.

75. *Verbal Model:* $\boxed{\text{Interest}} = \boxed{\text{Principal}} \cdot \boxed{\text{Rate}} \cdot \boxed{\text{Time}}$

Labels: Interest $= i$

Principal $= \$1000$

Rate $= 0.085$

Time $= 4$

Equation: $i = 1000 \cdot 0.085 \cdot 4$

$i = 340$

The interest is \$340.

77. *Verbal Model:* $\boxed{\text{Interest}} = \boxed{\text{Principal}} \cdot \boxed{\text{Rate}} \cdot \boxed{\text{Time}}$

Labels: Interest $= \$20{,}000$

Principal $= p$

Rate $= 0.095$

Time $= 4$

Equation: $20{,}000 = p \cdot 0.095 \cdot 4$

$\dfrac{20{,}000}{0.38} = p$

$52{,}631.58 \approx p$

The principal required is about \$52,631.58.

79. *Verbal Model:* $\boxed{\text{Interest}} = \boxed{\text{Principal}} \cdot \boxed{\text{Rate}} \cdot \boxed{\text{Time}}$

Labels: Interest $= 4700$

Principal 1 $= p$

Rate 1 $= 0.085$

Principal 2 $= 50{,}000 - p$

Rate 2 $= 0.10$

Time $= 1$

Equation: $4700 = 0.085p + 0.10(50{,}000 - p)$

$4700 = 0.085p + 5000 - 0.10p$

$-300 = -0.015p$

$\dfrac{-300}{-0.015} = p$

$20{,}000 = p$

$30{,}000 = 50{,}000 - p$

The smallest amount you can invest is \$30,000.

81. *Verbal Model:*

$\boxed{\text{Perimeter}} = 2 \cdot \boxed{\text{Width}} + 2 \cdot \boxed{\text{Length}}$

Labels: Perimeter $= 64$

Width $= \frac{3}{5}l$

Length $= l$

Equation: $64 = 2\left(\frac{3}{5}l\right) + 2l$

$64 = \frac{6}{5}l + 2l$

$64 = \frac{16}{5}l$

$20 = l$

The dimensions are 20 feet $\times$ 12 feet.

83. 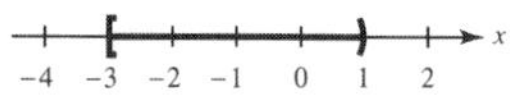

85. 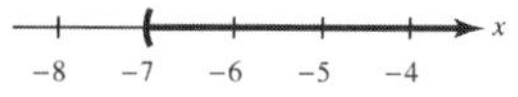

87. $x - 5 \le -1$

$x - 5 + 5 \le -1 + 5$

$x \le 4$

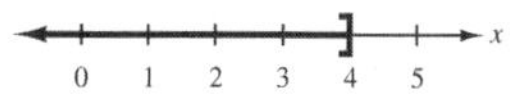

89. $-6x < -24$

$\dfrac{-6x}{-6} > \dfrac{-24}{-6}$

$x > 4$

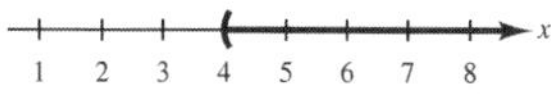

91. $5x + 3 > 18$

$5x > 15$

$x > 3$

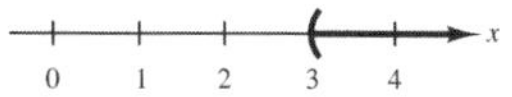

93. $8x + 1 \ge 10x - 11$

$8x - 10x + 1 \ge 10x - 10x - 11$

$-2x + 1 \ge -11$

$-2x + 1 - 1 \ge -11 - 1$

$-2x \ge -12$

$\dfrac{-2x}{-2} \le \dfrac{-12}{-2}$

$x \le 6$

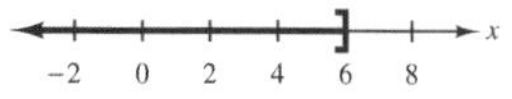

95. $\frac{1}{3} - \frac{1}{2}y < 12$

$2 - 3y < 72$

$-3y < 70$

$y > -\frac{70}{3}$

$-\frac{70}{3}$

−25 −24 −23 −22 y

97. $-4(3 - 2x) \le 3(2x - 6)$

$-12 + 8x \le 6x - 18$

$-12 + 8x - 6x \le 6x - 6x - 18$

$-12 + 2x \le -18$

$-12 + 12 + 2x \le -18 + 12$

$2x \le -6$

$\frac{2x}{2} \le \frac{-6}{2}$

$x \le -3$

−5 −4 −3 −2 −1 0 1 2 x

99. $-6 \le 2x + 8 < 4$

$-6 - 8 \le 2x + 8 - 8 < 4 - 8$

$-14 \le 2x < -4$

$\frac{-14}{2} \le \frac{2x}{2} < \frac{-4}{2}$

$-7 \le x < -2$

−7

−8 −6 −4 −2 0 x

101. $5 > \frac{x + 1}{-3} > 0$

$-15 < x + 1 < 0$

$-16 < x < -1$

−1

−20 −16 −12 −8 −4 0 x

103. $5x - 4 < 6$ and $3x + 1 > -8$

$5x - 4 + 4 < 6 + 4$ $\quad$ $3x + 1 - 1 > -8 - 1$

$5x < 10$ $\quad$ $3x > -9$

$\frac{5x}{5} < \frac{10}{5}$ $\quad$ $\frac{3x}{3} > \frac{-9}{3}$

$x < 2$ $\quad$ $x > -3$

$-3 < x < 2$

−3

−4 −2 0 2 x

105. *Verbal Model:* [Cost per minute] · [Number of minutes] ≥ [Money Remaining]

Labels: Cost per minute = 0.10

Number of minutes = x

Money remaining = 12.50

Equation: $0.10x \le 12.50$

$x \le 125$

You can talk for no more than 125 minutes.

107. $|x| = 6$

$x = 6$ or $x = -6$

109. $|4 - 3x| = 8$

$4 - 3x = 8$ or $4 - 3x = -8$

$4 - 4 - 3x = 8 - 4$ $\quad$ $4 - 4 - 3x = -8 - 4$

$-3x = 4$ $\quad$ $-3x = -12$

$\frac{-3x}{-3} = \frac{4}{-3}$ $\quad$ $\frac{-3x}{-3} = \frac{-12}{-3}$

$x = -\frac{4}{3}$ $\quad$ $x = 4$

111. $|5x + 4| - 10 = -6$

$|5x + 4| = 4$

$5x + 4 = 4$ or $5x + 4 = -4$

$5x + 4 - 4 = 4 - 4$ $\quad$ $5x + 4 - 4 = -4 - 4$

$5x = 0$ $\quad$ $5x = -8$

$\frac{5x}{5} = \frac{0}{5}$ $\quad$ $\frac{5x}{5} = \frac{-8}{5}$

$x = 0$ $\quad$ $x = -\frac{8}{5}$

113. $|3x - 4| = |x + 2|$

$3x - 4 = x + 2$ or $3x - 4 = -(x + 2)$

$2x = 6$ $\quad$ $3x - 4 = -x - 2$

$x = 3$ $\quad$ $4x = 2$

$x = \frac{2}{4} = \frac{1}{2}$

115. $|x - 4| > 3$

$x - 4 < -3$ or $x - 4 > 3$

$x < 1$ $\quad$ $x > 7$

117. $|3x| < 12$

$-12 < 3x < 12$

$-4 < x < 4$

119. $|2x - 7| < 15$

$$-15 < 2x - 7 < 15$$
$$-8 < 2x < 22$$
$$-4 < x < 11$$

121. $|b + 2| - 6 > 1$

$$|b + 2| > 7$$

$$b + 2 < -7 \quad \text{or} \quad b + 2 > 7$$
$$b < -9 \qquad\qquad b > 5$$

123. $(1, 5)$

$$1 < x < 5$$
$$1 - 3 < x - 3 < 5 - 3$$
$$-2 < x - 3 < 2$$
$$|x - 3| < 2$$

125. $|t - 78.3| \le 38.3$

$$-38.3 \le t - 78.3 \le 38.3$$
$$40 \le t \le 116.6$$

The minimum temperature is 40 degrees Fahrenheit and the maximum temperature is 116.6 degrees Fahrenheit.

116.6

0 20 40 60 80 100 120 t

Chapter Test for Chapter 2

1.
$$6x - 5 = 19$$
$$6x - 5 + 5 = 19 + 5$$
$$6x = 24$$
$$\frac{6x}{6} = \frac{24}{6}$$
$$x = 4$$

2.
$$5x - 6 = 7x - 12$$
$$5x - 7x - 6 = 7x - 7x - 12$$
$$-2x - 6 = -12$$
$$-2x - 6 + 6 = -12 + 6$$
$$-2x = -6$$
$$\frac{-2x}{-2} = \frac{-6}{-2}$$
$$x = 3$$

3. $15 - 7(1 - x) = 3(x + 8)$
$$15 - 7 + 7x = 3x + 24$$
$$8 + 7x = 3x + 24$$
$$8 + 7x - 3x = 3x + 24 - 3x$$
$$8 - 8 + 4x = 24 - 8$$
$$4x = 16$$
$$\frac{4x}{4} = \frac{16}{4}$$
$$x = 4$$

4.
$$\frac{2x}{3} = \frac{x}{2} + 4$$
$$6\left(\frac{2x}{3}\right) = \left(\frac{x}{2} + 4\right)6$$
$$4x = 3x + 24$$
$$4x - 3x = 3x + 24 - 3x$$
$$x = 24$$

5. *Verbal Model:* [Compared number] = [Percent] · [Base number]

Labels: Compared number $= a$

Percent $= 1.25$

Base number $= 3200$

Equation: $a = 1.25 \cdot 3200$

$a = 4000$

So, 125% of 3200 is 4000.

6. *Verbal Model:* [Compared number] = [Percent] · [Base number]

Labels: Compared number $= a$

Percent $= p$

Base number $= b$

Equation: $32 = p \cdot 8000$

$\frac{32}{8000} = p$

$0.004 = p$

So, 32 is 0.4% of 8000.

7. *Verbal Model:* [List price] − [Discount] = [Sale price]

Labels: List price $= x$

Discount $= 0.20x$

Sale price $= 8900$

Equation: $x - 0.20x = 8900$

$0.80x = 8900$

$x = 11{,}125$

The list price is \$11,125.

8. $\dfrac{\text{Total price}}{\text{Total units}} = \dfrac{\$2.49}{12 \text{ ounces}} = \dfrac{249}{1200} = \$0.2075$ per ounce

$\dfrac{\text{Total price}}{\text{Total units}} = \dfrac{\$2.99}{15 \text{ ounces}} = \dfrac{299}{1500} = \$0.199\overline{3}$ per ounce

The 15-ounce can is the better buy. It has a lower unit price.

9. *Verbal Model:* [Tax] = [Tax rate] · [Assessed value]

Labels: Tax $= 1650$

Tax rate $= p$

Assessed value $= 110{,}000$

Equation: $1650 = p \cdot 110{,}000$

$0.015 = p$

Verbal Model: [Tax] = [Tax rate] · [Assessed value]

Labels: Tax $= x$

Tax rate $= 0.015$

Assessed value $= 145{,}000$

Equation: $x = 0.015 \cdot 145{,}000$

$x = 2175$

The tax is \$2175.

10. *Verbal Model:* [Total bill] = [Cost of parts] + [Cost of labor]

Labels: Total bill $= 165$

Cost of parts $= 85$

Number of half hours of labor $= x$

Cost of labor $= 16x$

Equation: $165 = 85 + 16x$

$80 = 16x$

5 half hours $= x$

The repairs took $2\frac{1}{2}$ hours.

11. *Verbal Model:*

[Amount of food 1] + [Amount of food 2] = [Mixture]

Labels: Cost for food 1 $= 2.60$

Pounds of food 1 $= x$

Cost for food 2 $= 3.80$

Pounds of food 2 $= 40 - x$

Cost for mixture $= 3.35$

Pounds of mixture $= 40$

Equation: $2.60x + 3.80(40 - x) = 3.35(40)$

$-1.20x + 152 = 134$

$-1.20x = -18$

$x = 15$

There are 15 pounds at \$2.60 and 25 pounds at \$3.80.

12. *Verbal Model:*

[Distance of car 1] + 10 miles = [Distance of car 2]

Labels: Time $= x$

Distance of car 1 $= 40x$

Distance of car 2 $= 55x$

Equation: $40x + 10 = 55x$

$10 = 15x$

$\frac{10}{15} = x$

$\frac{2}{3} = x$

$\frac{2}{3}$ hour, or 40 minutes, must elapse.

13. *Verbal Model:* [Interest] = [Principal] · [Rate] · [Time]

Labels: Interest $= 300$

Principal $= p$

Rate $= 0.075$

Time $= 2$

Equation: $300 = p \cdot 0.075 \cdot 2$

$2000 = p$

The principal required is \$2000.

14. (a) $|3x - 6| = 9$

$$3x - 6 = 9 \quad \text{and} \quad -3x + 6 = 9$$
$$3x = 15 \qquad -3x = 3$$
$$\frac{3x}{3} = \frac{15}{3} \qquad \frac{-3x}{-3} = \frac{3}{-3}$$
$$x = 5 \qquad x = -1$$

(b) $|3x - 5| = |6x - 1|$

$$3x - 5 = 6x - 1 \quad \text{and} \quad -3x + 5 = 6x - 1$$
$$-4 = 3x \qquad 6 = 9x$$
$$\frac{-4}{3} = \frac{3x}{3} \qquad \frac{6}{9} = x$$
$$-\frac{4}{3} = x \qquad \frac{2}{3} = x$$

(c) $|9 - 4x| + 4 = 1$

$$|9 - 4x| = -3$$

There is no solution.

15. (a) $3x + 12 \geq -6$

$$3x \geq -18$$
$$x \geq -6$$

(b) $9 - 5x < 5 - 3x$

$$-2x < -4$$
$$x > 2$$

(c) $0 \leq \dfrac{1 - x}{4} < 2$

$$0 \leq 1 - x < 8$$

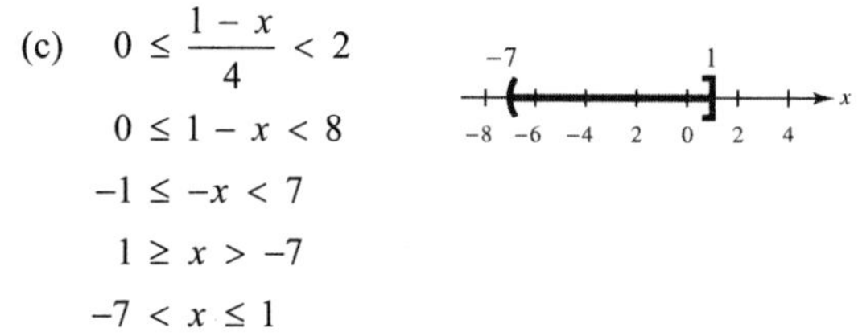

$$-1 \leq -x < 7$$
$$1 \geq x > -7$$
$$-7 < x \leq 1$$

(d) $-7 < 4(2 - 3x) \leq 20$

$$-7 < 8 - 12x \leq 20$$
$$-15 < -12x \leq 12$$
$$1 \leq x < \frac{5}{4}$$

16. $t \geq 8$

17. (a) $|x - 3| \leq 2$

$$-2 \leq x - 3 \leq 2$$
$$1 \leq x \leq 5$$

(b) $|5x - 3| > 12$

$$5x - 3 > 12 \quad \text{or} \quad 5x - 3 < -12$$
$$5x > 15 \qquad 5x < -9$$
$$x > 3 \qquad x < -\frac{9}{5}$$

(c) $\left|\dfrac{x}{4} + 2\right| < 0.2$

$$-0.2 < \frac{x}{4} + 2 < 0.2$$
$$-0.8 < x + 8 < 0.8$$
$$-8.8 < x < -7.2$$
$$-\frac{44}{5} < x < -\frac{36}{5}$$

18. *Verbal Model:* [Operating cost] $\leq 11{,}950$

Label: Number of miles $= m$

Equation: $0.37m + 2700 \leq 11{,}950$

$$0.37m \leq 9250$$
$$m \leq 25{,}000$$

The maximum number of miles is 25,000.

CHAPTER 3
Graphs and Functions

Section 3.1 The Rectangular Coordinate System 55

Section 3.2 Graphs of Equations 58

Section 3.3 Slope and Graphs of Linear Equations 64

Section 3.4 Equations of Lines 67

Mid-Chapter Quiz 71

Section 3.5 Graphs of Linear Inequalities 73

Section 3.6 Relations and Functions 77

Section 3.7 Graphs of Functions 79

Review Exercises 83

Chapter Test 88

CHAPTER 3
Graphs and Functions

Section 3.1 The Rectangular Coordinate System

1.

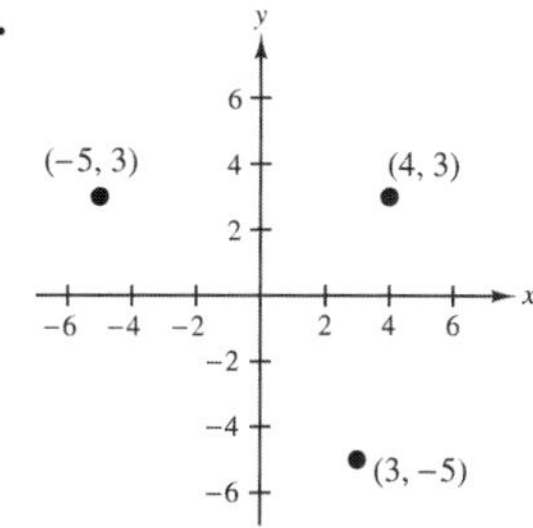

$(4, 3)$ is 4 units to the right of the vertical axis and 3 units above the horizontal axis.

$(-5, 3)$ is 5 units to the left of the vertical axis and 3 units above the horizontal axis.

$(3, -5)$ is 3 units to the right of the vertical axis and 5 units below the horizontal axis.

3.

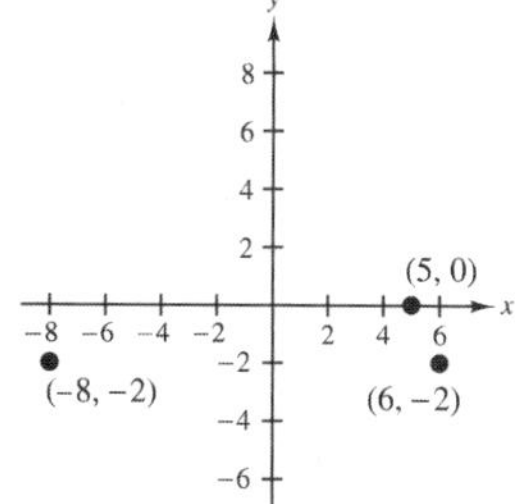

$(-8, -2)$ is 8 units to the left of the vertical axis and 2 units below the horizontal axis.

$(6, -2)$ is 6 units to the right of the vertical axis and 2 units below the horizontal axis.

$(5, 0)$ is 5 units to the right of the vertical axis and 0 units above or below the horizontal axis.

5.

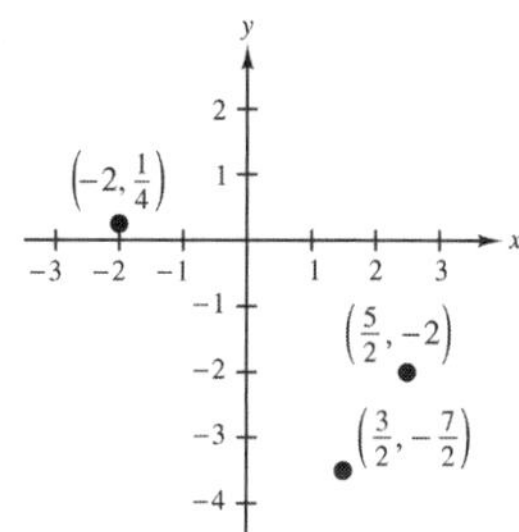

$\left(\frac{5}{2}, -2\right)$ is $\frac{5}{2}$ units to the right of the vertical axis and 2 units below the horizontal axis.

$\left(-2, \frac{1}{4}\right)$ is 2 units to the left of the vertical axis and $\frac{1}{4}$ unit above the horizontal axis.

$\left(\frac{3}{2}, -\frac{7}{2}\right)$ is $\frac{3}{2}$ units to the right of the vertical axis and $\frac{7}{2}$ units below the horizontal axis.

7.

Point	Position	Coordinates
A	2 units left, 4 units up	$(-2, 4)$
B	0 units right or left, 2 units down	$(0, -2)$
C	4 units right, 2 units down	$(4, -2)$

9.

Point	Position	Coordinates
A	4 units right, 2 units above	$(4, 2)$
B	1 unit left, 2 units below	$(-1, -2)$
C	0 units right or left, 0 units above or below	$(0, 0)$

11.

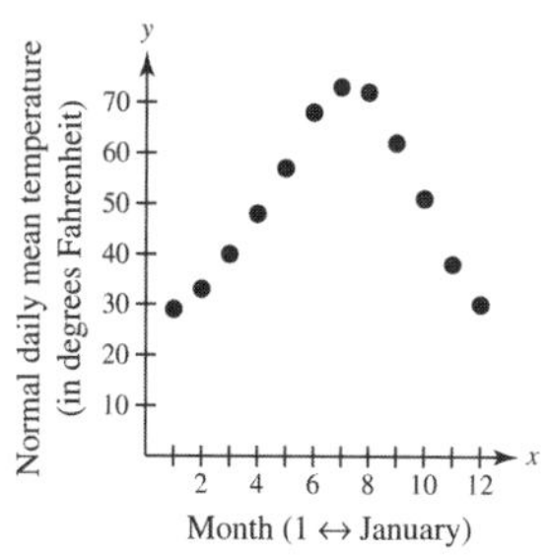

13.

Choose x	Calculate y from $y = 5x + 3$	Solution point
-2	$y = 5(-2) + 3 = -7$	$(-2, -7)$
0	$y = 5(0) + 3$	$(0, 3)$
2	$y = 5(2) + 3$	$(2, 13)$
4	$y = 5(4) + 3$	$(4, 23)$
6	$y = 5(6) + 3$	$(6, 33)$

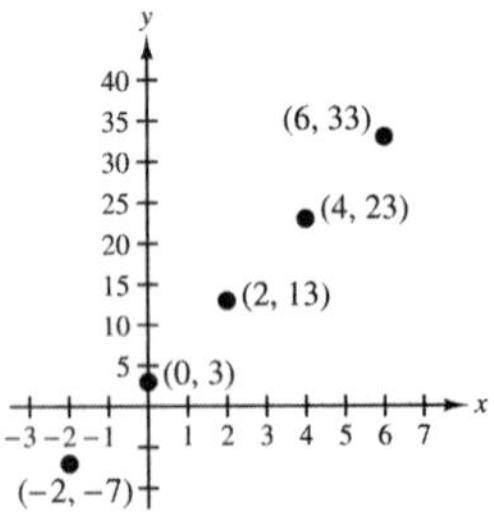

15.

Choose x	Calculate y from $y = \|2x - 7\| + 2$	Solution point
-4	$y = \|2(-4) - 7\| + 2$	$(-4, 17)$
0	$y = \|2(0) - 7\| + 2$	$(0, 9)$
3	$y = \|2(3) - 7\| + 2$	$(3, 3)$
5	$y = \|2(5) - 7\| + 2$	$(5, 5)$
10	$y = \|2(10) - 7\| + 2$	$(10, 15)$

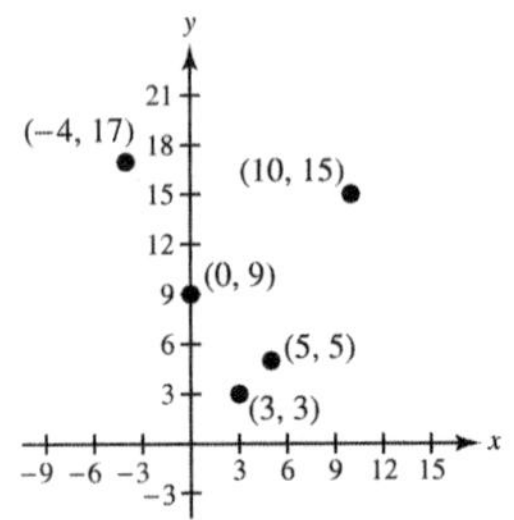

17. $4y - 2x + 1 = 0$

(a) $4(0) - 2(0) + 1 \stackrel{?}{=} 0$

$1 \neq 0$

Not a solution

(b) $4(0) - 2\left(\frac{1}{2}\right) + 1 \stackrel{?}{=} 0$

$0 - 1 + 1 \stackrel{?}{=} 0$

$0 = 0$

Solution

(c) $4\left(-\frac{7}{4}\right) - 2(-3) + 1 \stackrel{?}{=} 0$

$-7 + 6 + 1 \stackrel{?}{=} 0$

$0 = 0$

Solution

(d) $4\left(-\frac{3}{4}\right) - 2(1) + 1 \stackrel{?}{=} 0$

$-3 - 2 + 1 \stackrel{?}{=} 0$

$-4 \neq 0$

Not a solution

19. $x^2 + 3y = -5$

(a) $3^2 + 3(-2) \stackrel{?}{=} -5$

$9 - 6 \stackrel{?}{=} -5$

$3 \neq -5$

Not a solution

(b) $(-2)^2 + 3(-3) \stackrel{?}{=} -5$

$4 - 9 \stackrel{?}{=} -5$

$-5 = -5$

Solution

(c) $3^2 + 3(-5) \stackrel{?}{=} -5$

$9 - 15 \stackrel{?}{=} -5$

$-6 \neq -5$

Not a solution

(d) $4^2 + 3(-7) \stackrel{?}{=} -5$

$16 - 21 \stackrel{?}{=} -5$

$-5 = -5$

Solution

21. $d = \sqrt{(1-5)^2 + (3-6)^2}$

$= \sqrt{(-4)^2 + (-3)^2} = \sqrt{16+9} = \sqrt{25} = 5$

23. $d = \sqrt{(3-4)^2 + (7-5)^2}$

$= \sqrt{(-1)^2 + (2)^2} = \sqrt{1+4} = \sqrt{5}$

25. $d = \sqrt{(x_2 - x_1)^2 + (y_2 - y_1)^2}$

Let $(x, y) = (10, 10)$ and $(x_2, y_2) = (35, 40)$.

$d = \sqrt{(35-10)^2 + (40-10)^2}$

$= \sqrt{25^2 + 30^2}$

$= \sqrt{625 + 900}$

$= \sqrt{1525}$

$= 5\sqrt{61} \approx 39.05$

The pass is about 39.05 yards long.

27. Let $(x_1, y_1) = (-2, 0)$ and $(x_2, y_2) = (4, 8)$.

$\text{Midpoint} = \left(\frac{x_1 + x_2}{2}, \frac{y_1 + y_2}{2}\right)$

$= \left(\frac{-2+4}{2}, \frac{0+8}{2}\right)$

$= (1, 4)$

29. Let $(x_1, y_1) = (1, 6)$ and $(x_2, y_2) = (6, 3)$.

$\text{Midpoint} = \left(\frac{x_1 + x_2}{2}, \frac{y_1 + y_2}{2}\right)$

$= \left(\frac{1+6}{2}, \frac{6+3}{2}\right)$

$= \left(\frac{7}{2}, \frac{9}{2}\right)$

31. The graph of an ordered pair of real numbers is a point on a rectangular coordinate system.

33. Yes, the order of the points does not affect the computation of the midpoint.

35. $(-3, -5)$ is in Quadrant III.

37. $\left(-\frac{8}{9}, \frac{3}{4}\right)$ is in Quadrant II.

39. $(x, y), x > 0, y < 0$ is in Quadrant IV.

41. $(x, y), xy > 0$ is in Quadrant I or III.

43. $(-2, -1)$ shifted 2 units right and 5 units up $= (0, 4)$

$(-3, -4)$ shifted 2 units right and 5 units up $= (-1, 1)$

$(1, -3)$ shifted 2 units right and 5 units up $= (3, 2)$

45. $d = |5 - (-2)|$

$= |7|$

$= 7$

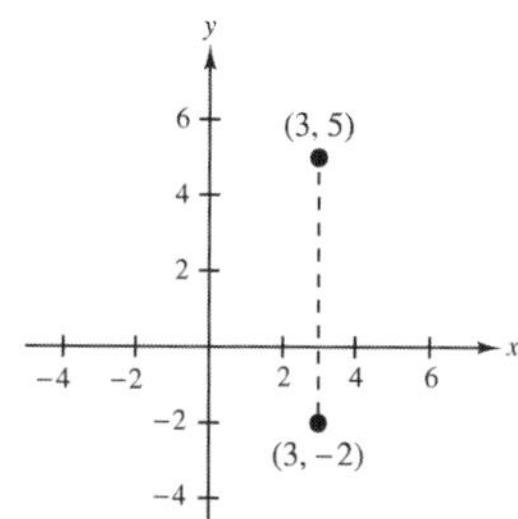

Vertical line

47. $d = |10 - 3|$

$= |7|$

$= 7$

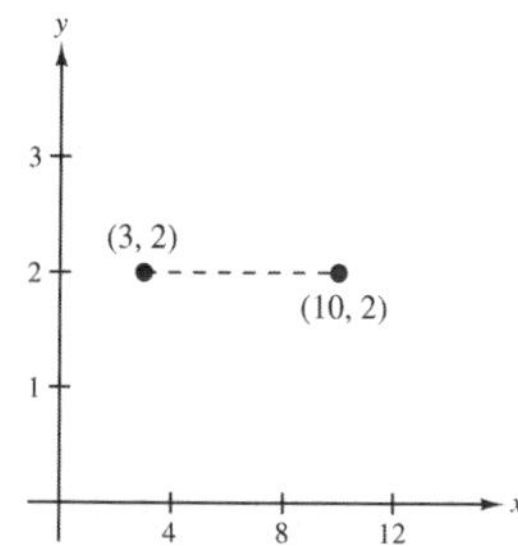

Horizontal line

49. $d_1 = \sqrt{(-2-0)^2 + (0-5)^2} = \sqrt{4+25} = \sqrt{29}$

$d_2 = \sqrt{(0-1)^2 + (5-0)^2} = \sqrt{1+25} = \sqrt{26}$

$d_3 = \sqrt{(-2-1)^2 + (0-0)^2} = \sqrt{9} = 3$

$P = \sqrt{29} + \sqrt{26} + 3 \approx 13.48$

51. Let the point (2009, 36.5) represent Apple's sales for the year 2009 and (2011, 108.2) represent the sales for the year 2011. The midpoint of these points is

$\text{Midpoint} = \left(\frac{x_1 + x_2}{2}, \frac{y_1 + y_2}{2}\right) = \left(\frac{2009 + 2011}{2}, \frac{36.5 + 108.2}{2}\right) = (2010, 72.35).$

Apple's net sales for the year 2010 is estimated to be $72.35 billion.

53. No. The distance between two points is always positive.

55. $(-3, 4)$ is not a solution point of $y = 4x + 15$ because

$4 \neq 4(-3) + 15$

$4 \neq 3.$

57.

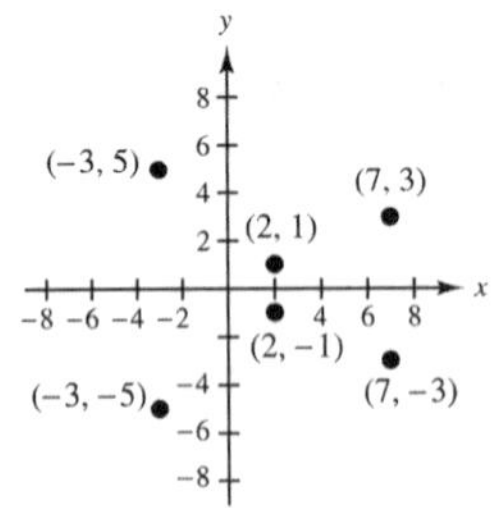

When the sign of the y-coordinate is changed, the point is on the opposite side of the x-axis from the original point.

59.
$$\frac{3}{4} = \frac{x}{28}$$
$$3 \cdot 28 = 4x$$
$$84 = 4x$$
$$21 = x$$

61.
$$\frac{a}{27} = \frac{4}{9}$$
$$9a = 4 \cdot 27$$
$$9a = 108$$
$$a = 12$$

63.
$$\frac{z + 1}{10} = \frac{z}{9}$$
$$9(z + 1) = 10z$$
$$9z + 9 = 10z$$
$$9 = z$$

65. Verbal model: [Discount] = [Discount rate] · [List price]

Labels: Discount $= 80 - 52 = 28$

List price $= 80$

Discount rate $= p$

Equation:
$$28 = p(80)$$
$$\frac{28}{80} = p$$
$$0.35 = p$$

The discount is \$28 and the discount rate is 35%.

67. Verbal model: [Sale price] = [List price] − [Discount]

Labels: Sale price $= x$

List price $= 112.50$

Discount $= 31.50$

Equation:
$$x = 112.50 - 31.50$$
$$x = 81$$

The sale price is \$81.

Verbal model: [Discount] = [Discount rate] · [List price]

Labels: Discount $= 31.50$

Discount price $= p$

List price $= 112.50$

Equation:
$$31.50 = p(112.50)$$
$$\frac{31.50}{112.50} = p$$
$$0.28 = p$$

The discount rate is 28%.

Section 3.2 Graphs of Equations

1.

x	-2	-1	0	1	2
$y = 3x$	-6	-3	0	3	6
Solution point	$(-2, -6)$	$(-1, -3)$	$(0, 0)$	$(1, 3)$	$(2, 6)$

3.

x	-2	-1	0	1	2
$y = 4 - x$	6	5	4	3	2
Solution point	$(-2, 6)$	$(-1, 5)$	$(0, 4)$	$(1, 3)$	$(2, 2)$

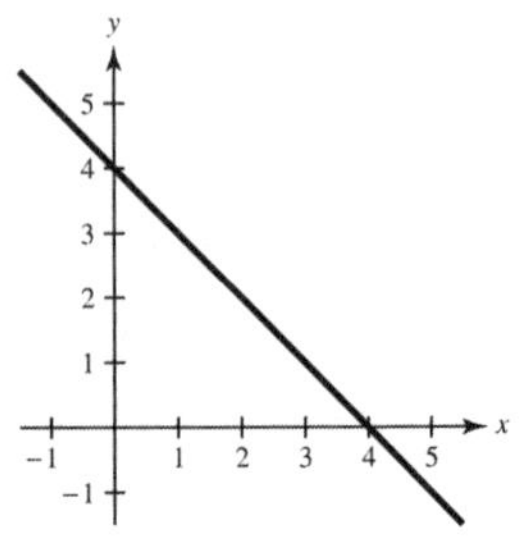

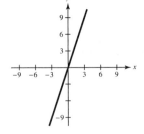

5. $2x - y = 3$

$-y = -2x + 3$

$y = 2x - 3$

x	-2	-1	0	1	2
$y = 2x - 3$	-7	-5	-3	-1	1
Solution point	$(-2, -7)$	$(-1, -5)$	$(0, -3)$	$(1, -1)$	$(2, 1)$

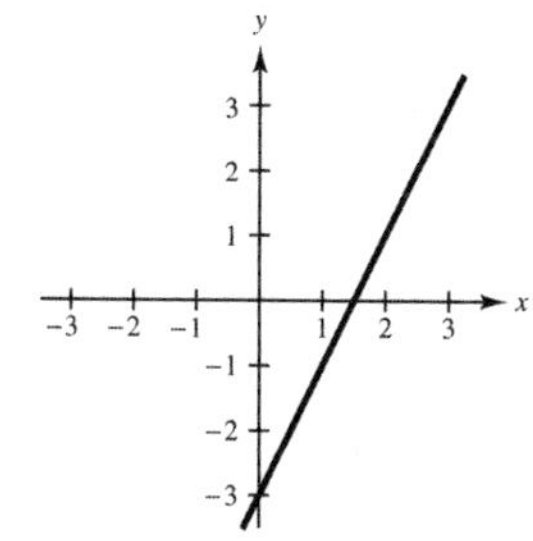

7. $3x + 2y = 2$

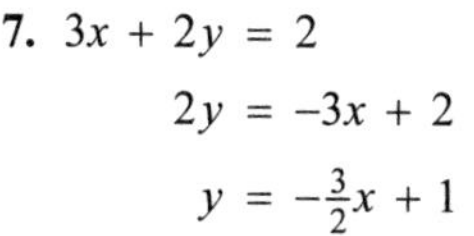

$2y = -3x + 2$

$y = -\frac{3}{2}x + 1$

x	-2	-1	0	1	2
$y = -\frac{3}{2}x + 1$	4	$\frac{5}{2}$	1	$-\frac{1}{2}$	-2
Solution point	$(-2, 4)$	$\left(-1, \frac{5}{2}\right)$	$(0, 1)$	$\left(1, -\frac{1}{2}\right)$	$(2, -2)$

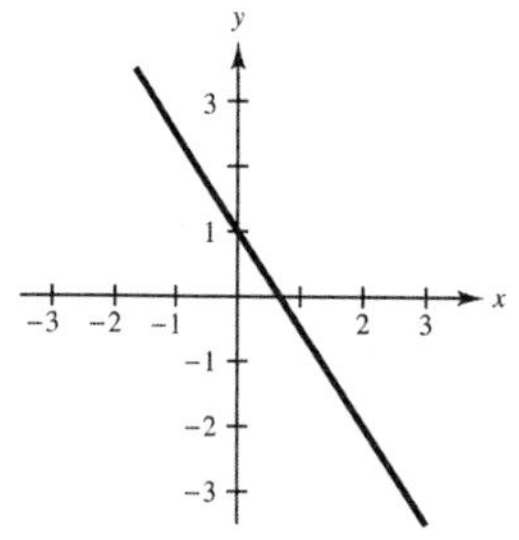

9.

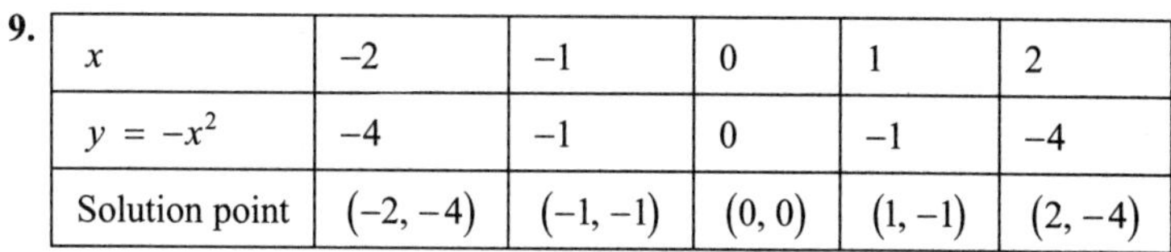

x	-2	-1	0	1	2
$y = -x^2$	-4	-1	0	-1	-4
Solution point	$(-2, -4)$	$(-1, -1)$	$(0, 0)$	$(1, -1)$	$(2, -4)$

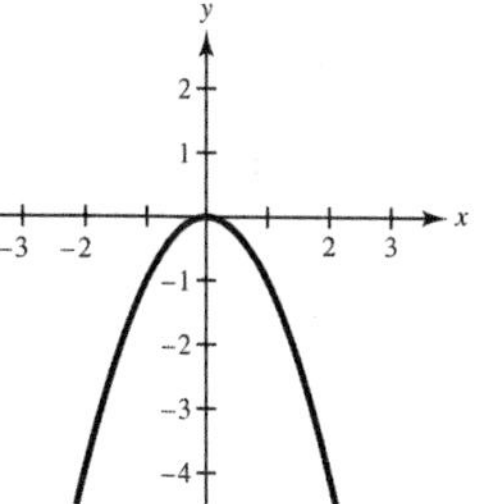

11.

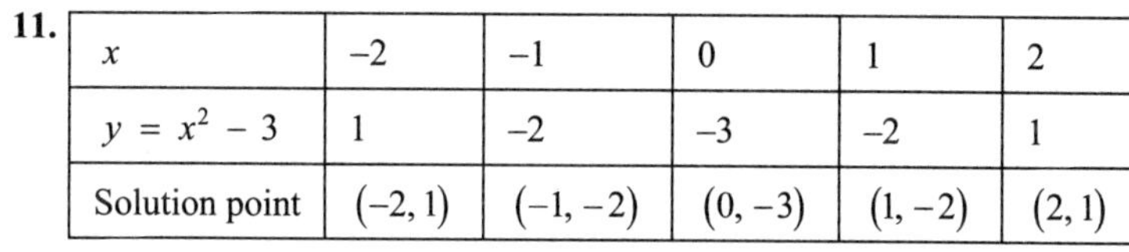

x	-2	-1	0	1	2
$y = x^2 - 3$	1	-2	-3	-2	1
Solution point	$(-2, 1)$	$(-1, -2)$	$(0, -3)$	$(1, -2)$	$(2, 1)$

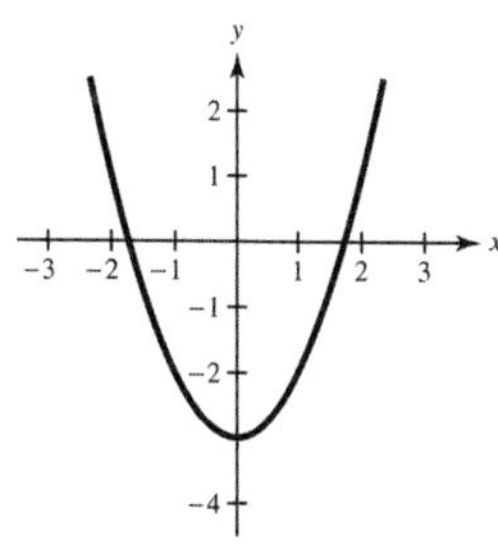

13. $-x^2 - 3x + y = 0$

$y = x^2 + 3x$

x	-2	-1	0	1	2
$y = x^2 + 3x$	-2	-2	0	4	10
Solution point	$(-2, -2)$	$(-1, -2)$	$(0, 0)$	$(1, 4)$	$(2, 10)$

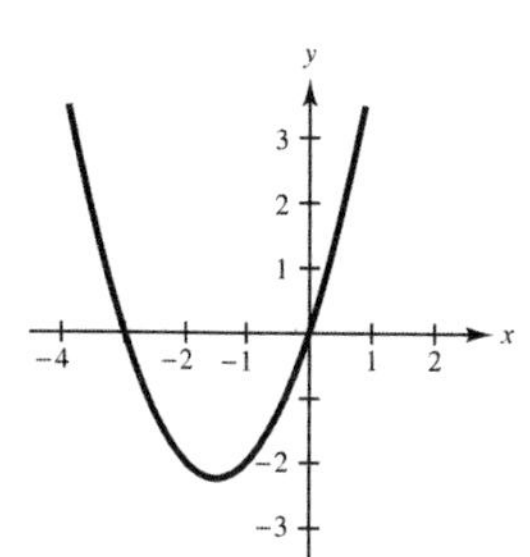

15. $x^2 - 2x - y = 1$

$$-y = -x^2 + 2x + 1$$

$$y = x^2 - 2x - 1$$

x	-2	-1	0	1	2
$y = x^2 - 2x - 1$	7	2	-1	-2	-1
Solution point	$(-2, 7)$	$(-1, 2)$	$(0, -1)$	$(1, -2)$	$(2, -1)$

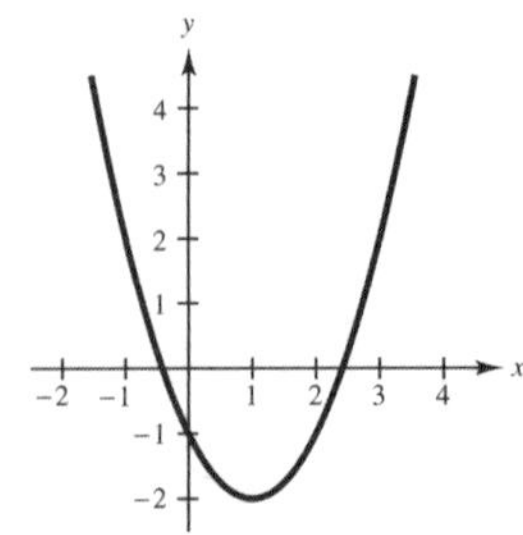

17.

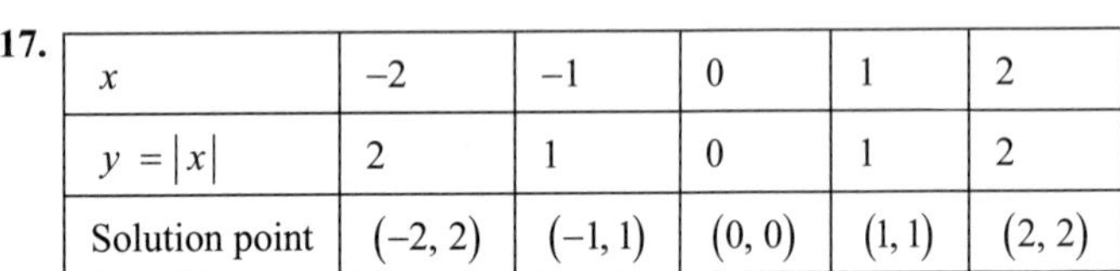

x	-2	-1	0	1	2
$y = \lvert x \rvert$	2	1	0	1	2
Solution point	$(-2, 2)$	$(-1, 1)$	$(0, 0)$	$(1, 1)$	$(2, 2)$

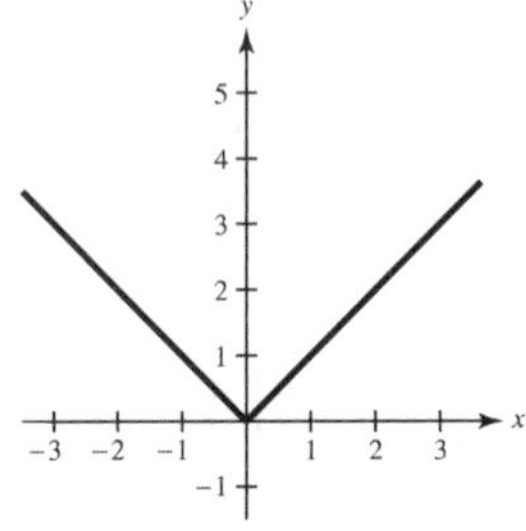

19.

x	-2	-1	0	1	2
$y = \lvert x \rvert + 3$	5	4	3	4	5
Solution point	$(-2, 5)$	$(-1, 4)$	$(0, 3)$	$(1, 4)$	$(2, 5)$

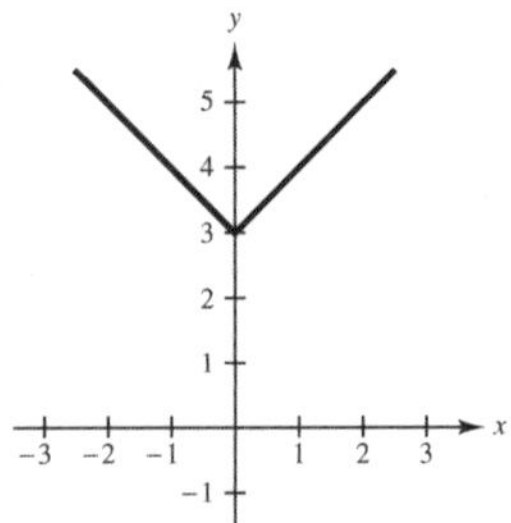

21.

x	-2	-1	0	1	2
$y = \lvert x + 3 \rvert$	1	2	3	4	5
Solution point	$(-2, 1)$	$(-1, 2)$	$(0, 3)$	$(1, 4)$	$(2, 5)$

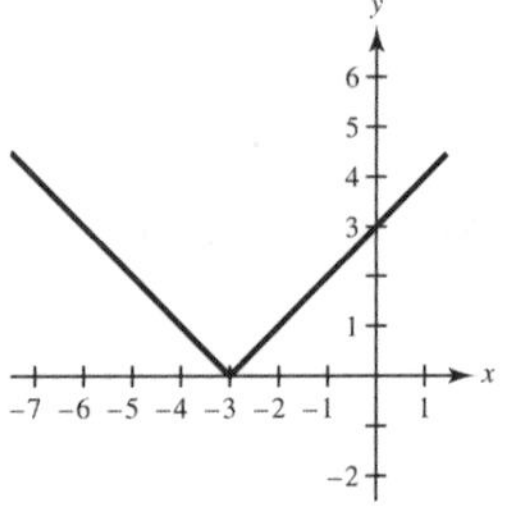

23. $y = 3 - x$

$y = 3 - 0$

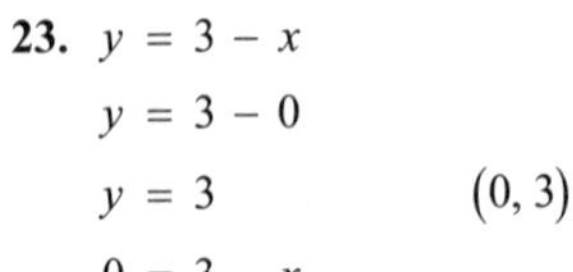

$y = 3$ $\quad (0, 3)$

$0 = 3 - x$

$x = 3$ $\quad (3, 0)$

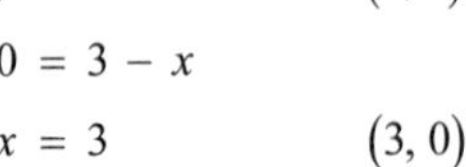

$y = 3 - 1$

$y = 2$ $\quad (1, 2)$

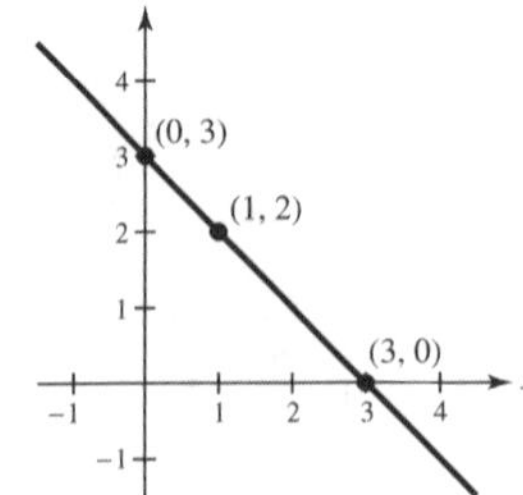

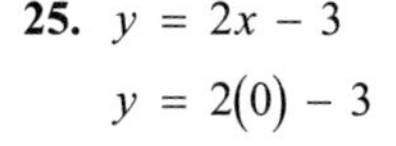

25.
$$\begin{aligned} y &= 2x - 3 \\ y &= 2(0) - 3 \\ y &= -3 && (0, -3) \\ 0 &= 2x - 3 \\ 3 &= 2x \\ \tfrac{3}{2} &= x && \left(\tfrac{3}{2}, 0\right) \\ y &= 2(3) - 3 \\ y &= 3 && (3, 3) \end{aligned}$$

27.
$$\begin{aligned} 4x + y &= 3 \\ 4(0) + y &= 3 \\ y &= 3 && (0, 3) \\ 4x + 0 &= 3 \\ 4x &= 3 \\ x &= \tfrac{3}{4} && \left(\tfrac{3}{4}, 0\right) \\ 4(1) + y &= 3 \\ y &= -1 && (1, -1) \end{aligned}$$

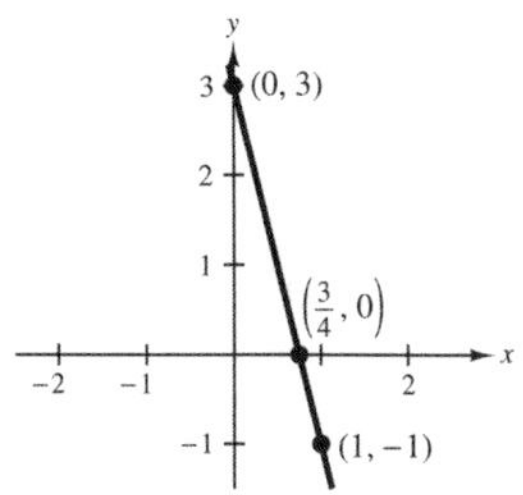

29.
$$\begin{aligned} 2x - 3y &= 6 \\ 2(0) - 3y &= 6 \\ -3y &= 6 \\ y &= -2 && (0, -2) \\ 2x - 3(0) &= 6 \\ 2x &= 6 \\ x &= 3 && (3, 0) \\ 2(1) - 3y &= 6 \\ -3y &= 4 \\ y &= -\tfrac{4}{3} && \left(1, -\tfrac{4}{3}\right) \end{aligned}$$

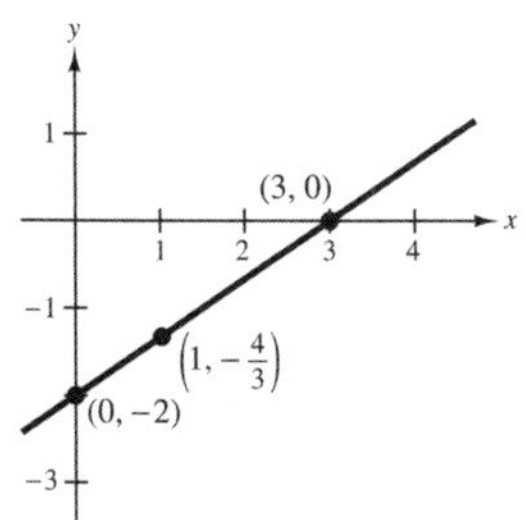

31.
$$\begin{aligned} 3x + 4y &= 12 \\ 3(0) + 4y &= 12 \\ 0 + 4y &= 12 \\ 4y &= 12 \\ y &= 3 && (0, 3) \\ 3x + 4(0) &= 12 \\ 3x + 0 &= 12 \\ 3x &= 12 \\ x &= 4 && (4, 0) \\ 3(1) + 4y &= 12 \\ 3 + 4y &= 12 \\ 4y &= 9 \\ y &= \tfrac{9}{4} && \left(1, \tfrac{9}{4}\right) \end{aligned}$$

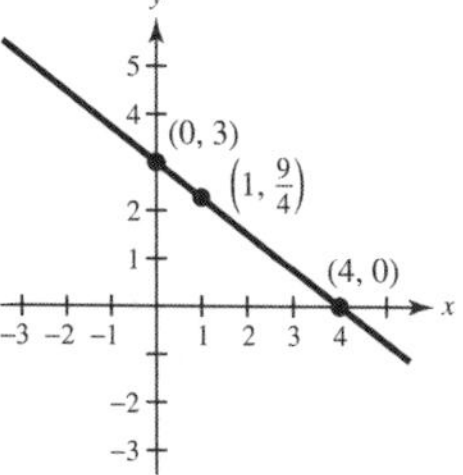

33.
$$\begin{aligned} x + 5y &= 10 \\ 0 + 5y &= 10 \\ y &= 2 && (0, 2) \\ x + 5(0) &= 10 \\ x &= 10 && (10, 0) \\ 5 + 5y &= 10 \\ 5y &= 5 \\ y &= 1 && (5, 1) \end{aligned}$$

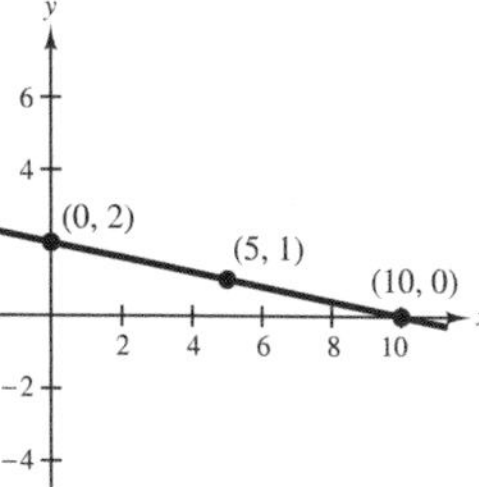

35. $5x - y = 10$

$5(0) - y = 10$

$0 - y = 10$

$-y = 10$

$y = -10$ $(0, -10)$

$5x - (0) = 10$

$5x = 10$

$x = 2$ $(2, 0)$

$5(1) - y = 10$

$5 - y = 10$

$-y = 5$

$y = -5$ $(1, -5)$

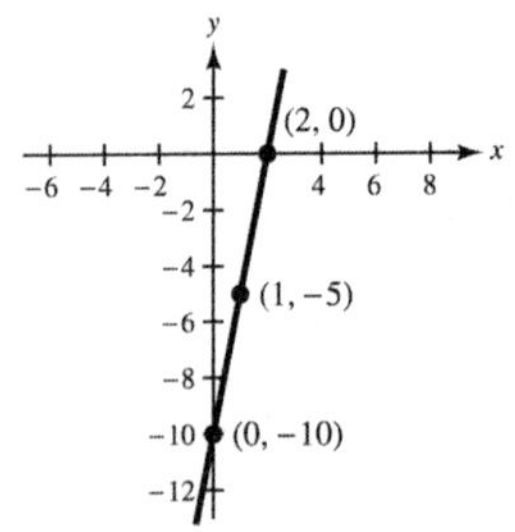

37. (a) The annual depreciation is

$$\frac{40{,}000 - 5000}{7} = 5000.$$

$40{,}000 - 5000(1) = 35{,}000$

$40{,}000 - 5000(2) = 30{,}000$

$40{,}000 - 5000(3) = 25{,}000$

So, $y = 40{,}000 - 5000t,\ 0 \le t \le 7$.

(b)

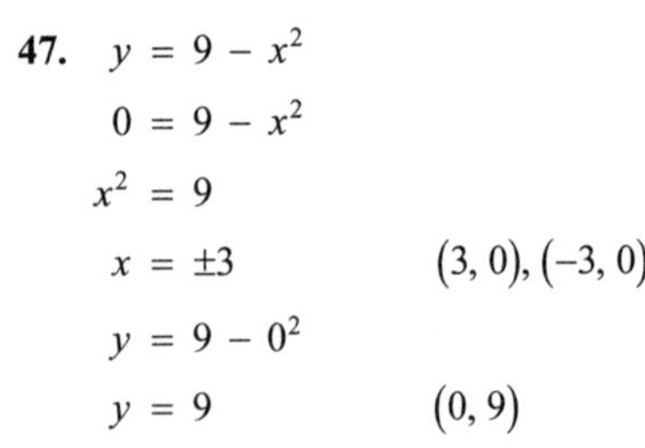

(c) $y = 40{,}000 - 5000(0)$

$y = 40{,}000,\ (0, 40{,}000)$

The y-intercept represents the value of the delivery van when purchased.

39. (a)

x	0	3	6	9	12
F	0	4	8	12	16

(b)

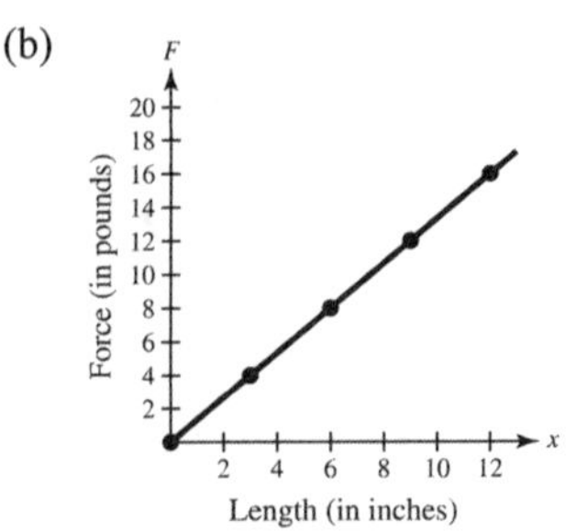

41. To complete the graph of the equation, connect the points with a smooth curve.

43. False. To find the x-intercept, substitute 0 for y in the equation and solve for x.

45. $y = x^2 - 9$

$0 = x^2 - 9$

$0 = (x - 3)(x + 3)$

$x = \pm 3$ $(3, 0), (-3, 0)$

$y = 0^2 - 9$

$y = -9$ $(0, -9)$

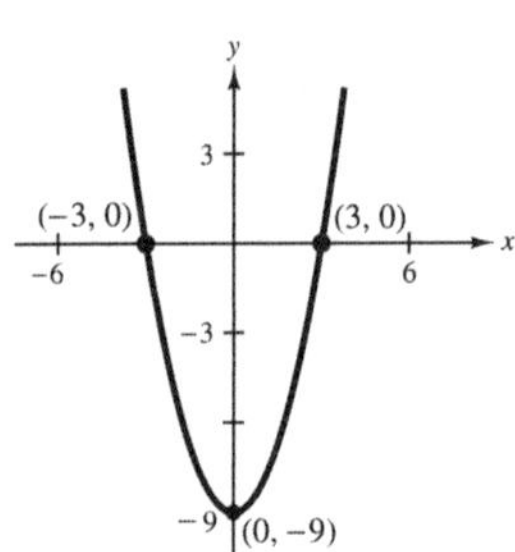

47. $y = 9 - x^2$

$0 = 9 - x^2$

$x^2 = 9$

$x = \pm 3$ $(3, 0), (-3, 0)$

$y = 9 - 0^2$

$y = 9$ $(0, 9)$

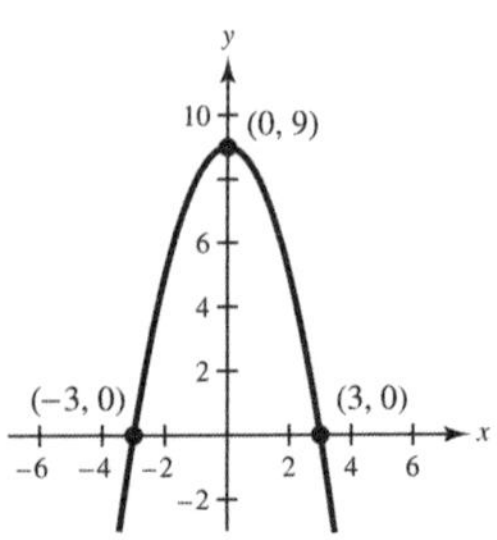

49. $y = x(x - 2)$

$y = 0^2 - 2(0)$

$y = 0 \quad (0, 0)$

$0 = x^2 - 2x$

$0 = x(x - 2)$

$x = 0, 2 \quad (0, 0), (2, 0)$

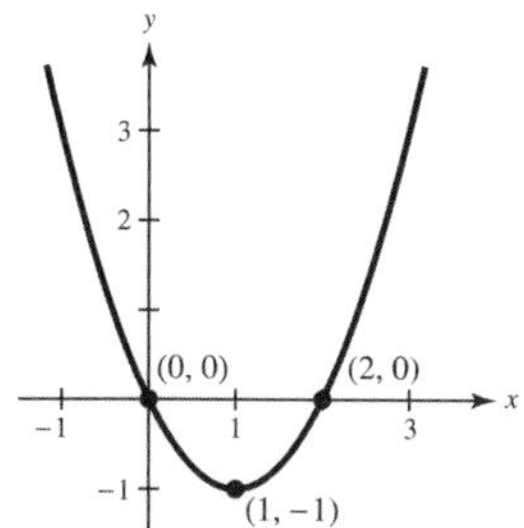

51. $y = -x(x + 4)$

$y = -(0)(0 + 4)$

$y = 0 \quad (0, 0)$

$0 = -x(x + 4)$

$x = 0, -4 \quad (0, 0), (-4, 0)$

$y = -(-2)(-2 + 4)$

$y = 2(2)$

$y = 4 \quad (-2, 4)$

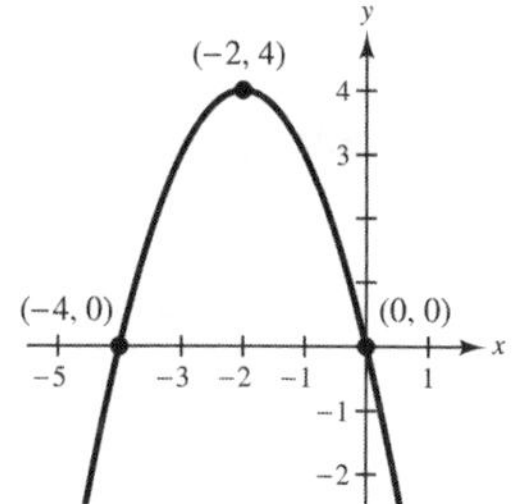

53. $y = |x + 2|$

$= 2 \quad (0, 2)$

$y = |x + 2|$

$0 = x + 2$

$-2 = x \quad (-2, 0)$

$y = |-4 + 2|$

$= 2 \quad (-4, 2)$

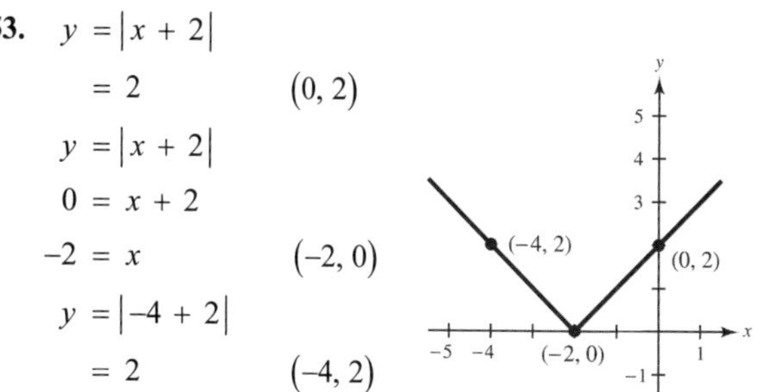

55. $y = |x - 3|$

$0 = |x - 3|$

$x = 3 \quad (3, 0)$

$y = |0 - 3|$

$y = |-3|$

$y = 3 \quad (0, 3)$

$y = |6 - 3|$

$y = |3|$

$y = 3 \quad (6, 3)$

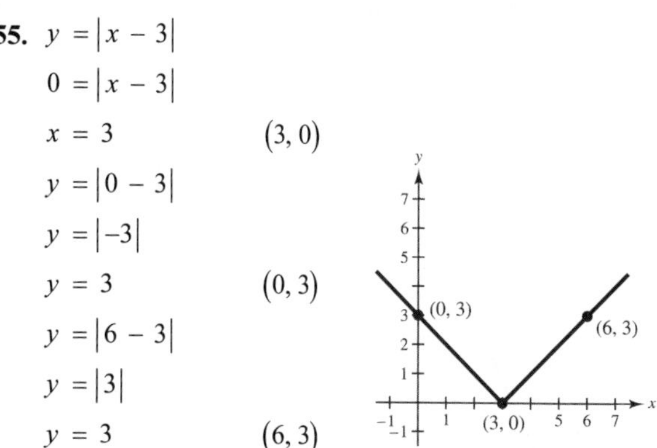

57. The scales on the y-axes are different. From graph (a) it appears that sales have not increased. From graph (b) it appears that sales have increased dramatically.

59.

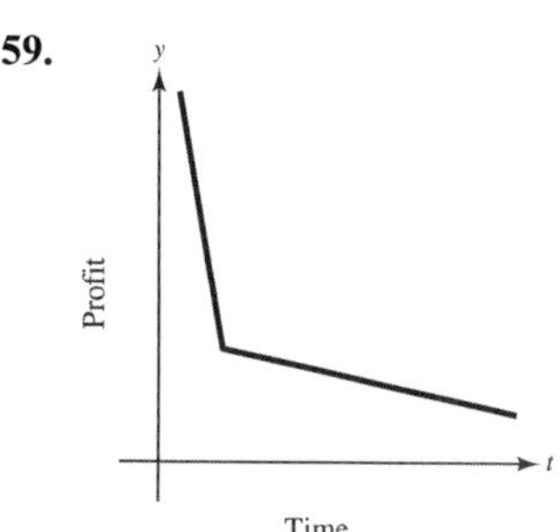

61. A horizontal line has no x-intercepts unless $y = 0$. The y-intercept is $(0, b)$, where b is any real number. So, a horizontal line has one y-intercept.

63. $\frac{1}{7}$

65. $\frac{5}{4}$

67. $x - 8 = 0$

$x = 8$

Check:

$8 - 8 \stackrel{?}{=} 0$

$0 = 0$

69. $4x + 15 = 23$

$4x = 8$

$x = 2$

Check:

$4(2) + 15 \stackrel{?}{=} 23$

$8 + 15 = 23$

71.

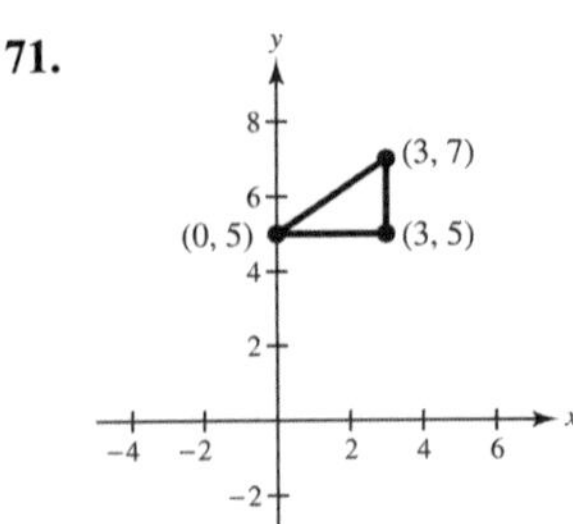

Section 3.3 Slope and Graphs of Linear Equations

1. $m = \dfrac{8 - 0}{4 - 0} = \dfrac{8}{4} = 2$

The slope is 2.

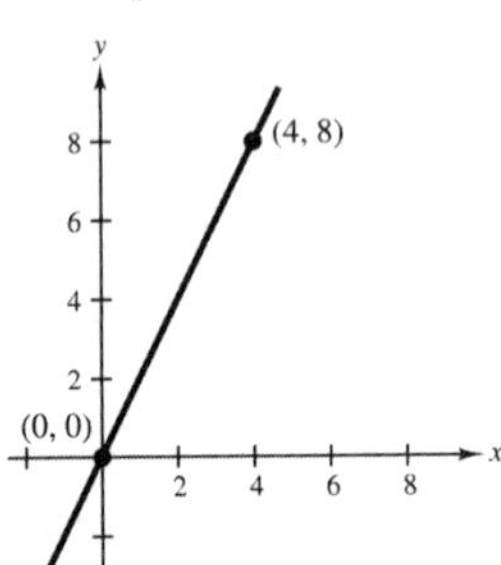

3. $m = \dfrac{5 - 3}{-2 - (-4)} = \dfrac{2}{2} = 1$

The slope is 1.

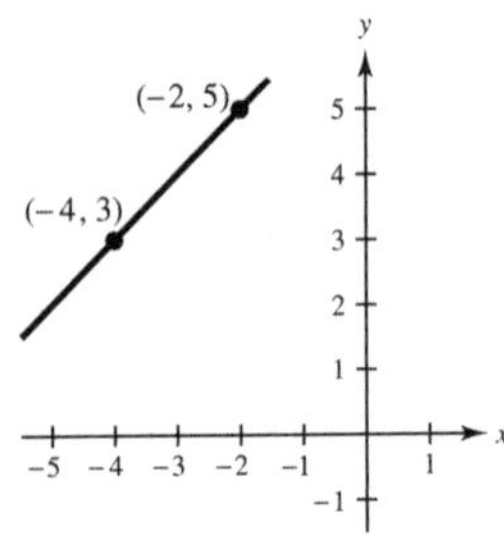

5. $m = \dfrac{10 - 7}{-2 - 3} = \dfrac{3}{-5} = -\dfrac{3}{5}$

The slope is $-\dfrac{3}{5}$.

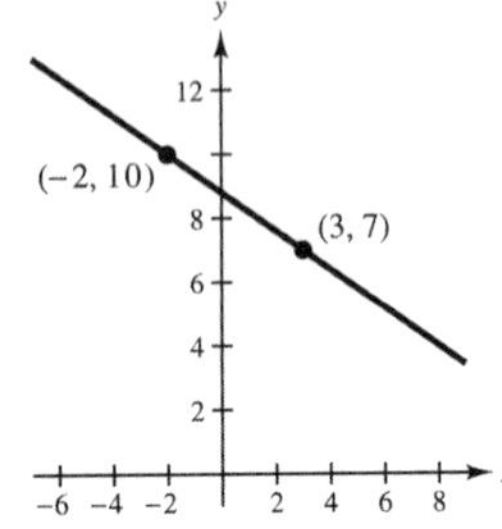

7. $m = \dfrac{-\frac{5}{2} - 2}{5 - \frac{3}{4}} = \dfrac{-\frac{9}{2}}{\frac{17}{4}} = -\dfrac{9}{2} \cdot \dfrac{4}{17} = \dfrac{-9 \cdot 2}{17} = -\dfrac{18}{17}$

The slope is $-\dfrac{18}{17}$.

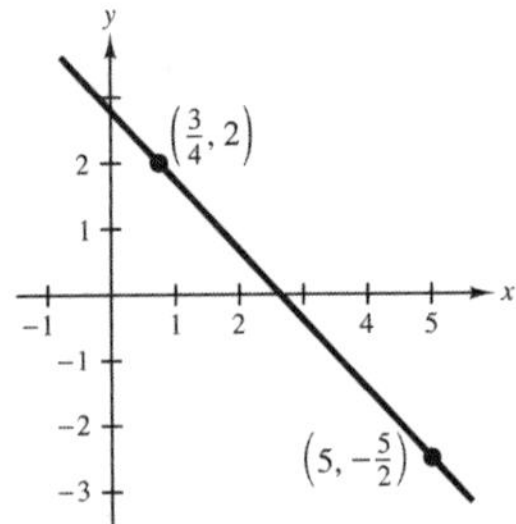

9. The slope of the line through (3, 2) and (3, 4) is undefined.

$\dfrac{4 - 2}{3 - 3} = \dfrac{2}{0}$

Because division by zero is not defined, the slope of a vertical line is not defined.

11. The slope of the line through (0, 0) and (5, 2) is

$m = \dfrac{2 - 0}{5 - 0}$

$= \dfrac{2}{5}.$

13. The slope of the line through (2, 4) and (3, 2) is

$m = \dfrac{2 - 4}{3 - 2}$

$= -\dfrac{2}{1}$

$= -2.$

15. (a) $m = \frac{3}{4} \Rightarrow L_3$

(b) $m = 0 \Rightarrow L_2$

(c) $m = -3 \Rightarrow L_1$

17. $y = 2x - 1$

slope $= 2$

y-intercept $= (0, -1)$

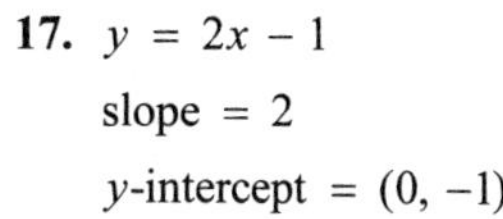

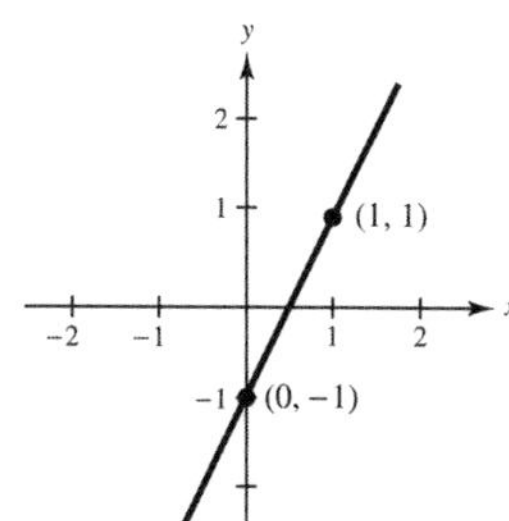

19. $y = -x - 2$

slope $= -1$

y-intercept $= (0, -2)$

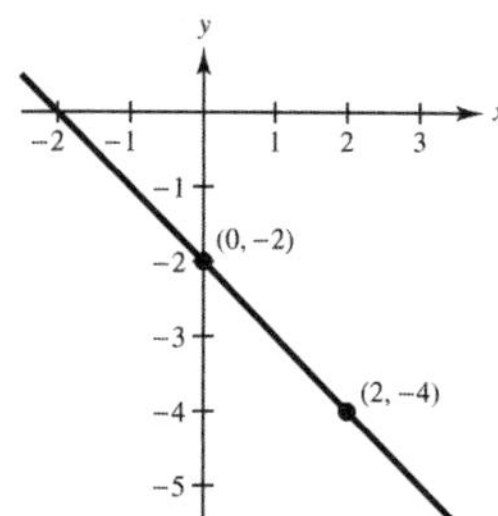

21. $y = -\frac{1}{2}x + 4$

slope $= -\frac{1}{2}$

y-intercept $= (0, 4)$

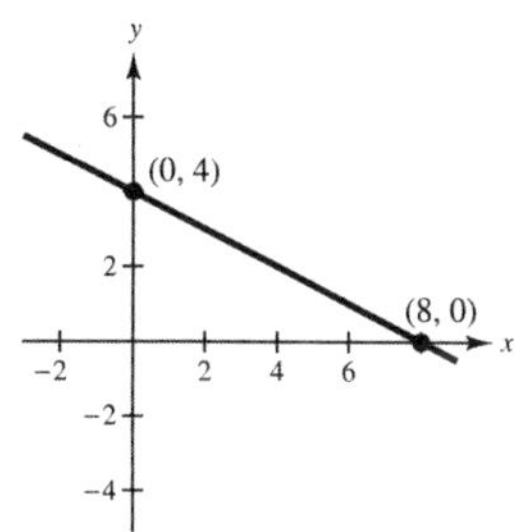

23. $y = 3x - 2$

$m = 3$; y-intercept $= (0, -2)$

25. $4x - 6y = 24$

$$-6y = -4x + 24$$

$$y = \frac{2}{3}x - 4$$

$m = \frac{2}{3}$; y-intercept $= (0, -4)$

27. $x + y = 0$

$$y = -x$$

slope $= -1$

y-intercept $= 0$

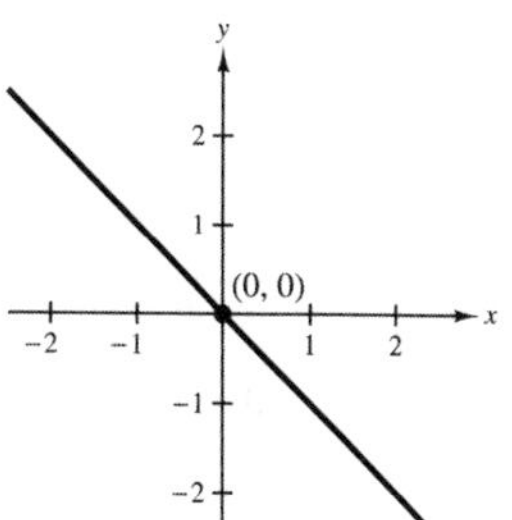

29. $3x - y - 2 = 0$

$$-y = -3x + 2$$

$$y = 3x - 2$$

slope $= 3$

y-intercept $= -2$

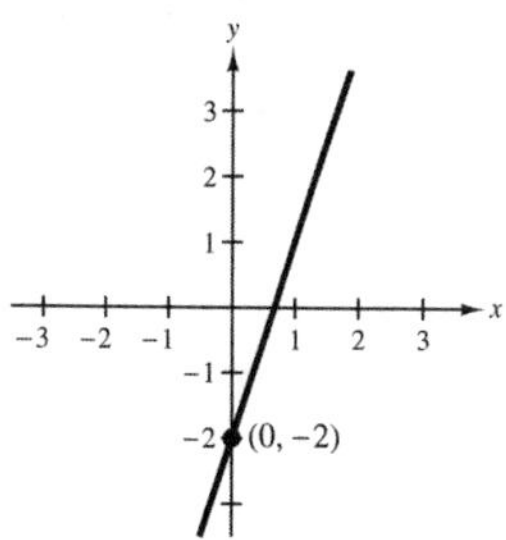

31. $3x + 2y - 2 = 0$

$$2y = -3x + 2$$

$$y = -\frac{3}{2}x + 1$$

slope $= -\frac{3}{2}$

y-intercept $= 1$

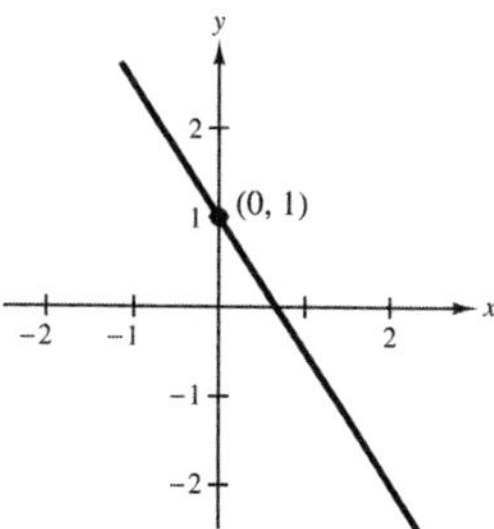

33.

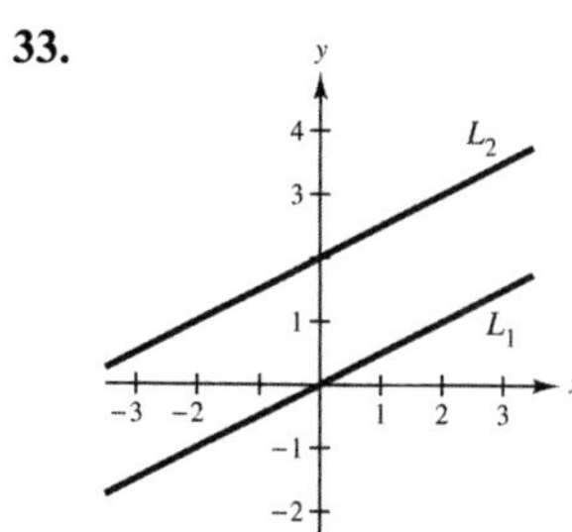

35. $L_1: y = \frac{1}{2}x - 2$

$L_2: y = \frac{1}{2}x + 3$

$m_1 = \frac{1}{2}$ and $m_2 = \frac{1}{2}$

$m_1 = m_2$ so the lines are parallel.

37. $L_1: y = \frac{3}{4}x - 3$

$L_2: y = -\frac{4}{3}x + 1$

$m_1 = \frac{3}{4}$ and $m_2 = -\frac{4}{3}$

$m_1 \cdot m_2 = -1$ so the lines are perpendicular.

39. $L_1: y = -4x + 6$

$L_2: y = -\frac{1}{4}x - 1$

$m_1 = -4$ and $m_2 = -\frac{1}{4}$

$m_1 \neq m_2$ and $m_1 \cdot m_2 \neq -1$, so the lines are neither parallel nor perpendicular.

41. Let w represent hourly wages and let t represent the year. The two given data points are represented by (t_1, w_1) and (t_2, w_2).

$(t_1, w_1) = (2005, 17.05)$

$(t_2, w_2) = (2010, 20.43)$

Use the formula for slope to find the average rate of change.

$$\begin{aligned}\text{Rate of change} &= \frac{w_2 - w_1}{t_2 - t_1}\\ &= \frac{20.43 - 17.05}{2010 - 2005}\\ &= 0.676\end{aligned}$$

From 2005 to 2010, the average rate of change in the hourly wage of health service employees was about \$0.68.

43. The rise is the numerator of the fraction $-\frac{2}{3}$, so the rise is -2. The run is the denominator of the fraction $-\frac{2}{3}$, so the run is 3.

45. A line with a negative slope falls from the left to right.

47. $m = \dfrac{\frac{1}{4} - \frac{1}{8}}{\frac{3}{4} - \frac{-3}{2}} \cdot \dfrac{8}{8} = \dfrac{2 - 1}{6 + 12} = \dfrac{1}{18}$ Line rises.

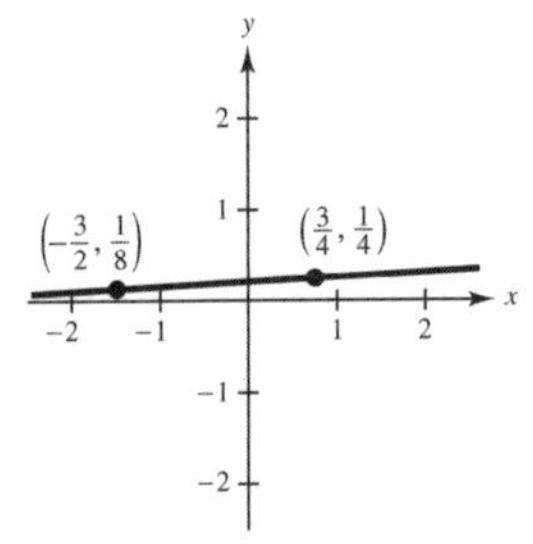

49. $m = \dfrac{6 - (-1)}{-4.2 - 4.2} = \dfrac{7}{-8.4} = -\dfrac{70}{84} = -\dfrac{5}{6}$ Line falls.

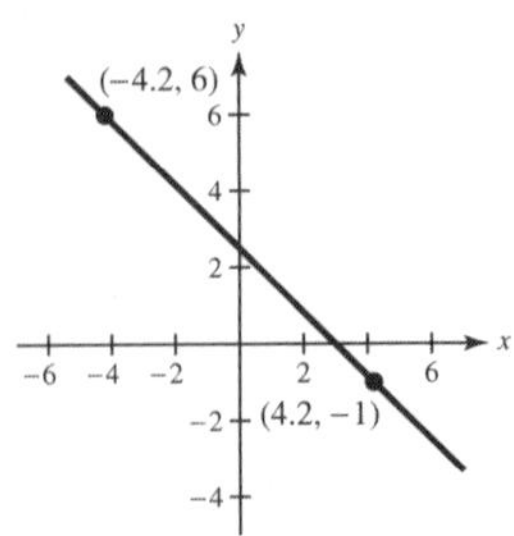

51.
$$\begin{aligned}\frac{-2}{3} &= \frac{7 - 5}{x - 4}\\ -2(x - 4) &= 6\\ -2x + 8 &= 6\\ -2x &= -2\\ x &= 1\end{aligned}$$

53.
$$\begin{aligned}\frac{3}{2} &= \frac{3 - y}{9 - (-3)}\\ 3(12) &= 2(3 - y)\\ 36 &= 6 - 2y\\ 30 &= -2y\\ -15 &= y\end{aligned}$$

55. $L_1: m_1 = \dfrac{8 - 4}{2 - 0} = \dfrac{4}{2} = 2$

$L_2: m_2 = \dfrac{5 - (-1)}{3 - 0} = \dfrac{6}{3} = 2$

$m_1 = m_2$ so the lines are parallel.

57. $L_1: m_1 = \dfrac{-2 - 2}{6 - 0} = \dfrac{-4}{6} = -\dfrac{2}{3}$

$L_2: m_2 = \dfrac{4 - 0}{8 - 2} = \dfrac{4}{6} = \dfrac{2}{3}$

$m_1 \neq m_2$ and $m_1 \cdot m_2 \neq -1$, so the lines are neither parallel nor perpendicular.

59. $3x - 5y - 15 = 0$

$3(0) - 5y - 15 = 0$

$-5y = 15$

$y = -3 \quad (0, -3)$

$3x - 5(0) - 15 = 0$

$3x = 15$

$x = 5 \quad (5, 0)$

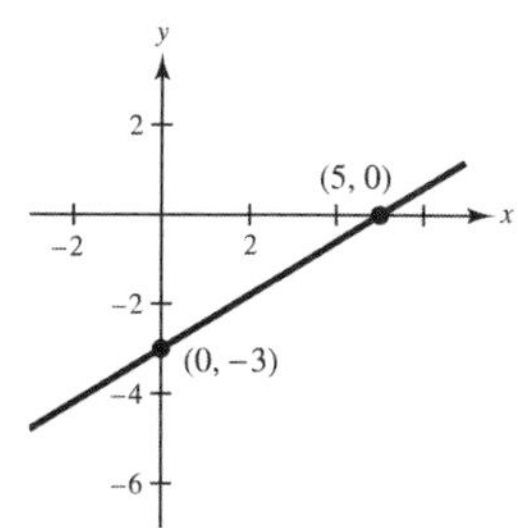

61. $-4x - 2y + 16 = 0$

$-4x - 2(0) + 16 = 0$

$-4x = -16$

$x = 4 \quad (4, 0)$

$-4(0) - 2y + 16 = 0$

$-2y = -16$

$y = 8 \quad (0, 8)$

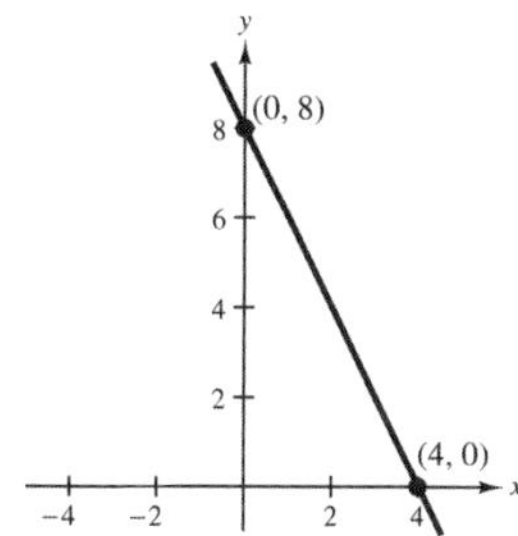

63. $-\dfrac{8}{100} = \dfrac{-2000}{x}$

$-8x = -200{,}000$

$x = 25{,}000$

The change in your horizontal position is 25,000 feet.

65. $\dfrac{3}{4} = \dfrac{h}{15}$

$45 = 4h$

$\dfrac{45}{4} = h$

The maximum height in the attic is $\dfrac{45}{4}$ feet $= 11.25$ feet.

67. Yes, any pair of points on a line can be used to calculate the slope of the line. When different pairs of points are selected, the change in y and the change in x are the lengths of the sides of similar triangles. Corresponding sides of similar triangles are proportional.

69. Two lines with undefined slope are parallel. Because the slopes are undefined, the lines are vertical lines, so they are parallel.

71. No, it is not possible for two lines with positive slopes to be perpendicular to each other. Their slopes must be negative reciprocals of each other.

73. $x < 0$

75. $85 \le z \le 100$

77. $|x| = 8$

$x = -8$ and $x = 8$

79. $|4h| = 24$

$4h = -24$ or $4h = 24$

$h = -6 \qquad h = 6$

81. $|x + 4| = 5$

$x + 4 = -5$ or $x + 4 = 5$

$x = -9 \qquad x = 1$

83. $|6b + 8| = -2b$

$6b + 8 = -2b$ or $6b + 8 = 2b$

$8 = -8b \qquad 8 = -4b$

$-1 = b \qquad -2 = b$

Section 3.4 Equations of Lines

1. $y - 0 = -\frac{1}{2}(x - 0)$

$y = -\frac{1}{2}x$

3. $y + 4 = 3(x - 0)$

$y + 4 = 3x$

$y = 3x - 4$

5. $y - 6 = -\frac{3}{4}(x - 0)$

$y - 6 = -\frac{3}{4}x$

$y = -\frac{3}{4}x + 6$

7. $y - 8 = -2[x - (-2)]$

$y - 8 = -2(x + 2)$

$y - 8 = -2x - 4$

$y = -2x + 4$

9. $y - (-7) = \frac{5}{4}[x - (-4)]$

$y + 7 = \frac{5}{4}(x + 4)$

$y + 7 = \frac{5}{4}x + 5$

$y = \frac{5}{4}x - 2$

11. Let $(x_1, y_1) = (0, 0)$ and let $(x_2, y_2) = (5, 2)$.

$m = \frac{2 - 0}{5 - 0} = \frac{2}{5}$

$y - y_1 = m(x - x_1)$

$y - 0 = \frac{2}{5}(x - 0)$

$y = \frac{2}{5}x$

$5y = 2x$

$-2x + 5y = 0$

The general form of the equation of the line is $-2x + 5y = 0$.

13. Let $(x_1, y_1) = (1, 4)$ and let $(x_2, y_2) = (5, 6)$.

$m = \frac{6 - 4}{5 - 1}$

$= \frac{1}{2}$

$y - y_1 = m(x - x_1)$

$y - 4 = \frac{1}{2}(x - 1)$

$2y - 8 = x - 1$

$-x + 2y - 7 = 0$

The general form of the equation of the line is $-x + 2y - 7 = 0$.

15. Let $(x_1, y_1) = \left(\frac{3}{2}, 3\right)$ and let $(x_2, y_2) = \left(\frac{9}{2}, 4\right)$.

$m = \frac{4 - 3}{\frac{9}{2} - \frac{3}{2}}$

$= \frac{1}{\frac{6}{2}}$

$= \frac{1}{3}$

$y - y_1 = m(x - x_1)$

$y - 3 = \frac{1}{3}\left(x - \frac{3}{2}\right)$

$y - 3 = \frac{1}{3}x - \frac{1}{2}$

$6y - 18 = 2x - 3$

$-2x + 6y - 15 = 0$

$2x - 6y + 15 = 0$

The general form of the equation of the line is $2x - 6y + 15 = 0$.

17. $x = -1$ because every x-coordinate is -1.

19. $y = -5$ because every y-coordinate is -5.

21. $x = -7$ because both points have an x-coordinate of -7.

23. Because the line is horizontal and passes through the point $(0, -2)$, every point on the line has a y-coordinate of -2. So, an equation of the line is $y = -2$.

25. $6x - 2y = 3$ slope $= 3$

$-2y = -6x + 3$

$y = 3x - \frac{3}{2}$

(a) $y - 1 = 3(x - 2)$

$y - 1 = 3x - 6$

$y = 3x - 5$

(b) $y - 1 = -\frac{1}{3}(x - 2)$

$y - 1 = -\frac{1}{3}x + \frac{2}{3}$

$y = -\frac{1}{3}x + \frac{2}{3} + \frac{3}{3}$

$y = -\frac{1}{3}x + \frac{5}{3}$

27. $5x + 4y = 24$

$$4y = -5x + 24$$

$$y = -\tfrac{5}{4}x + 6 \qquad \text{slope} = -\tfrac{5}{4}$$

(a) $y - 4 = -\tfrac{5}{4}[x - (-5)]$

$$y - 4 = -\tfrac{5}{4}(x + 5)$$

$$y - 4 = -\tfrac{5}{4}x - \tfrac{25}{4}$$

$$y = -\tfrac{5}{4}x - \tfrac{25}{4} + \tfrac{16}{4}$$

$$y = -\tfrac{5}{4}x - \tfrac{9}{4}$$

(b) $y - 4 = \tfrac{4}{5}[x - (-5)]$

$$y - 4 = \tfrac{4}{5}(x + 5)$$

$$y - 4 = \tfrac{4}{5}x + 4$$

$$y = \tfrac{4}{5}x + 8$$

29. $4x - y - 3 = 0$

$$-y = -4x + 3$$

$$y = 4x + 3$$

$$\text{slope} = 4$$

(a) $y - (-3) = 4(x - 5)$

$$y + 3 = 4x - 20$$

$$y = 4x - 23$$

(b) $y - (-3) = -\tfrac{1}{4}(x - 5)$

$$y + 3 = -\tfrac{1}{4}x + \tfrac{5}{4}$$

$$y = -\tfrac{1}{4}x - \tfrac{7}{4}$$

31. $(5, 1500), (6, 1560)$

$$m = \frac{1560 - 1500}{6 - 5} = \frac{60}{1} = 60$$

$$N - 1500 = 60(t - 5)$$

$$N - 1500 = 60t - 300$$

$$N = 60t + 1200$$

33. $m = \dfrac{6000 - 5000}{50 - 0} = \dfrac{1000}{50} = 20$

$$C - 5000 = 20(x - 0)$$

$$C = 20x + 5000$$

$$C = 20(400) + 5000$$

$$= 13{,}000$$

The cost is \$13,000.

35. $y - y_1 = m(x - x_1)$

$$y - 1 = 4(x - 1)$$

The slope is 4. The line passes through $(1, 1)$.

37. Yes. The horizontal line passing through $(-4, 5)$ is $y = 5$.

39. $\dfrac{x}{3} + \dfrac{y}{2} = 1$

41. $\dfrac{x}{\frac{-5}{6}} + \dfrac{y}{\frac{-7}{3}} = 1$

$$-\frac{6x}{5} - \frac{3y}{7} = 1$$

43. $m = \dfrac{200{,}000 - 500{,}000}{2 - 5} = \dfrac{-300{,}000}{-3} = 100{,}000$

$$S - 500{,}000 = 100{,}000(t - 5)$$

$$S - 500{,}000 = 100{,}000t - 500{,}000$$

$$S = 100{,}000t$$

$$S = 100{,}000(6) = 600{,}000$$

The total sales for the sixth year are \$600,000.

45. (a) $(0, 7400), (4, 1500)$

$$m = \frac{7400 - 1500}{0 - 4} = \frac{5900}{-4} = -1475$$

$$V - 7400 = -1475(t - 0)$$

$$V - 7400 = -1475t$$

$$V = -1475t + 7400$$

(b) $V = -1475(2) + 7400$

$$V = -2950 + 7400$$

$$V = 4450$$

The photocopier has a value of \$4450 after 2 years.

47. $(0, 0), (40, 5)$

$$m = \frac{5 - 0}{40 - 0} = \frac{5}{40} = \frac{1}{8}$$

$$y - 0 = \frac{1}{8}(x - 0)$$

$$y = \frac{1}{8}x$$

$$8y = x$$

$$x - 8y = 0$$

Depth of water $= 9 - y$

Distance from deep end	0	8	16	24	32	40
Depth of water	9	8	7	6	5	4

(a) $9 - y$

$$9 = 9 - y$$
$$0 = y$$
$$x - 8(0) = 0$$
$$x = 0$$

(b) $9 - y$

$$8 = 9 - y$$
$$-1 = -y$$
$$1 = y$$
$$x - 8(1) = 0$$
$$x = 8$$

(c) $9 - y$

$$7 = 9 - y$$
$$-2 = -y$$
$$2 = y$$
$$x - 8(2) = 0$$
$$x = 16$$

(d) $9 - y$

$$6 = 9 - y$$
$$-3 = -y$$
$$3 = y$$
$$x - 8(3) = 0$$
$$x = 24$$

(e) $9 - y$

$$5 = 9 - y$$
$$-4 = -y$$
$$4 = y$$
$$x - 8(4) = 0$$
$$x = 32$$

(f) $9 - y$

$$4 = 9 - y$$
$$-5 = -y$$
$$5 = y$$
$$x - 8(5) = 0$$
$$x = 40$$

49. Point-slope form: $y - y_1 = m(x - x_1)$

Slope-intercept form: $y = mx + b$

General form: $ax + by + c = 0$

51. The variable y is missing in the equation of a vertical line because any point on a vertical line is independent of y.

53.

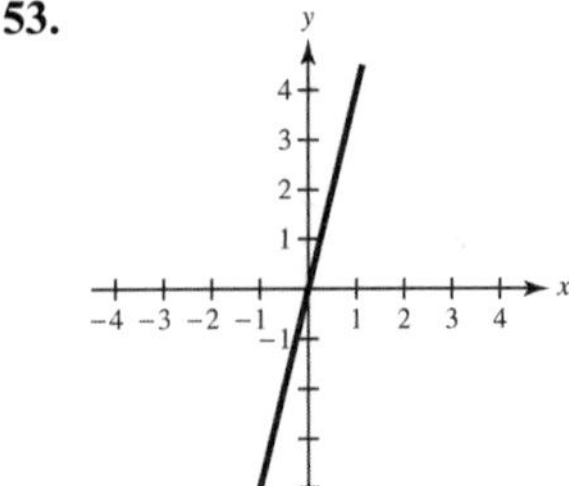

55.

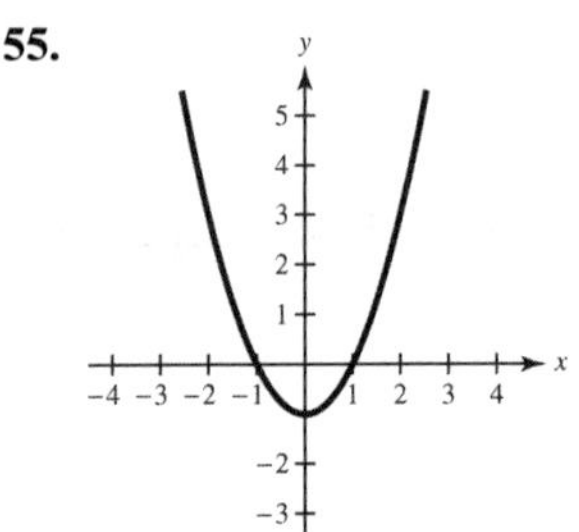

57. $\dfrac{4 - 2}{a - 1} = 2$

$$\frac{2}{a - 1} = 2$$
$$2 = 2(a - 1)$$
$$2 = 2a - 2$$
$$4 = 2a$$
$$2 = a$$

59. $\dfrac{3 - a}{-2 - (-4)} = \dfrac{1}{2}$

$$2(3 - a) = 1(2)$$
$$6 - 2a = 2$$
$$-2a = -4$$
$$a = 2$$

61. $\dfrac{a - 0}{0 - 5} = -\dfrac{3}{5}$

$$5a = 15$$
$$a = 3$$

63. $\dfrac{a - (-2)}{-1 - (-7)} = -1$

$$a + 2 = -6$$
$$a = -8$$

Mid-Chapter Quiz for Chapter 3

1. Quadrant I or II. Because x can be any real number and y is 4, the point $(x, 4)$ can only be located in quadrants in which the y-coordinate is positive.

2. $4x - 3y = 10$

(a) $4(2) - 3(1) \stackrel{?}{=} 10$

$8 - 3 \stackrel{?}{=} 10$

$5 \neq 10$ Not a solution

(b) $4(1) - 3(-2) \stackrel{?}{=} 10$

$4 + 6 \stackrel{?}{=} 10$

$10 = 10$ Solution

(c) $4(2.5) - 3(0) \stackrel{?}{=} 10$

$10 - 0 \stackrel{?}{=} 10$

$10 = 10$ Solution

(d) $4(2) - 3\left(-\frac{2}{3}\right) \stackrel{?}{=} 10$

$8 + 2 \stackrel{?}{=} 10$

$10 = 10$ Solution

3.

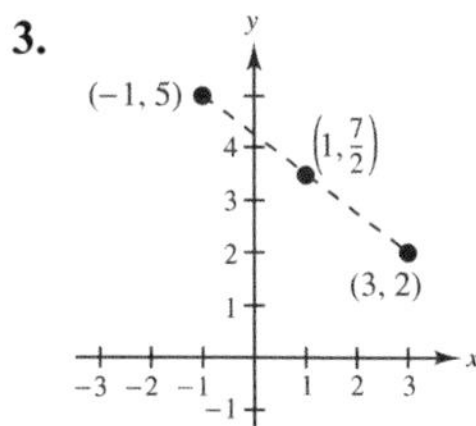

$$d = \sqrt{(-1-3)^2 + (5-2)^2}$$
$$= \sqrt{16+9}$$
$$= \sqrt{25}$$
$$= 5$$
$$M = \left(\frac{-1+3}{2}, \frac{5+2}{2}\right) = \left(1, \frac{7}{2}\right)$$

4.

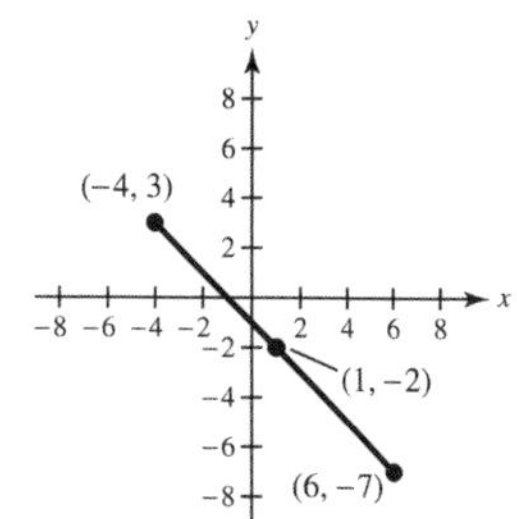

$$d = \sqrt{[6-(-4)]^2 + (-7-3)^2}$$
$$= \sqrt{10^2 + (-10)^2}$$
$$= \sqrt{100+100}$$
$$= \sqrt{200} = \sqrt{2 \cdot 100} = 10\sqrt{2}$$
$$M = \left(\frac{-4+6}{2}, \frac{3+(-7)}{2}\right) = \left(\frac{2}{2}, \frac{-4}{2}\right) = (1, -2)$$

5. $3x + y - 6 = 0$

$3(0) + y - 6 = 0$

$y = 6$ $(0, 6)$

$3x + 0 - 6 = 0$

$3x = 6$

$x = 2$ $(2, 0)$

$3(1) + y - 6 = 0$

$y = 3$ $(1, 3)$

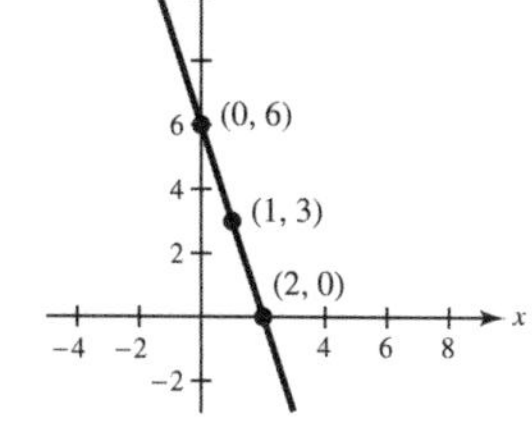

6. $y = 6x - x^2$

$y = 6(0) - 0^2$

$= 0$ $(0, 0)$

$y = 6(6) - 6^2$

$= 0$ $(6, 0)$

$y = 6(3) - 3^2$

$= 18 - 9$

$= 9$ $(3, 9)$

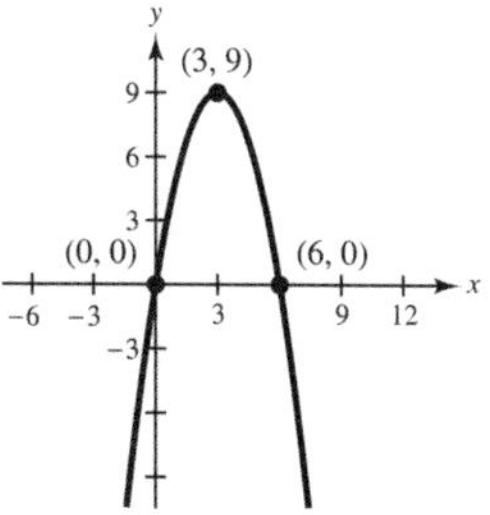

7. $y = |x - 2| - 3$

$y = |0 - 2| - 3$

$= -1$ $(0, -1)$

$y = |5 - 2| - 3$

$= 0$ $(5, 0)$

$y = |2 - 2| - 3$

$= -3$ $(2, -3)$

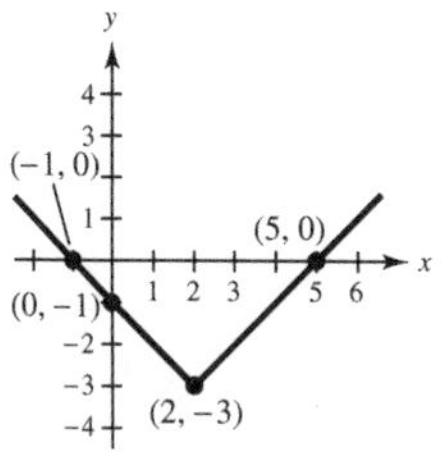

8. $m = \dfrac{8 - 8}{7 - (-3)} = \dfrac{0}{10} = 0$ Line is horizontal.

9. $m = \dfrac{5 - 0}{6 - 3} = \dfrac{5}{3}$ Line rises.

10. $m = \dfrac{-1 - 7}{4 - (-2)} = \dfrac{-8}{6} = -\dfrac{4}{3}$

Line falls.

11. $3x + 6y = 6$

$$6y = -3x + 6$$

$$y = -\frac{1}{2}x + 1$$

$m = -\frac{1}{2}$; y-intercept $= (0, 1)$

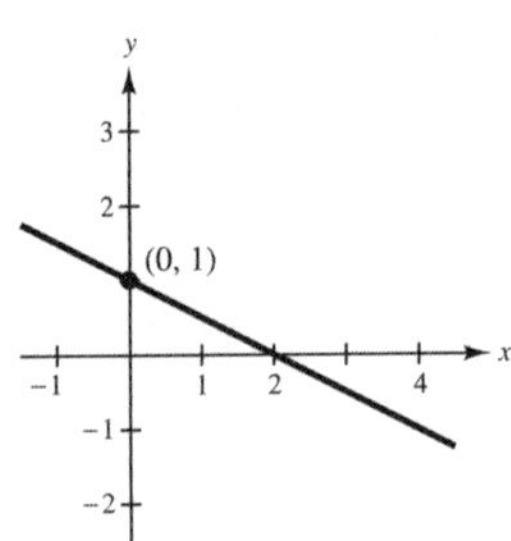

12. $6x - 4y = 12$

$$-4y = -6x + 12$$

$$y = \frac{3}{2}x - 3$$

$m = \frac{3}{2}$

y-intercept $= (0, -3)$

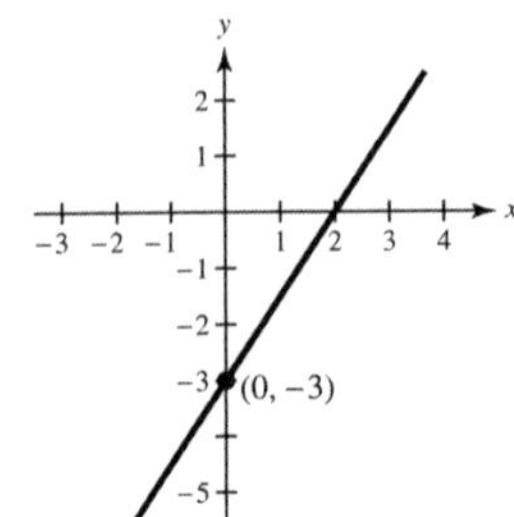

13. $y = 3x + 2$; $y = -\frac{1}{3}x - 4$

$m_1 = 3$

$m_2 = -\frac{1}{3}$

$m_1 \cdot m_2 = -1$

The lines are perpendicular.

14. L_1: $m_1 = \dfrac{(-9) - 3}{(-2) - 4} = \dfrac{-12}{-6} = 2 \Rightarrow y - 3 = 2(x - 4) \Rightarrow y = 2x - 5$

L_2: $m_2 = \dfrac{5 - (-5)}{5 - 0} = \dfrac{10}{5} = 2 \Rightarrow y = 2x - 5$

The lines are neither parallel nor perpendicular because they are the same line.

15.
$$y - (-1) = \frac{1}{2}(x - 6)$$
$$y + 1 = \frac{1}{2}(x - 6)$$
$$2(y + 1) = x - 6$$
$$2y + 2 = x - 6$$
$$x - 2y - 8 = 0$$

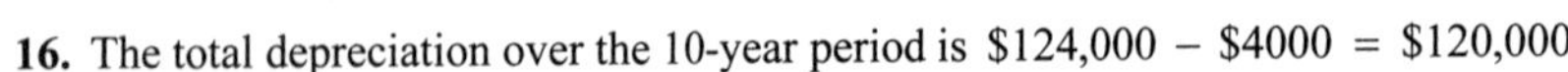

16. The total depreciation over the 10-year period is $124,000 − $4000 = $120,000.

The annual depreciation is $\dfrac{\$120{,}000}{10} = \$12{,}000$.

$124,000 − (2)$12,000 = $100,000

$124,000 − (3)$12,000 = $88,000

$124,000 − (4)$12,000 = $76,000

So, $y = 124{,}000 - 12{,}000t$, $0 \le t \le 10$.

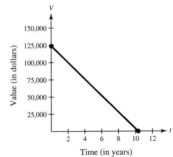

Section 3.5 Graphs of Linear Inequalities

1. $x - 2y < 4$

(a) $0 - 2(0) \overset{?}{<} 4$

$0 < 4$

$(0, 0)$ *is* a solution.

(b) $2 - 2(-1) \overset{?}{<} 4$

$2 + 2 < 4$

$4 \not< 4$

$(2, -1)$ *is not* a solution.

(c) $3 - 2(4) \overset{?}{<} 4$

$3 - 8 < 4$

$-5 < 4$

$(3, 4)$ *is* a solution.

(d) $5 - 2(1) \overset{?}{<} 4$

$5 - 2 < 4$

$3 < 4$

$(5, 1)$ *is* a solution.

3. $3x + y \ge 10$

(a) $3(1) + 3 \overset{?}{\ge} 10$

$9 \not\ge 10$

$(1, 3)$ *is not* a solution.

(b) $3(-3) + 1 \overset{?}{\ge} 10$

$-8 \not\ge 10$

$(-3, 1)$ *is not* a solution.

(c) $3(3) + 1 \overset{?}{\ge} 10$

$10 \ge 10$

$(3, 1)$ *is* a solution.

(d) $3(2) + 15 \overset{?}{\ge} 10$

$21 \ge 10$

$(2, 15)$ *is* a solution.

5. $y > 0.2x - 1$

(a) $2 \overset{?}{>} 0.2(0) - 1$

$2 > -1$

$(0, 2)$ *is* a solution.

(b) $0 \overset{?}{>} 0.2(6) - 1$

$0 > 0.2$

$(6, 0)$ *is not* a solution.

(c) $-1 \overset{?}{>} 0.2(4) - 1$

$-1 \not> -0.2$

$(4, -1)$ *is not* a solution.

(d) $7 \overset{?}{>} 0.2(-2) - 1$

$7 > -1.4$

$(-2, 7)$ *is* a solution.

7. $x \ge 6$

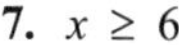

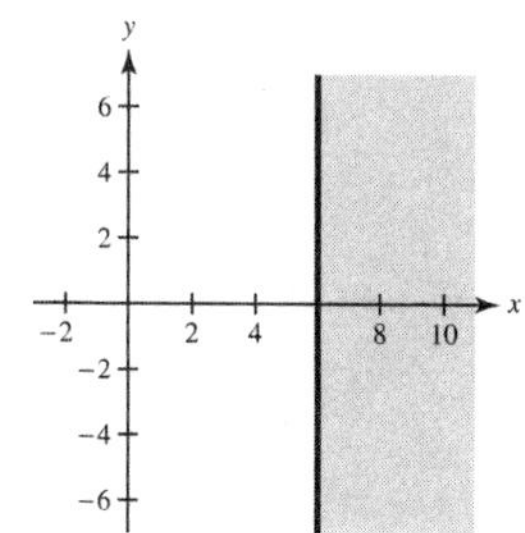

9. $y < 5$

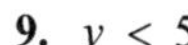

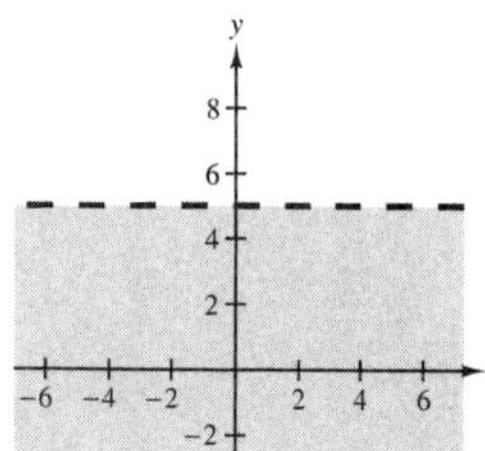

11. $y > \frac{1}{2}x$

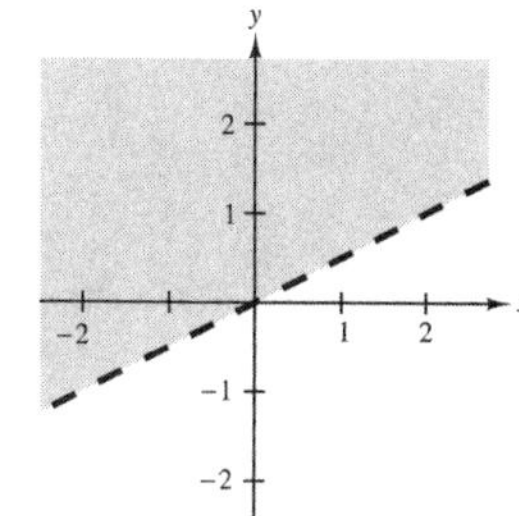

13. $y \geq -2$; (b)

14. $x < -2$; (a)

15. $x - y < 0$; (d)

16. $x - y > 0$; (e)

17. $x + y < 4$; (f)

18. $x + y \leq 4$; (c)

19. $y \geq 3 - x$

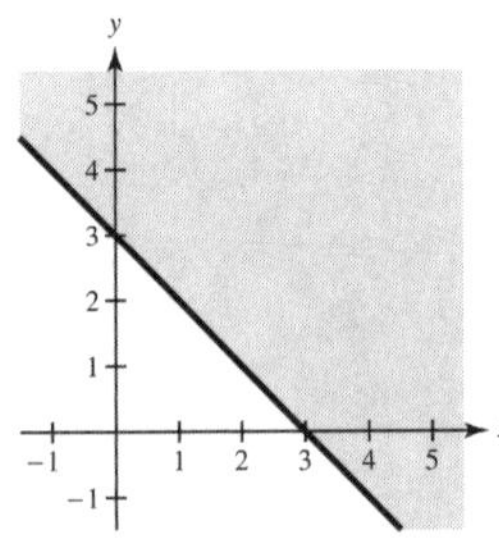

21. $y \leq x + 2$

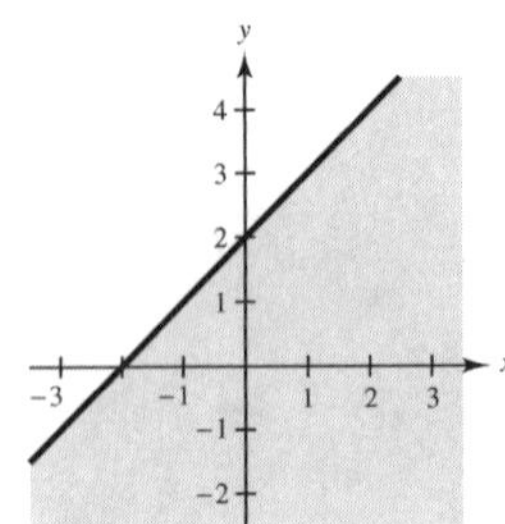

23. $x + y \geq 4$

$$y \geq -x + 4$$

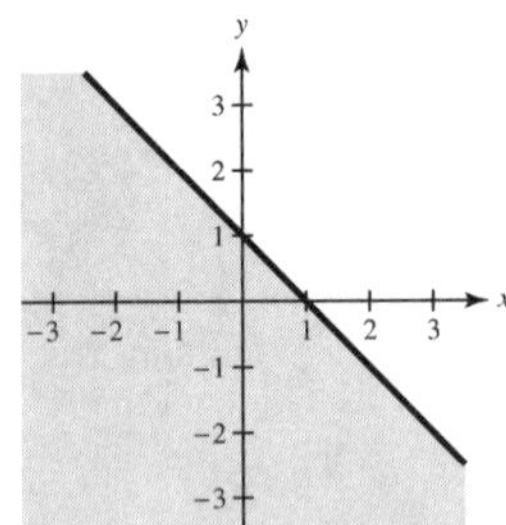

25. Because the origin $(0, 0)$ does not satisfy the inequality, the graph consists of the half-plane lying below the line.

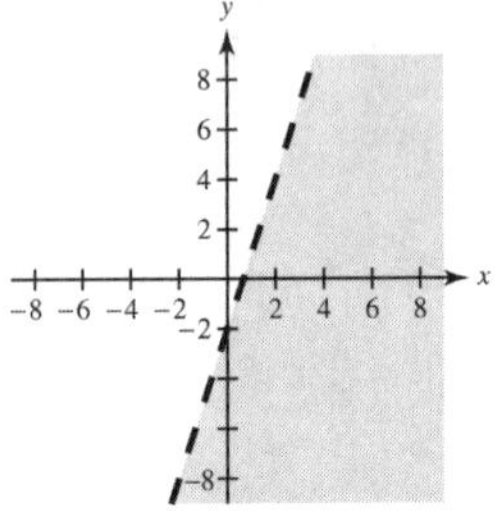

27. $2x + y \geq 4$

$$y \geq -2x + 4$$

Because the origin $(0, 0)$ does not satisfy the inequality, the graph consists of the half-plane lying on or above the line.

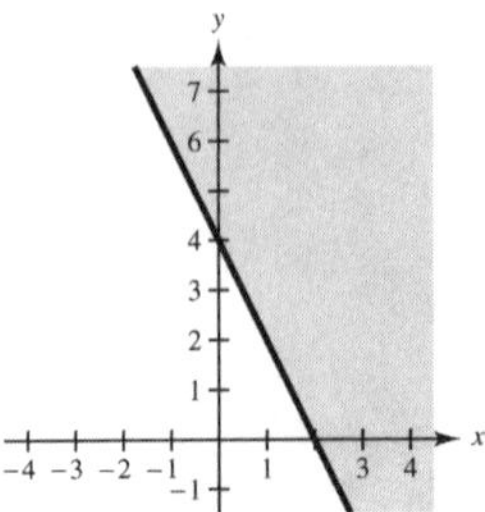

29. $-x + 2y < 4$

$$2y < x + 4$$

$$y < \frac{1}{2}x + 2$$

Because the origin $(0, 0)$ satisfies the inequality, the graph consists of the half-plane lying below the line.

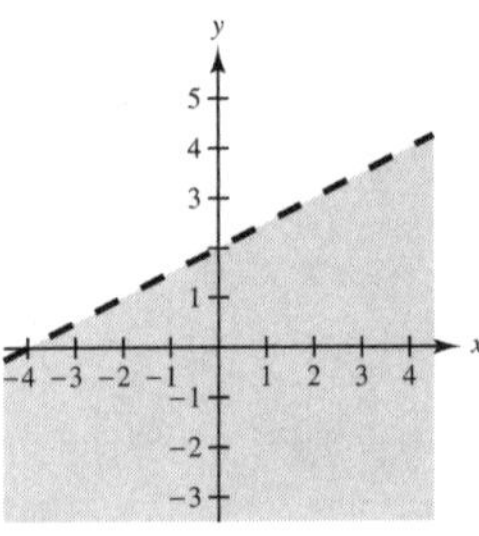

31. $\frac{1}{2}x + y > 6$

$$y > -\frac{1}{2}x + 6$$

Because the origin $(0, 0)$ does not satisfy the inequality, the graph consists of the half-plane lying above the line.

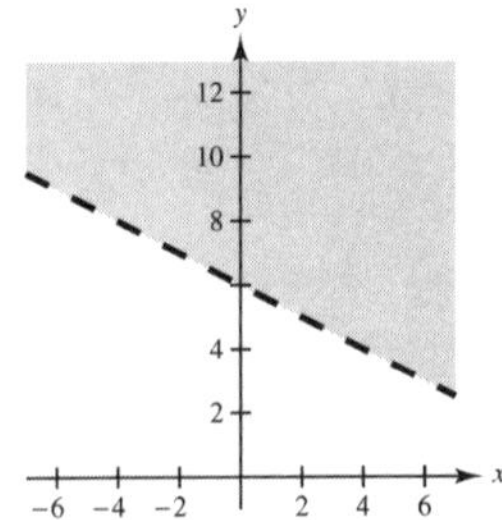

33. $x - 4y \le 8$

$-4y \le -x + 8$

$y \ge \frac{1}{4}x - 2$

Because the origin (0, 0) satisfies the inequality, the graph consists of the half-plane lying on or above the line.

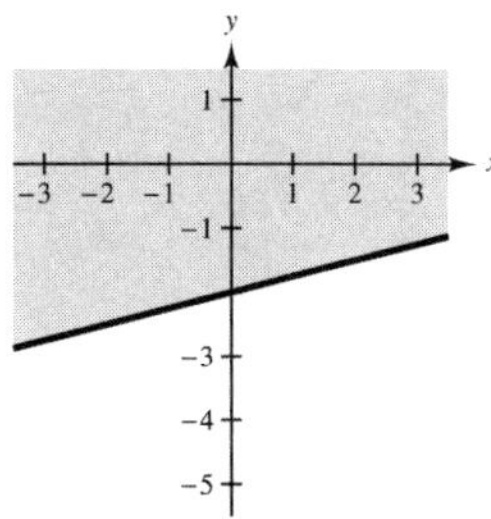

35. $3x + 2y \ge 2$

$2y \ge -3x + 2$

$y \ge -\frac{3}{2}x + 1$

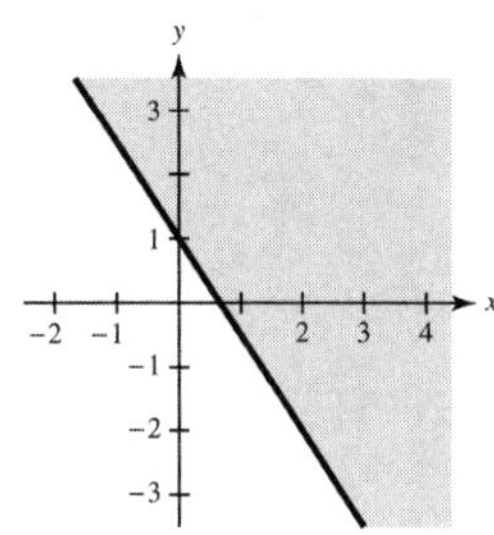

37. $5x + 4y < 20$

$4y < -5x + 20$

$y < -\frac{5}{4}x + 5$

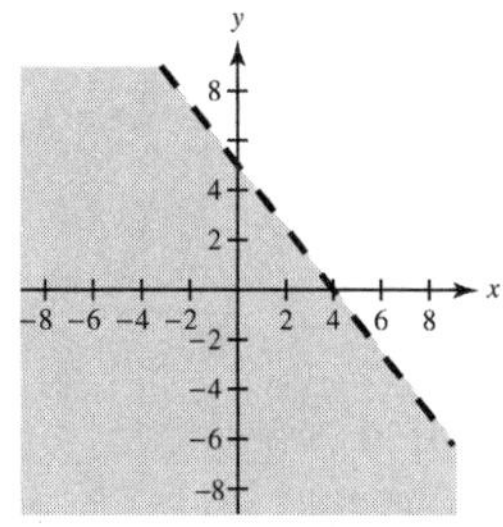

39. $x - 3y - 9 < 0$

$-3y < -x + 9$

$y > \frac{1}{3}x - 3$

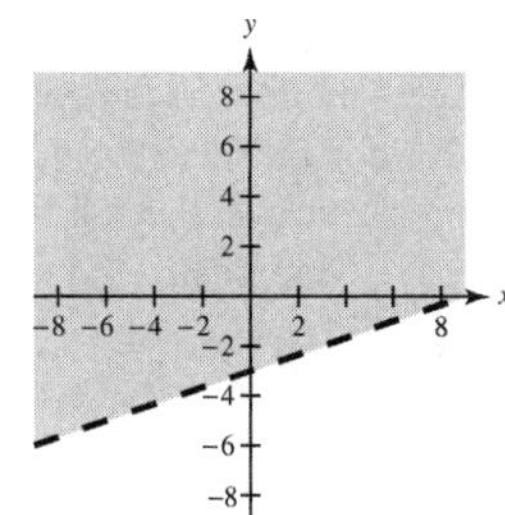

41. $3x - 2 \le 5x + y$

$-y \le 2x + 2$

$y \ge -2x - 2$

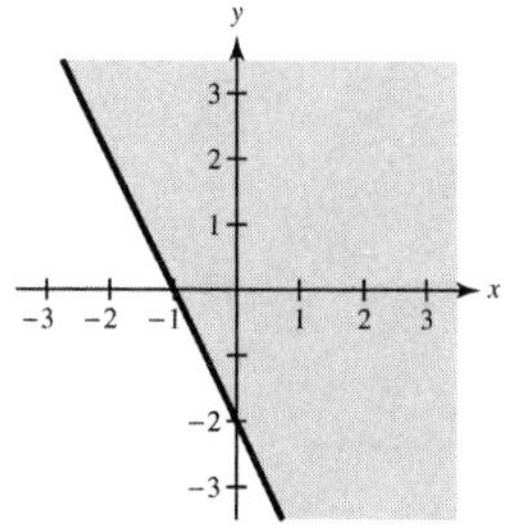

43. $0.2x + 0.3y < 2$

$2x + 3y < 20$

$3y < -2x + 20$

$y < -\frac{2}{3}x + \frac{20}{3}$

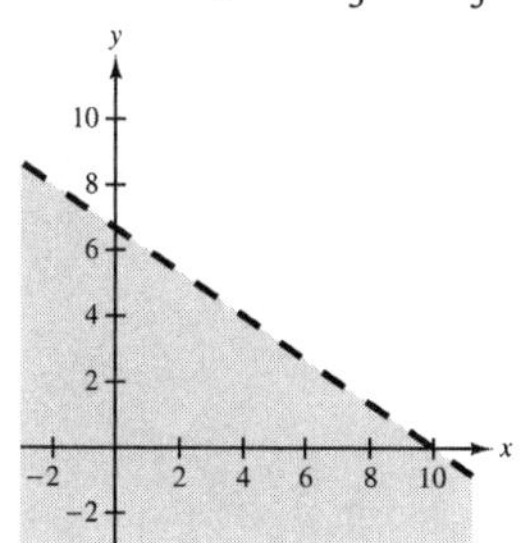

45. $y - 1 > -\frac{1}{2}(x - 2)$

$y - 1 = -\frac{1}{2}x + 1$

$y = -\frac{1}{2}x + 2$

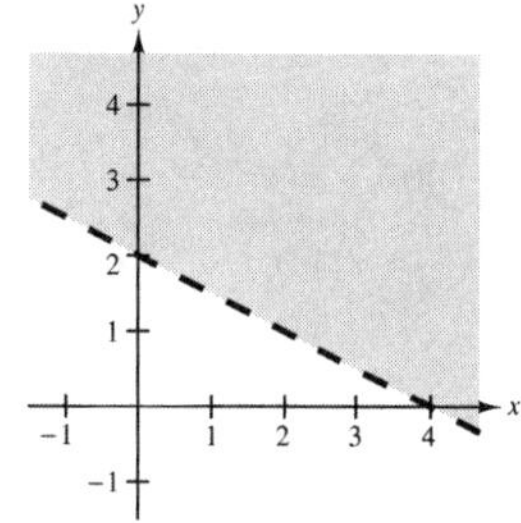

47. $\frac{x}{3} + \frac{y}{4} \le 1$

$4x + 3y \le 12$

$3y \le -4x + 12$

$y \le -\frac{4}{3}x + 4$

49. (a) $11x + 9y \geq 240$

$$9y \geq -11x + 240$$

$$y \geq -\tfrac{11}{9}x + \tfrac{80}{3}; \quad x \geq 0, y \geq 0$$

(b)

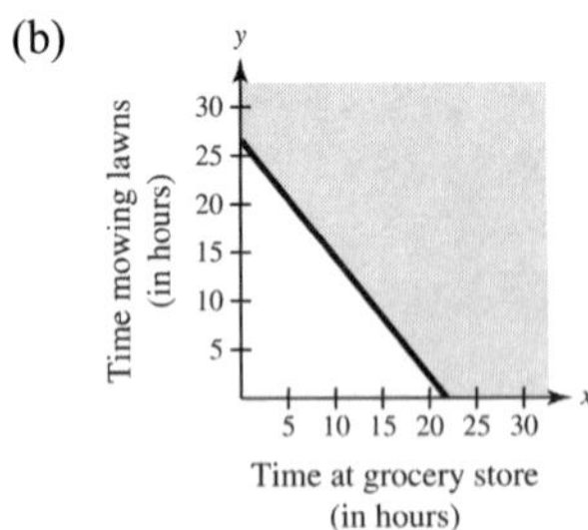

(x, y): $(2, 25)$, $(4, 22)$, $(10, 15)$

51. False. Another way to determine if $(2, 4)$ is a solution of the inequality is to substitute $x = 2$ and $y = 4$ into the inequality $2x + 3y > 12$.

53. The solution of $x - y > 1$ does not include the points on the line $x - y = 1$. The solution of $x - y \geq 1$ does include the points on the line $x - y = 1$.

55. $m = \dfrac{2 - 5}{3 + 1} = -\dfrac{3}{4}$

$$y - 2 > -\frac{3}{4}(x - 3)$$

$$4y - 8 > -3x + 9$$

$$3x + 4y > 17$$

57. $y < 2$

59. (a) $10x + 15y \leq 1000$

$$15y \leq -10x + 1000$$

$$y \leq -\tfrac{2}{3}x + \tfrac{200}{3}, \quad x \geq 0, y \geq 0$$

(b)

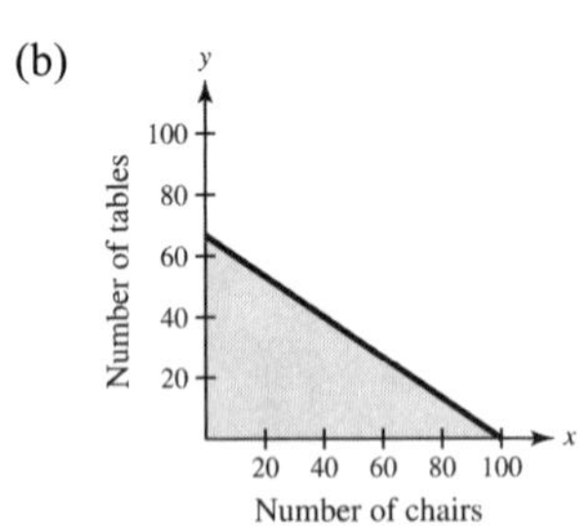

61. (a) *Verbal model:*

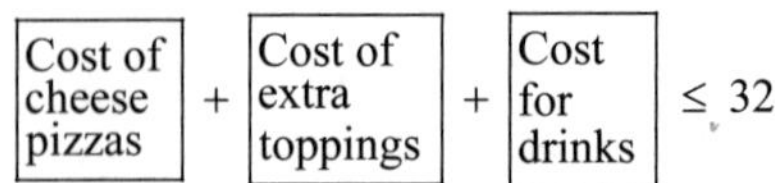

Labels: Cost of cheese pizzas $= 2(10) = \$20$

Cost for extra toppings $= 0.60x$ (dollars)

Cost for drinks $= y$ (dollars)

Inequality: $20 + 0.60x + y \leq 32$

$$0.60x + y \leq 12, \quad x \geq 0, y \geq 0$$

(b)

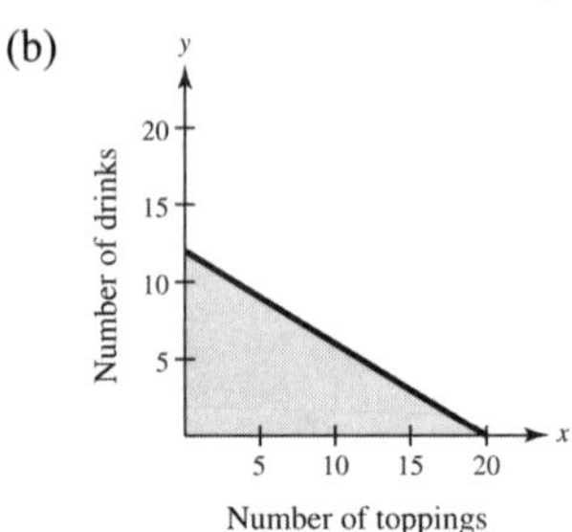

(c) $(6, 6)$

$$0.60(6) + 6 \overset{?}{\leq} 12$$

$$3.6 + 6 \overset{?}{\leq} 12$$

$$9.6 \leq 12$$

$(6, 6)$ is a solution of the inequality.

63. (a) $2w + t \geq 70$

$$t \geq -2w + 70, \quad x \geq 0, y \geq 0$$

(b)

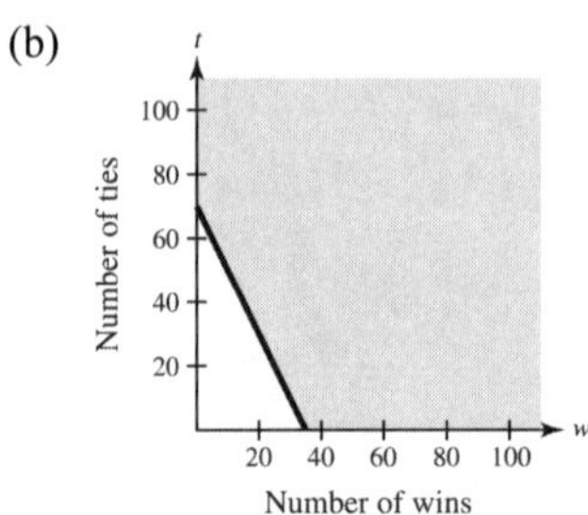

(w, t): $(10, 50)$, $(20, 30)$, $(30, 10)$

65. On the real number line, the solution of $x \leq 3$ is an unbounded interval.

On a rectangular coordinate system, the solution of $x \leq 3$ is a half-plane.

67. To write a double inequality such that the solution is the graph of a line, use the same real number as the bounds of the inequality, and use $\leq$ for one inequality symbol and $\leq$ for the other symbol. For example, the solution of $2 \leq x + y \leq 2$ is $x + y = 2$. Graphically, $x + y = 2$ is a line in the plane.

69. The boundary lines of the inequalities are parallel lines. Therefore, $a = c$. Because every point in the solution set of $y \le ax + b$ is a solution of $y < cx + d$, the inequality $y \le ax + b$ lies in the half-plane of $y < cx + d$. Therefore, $y = ax + b$ lies under $y = cx + d$ and $b < d$ (assuming $b > 0$ and $d > 0$).

71. $|x + 2| = 3$

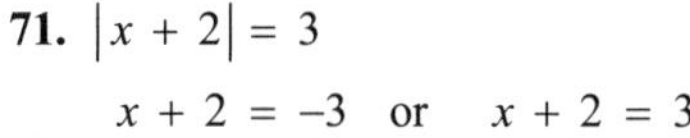

$x + 2 = -3$ or $x + 2 = 3$

73. $|8 - 3x| = 10$

$8 - 3x = -10$ or $8 - 3x = 10$

75.

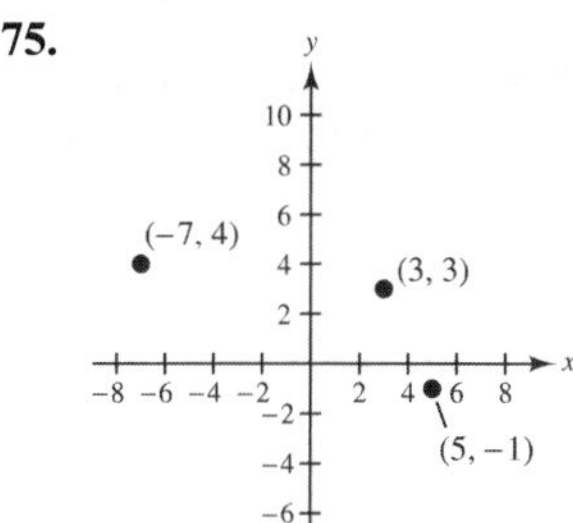

77.

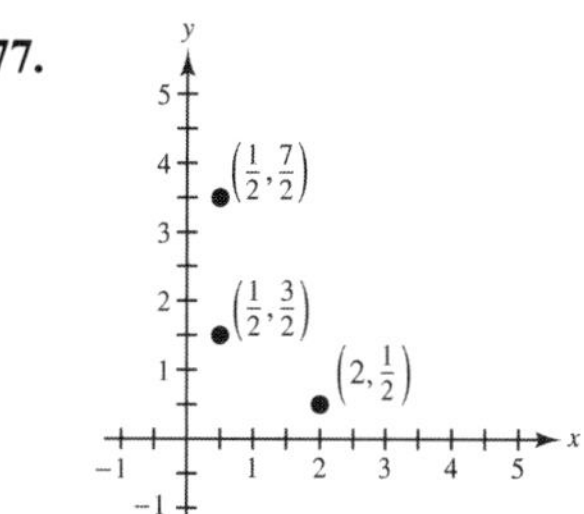

79. $2x - 3y = -6$

$$-3y = -6 - 2x$$

$$y = \tfrac{2}{3}x + 2$$

(a) $y - y_1 = m(x - x_1)$

$$y - 7 = \tfrac{2}{3}(x - 5)$$

$$y = \tfrac{2}{3}x + \tfrac{11}{3}$$

(b) $y - y_1 = m(x - x_1)$

$$y - 7 = -\tfrac{3}{2}(x - 5)$$

$$y = -\tfrac{3}{2}x + \tfrac{29}{2}$$

Section 3.6 Relations and Functions

1. Domain $= \{-2, 0, 1\}$

Range $= \{-1, 0, 1, 4\}$

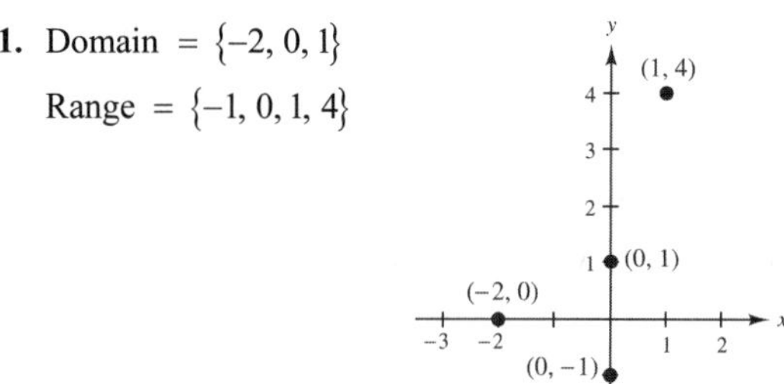

3. Domain $= \{0, 2, 4, 5, 6\}$

Range $= \{-3, 0, 5, 8\}$

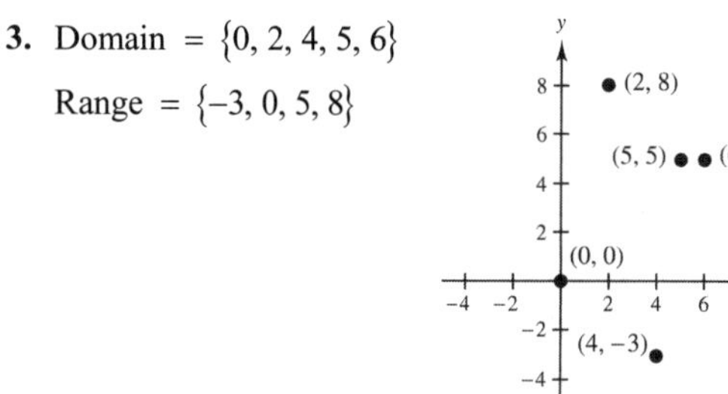

5. $(1, 1), (2, 8), (3, 27), (4, 64), (5, 125), (6, 216), (7, 343)$

7. {(2008, Philadelphia Phillies), (2009, New York Yankees), (2010, San Francisco Giants), (2011, St. Louis Cardinals)}

9. $\{(3, 9), (1, 3), (2, 6), (8, 24), (7, 21)\}$

11. (a) Yes, this relation is a function because each number in the domain is paired with exactly one number in the range.

(b) No, this relation is not a function because the 1 in the domain is paired with 2 different numbers in the range.

(c) Yes, this relation is a function because each number in the domain is paired with exactly one number in the range.

(d) No, this relation is not a function because each number in the domain is not paired with a number.

13. $x^2 + y^2 = 25$

$0^2 + 5^2 \stackrel{?}{=} 25$ $\qquad$ $0^2 + (-5)^2 \stackrel{?}{=} 25$

$25 = 25$ $\qquad$ $25 = 25$

There are two values of y associated with one value of x, which implies y is not a function of x.

15. $|y| = x + 2$

$|3| \stackrel{?}{=} 1 + 2$ $\qquad$ $|-3| \stackrel{?}{=} 1 + 2$

$3 = 3$ $\qquad$ $3 = 3$

There are two values of y associated with one value of x, which implies y is not a function of x.

17. $y^2 = x$

$y = \sqrt{x}$ or $-\sqrt{x}$

Because $(4, 2)$ and $(4, -2)$ are both solutions, y is not a function of x.

19. $y = |x|$

$y = |-3| = 3$

Because one value of x corresponds to one value of y, y is a function of x.

21. $f(x) = 12x - 7$

(a) $f(3) = 12(3) - 7 = 29$

(b) $f\left(\frac{3}{2}\right) = 12\left(\frac{3}{2}\right) - 7 = 11$

(c) $f(a) + f(1) = \left[12(a) - 7\right] + \left[12(1) - 7\right]$
$= 12a - 7 + 12 - 7$
$= 12a - 2$

(d) $f(a + 1) = 12(a + 1) - 7$
$= 12a + 12 - 7$
$= 12a + 5$

23. $f(x) = \dfrac{3x}{x - 5}$

(a) $f(0) = \dfrac{3(0)}{0 - 5} = 0$

(b) $f\left(\dfrac{5}{3}\right) = \dfrac{3\left(\frac{5}{3}\right)}{\frac{5}{3} - 5} \cdot \dfrac{3}{3} = \dfrac{15}{5 - 15} = \dfrac{15}{-10} = -\dfrac{3}{2}$

(c) $f(2) - f(-1) = \left[\dfrac{3(2)}{2 - 5}\right] - \left[\dfrac{3(-1)}{-1 - 5}\right]$
$= \dfrac{6}{-3} - \dfrac{-3}{-6} = -2 - \dfrac{1}{2} = -\dfrac{5}{2}$

(d) $f(x + 4) = \dfrac{3(x + 4)}{x + 4 - 5} = \dfrac{3x + 12}{x - 1}$

25. The domain of $f(x) = x^2 + x - 2$ is all real numbers x.

27. The domain of $f(t) = \dfrac{t + 3}{t(t + 2)}$ is all real numbers t such that $t(t + 2) \neq 0$. So, $t \neq 0$ and $t \neq -2$.

29. The domain of $g(x) = \sqrt{x + 4}$ is all real numbers x such that $x + 4 \geq 0$. So, $x \geq -4$.

31. Domain: All real numbers r such that $r > 0$

Range: All real numbers C such that $C > 0$

33. Domain: All real numbers r such that $r > 0$

Range: All real numbers A such that $A > 0$

35. The domain of a function is the set of inputs of the function, or the set of all the first components of the ordered pairs. The range of a function is the set of outputs of the function, or the set of all the second components of the ordered pairs.

37. Negative numbers are excluded from the domain because you cannot produce a negative number of video games.

39. No, this relation is not a function because -1 in the domain is matched to 2 numbers (6 and 7) in the range.

41. Yes, this relation is a function as each number in the domain is matched to exactly one number in the range.

43. $g(x) = 2 - 4x + x^2$

(a) $g(4) = 2 - 4(4) + 4^2 = 2 - 16 + 16 = 2$

(b) $g(0) = 2 - 4(0) + 0^2 = 2$

(c) $g(2y) = 2 - 4(2y) + (2y)^2 = 2 - 8y + 4y^2$

(d) $g(4) + g(6) = \left[2 - 4(4) + 4^2\right] + \left[2 - 4(6) + 6^2\right]$
$= (2 - 16 + 16) + (2 - 24 + 36)$
$= 2 + 14 = 16$

45. The domain of $f(x) = \sqrt{2x - 1}$ is all real numbers x such that $2x - 1 \geq 0$. So, $x \geq \frac{1}{2}$.

47. The domain of $f(t) = |t - 4|$ is all real numbers t.

49. *Verbal Model:*

[Total cost] = [Variable costs] + [Fixed costs]

Labels: Total cost $= C(x)$

Variable costs $= 3.25x$

Fixed costs $= 495$

Number of units $= x$

Function: $C(x) = 3.25x + 495, x > 0$

51. *Verbal Model:*

[Volume] = [Length] · [Width] · [Height]

Labels: Volume $= V$

Length $= (24 - 2x)$

Width $= (24 - 2x)$

Height $= x$

Function: $V = x(24 - 2x)^2, x > 0$

53. $S(L) = \dfrac{128{,}160}{L}$

(a) $S(12) = \dfrac{128{,}160}{12} = 10{,}680$ pounds

(b) $S(16) = \dfrac{128{,}160}{16} = 8010$ pounds

55. (a) $w(20) = 12(20) = \$240$

$w(35) = 12(35) = \$420$

$w(50) = 18(50 - 40) + 480 = \$660$

$w(55) = 18(55 - 40) + 480 = \$750$

(b) When the worker works between 0 and 40 hours, the wage is \$12 per hour. When more than 40 hours are worked, the worker earns time-and-a-half for the additional hours.

57. Because the prices of shoes vary, the store may sell more or fewer shoes depending on the prices. So, the use of the word function is not mathematically correct.

59. If set B is a function, no element of the domain is matched with two different elements in the range. A subset of B would have fewer elements in the domain, and therefore, a subset of B would be a function.

61. Statement: The distance driven on a trip is a function of the number of hours spent in the car.

Domain: The number of hours in the car

Range: The distance driven

The function has an infinite number of ordered pairs.

63. Multiplication Property of Zero

65. Distributive Property

67.

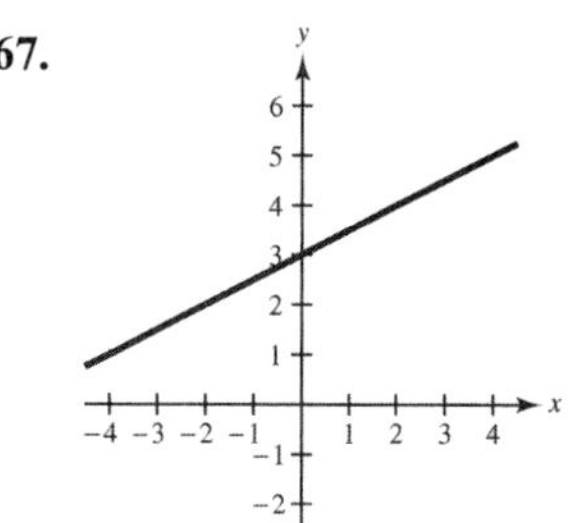

69.

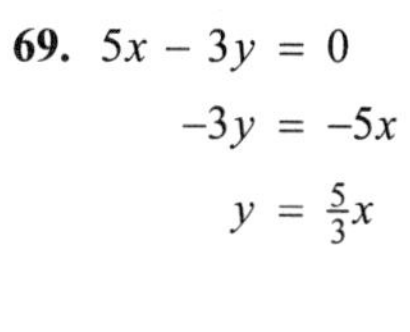

$5x - 3y = 0$

$-3y = -5x$

$y = \frac{5}{3}x$

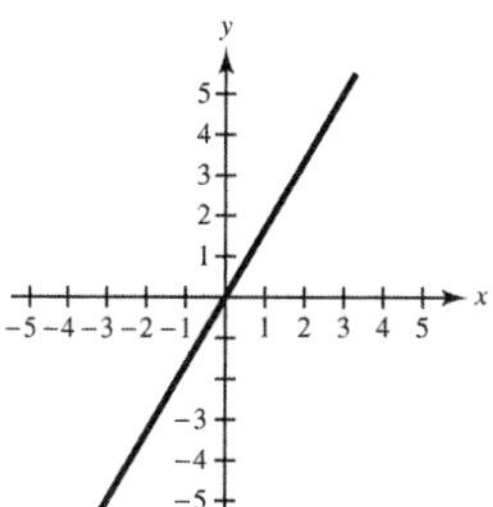

71.

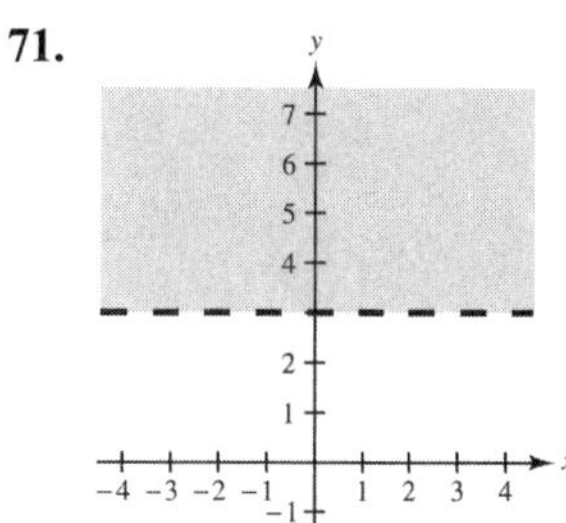

73.

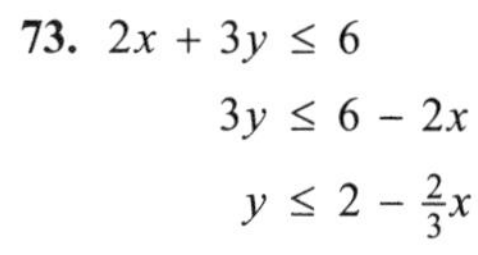

$2x + 3y \le 6$

$3y \le 6 - 2x$

$y \le 2 - \frac{2}{3}x$

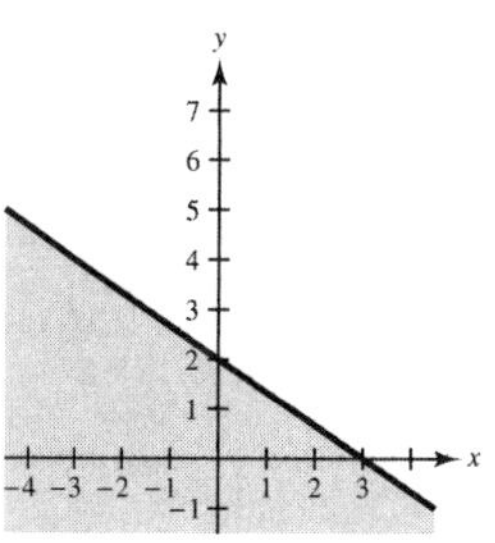

Section 3.7 Graphs of Functions

1. $f(x) = 2x - 7$

x	0	1	2	3	4	5
$f(x)$	−7	−5	−3	−1	1	3

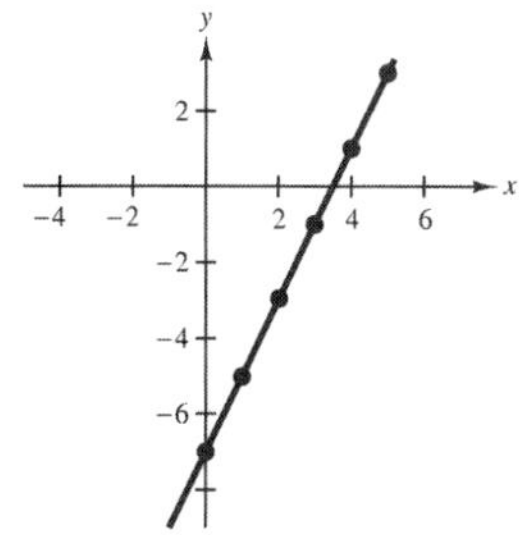

3. $h(x) = -(x - 1)^2$

x	−1	0	1	2	3
$h(x)$	−4	−1	0	−1	−4

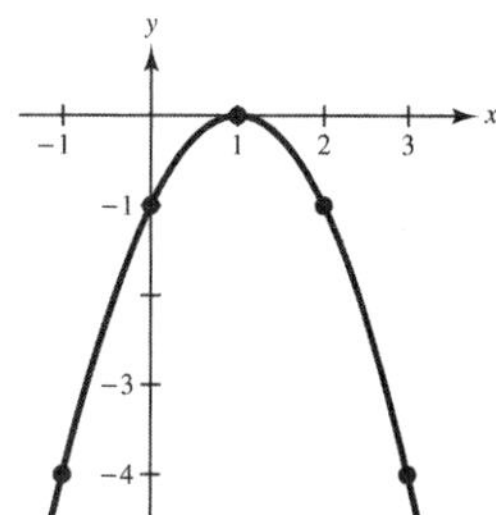

5. $f(x) = |x + 3|$

x	-6	-3	0	3	6
$f(x)$	3	0	3	6	9

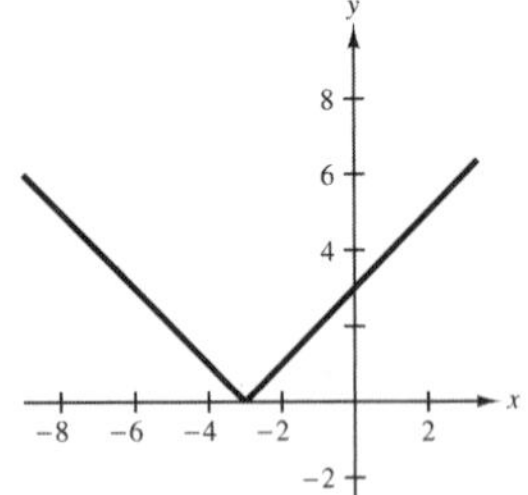

7.

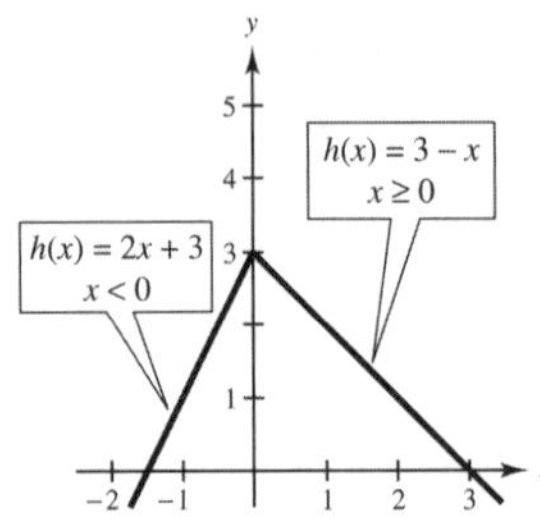

Domain: $-\infty < x < \infty$

Range: $\infty < y \le 3$

9.

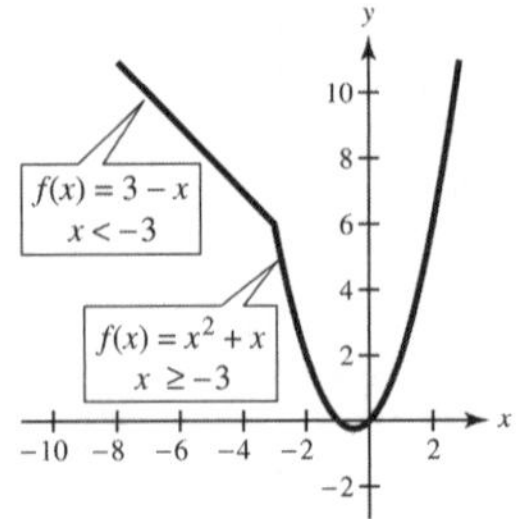

Domain: $-\infty < x < \infty$

Range: $-\frac{1}{4} \le y < \infty$

11.

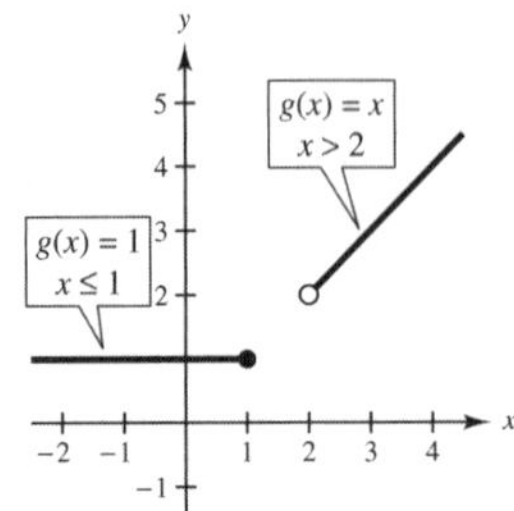

Domain: $-\infty < x \le 1$ and $2 < x < \infty$

Range: $y = 1$ and $2 < y < \infty$

13. Basic function: $f(x) = |x|$

The graph is shifted 1 unit downward.

15. Basic function: $f(x) = c$

The constant c is 7.

17. Basic function: $f(x) = x^2$

The graph is upside down and shifted 1 unit to the right and 1 unit downward.

19. Yes, $y = \frac{1}{3}x^3$ passes the Vertical Line Test and is a function of x.

21. No, y is not a function of x by the Vertical Line Test.

23. (a) Vertical shift 2 units upward

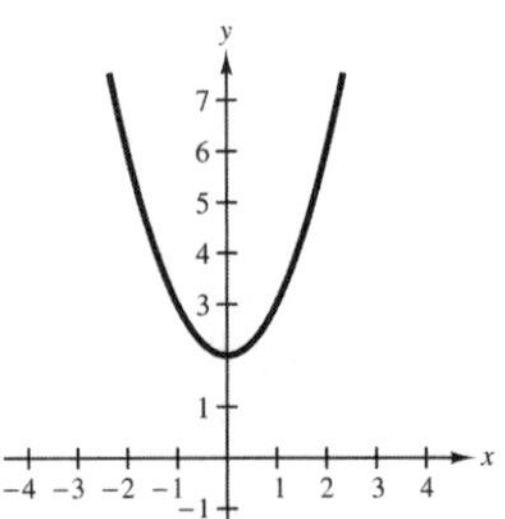

(b) Vertical shift 4 units downward

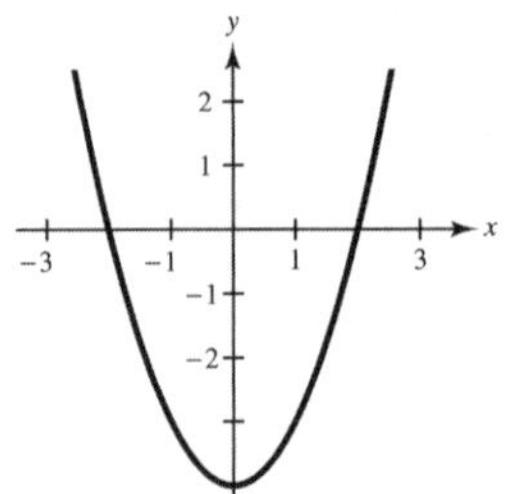

(c) Horizontal shift 2 units to the left

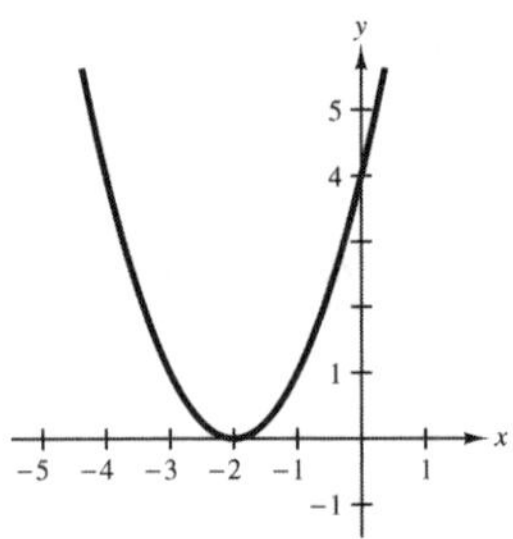

(d) Horizontal shift 4 units to the right

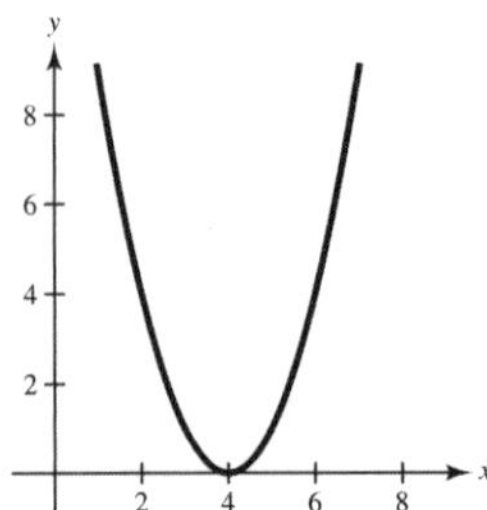

25. (a) Vertical shift 3 units upward

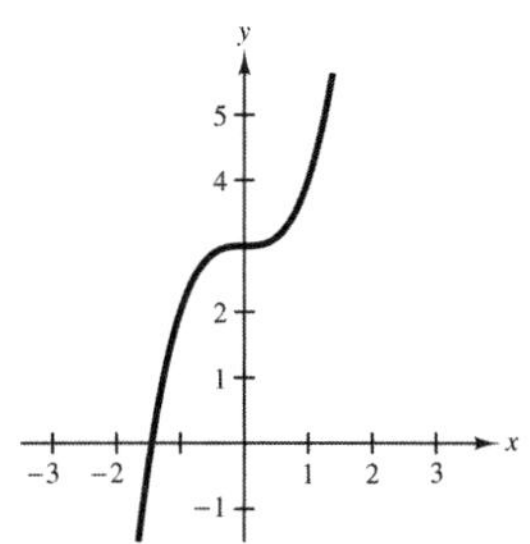

(b) Vertical shift 5 units downward

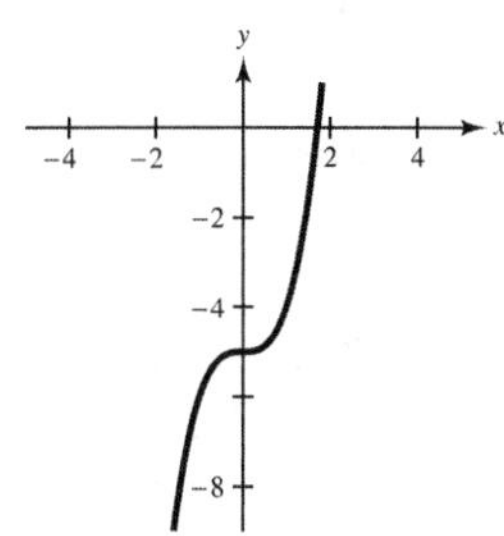

(c) Horizontal shift 3 units to the right

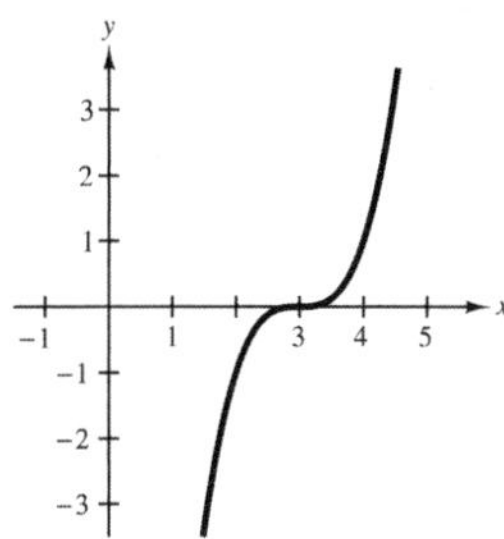

(d) Horizontal shift 2 units to the left

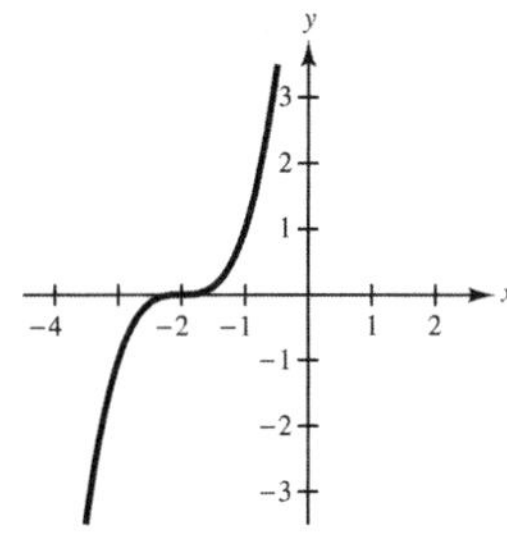

27. $y = x^2$; (d)

28. $y = -x^2$; (a)

29. $y = (-x)^2$; (d)

30. $y = x^3$; (b)

31. $y = -x^3$; (c)

32. $y = (-x)^3$; (c)

33. The basic function related to $h(x) = (x - 1)^2 + 2$ is $f(x) = x^2$.

35. The part of $h(x) = (x - 1)^2 + 2$ that indicates a vertical shift is $(x - 1)^2 + 2$. The expression is of the form $f(x) + c$.

37. 

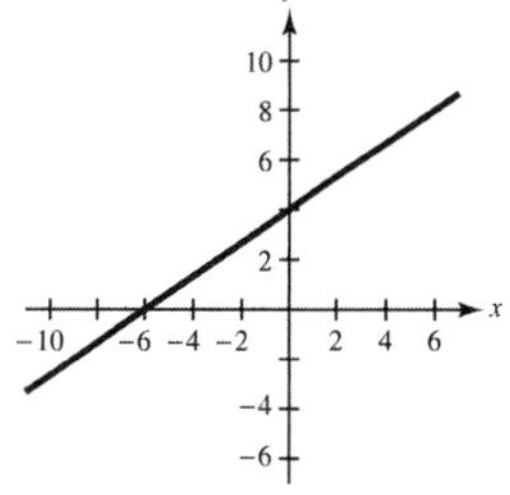

y is a function of x.

39.

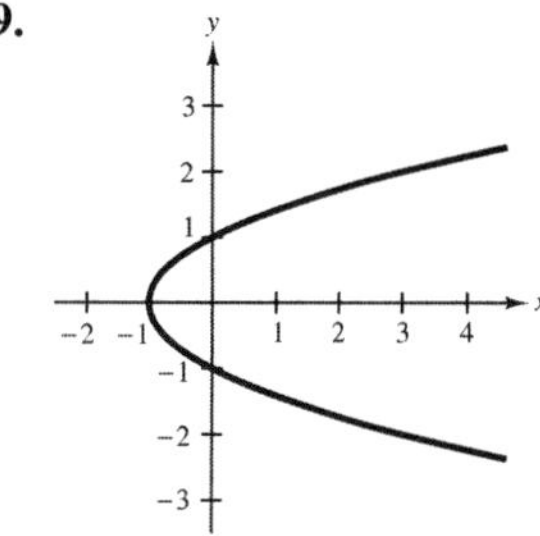

y is not a function of x.

41. Basic function: $y = x^3$

Transformation: Horizontal shift 2 units right

Equation: $y = (x - 2)^3$

43. Basic function: $y = x^2$

Transformation: Reflection in the x-axis, horizontal shift 1 unit left, vertical shift 1 unit upward

Equation: $y = -(x + 1)^2 + 1$

45. (a) $y = f(x) + 2$

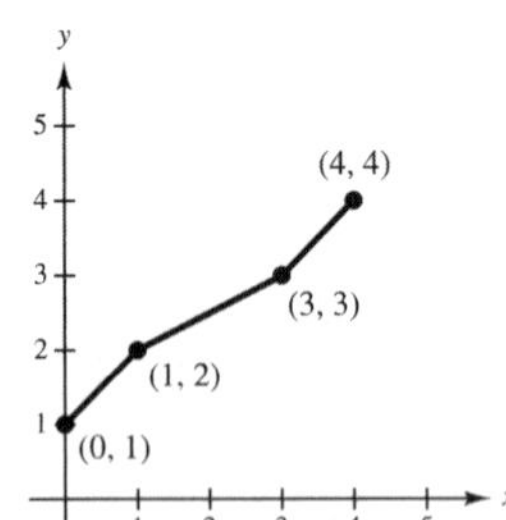

(b) $y = -f(x)$

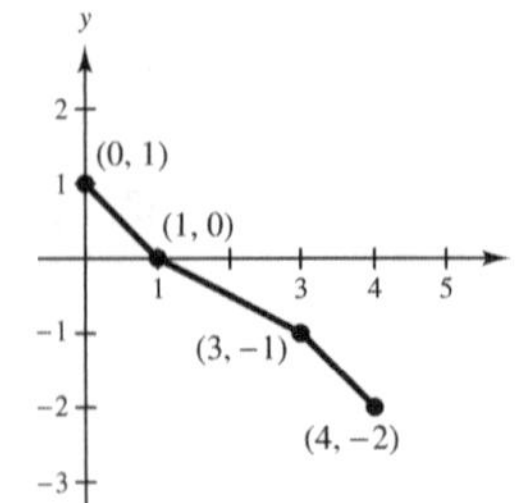

(c) $y = f(x - 2)$

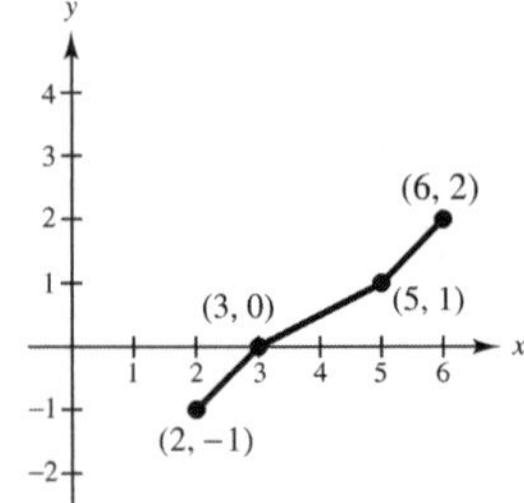

(d) $y = f(x + 2)$

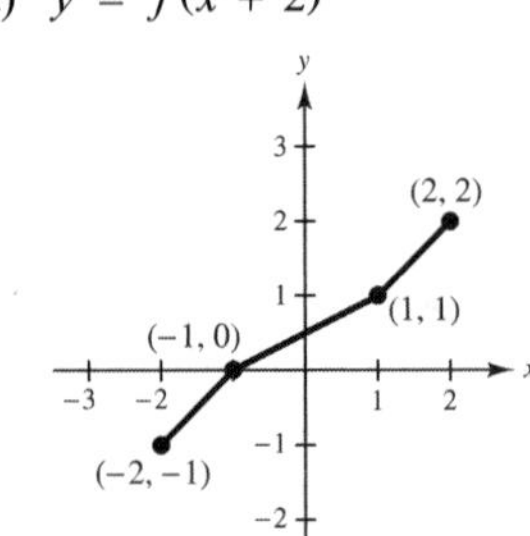

(e) $y = f(x) - 1$

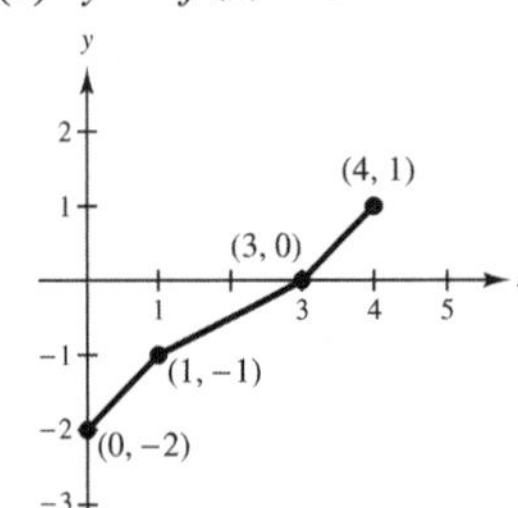

(f) $y = f(-x)$

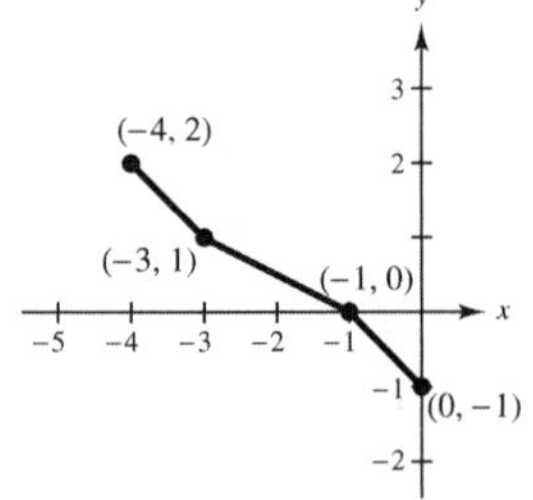

47. (a)

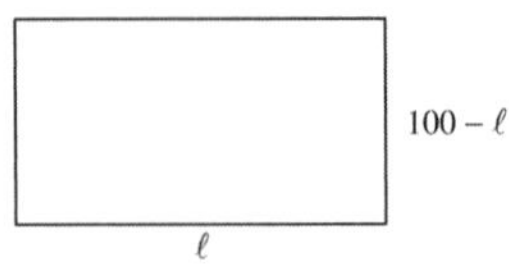

$A = l \cdot w$

$A = l(100 - l)$

Let l = length.

$100 - l$ = width

$P = 2l + 2w$

$200 = 2l + 2w$

$100 = l + w$

$100 - l = w$

(b)

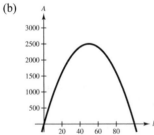

(c) When $l = 50$, the largest value of A is 2500. $(50, 2500)$ is the highest point on the graph of A giving the largest value of the function. The figure is a square.

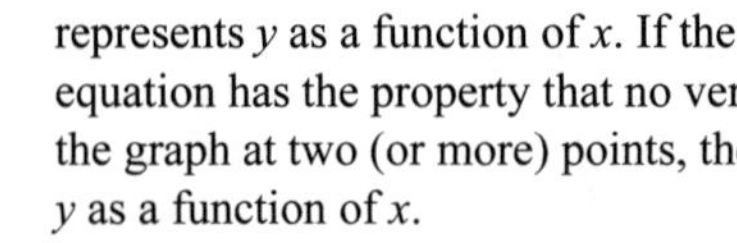

49. Use the Vertical Line Test to determine if an equation represents y as a function of x. If the graph of an equation has the property that no vertical line intersects the graph at two (or more) points, the equation represents y as a function of x.

51. The graph of $g(x) = f(-x)$ is a reflection in the y-axis of the graph of $f(x)$.

53. The sum of four times a number and 1

55. The ratio of two times a number and 3

57. $2x + y = 4$

$y = 4 - 2x$

59. $-4x + 3y + 3 = 0$

$3y = 4x - 3$

$y = \frac{4}{3}x - 1$

61. $y < 2x + 1$

$1 \stackrel{?}{<} 2(0) + 1$

$1 \not< 1$; $(0, 1)$ is not a solution.

63. $2x - 3y > 2y$

$2(6) - 3(2) \stackrel{?}{>} 2(2)$

$12 - 6 \stackrel{?}{>} 4$

$6 > 4$; $(6, 2)$ is a solution.

Review Exercises for Chapter 3

1. 

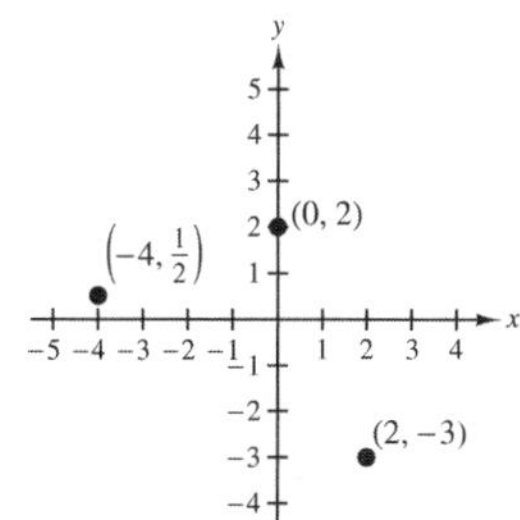

3. Quadrant I

5. Quadrant I or IV

7. (a) $(2, 3)$: $3 \stackrel{?}{=} 4 - \frac{1}{2}(2)$

$3 \stackrel{?}{=} 4 - 1$

$3 = 3$

$(2, 3)$ is a solution.

(b) $(-1, 5)$: $5 \stackrel{?}{=} 4 - \frac{1}{2}(-1)$

$5 \stackrel{?}{=} 4 + \frac{1}{2}$

$5 \neq 4\frac{1}{2}$

$(-1, 5)$ is not a solution.

(c) $(-6, 1)$: $1 \stackrel{?}{=} 4 - \frac{1}{2}(-6)$

$1 \stackrel{?}{=} 4 + 3$

$1 \neq 7$

$(-6, 1)$ is not a solution.

(d) $(8, 0)$: $0 \stackrel{?}{=} 4 - \frac{1}{2}(8)$

$0 \stackrel{?}{=} 4 - 4$

$0 = 0$

$(8, 0)$ is a solution.

9. $d = \sqrt{(4-4)^2 + (3-8)^2}$

$= \sqrt{0 + 25}$

$= \sqrt{25}$

$= 5$

11. $d = \sqrt{(-5-1)^2 + (-1-2)^2}$

$= \sqrt{36 + 9}$

$= \sqrt{45}$

$= 3\sqrt{5}$

13. $M = \left(\frac{1+7}{2}, \frac{4+2}{2}\right) = \left(\frac{8}{2}, \frac{6}{2}\right) = (4, 3)$

15. $M = \left(\frac{5+(-3)}{2}, \frac{-2+(5)}{2}\right) = \left(\frac{2}{2}, \frac{3}{2}\right) = \left(1, \frac{3}{2}\right)$

17. $3y - 2x - 3 = 0$

$3y - 2(0) - 3 = 0$

$3y = 3$

$y = 1 \quad (0, 1)$

$3(0) - 2x - 3 = 0$

$-2x = 3$

$x = -\frac{3}{2} \quad \left(-\frac{3}{2}, 0\right)$

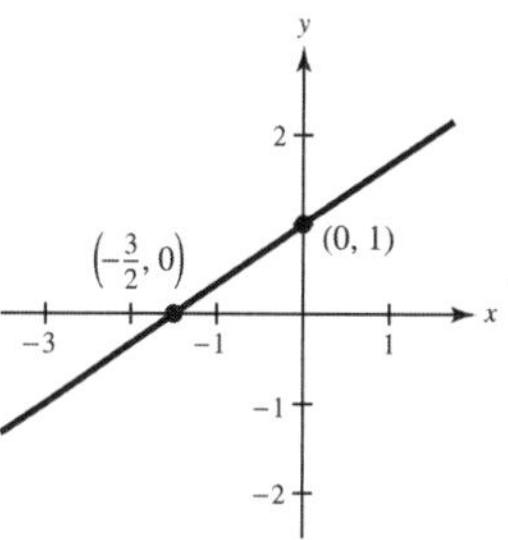

19. $y = x^2 - 1$

$y = 0^2 - 1$

$= -1 \quad (0, -1)$

$0 = x^2 - 1$

$0 = (x-1)(x+1)$

$x = 1, x = -1 \quad (1, 0), (-1, 0)$

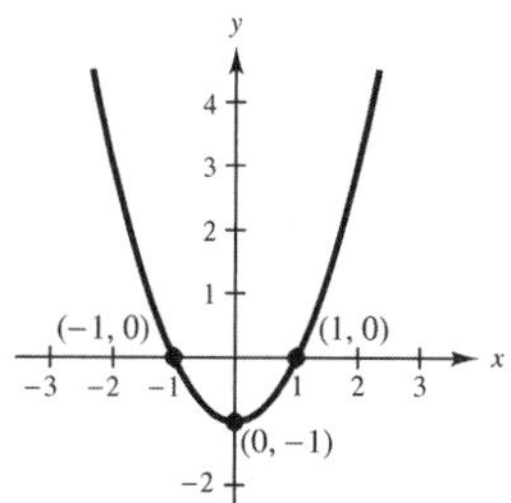

21. $y = |x| - 2$

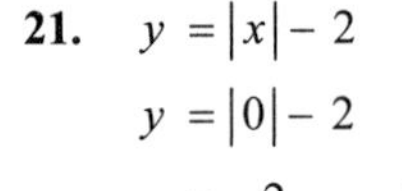

$y = |0| - 2$

$= -2 \qquad (0, -2)$

$0 = |x| - 2$

$2 = |x|$

$\pm 2 = x \qquad (2, 0), (-2, 0)$

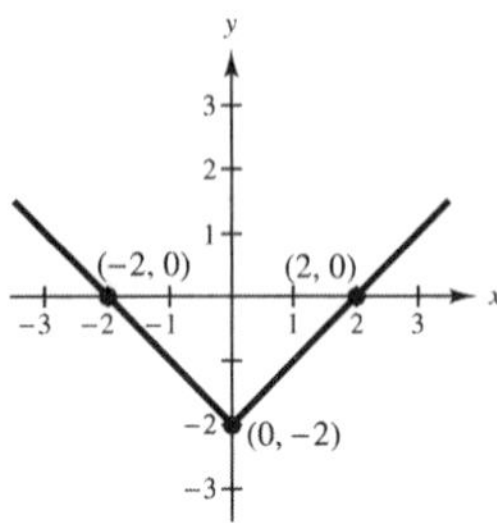

23. $8x - 2y = -4$

y-intercept: $8(0) - 2y = -4$

$-2y = -4$

$y = 2 \qquad (0, 2)$

x-intercept: $8x - 2(0) = -4$

$8x = -4$

$x = -\frac{1}{2} \qquad \left(-\frac{1}{2}, 0\right)$

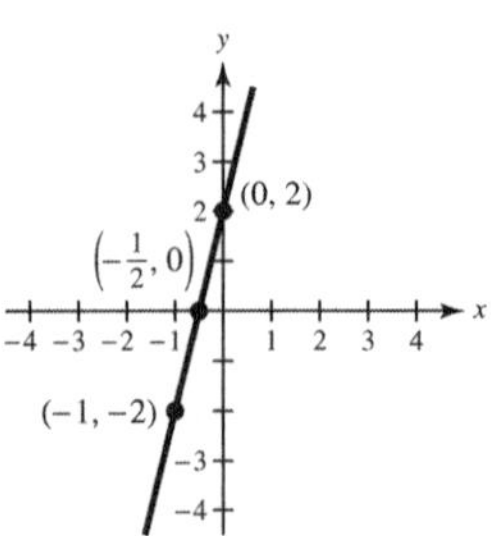

25. $y = 5 - |x|$

y-intercept: $y = 5 - |x|$

$y = 5 \qquad (0, 5)$

x-intercepts: $0 = 5 - |x|$

$|x| = 5$

$x = -5, x = 5 \qquad (-5, 0), (5, 0)$

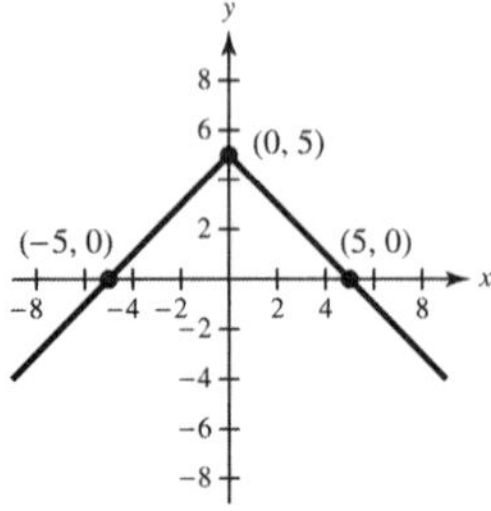

27. $y = |2x + 1| - 5$

y-intercept: $y = |2(0) + 1| - 5$

$= 1 - 5$

$= -4 \qquad (0, -4)$

x-intercepts: $0 = |2x + 1| - 5$

$5 = |2x + 1|$

$5 = 2x + 1$ or $-5 = 2x + 1$

$4 = 2x \qquad -6 = 2x$

$2 = x \qquad -3 = x$

$(2, 0), (-3, 0)$

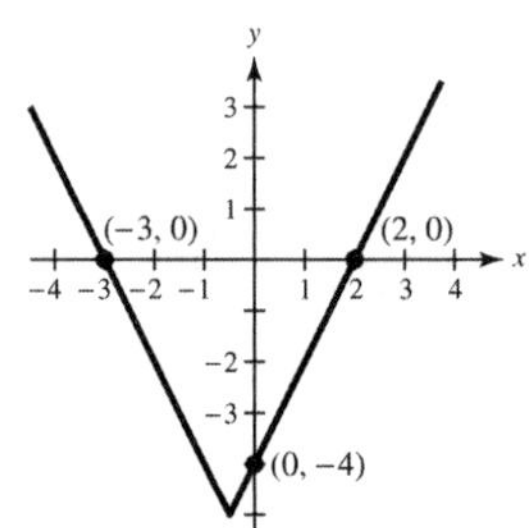

29. (a) The annual depreciation is

$$\frac{35{,}000 - 15{,}000}{5} = 4000.$$

$35{,}000 - 4000(1) = 31{,}000$

$35{,}000 - 4000(2) = 27{,}000$

$35{,}000 - 4000(3) = 23{,}000$

So, $y = 35{,}000 - 4000t$.

(b)

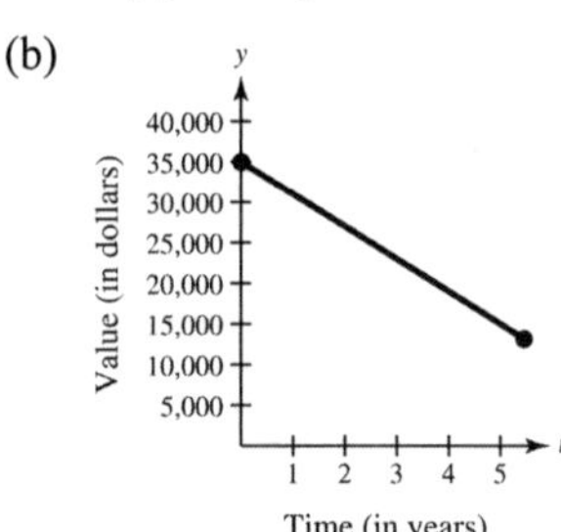

(c) $y = 35{,}000 - 4000(0)$

$y = 35{,}000$

$(0, 35{,}000)$

The y-intercept represents the value of the SUV when purchased.

31. $m = \dfrac{-1 - 2}{-3 - 4} = \dfrac{-3}{-7} = \dfrac{3}{7}$

33. $m = \dfrac{\frac{3}{4} - \frac{3}{4}}{4 - (-6)} = \dfrac{0}{10} = 0$

35. $m = \dfrac{0 - 6}{8 - 0} = \dfrac{-6}{8} = \dfrac{-3}{4} = -\dfrac{3}{4}$

37. $5x - 2y - 4 = 0$

$-2y = -5x + 4$

$y = \frac{5}{2}x - 2$

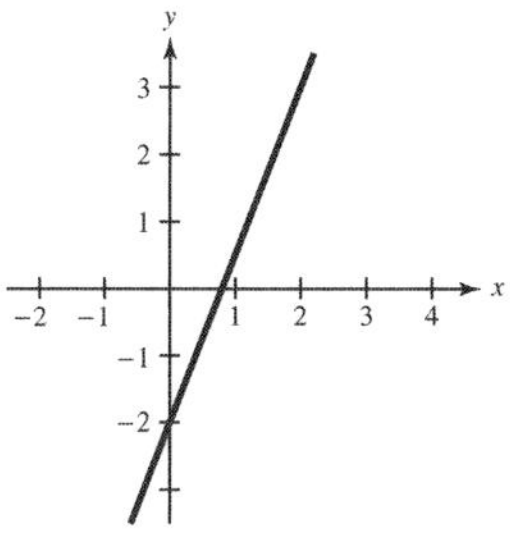

39. $6x + 2y + 5 = 0$

$2y = -6x - 5$

$y = -3x - \frac{5}{2}$

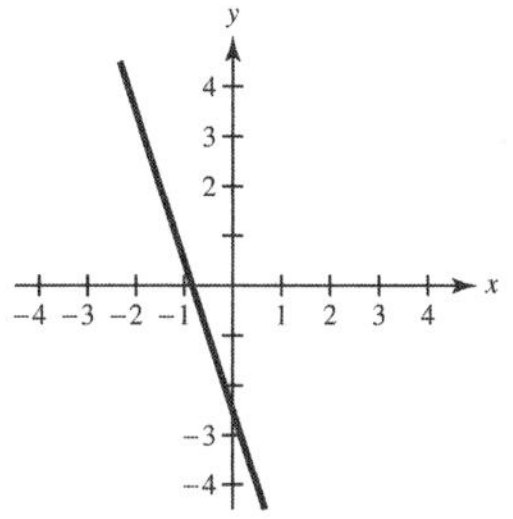

41. $L_1: y = \frac{3}{2}x + 1$

$L_2: y = \frac{2}{3}x - 1$

$m_1 = \frac{3}{2}, m_2 = \frac{2}{3}$

$m_1 \neq m_2, m_1 \cdot m_2 \neq -1$

The lines are neither parallel nor perpendicular.

43. $L_1: y = \frac{3}{2}x - 2$

$L_2: y = -\frac{2}{3}x + 1$

$m_1 = \frac{3}{2}, m_2 = -\frac{2}{3}$

$m_1 \cdot m_2 = -1$

The lines are perpendicular.

45. Let p represent the price of milk and let t represent the year. The two data points are (t_1, p_1) and (t_2, p_2).

$(t_1, p_1) = (2000, 2.79)$

$(t_2, p_2) = (2010, 3.32)$

The slope formula can be used to find the average rate of change.

$m = \dfrac{3.32 - 2.79}{2010 - 2000} = 0.053$

From 2000 to 2010, the average rate of change of the price of a gallon of whole milk is \$0.053.

47. $m = -4;\ (x_1, y_1) = (1, -4)$

$y - y_1 = m(x - x_1)$

$y - (-4) = -4(x - 1)$

$y + 4 = -4x + 4$

$y = -4x$

49. $m = \frac{1}{4};\ (x_1, y_1) = (-6, 5)$

$y - y_1 = m(x - x_1)$

$y - 5 = \frac{1}{4}(x - (-6))$

$y - 5 = \frac{1}{4}(x + 6)$

$y - 5 = \frac{1}{4}x + \frac{3}{2}$

$y = \frac{1}{4}x + \frac{13}{2}$

51. $(x_1, y_1) = (-6, 0);\ (x_2, y_2) = (0, -3)$

$m = \dfrac{-3 - 0}{0 - (-6)}$

$= -\dfrac{1}{2}$

$y - y_1 = m(x - x_1)$

$y - 0 = -\dfrac{1}{2}(x - (-6))$

$y = -\dfrac{1}{2}(x + 6)$

$2y = -(x + 6)$

$2y = -x - 6$

$x + 2y + 6 = 0$

53. $(x_1, y_1) = (-2, -3);\ (x_2, y_2) = (4, 6)$

$m = \dfrac{6 - (-3)}{4 - (-2)}$

$= \dfrac{3}{2}$

$y - y_1 = m(x - x_1)$

$y - (-3) = \dfrac{3}{2}(x - (-2))$

$y + 3 = \dfrac{3}{2}(x + 2)$

$2y + 6 = 3(x + 2)$

$2y + 6 = 3x + 6$

$-3x + 2y = 0$

$3x - 2y = 0$

55. $y = -9$

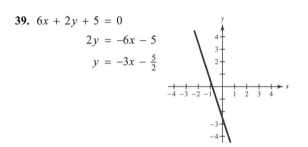

57. $x = -5$

59. $3x + y = 2$

$$y = -3x + 2$$

(a) $y + \frac{4}{5} = -3\left(x - \frac{3}{5}\right)$

$$y + \frac{4}{5} = -3x + \frac{9}{5}$$

$$y = -3x + 1$$

(b) $y + \frac{4}{5} = \frac{1}{3}\left(x - \frac{3}{5}\right)$

$$y + \frac{4}{5} = \frac{1}{3}x - \frac{3}{15}$$

$$y = \frac{1}{3}x - 1$$

61. (a) Let $(t_1, S_1) = (7, 32{,}000)$. Because the annual salary increased by about \$1050 per year, the slope is $m = 1050$.

$$S - S_1 = m(t - t_1)$$

$$S - 32{,}000 = 1050(t - 7)$$

$$S - 32{,}000 = 1050t - 7350$$

$$S = 1050t + 24{,}650$$

The equation that represents the salary is $S = 1050t + 24{,}650$.

(b) For $t = 15$, the equation $S = 1050t + 24{,}650$ has the value $S = 1050(15) + 24{,}650 = 40{,}400$. Your annual salary in the year 2015 is predicted to be \$40,400.

(c) For $t = 10$, the equation $S = 1050t + 24{,}650$ has the value $S = 1050(10) + 24{,}650 = 35{,}150$. Your annual salary in the year 2010 is estimated to be \$35,150.

63. $5x - 8y \geq 12$

(a) $5(-1) - 8(2) \overset{?}{\geq} 12$

$-5 - 16 \geq 12$

$-21 \not\geq 12$ Not a solution

(b) $5(3) - 8(-1) \overset{?}{\geq} 12$

$15 + 8 \geq 12$

$23 \geq 12$ Solution

(c) $5(4) - 8(0) \overset{?}{\geq} 12$

$20 - 0 \geq 12$

$20 \geq 12$ Solution

(d) $5(0) - 8(3) \overset{?}{\geq} 12$

$0 - 24 \geq 12$

$-24 \not\geq 12$ Not a solution

65. $y > -2$

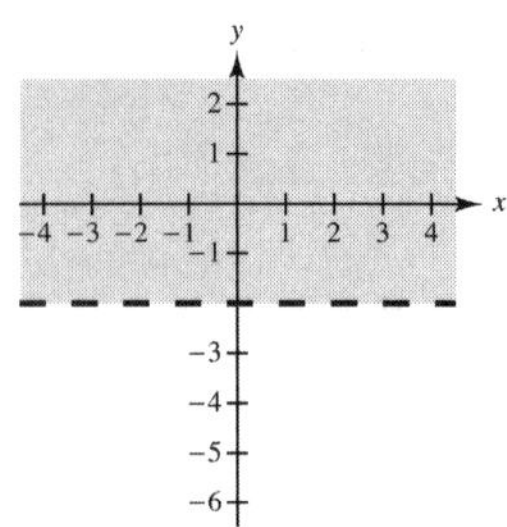

67. $x - 2 \geq 0$

$$x \geq 2$$

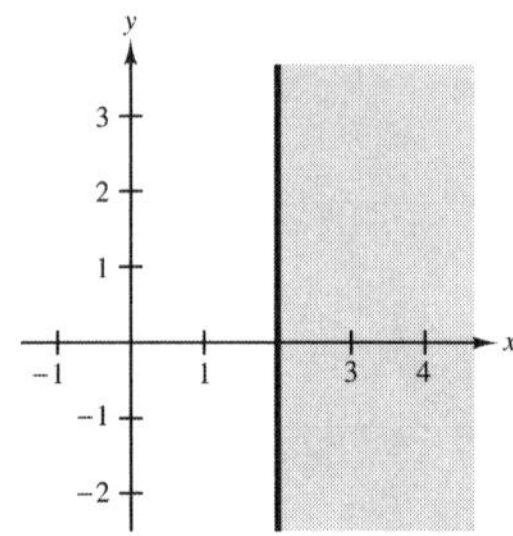

69. $2x + y < 1$

$$y < -2x + 1$$

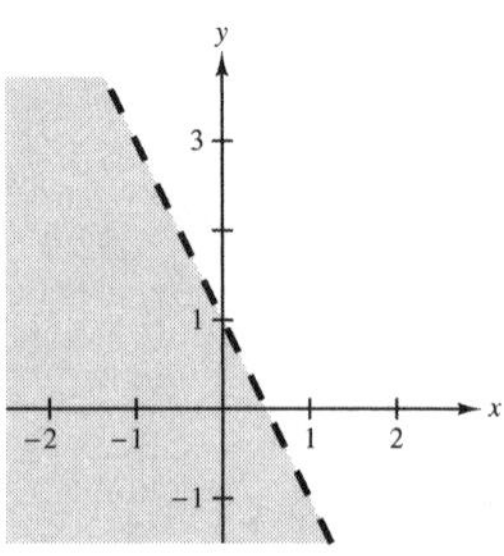

71. $-(x - 1) \leq 4y - 2$

$-x + 1 \leq 4y - 2$

$-x + 3 \leq 4y$

$-\frac{1}{4}x + \frac{3}{4} \leq y$

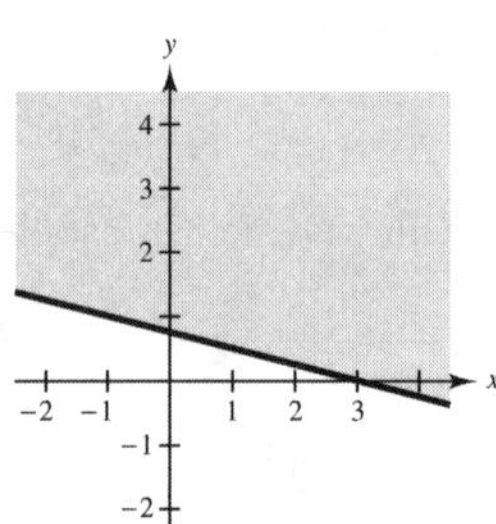

73. (a) $P = 2x + 2y$

$2x + 2y \leq 800$

$x + y \leq 400, \quad x \geq 0, y \geq 0$

(b)

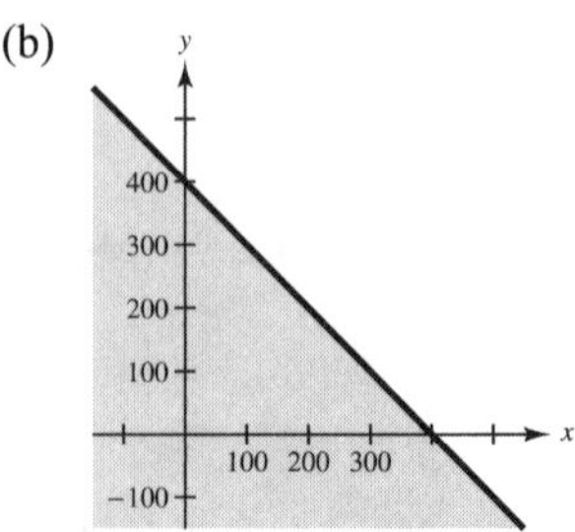

75. Domain: $\{-3, -1, 0, 1\}$

Range: $\{0, 1, 4, 5\}$

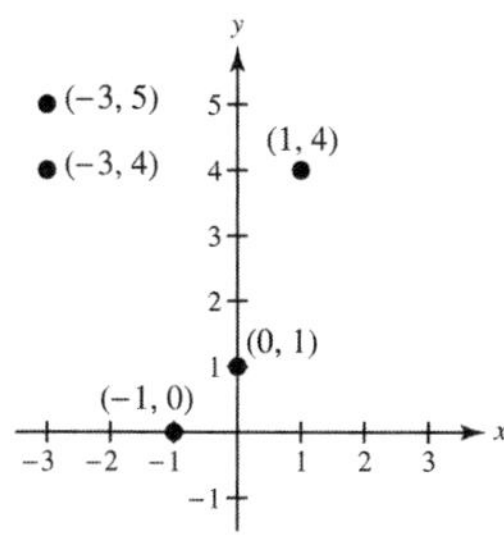

77. No, this relation is not a function because the 8 in the domain is matched with two numbers (1 and 2) in the range.

79. Yes, this relation is a function because each number in the domain is matched with only one number in the range.

81. $f(t) = \sqrt{5 - t}$

(a) $f(-4) = \sqrt{5 - (-4)} = \sqrt{9} = 3$

(b) $f(5) = \sqrt{5 - 5} = 0$

(c) $f(3) = \sqrt{5 - 3} = \sqrt{2}$

(d) $f(5z) = \sqrt{5 - 5z}$

83. $f(x) = \begin{cases} -3x, & x \le 0 \\ 1 - x^2, & x > 0 \end{cases}$

(a) $f(3) = 1 - 3^2 = -8$

(b) $f\left(-\frac{2}{3}\right) = -3\left(-\frac{2}{3}\right) = 2$

(c) $f(0) = -3(0) = 0$

(d) $f(4) - f(3) = (1 - 4^2) - (1 - 3^2)$
$= -15 - (-8) = -15 + 8 = -7$

85. $h(x) = 4x^2 - 7$

Domain: $-\infty < x < \infty$

87. $f(x) = \sqrt{3x + 10}$

$3x + 10 \ge 0$

$3x \ge -10$

$x \ge -\frac{10}{3}$

Domain: $x \ge -\frac{10}{3}$

Range: $f(x) \ge 0$

89. *Verbal model:* [Perimeter] = 2[Length] + 2[Width]

$150 = 2\text{Length} + 2x$

$\frac{150 - 2x}{2} = \text{Length}$

$75 - x = \text{Length}$

Verbal model: [Area] = [Length] · [Width]

Labels: Area $= A$

Length $= 75 - x$

Width $= x$

Function: $A = (75 - x)x$

Domain: $0 < x < \frac{75}{2}$

91. $y = 4 - (x - 3)^2$

x	0	1	2	3	4	5
$f(x)$	–5	0	3	4	3	0

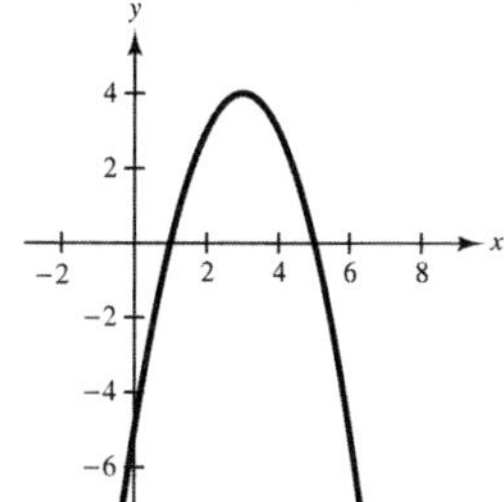

93. $g(x) = 6 - 3x, \ -2 \le x \le 4$

x	–2	–1	0	1	2	4
$g(x)$	12	9	6	3	0	–6

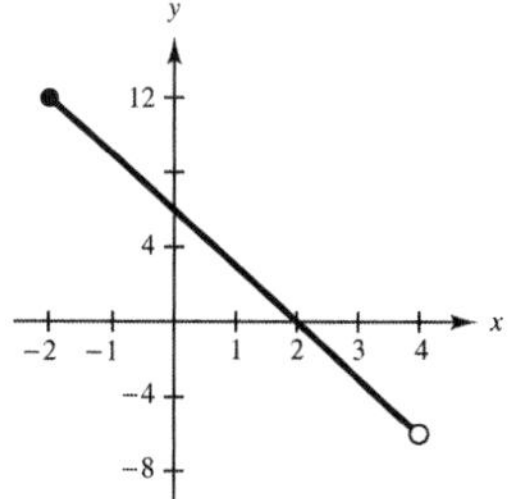

95.

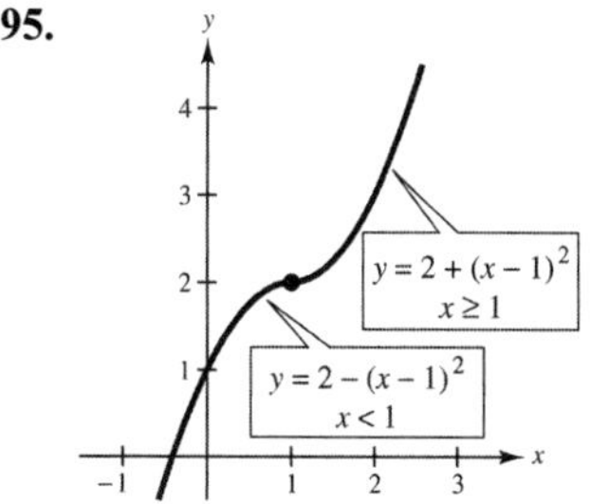

Domain: $-\infty < x < \infty$

Range: $-\infty < y < \infty$

97. $f(x) = -x^2 + 2$ (c)

98. $f(x) = |x| - 3$ (a)

99. $f(x) = -\sqrt{x}$ (b)

100. $f(x) = (x - 2)^3$ (d)

101. No, y is not a function of x.

103. $h(x) = -\sqrt{x}$ is a reflection in the x-axis of $f(x) = \sqrt{x}$.

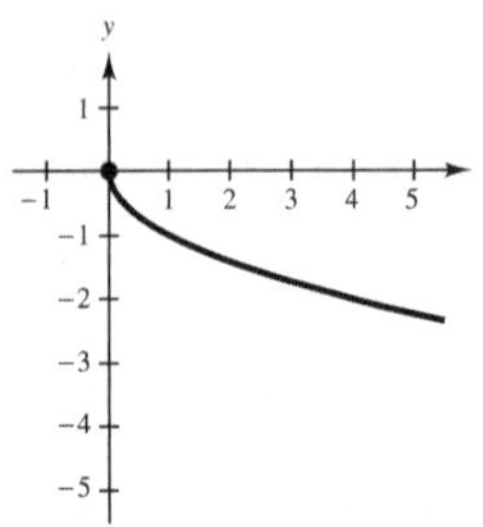

105. $h(x) = \sqrt{x + 3}$

Horizontal shift 3 units to the left

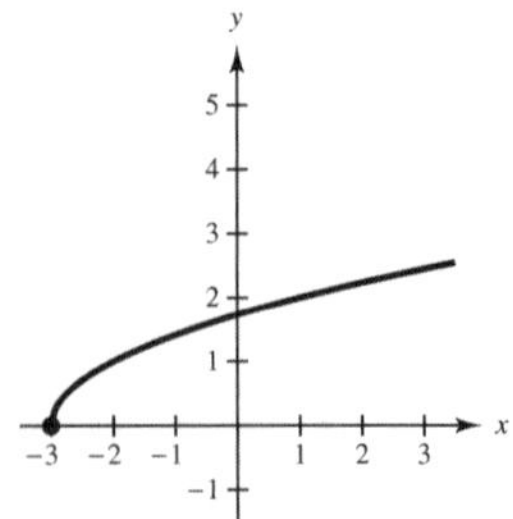

Chapter Test for Chapter 3

1. Quadrant IV

2. 

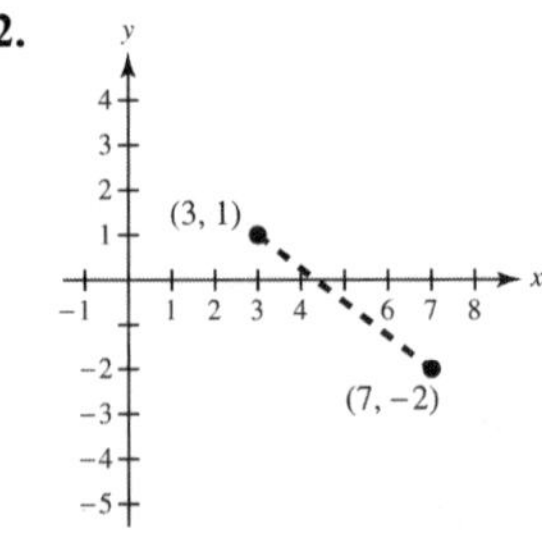

$$d = \sqrt{(7-3)^2 + (-2-1)^2} = \sqrt{16+9} = \sqrt{25} = 5$$

$$M = \left(\frac{7+3}{2}, \frac{-2+1}{2}\right) = \left(\frac{10}{2}, \frac{-1}{2}\right) = \left(5, -\frac{1}{2}\right)$$

3. $y = -3(x + 1)$

y-intercept: $y = -3(0 + 1)$

$y = -3, \ (0, -3)$

x-intercept: $0 = -3(x + 1)$

$0 = -3x - 3$

$3 = -3x$

$-1 = x, \ (-1, 0)$

4.

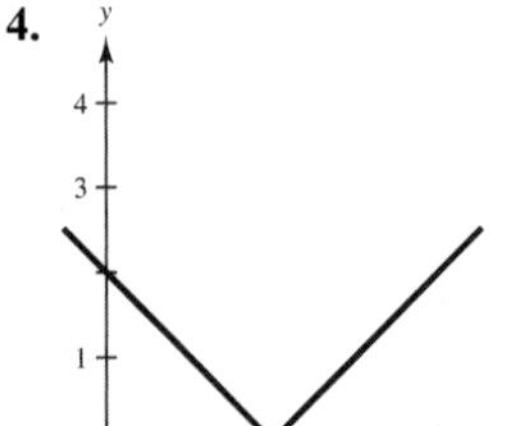

5. (a) $m = \dfrac{3 - 7}{2 + 4} = -\dfrac{4}{6} = -\dfrac{2}{3}$

(b) $m = \dfrac{6 + 2}{3 - 3} = \dfrac{8}{0} =$ undefined

6.

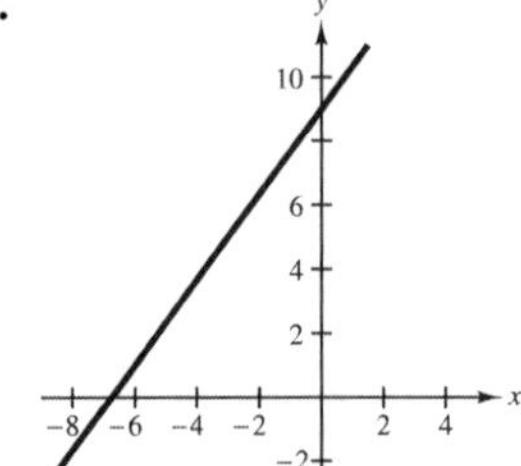

7. $2x + 5y = 10$

$2(0) + 5y = 10$

$5y = 10$

$y = 2 \quad (0, 2)$

$2x + 5(0) = 10$

$2x = 10$

$x = 5 \quad (5, 0)$

8. $m = \dfrac{3 - (-6)}{8 - 2} = \dfrac{9}{6} = \dfrac{3}{2}$

$$y - 3 = \frac{3}{2}(x - 8)$$

$$2(y - 3) = 3(x - 8)$$

$$2y - 6 = 3x - 24$$

$$-3x + 2y + 18 = 0$$

$$3x - 2y - 18 = 0$$

9. $x = -2$

$x + 2 = 0$

10. $3x - 5y = 4$

$-5y = -3x + 4$

$y = \frac{-3x}{-5} + \frac{4}{-5}$

$y = \frac{3}{5}x - \frac{4}{5}$

slope $= \frac{3}{5}$

(a) $y - 3 = \frac{3}{5}(x + 2)$

$y - 3 = \frac{3}{5}x + \frac{6}{5}$

$y = \frac{3}{5}x + \frac{21}{5}$

(b) $y - 3 = -\frac{5}{3}(x + 2)$

$y - 3 = -\frac{5}{3}x - \frac{10}{3}$

$y = -\frac{5}{3}x - \frac{1}{3}$

11. $x + 4y \le 8$

$4y \le -x + 8$

$y \le -\frac{1}{4}x + 2$

12. No, $y^2(4 - x) = x^3$ is not a function of x because the graph does not pass the Vertical Line Test.

13. (a) The relation is a function because each x number is matched with exactly one y number.

(b) The relation is not a function because 0 is matched with two numbers, 0 and -4.

14. (a) $g(2) = \frac{2}{2 - 3} = -2$

(b) $g\left(\frac{7}{2}\right) = \frac{\frac{7}{2}}{\frac{7}{2} - 3} = \frac{7}{7 - 6} = 7$

(c) $g(x + 2) = \frac{x + 2}{(x + 2) - 3} = \frac{x + 2}{x - 1}$

15. (a) $h(t) = \sqrt{9 - t}$

$9 - t \ge 0$

$-t \ge -9$

$t \le 9$

Domain: All real values of t such that $t \le 9$

(b) $f(x) = \frac{x + 1}{x - 4}$

$x - 4 \ne 0$

$x \ne 4$

Domain: All real values of x such that $x \ne 4$

16.

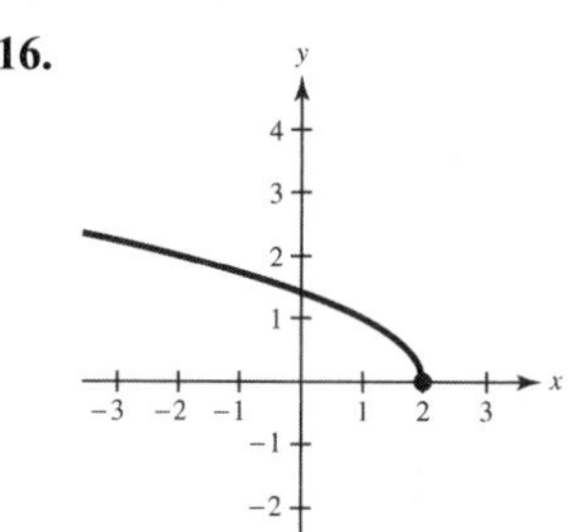

17. $g(x) = -(x - 2)^2 + 1$ is a reflection in the x-axis, horizontal shift 2 units to the right and a vertical shift 1 unit upward.

18. $(0, \$26{,}000), (4, \$10{,}000)$

$m = \frac{10{,}000 - 26{,}000}{4 - 0} = \frac{-16{,}000}{4} = -4000$

$V - 26{,}000 = -4000(t - 0)$

$V = -4000t + 26{,}000$

$16{,}000 = -4000t + 26{,}000$

$-10{,}000 = -4000t$

$\frac{-10{,}000}{-4000} = t$

$2.5 = \frac{5}{2} = t$

After 2.5 years, the car will be worth \$16,000.

19. (a) $y = |x - 2|$

(b) $y = |x| - 2$

(c) $y = -|x| + 2$

CHAPTER 4
Systems of Equations and Inequalities

Section 4.1 Systems of Equations 91

Section 4.2 Linear Systems in Two Variables 97

Section 4.3 Linear Systems in Three Variables 101

Mid-Chapter Quiz 106

Section 4.4 Matrices and Linear Systems 108

Section 4.5 Determinants and Linear Systems 117

Section 4.6 Systems of Linear Inequalities 125

Review Exercises 131

Chapter Test 139

Cumulative Test 142

CHAPTER 4
Systems of Equations and Inequalities

Section 4.1 Systems of Equations

1. (a) $(1, 4)$

$$1 + 2(4) \stackrel{?}{=} 9$$
$$9 = 9$$

$$-2(1) + 3(4) \stackrel{?}{=} 10$$
$$-2 + 12 \stackrel{?}{=} 10$$
$$10 = 10$$

Solution

(b) $(3, -1)$

$$3 + 2(-1) \stackrel{?}{=} 9$$
$$3 - 2 \stackrel{?}{=} 9$$
$$1 \neq 9$$

Not a solution

3. (a) $(-3, 2)$

$$-2(-3) + 7(2) \stackrel{?}{=} 46$$
$$6 + 14 \stackrel{?}{=} 46$$
$$20 \neq 46$$

Not a solution

(b) $(-2, 6)$

$$-2(-2) + 7(6) \stackrel{?}{=} 46$$
$$4 + 42 \stackrel{?}{=} 46$$
$$46 = 46$$

$$3(-2) + 6 \stackrel{?}{=} 0$$
$$-6 + 6 \stackrel{?}{=} 0$$
$$0 = 0$$

Solution

5.

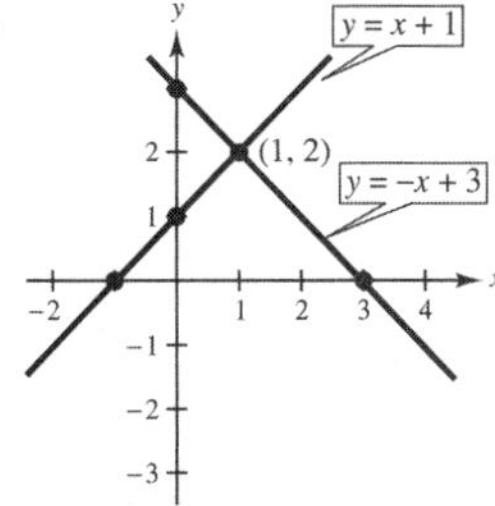

The point of intersection is $(1, 2)$.

7. $x - y = 2$ $\quad\quad$ $x + y = 2$

$$-y = -x + 2 \quad\quad y = -x + 2$$
$$y = x - 2$$

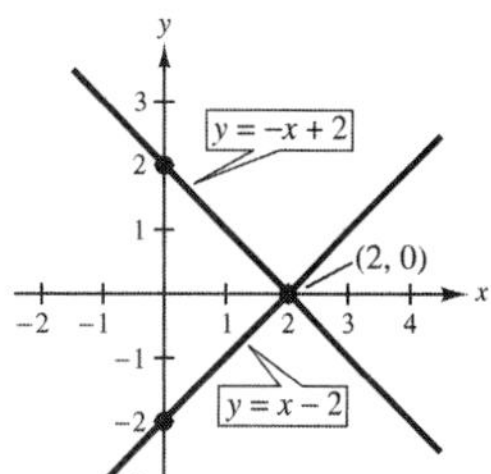

The point of intersection is $(2, 0)$.

9. Solve the first equation for y.

$$3x - 4y = 5$$
$$-4y = -3x + 5$$
$$y = \frac{3}{4}x - \frac{5}{4}$$

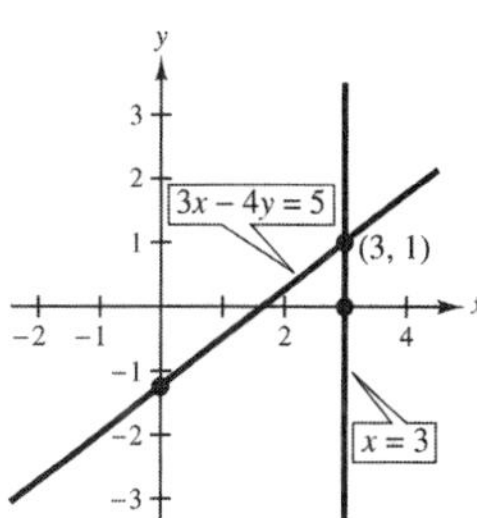

The point of intersection is $(3, 1)$.

11. $4x + 5y = 20$ $\quad\quad$ $\frac{4}{5}x + y = 4$

$$5y = -4x + 20 \quad\quad y = -\frac{4}{5}x + 4$$
$$y = -\frac{4}{5}x + 4$$

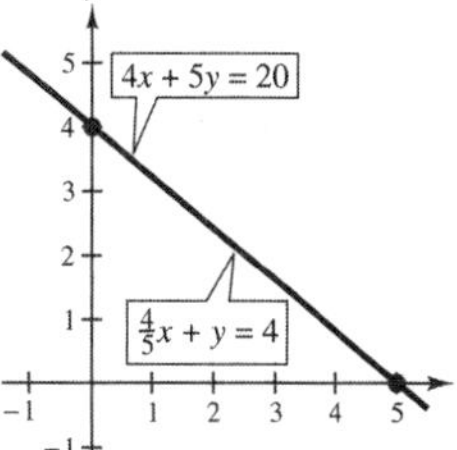

The lines representing the two equations are the same, so the system has infinitely many solutions.

13. $y = 2$

Substitute into second equation.

$$\begin{aligned} x - 6(2) &= -6 \\ x - 12 &= -6 \\ x &= 6 \end{aligned}$$

The solution is $(6, 2)$.

15. Solve for x in the first equation.

$$x = 2y$$

Substitute into second equation.

$$\begin{aligned} 3(2y) + 2y &= 8 \\ 6y + 2y &= 8 \\ 8y &= 8 \\ y &= 1 \end{aligned}$$

$$\begin{aligned} x &= 2(1) \\ &= 2 \end{aligned}$$

The solution is $(2, 1)$.

17. Solve for y in the first equation.

$$y = 3 - x$$

Substitute into second equation.

$$\begin{aligned} 2x - (3 - x) &= 0 \\ 2x - 3 + x &= 0 \\ 3x &= 3 \\ x &= 1 \end{aligned}$$

$$\begin{aligned} y &= 3 - 1 \\ y &= 2 \end{aligned}$$

The solution is $(1, 2)$.

19. Solve for x in the first equation.

$$x = 2 - y$$

Substitute into second equation.

$$\begin{aligned} 2 - y - 4y &= 12 \\ -5y &= 10 \\ y &= -2 \end{aligned}$$

$$\begin{aligned} x &= 2 - (-2) \\ x &= 4 \end{aligned}$$

The solution is $(4, -2)$.

21. Solve for x in the second equation.

$$x = -7 + 7y$$

Substitute into first equation.

$$\begin{aligned} -7 + 7y + 6y &= 19 \\ 13y &= 26 \\ y &= 2 \end{aligned}$$

$$x = -7 + 7(2) = 7$$

The solution is $(7, 2)$.

23. Solve for y in the first equation.

$$\begin{aligned} 5y &= -8x + 100 \\ y &= -\tfrac{8}{5}x + 20 \end{aligned}$$

Substitute into second equation.

$$\begin{aligned} 9x - 10\left(-\tfrac{8}{5}x + 20\right) &= 50 \\ 9x + 16x - 200 &= 50 \\ 25x &= 250 \\ x &= 10 \end{aligned}$$

$$\begin{aligned} y &= -\tfrac{8}{5}(10) + 20 \\ y &= 4 \end{aligned}$$

The solution is $(10, 4)$.

25. Solve for y in the first equation.

$$y = -3x + 8$$

Substitute into second equation.

$$\begin{aligned} 3x + (-3x + 8) &= 6 \\ 8 &\neq 6 \end{aligned}$$

This system of equations has no solution.

27. Solve for y in the first equation.

$$y = -\tfrac{4}{3}x + 1$$

Substitute into second equation.

$$\begin{aligned} 12x + 9\left(-\tfrac{4}{3}x + 1\right) &= 6 \\ 12x - 12x + 9 &= 6 \\ 9 &\neq 6 \end{aligned}$$

This system of equations has no solution.

29. Solve for x in the first equation.

$$x = 4y - 7$$

Substitute into second equation.

$$\begin{aligned} 3(4y - 7) - 12y &= -21 \\ 12y - 21 - 12y &= -21 \\ -21 &= -21 \end{aligned}$$

This system of equations has infinitely many solutions.

31. *Verbal Model:* [Amount of hay 1] + [Amount of hay 2] = 100

$125 · [Amount of hay 1] + $75 · [Amount of hay 2] = $90 · 100

Labels: Amount of hay 1 = x

Amount of hay 2 = y

System:

$$x + y = 100$$
$$125x + 75y = 90(100)$$

Solve for x in the first equation.

$x = 100 - y$

Substitute into second equation.

$$125(100 - y) + 75y = 9000$$
$$12{,}500 - 125y + 75y = 9000$$
$$-50y = -3500$$
$$y = 70$$

$x = 100 - 70$

$x = 30$

30 tons at $125 per ton and 70 tons at $75 per ton should be used.

33. *Verbal Model:* [Total cost] = [Cost per unit] · [Number of units] + [Initial cost]

[Total revenue] = [Price per unit] · [Number of units]

Labels: Total cost = C

Cost per unit = 1.20

Number of units = x

Initial cost = 8000

Total revenue = R

Price per unit = 2.00

System:

$$C = 1.20x + 8000$$
$$R = 2.00x$$

Break-even point occurs when $R = C$ so

$$1.20x + 8000 = 2.00x$$
$$8000 = 0.80x$$
$$10{,}000 = x.$$

10,000 candy bars must be sold.

35. After finding a value for one of the variables, substitute this value back into one of the original equations. This is called back-substitution.

37. When solving a system of linear equations by the method of substitution, if you obtain a true result such as $15 = 15$, then the system of linear equations has infinitely many solutions.

39. Solve for x in the first equation.

$$4x = -15 + 14y$$
$$x = \frac{-15 + 14y}{4}$$

Substitute into second equation.

$$18\left(\frac{-15 + 14y}{4}\right) - 12y = 9$$
$$18(-15 + 14y) - 48y = 36$$
$$-270 + 252y - 48y = 306$$
$$204y = 306$$
$$y = \frac{3}{2}$$

$$x = \frac{-15 + 14\left(\frac{3}{2}\right)}{4} = \frac{-15 + 21}{4} = \frac{3}{2}$$

The solution is $\left(\frac{3}{2}, \frac{3}{2}\right)$.

41. Solve for y in the second equation.

$$y = -x + 20$$

Substitute into first equation.

$$\tfrac{1}{5}x + \tfrac{1}{2}(-x + 20) = 8$$
$$\tfrac{1}{5}x - \tfrac{1}{2}x + 10 = 8$$
$$\tfrac{1}{5}x - \tfrac{1}{2}x = -2$$
$$2x - 5x = -20$$
$$-3x = -20$$
$$x = \tfrac{20}{3}$$

$$y = -\tfrac{20}{3} + 20$$
$$y = \tfrac{40}{3}$$

The solution is $\left(\frac{20}{3}, \frac{40}{3}\right)$.

43. Solve for x in the first equation.

$$8\left(\tfrac{1}{8}x + \tfrac{1}{2}y\right) = (1)8$$
$$x + 4y = 8$$
$$x = 8 - 4y$$

Substitute into second equation.

$$5\left(\tfrac{3}{5}x + y\right) = \left(\tfrac{3}{5}\right)5$$
$$3x + 5y = 3$$
$$3(8 - 4y) + 5y = 3$$
$$24 - 12y + 5y = 3$$
$$-7y = -21$$
$$y = 3$$

$$x = 8 - 4(3)$$
$$x = 8 - 12$$
$$x = -4$$

The solution is $(-4, 3)$.

45. Answers will vary.

$$\begin{cases} 2x - 3y = -7 \\ x + y = 9 \end{cases}$$

47. Answers will vary.

$$\begin{cases} 2x + y = -1 \\ 4x + 2y = 7 \end{cases}$$

49. $3y = ax + b$; $\qquad y = \frac{2}{3}x + 1$

$$y = \frac{a}{3}x + \frac{b}{3}$$

$$\frac{a}{3} = \frac{2}{3} \qquad \frac{b}{3} = 1$$
$$3a = 6 \qquad b = 3$$
$$a = 2$$

51. $\frac{3}{4}x - ay + b = 0$

$$-ay = -\frac{3}{4}x - b$$
$$y = \frac{-3x}{-4a} - \frac{b}{-a}$$
$$y = \frac{3}{4a}x + \frac{b}{a}$$

$$6x - 8y - 48 = 0$$
$$-8y = -6x + 48$$
$$y = \frac{3}{4}x - 6$$

$$\frac{3}{4a} = \frac{3}{4} \qquad \frac{b}{a} = -6$$
$$12 = 12a \qquad b = -6a$$
$$1 = a \qquad = -6(1) = -6$$

53. *Verbal Model:* $\boxed{\text{Larger number}} + \boxed{2} \cdot \boxed{\text{Smaller number}} = \boxed{61}$

$\boxed{\text{Larger number}} - \boxed{\text{Smaller number}} = \boxed{7}$

Labels: Larger number $= x$

Smaller number $= y$

System: $x + 2y = 61$

$x - y = 7$

Solve for x in the second equation.

$x = y + 7$

Substitute into first equation.

$y + 7 + 2y = 61$

$3y + 7 = 61$

$3y = 54$

$y = 18$

$x = 18 + 7$

$x = 25$

$(25, 18)$

55. *Verbal Model:* $2 \cdot \boxed{\text{Smaller number}} - \boxed{\text{Larger number}} = 13$

$\boxed{\text{Smaller number}} + 2 \cdot \boxed{\text{Larger number}} = 114$

Labels: Larger number $= x$

Smaller number $= y$

System: $2y - x = 13$

$y + 2x = 114$

Solve for x in the first equation.

$x = 2y - 13$

Substitute into second equation.

$y + 2(2y - 13) = 114$

$y + 4y - 26 = 114$

$5y = 140$

$y = 28$

$x = 2(28) - 13$

$= 56 - 13$

$x = 43$

$(28, 43)$

57. *Verbal Model:* $2 \cdot \boxed{\text{Length}} + 2 \cdot \boxed{\text{Width}} = 68$

$\boxed{\text{Length}} = \frac{7}{10} \cdot \boxed{\text{Width}}$

Labels: Length $= x$

Width $= y$

System: $2x + 2y = 68$

$x = \frac{7}{10}y$

Substitute into first equation.

$2\left(\frac{7}{10}y\right) + 2y = 68$

$\frac{7}{5}y + 2y = 68$

$7y + 10y = 340$

$17y = 340$

$y = 20$

$x = \frac{7}{10}(20) = 7(2) = 14$

14 yards × 20 yards

59. *Verbal Model:* [Amount in 8.5% bond] + [Amount in 10% bond] = 12,000

8.5% · [Amount in 8.5% bond] + 10% · [Amount in 10% bond] = 1140

Labels: Amount in 8.5% bond $= x$

Amount in 10% bond $= y$

System:

$$x + y = 12{,}000$$
$$0.085x + 0.10y = 1140$$

Solve for x in the first equation.

$x = 12{,}000 - y$

Substitute into second equation.

$$0.085(12{,}000 - y) + 0.10y = 1140$$
$$1020 - 0.085y + 0.10y = 1140$$
$$0.015y = 120$$
$$y = 8000$$

$x = 12{,}000 - 8000$

$x = 4000$

\$4000 is at 8.5% and \$8000 is at 10%.

61. The graphical method usually yields approximate solutions.

63. The system will have exactly one solution because the system is consistent.

65. Pick any coefficient for each variable, then substitute the desired solution for the corresponding variables to find the right-hand side of the equation. Repeat this process for the second equation in the system.

Example:

Solution: $(1, 4)$

Equation 1: $-x + 3y$

$-(1) + 3(4) = 11$

Equation 2: $2x - y$

$2(1) - 4 = -2$

System: $\begin{cases} -x + 3y = 11 \\ 2x - y = -2 \end{cases}$

67. $6x - 2 = -7$

$6x = -5$

$x = -\frac{5}{6}$

Check: $6\left(-\frac{5}{6}\right) - 2 \stackrel{?}{=} -7$

$-5 - 2 \stackrel{?}{=} -7$

$-7 = -7$

69. $4x + 21 = 4(x + 5)$

$4x + 21 = 4x + 20$

$21 \neq 20$

71.

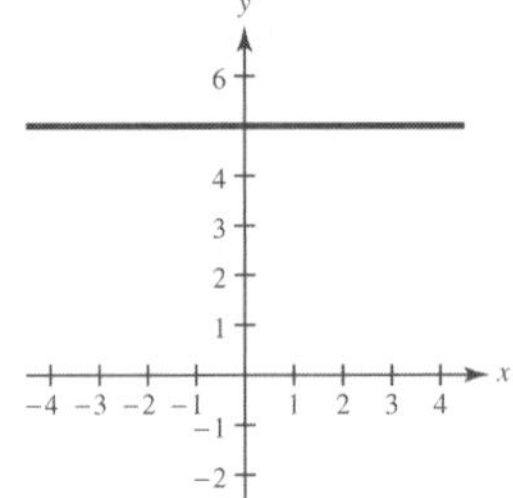

73.

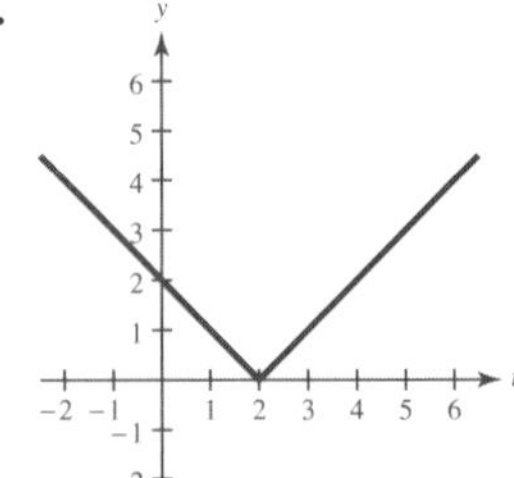

75. $|y| = x + 4$

$|7| = 3 + 4$

$7 = 7$

$|-7| = 3 + 4$

$7 = 7$

There are two values of y associated with one value of x. So, y is not a function of x.

Section 4.2 Linear Systems in Two Variables

1. $$\begin{aligned} 2x + y &= 4 \\ x - y &= 2 \\ \hline 3x \quad &= 6 \\ x \quad &= 2 \end{aligned}$$

$$\begin{aligned} 2 - y &= 2 \\ -y &= 0 \\ y &= 0 \end{aligned}$$

The solution is $(2, 0)$.

3. $$\begin{aligned} -x + 2y &= 1 \\ x - y &= 2 \\ \hline y &= 3 \end{aligned}$$

$$\begin{aligned} x - 3 &= 2 \\ x &= 5 \end{aligned}$$

The solution is $(5, 3)$.

5. $$\begin{aligned} 6x - 6y = 25 &\Rightarrow 6x - 6y = 25 \\ 3y = 11 &\Rightarrow \quad 6y = 22 \\ \hline & \quad 6x = 47 \\ & \quad x = \tfrac{47}{6} \end{aligned}$$

$$\begin{aligned} 3y &= 11 \\ y &= \tfrac{11}{3} \end{aligned}$$

The solution is $\left(\frac{47}{6}, \frac{11}{3}\right)$.

7. $$\begin{aligned} x + 7y &= -6 \\ x &= -7y - 6 \end{aligned}$$

$$\begin{aligned} (-7y - 6) - 5y &= 18 \\ -7y - 6 - 5y &= 18 \\ -12y &= 24 \\ y &= -2 \end{aligned}$$

$$\begin{aligned} x + 7(-2) &= -6 \\ x - 14 &= -6 \\ x &= 8 \end{aligned}$$

The solution is $(8, -2)$.

9. $$\begin{aligned} 5x + 2y = 7 &\Rightarrow 5x + 2y = 7 \\ 3x - y = 13 &\Rightarrow 6x - 2y = 26 \\ \hline & \quad 11x = 33 \\ & \quad x = 3 \end{aligned}$$

$$\begin{aligned} 3(3) - y &= 13 \\ -y &= 4 \\ y &= -4 \end{aligned}$$

The solution is $(3, -4)$.

11. $$\begin{aligned} x - 3y = 2 &\Rightarrow -3x + 9y = -6 \\ 3x - 7y = 4 &\Rightarrow 3x - 7y = 4 \\ \hline & \quad 2y = -2 \\ & \quad y = -1 \end{aligned}$$

$$\begin{aligned} x - 3(-1) &= 2 \\ x &= -1 \end{aligned}$$

The solution is $(-1, -1)$.

13. $$\begin{aligned} 4x + 3y = -10 &\Rightarrow 4x + 3y = -10 \\ 3x - y = -14 &\Rightarrow 9x - 3y = -42 \\ \hline & \quad 13x = -52 \\ & \quad x = -4 \end{aligned}$$

$$\begin{aligned} 3(-4) - y &= -14 \\ -y &= -2 \\ y &= 2 \end{aligned}$$

The solution is $(-4, 2)$.

15. $$\begin{aligned} x + y = 0 &\Rightarrow 3x + 3y = 0 \\ -3x - 3y = 0 &\Rightarrow -3x - 3y = 0 \\ \hline & \quad 0 = 0 \end{aligned}$$

Infinitely many solutions

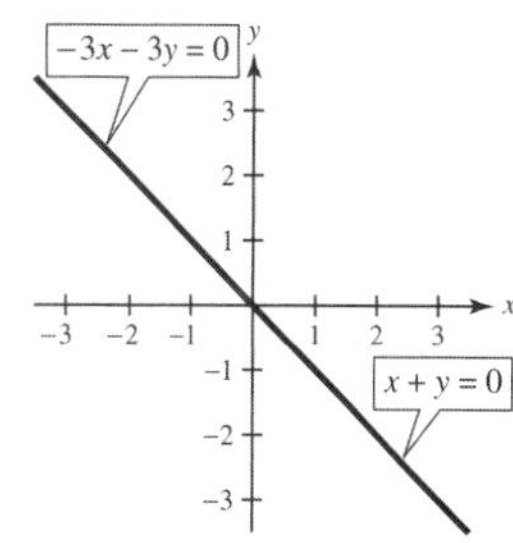

17. $$\begin{aligned} 12x - 5y = 2 &\Rightarrow 24x - 10y = 4 \\ -24x + 10y = 6 &\Rightarrow -24x + 10y = 6 \\ \hline & \quad 0 \neq 10 \end{aligned}$$

There is no solution.

19. $$\begin{aligned} -2x + 3y = 9 &\Rightarrow -6x + 9y = 27 \\ 6x - 9y = -27 &\Rightarrow 6x - 9y = -27 \\ \hline & \quad 0 = 0 \end{aligned}$$

There are infinitely many solutions.

21. $$\begin{aligned} 4x - 8y = 36 &\Rightarrow 12x - 24y = 108 \\ 3x - 6y = 15 &\Rightarrow 12x - 24y = 60 \\ \hline & \quad 0 \neq 48 \end{aligned}$$

There is no solution.

23. *Verbal Model:* [Band 4-hour rate] + [Band additional hour rate] · [Number of additional hours] = [Total cost]

[DJ 4-hour rate] + [DJ additional hour rate] · [Number of additional hours] = [Total cost]

Labels: Band 4-hour rate = 500

Band additional hour rate = 50

DJ 4-hour rate = 300

DJ additional hour rate = 75

Number of additional hours = x

Total cost = y

System:

$$\begin{aligned} 500 + 50x &= y \\ 300 + 75x &= y \\ \hline 200 - 25x &= 0 \\ -25x &= -200 \\ x &= 8 \end{aligned}$$

After 8 additional hours, the cost for the DJ will exceed the cost of the band. So, the total time will be 12 hours.

25. *Verbal Model:* [Plane speed (still air)] − [Speed of air] = [Speed into head wind]

[Plane speed (still air)] + [Speed of air] = [Speed into head wind]

Labels: Plane speed = x

Speed of air = y

System:

$$x - y = \frac{1800}{3.6} \Rightarrow x - y = 500$$

$$x + y = \frac{1800}{3} \Rightarrow x + y = 600$$

$$\begin{aligned} 2x &= 1100 \\ x &= 550 \end{aligned}$$

$$\begin{aligned} 550 - y &= 500 \\ -y &= -50 \\ y &= 50 \end{aligned}$$

The speed of the plane is 550 miles per hour. The wind speed is 50 miles per hour.

27. *Verbal Model:* [Distance] = [Rate] · [Time]

Time at 55 mph = x

Labels: D_1 = distance at 40 mph for 2 hours + at 55 mph for x hours

D_2 = distance at 50 mph for $(2 + x)$ hours

System: $D_1 = D_2$

$$\begin{aligned} 40(2) + 55(x) &= 50(2 + x) \\ 80 + 55x &= 100 + 50x \\ 5x &= 20 \\ x &= 4 \end{aligned}$$

The van must travel 4 hours longer.

29. (a)

$$\begin{aligned} 5m + 3b &= 7 \\ -3m - 3b &= -4 \\ \hline 2m &= 3 \\ m &= \tfrac{3}{2} \end{aligned}$$

$$\begin{aligned} 5\left(\tfrac{3}{2}\right) + 3b &= 7 \\ 15 + 6b &= 14 \\ 6b &= -1 \\ b &= -\tfrac{1}{6} \end{aligned}$$

$$y = \tfrac{3}{2}x - \tfrac{1}{6}$$

(b)

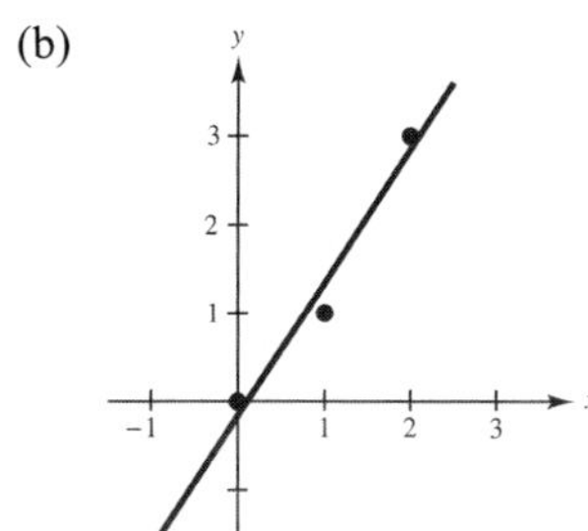

31. To eliminate the y variable in a system, add the two equations together. When the y-terms with opposite coefficients are added together, the sum is zero.

33. An inconsistent system of linear equations is a system that has no solution.

35.

$$\begin{aligned} 2x - y &= 20 \\ -x + y &= -5 \\ \hline x &= 15 \end{aligned}$$

$$\begin{aligned} -15 + y &= -5 \\ y &= 10 \end{aligned}$$

The solution is $(15, 10)$.

37.

$$\begin{aligned} y = 5x - 3 &\Rightarrow \quad y = 5x - 3 \\ y = -2x + 11 &\Rightarrow -y = 2x - 11 \\ \hline & \quad\; 0 = 7x - 14 \\ & \quad\; 14 = 7x \\ & \quad\; 2 = x \end{aligned}$$

$$\begin{aligned} y &= 5(2) - 3 \\ y &= 10 - 3 \\ y &= 7 \end{aligned}$$

The solution is $(2, 7)$.

39.

$$\begin{aligned} \tfrac{3}{2}x + 2y &= 12 \\ \tfrac{1}{4}x + y &= 4 \\ y &= 4 - \tfrac{1}{4}x \end{aligned}$$

$$\begin{aligned} \tfrac{3}{2}x + 2\left(4 - \tfrac{1}{4}x\right) &= 12 \\ \tfrac{3}{2}x + 8 - \tfrac{1}{2}x &= 12 \\ x &= 4 \end{aligned}$$

$$\begin{aligned} y &= 4 - \tfrac{1}{4}(4) \\ &= 4 - 1 = 3 \end{aligned}$$

The solution is $(4, 3)$.

41.

$$\begin{aligned} 2u + 3v = 8 &\Rightarrow -6u - 9v = -24 \\ 3u + 4v = 13 &\Rightarrow \quad 6u + 8v = 26 \\ \hline & \qquad\qquad -v = 2 \\ & \qquad\qquad\; v = -2 \end{aligned}$$

$$\begin{aligned} 2u + 3(-2) &= 8 \\ 2u &= 14 \\ u &= 7 \end{aligned}$$

The solution is $(7, -2)$.

43.

$$\begin{aligned} 4x - 5y = 3 &\Rightarrow -5y = -4x + 3 \Rightarrow y = \tfrac{4}{5}x - \tfrac{3}{5} \\ -8x + 10y = -6 &\Rightarrow 10y = 8x - 6 \Rightarrow y = \tfrac{4}{5}x - \tfrac{3}{5} \end{aligned}$$

Many solutions $\Rightarrow$ consistent

45.

$$\begin{aligned} -2x + 5y = 3 &\Rightarrow 5y = 2x + 3 \Rightarrow y = \tfrac{2}{5}x + \tfrac{3}{5} \\ 5x + 2y = 8 &\Rightarrow 2y = -5x + 8 \Rightarrow y = -\tfrac{5}{2}x + 4 \end{aligned}$$

One solution $\Rightarrow$ consistent

47. *Verbal Model:* $12\left(\boxed{\text{Cost of regular gasoline}}\right) + 8\left(\boxed{\text{Cost of premium gasoline}}\right) = \boxed{76.48}$

$\boxed{\text{Cost of premium gasoline}} = \boxed{0.11} + \boxed{\text{Cost of regular gasoline}}$

Labels: Cost of regular gasoline $= x$

Cost of premium gasoline $= y$

System:

$$12x + 8y = 76.48$$
$$y = 0.11 + x$$

$$12x + 8(0.11 + x) = 76.48$$
$$12x + 0.88 + 8x = 76.48$$
$$20x = 75.60$$
$$x = 3.78$$
$$y = 0.11 + 3.78 = 3.89$$

Regular unleaded gasoline costs \$3.78 per gallon and premium unleaded gasoline costs \$3.89 per gallon.

49. *Verbal Model:* $\boxed{\text{Number of liters Solution 1}} + \boxed{\text{Number of liters Solution 2}} = 30$

$\boxed{\text{Value of Solution 1}} + \boxed{\text{Value of Solution 2}} = 30(0.46)$

Labels: Number of liters Solution 1 $= x$

Number of liters Solution 2 $= y$

System:

$$x + y = 30$$
$$0.40x + 0.70y = 30(0.46)$$
$$x = 30 - y$$

$$x = 30 - y$$

$$0.40(30 - y) + 0.70y = 30(0.46)$$
$$40(30 - y) + 70y = 30(46)$$
$$1200 - 40y + 70y = 1380$$
$$30y = 180$$
$$y = 6$$
$$x = 30 - 6 = 24$$

There are 24 liters of 40% solution and 6 liters of 70% solution.

51. When solving a system by elimination, you can recognize that it has infinitely many solutions when adding a nonzero multiple of one equation to another equation to eliminate a variable, and you get $0 = 0$ for the result.

53. (1) Obtain coefficients for x (or y) that differ only in sign by multiplying all terms of one or both equations by suitable chosen constants.

(2) Add the equations to eliminate one variable, and solve the resulting equation.

(3) Back-substitute the value obtained in Step (2) in either of the original equations and solve for the other variable.

(4) Check your solution in both of the original equations.

55. $m = \dfrac{2 - 0}{4 - 0} = \dfrac{2}{4} = \dfrac{1}{2}$

$$y - 0 = \frac{1}{2}(x - 0)$$
$$y = \frac{1}{2}x$$
$$2y = x$$
$$x - 2y = 0$$

57. $m = \dfrac{2 - 2}{5 - (-1)} = \dfrac{0}{6} = 0$

$$y - 2 = 0[x - (-1)]$$
$$y - 2 = 0$$

59. This set of ordered pairs does represent a function. No input value is matched with two output values.

61. This set of ordered pairs does not represent a function. The input value 3 is matched with two different output values, –2 and –4.

63. (a)

$$3x - 4y = 10 \qquad 2x + 6y = -2$$
$$3(2) - 4(-1) \stackrel{?}{=} 10 \qquad 2(2) + 6(-1) \stackrel{?}{=} -2$$
$$6 + 4 \stackrel{?}{=} 10 \qquad 4 - 6 \stackrel{?}{=} -2$$
$$10 = 10 \qquad -2 = -2$$

$(2, -1)$ is a solution.

(b)

$$3(-1) - 4(0) \stackrel{?}{=} 10 \qquad 2(-1) + 6(0) \stackrel{?}{=} -2$$
$$-3 - 0 \stackrel{?}{=} 10 \qquad -2 + 0 \stackrel{?}{=} -2$$
$$-3 \neq 10 \qquad -2 = -2$$

$(-1, 0)$ is not a solution.

Section 4.3 Linear Systems in Three Variables

1.

$$3y - (-5) = 2$$
$$3y = -3$$
$$y = -1$$
$$x - 2(-1) + 4(-5) = 4$$
$$x + 2 - 20 = 4$$
$$x - 18 = 4$$
$$x = 22$$

The solution is $(22, -1, -5)$.

3.

$$3 + z = 2$$
$$z = -1$$
$$x - 2(3) + 4(-1) = 4$$
$$x - 6 - 4 = 4$$
$$x - 10 = 4$$
$$x = 14$$

The solution is $(14, 3, -1)$.

5.

$$\begin{cases} x + z = 4 \\ y = 2 \\ 4x + z = 7 \end{cases}$$

$$\begin{cases} x + z = 4 \\ y = 2 \\ -3z = -9 \end{cases}$$

$$\begin{cases} x + z = 4 \\ y = 2 \\ z = 3 \end{cases}$$

$$\begin{cases} x = 1 \\ y = 2 \\ z = 3 \end{cases}$$

The solution is $(1, 2, 3)$.

7.

$$\begin{cases} x + y + z = 6 \\ 2x - y + z = 3 \\ 3x - z = 0 \end{cases}$$

$$\begin{cases} x + y + z = 6 \\ -3y - z = -9 \\ -3y - 4z = -18 \end{cases}$$

$$\begin{cases} x + y + z = 6 \\ y + \frac{1}{3}z = 3 \\ -3y - 4z = -18 \end{cases}$$

$$\begin{cases} x + y + z = 6 \\ y + \frac{1}{3}z = 3 \\ -3z = -9 \end{cases}$$

$$\begin{cases} x + y + z = 6 \\ y + \frac{1}{3}z = 3 \\ z = 3 \end{cases}$$

$$y + \tfrac{1}{3}(3) = 3$$
$$y = 2$$
$$x + 2 + 3 = 6$$
$$x = 1$$

The solution is $(1, 2, 3)$.

9. $\begin{cases} x + y + z = -3 \\ 4x + y - 3z = 11 \\ 2x - 3y + 2z = 9 \end{cases}$

$\begin{cases} x + y + z = -3 \\ -3y - 7z = 23 \\ -5y = 15 \end{cases}$

$\begin{cases} x + y + z = -3 \\ y + \frac{7}{3}z = -\frac{23}{3} \\ y = -3 \end{cases}$

$\begin{cases} x + y + z = -3 \\ y + \frac{7}{3}z = -\frac{23}{3} \\ -\frac{7}{3}z = \frac{14}{3} \end{cases}$

$\begin{cases} x + y + z = -3 \\ y + \frac{7}{3}z = -\frac{23}{3} \\ z = -2 \end{cases}$

$y = -3$

$x + (-3) + (-2) = -3$

$x - 5 = -3$

$x = 2$

The solution is $(2, -3, -2)$.

11. $\begin{cases} x + 6y + 2z = 9 \\ 3x - 2y + 3z = -1 \\ 5x - 5y + 2z = 7 \end{cases}$

$\begin{cases} x + 6y + 2z = 9 \\ -20y - 3z = -28 \\ -35y - 8z = -38 \end{cases}$

$\begin{cases} x + 6y + 2z = 9 \\ -20y - 3z = -28 \\ -\frac{11}{4}z = 11 \end{cases}$

$\begin{cases} x + 6y + 2z = 9 \\ y + \frac{3}{20}z = \frac{7}{5} \\ z = -4 \end{cases}$

$y + \frac{3}{20}(-4) = \frac{7}{5}$

$y - \frac{3}{5} = \frac{7}{5}$

$y = \frac{10}{5}$

$y = 2$

$x + 6(2) + 2(-4) = 9$

$x + 12 - 8 = 9$

$x = 5$

The solution is $(5, 2, -4)$.

13. $\begin{cases} 6y + 4z = -12 \\ 3x + 3y = 9 \\ 2x - 3z = 10 \end{cases}$

$\begin{cases} x + y = 3 \\ 6y + 4z = -12 \\ 2x - 3z = 10 \end{cases}$

$\begin{cases} x + y = 3 \\ 6y + 4z = -12 \\ -2y - 3z = 4 \end{cases}$

$\begin{cases} x + y = 3 \\ y + \frac{2}{3}z = -2 \\ -2y - 3z = 4 \end{cases}$

$\begin{cases} x + y = 3 \\ y + \frac{2}{3}z = -2 \\ -\frac{5}{3}z = 0 \end{cases}$

$\begin{cases} x + y = 3 \\ y + \frac{2}{3}z = -2 \\ z = 0 \end{cases}$

$y + \frac{2}{3}(0) = -2$

$y = -2$

$x + (-2) = 3$

$x = 5$

The solution is $(5, -2, 0)$.

15. $\begin{cases} 2x + y + 3z = 1 \\ 2x + 6y + 8z = 3 \\ 6x + 8y + 18z = 5 \end{cases}$

$\begin{cases} 2x + y + 3z = 1 \\ 5y + 5z = 2 \\ 5y + 9z = 2 \end{cases}$

$\begin{cases} 2x + y + 3z = 1 \\ 5y + 5z = 2 \\ 4z = 0 \end{cases}$

$\begin{cases} x + \frac{1}{2}y + \frac{3}{2}z = \frac{1}{2} \\ y + z = \frac{2}{5} \\ z = 0 \end{cases}$

$y + 0 = \frac{2}{5}$

$y = \frac{2}{5}$

$x + \frac{1}{2}\left(\frac{2}{5}\right) + \frac{3}{2}(0) = \frac{1}{2}$

$x + \frac{1}{5} = \frac{1}{2}$

$x = \frac{5}{10} - \frac{2}{10} = \frac{3}{10}$

The solution is $\left(\frac{3}{10}, \frac{2}{5}, 0\right)$.

17. $\begin{cases} x + 2y + 6z = 5 \\ -x + y - 2z = 3 \\ x - 4y - 2z = 1 \end{cases}$

$\begin{cases} x + 2y + 6z = 5 \\ 3y + 4z = 8 \\ x - 4y - 2z = 1 \end{cases}$

$\begin{cases} x + 2y + 6z = 5 \\ 3y + 4z = 8 \\ -6y - 8z = -4 \end{cases}$

$\begin{cases} x + 2y + 6z = 5 \\ 3y + 4z = 8 \\ 0 = 12 \end{cases}$

There is no solution.

19. $\begin{cases} y + z = 5 \\ 2x + 4z = 4 \\ 2x - 3y = -14 \end{cases}$

$\begin{cases} y + z = 5 \\ 2x + 4z = 4 \\ -3y - 4z = -18 \end{cases}$

$\begin{cases} y + z = 5 \\ 2x + 4z = 4 \\ -z = -3 \end{cases}$

$\begin{cases} y + z = 5 \\ 2x + 4z = 4 \\ z = 3 \end{cases}$

$y + 3 = 5$

$y = 2$

$2x + 4(3) = 4$

$2x + 12 = 4$

$2x = -8$

$x = -4$

The solution is $(-4, 2, 3)$.

21. $\begin{cases} 2x + z = 1 \\ 5y - 3z = 2 \\ 6x + 20y - 9z = 11 \end{cases}$

$\begin{cases} x + \frac{1}{2}z = \frac{1}{2} \\ 5y - 3z = 2 \\ 6x + 20y - 9z = 11 \end{cases}$

$\begin{cases} x + \frac{1}{2}z = \frac{1}{2} \\ 5y - 3z = 2 \\ 20y - 12z = 8 \end{cases}$

$\begin{cases} x + \frac{1}{2}z = \frac{1}{2} \\ y - \frac{3}{5}z = \frac{2}{5} \\ 20y - 12z = 8 \end{cases}$

$\begin{cases} x + \frac{1}{2}z = \frac{1}{2} \\ y - \frac{3}{5}z = \frac{2}{5} \\ 0 = 0 \end{cases}$

$y = \frac{3}{5}z + \frac{2}{5}$

$x + \frac{1}{2}z = \frac{1}{2}$

$x = \frac{1}{2} - \frac{1}{2}z$

Let $z = a$; $\left(\frac{1}{2} - \frac{1}{2}a, \frac{3}{5}a + \frac{2}{5}, a\right)$.

23. $\begin{cases} x + 4y - 2z = 2 \\ -3x + y + z = -2 \\ 5x + 7y - 5z = 6 \end{cases}$

$\begin{cases} x + 4y - 2z = 2 \\ 13y - 5z = 4 \\ -13y + 5z = -4 \end{cases}$

$\begin{cases} x + 4y - 2z = 2 \\ y - \frac{5}{13}z = \frac{4}{13} \\ -13y + 5z = -4 \end{cases}$

$\begin{cases} x + 4y - 2z = 2 \\ y - \frac{5}{13}z = \frac{4}{13} \\ 0 = 0 \end{cases}$

$y = \frac{5}{13}z + \frac{4}{13}$

$x + 4\left(\frac{5}{13}z + \frac{4}{13}\right) - 2z = 2$

$x + \frac{20}{13}z + \frac{16}{13} - \frac{26}{13}z = \frac{26}{13}$

$x - \frac{6}{13}z = \frac{10}{13}$

$x = \frac{6}{13}z + \frac{10}{13}$

Let $z = a$; $\left(\frac{6}{13}a + \frac{10}{13}, \frac{5}{13}a + \frac{4}{13}, a\right)$.

25. $\begin{cases} x + 2y - 7z = -4 \\ 2x + y + z = 13 \\ 3x + 9y - 36z = -33 \end{cases}$

$\begin{cases} x + 2y - 7z = -4 \\ -3y + 15z = 21 \\ 3y - 15z = -21 \end{cases}$

$\begin{cases} x + 2y - 7z = -4 \\ -3y + 15z = 21 \\ 0 = 0 \end{cases}$

$y = 5z - 7$

$x + 2(5z - 7) - 7z = -4$

$x + 10z - 14 - 7z = -4$

$x = -3z + 10$

Let $z = a$; $(-3a + 10, 5a - 7, a)$.

27. $\begin{cases} 128 = \frac{1}{2}a(1)^2 + v_0(1) + s_0 \\ 80 = \frac{1}{2}a(2)^2 + v_0(2) + s_0 \\ 0 = \frac{1}{2}a(3)^2 + v_0(3) + s_0 \end{cases}$

$\begin{cases} 128 = \frac{1}{2}a + v_0 + s_0 \\ 80 = 2a + 2v_0 + s_0 \\ 0 = \frac{9}{2}a + 3v_0 + s_0 \end{cases}$

$\begin{cases} 256 = a + 2v_0 + 2s_0 \\ 80 = 2a + 2v_0 + s_0 \\ 0 = \frac{9}{2}a + 3v_0 + s_0 \end{cases}$

$\begin{cases} 256 = a + 2v_0 + 2s_0 \\ -432 = -2v_0 - 3s_0 \\ -1152 = -6v_0 - 8s_0 \end{cases}$

$\begin{cases} 256 = a + 2v_0 + 2s_0 \\ 216 = v_0 + \frac{3}{2}s_0 \\ -1152 = -6v_0 - 8s_0 \end{cases}$

$\begin{cases} 256 = a + 2v_0 + 2s_0 \\ 216 = v_0 + \frac{3}{2}s_0 \\ 144 = + s_0 \end{cases}$

$216 = v_0 + \frac{3}{2}(144)$

$0 = v_0$

$256 = a + 0 + 288$

$-32 = a$

$s = -16t^2 + 144$

29. The process of Gaussian elimination can be used to solve a system of linear equations by converting the system to an equivalent system in row-echelon form. To find the solution, back-substitution is used on the system in row-echelon form.

31. *Sample answer:*

$\begin{cases} x + 2y - 3z = 13 \\ y + 4z = -11 \\ z = -3 \end{cases}$

33. Yes. The first equation was multiplied by –2 and added to the second equation. Then the first equation was multiplied by –3 and added to the third equation.

35. $\begin{aligned} x + y + z &= 3 \\ 2x + y + 2z &= 9 \\ x - 2z &= 0 \end{aligned}$ or $\begin{aligned} x + 2y - z &= -4 \\ y + 2z &= 1 \\ 3x + y + 3z &= 15 \end{aligned}$

Many correct answers. Write equations so that $(4, -3, 2)$ satisfies each equation.

37. Let x = measure of first angle, y = measure of second angle, and z = measure of third angle.

$\begin{cases} x + y + z = 180 \\ x + y = 2z \\ y = z - 28 \end{cases}$

$\begin{cases} x + y + z = 180 \\ x + y - 2z = 0 \\ y - z = -28 \end{cases}$

$$\begin{array}{r} x + y + z = 180 \\ -x - y + 2z = 0 \\ \hline 3z = 180 \\ z = 60 \end{array}$$

$y - 60 = -28$

$y = 32$

$x + 32 - 2(60) = 0$

$x + 32 - 120 = 0$

$x = 88$

The measures of the three angles are 32°, 88°, and 60°.

39. Let x = pounds of vanilla coffee, y = pounds of hazelnut coffee, z = pounds of French roast coffee.

$$\begin{cases} x + y + z = 10 \\ 6x + 6.5y + 7z = 66 \\ y = z \end{cases}$$

$$\begin{cases} -6x - 6y - 6z = -60 \\ 6x + 6.5y + 7z = 66 \\ y - z = 0 \end{cases}$$

$$\begin{aligned} 0.5y + z &= 6 \\ y - z &= 0 \\ \hline 1.5y &= 6 \\ y &= 4 \end{aligned}$$

$$z = 4$$

$$\begin{aligned} x + 4 + 4 &= 10 \\ x &= 2 \end{aligned}$$

The package contains 2 pounds of vanilla coffee, 4 pounds of hazelnut coffee, and 4 pounds of French roast coffee.

41. $$\begin{cases} 0.40x + 0.30y + 0.50z = 30 \\ 0.20x + 0.25y + 0.25z = 17 \\ 0.10x + 0.15y + 0.25z = 10 \end{cases}$$

$$\begin{cases} x + 0.75y + 1.25z = 75 \\ 0.20x + 0.25y + 0.25z = 17 \\ 0.10x + 0.15y + 0.25z = 10 \end{cases}$$

$$\begin{cases} x + 0.75y + 1.25z = 75 \\ 0.1y = 2 \\ 0.075y + .125z = 2.5 \end{cases}$$

$$\begin{cases} x + 0.75y + 1.25z = 75 \\ y = 20 \\ 0.075y + 0.125z = 2.5 \end{cases}$$

$$\begin{cases} x + 1.25z = 60 \\ y = 20 \\ 0.125z = 1 \end{cases}$$

$$\begin{cases} x + 1.25z = 60 \\ y = 20 \\ z = 8 \end{cases}$$

$$\begin{cases} x = 50 \\ y = 20 \\ z = 8 \end{cases}$$

String: 50, Wind: 20, Percussion: 8

43. The solution is apparent because the row-echelon form is

$$\begin{cases} x = 1 \\ y = -3. \\ z = 4 \end{cases}$$

45. Three planes have no point in common when two of the planes are parallel and the third plane intersects the other two planes.

47. The graphs are three planes with three possible situations. If all three planes intersect at one point, there is one solution. If all three planes intersect in one line, there is an infinite number of solutions. If each pair of planes intersects in a line, but the three lines of intersection are all parallel, there is no solution.

49. Terms: $3x$, 2

Coefficients: 3, 2

51. Terms: $14t^5$, $-t$, 25

Coefficients: 14, −1, 25

53. $$\begin{aligned} 2x + 3y = 17 &\Rightarrow 8x + 12y = 68 \\ 4y = 12 &\Rightarrow \quad\;\; -12y = -36 \\ & \overline{8x \qquad\quad = 32} \\ & x \qquad\quad\; = 4 \end{aligned}$$

$$\begin{aligned} 2(4) + 3y &= 17 \\ 3y &= 9 \\ y &= 3 \end{aligned}$$

The solution is $(4, 3)$.

55. $$\begin{aligned} 3x - 4y &= -30 \\ 5x + 4y &= 14 \\ \hline 8x &= -16 \\ x &= -2 \end{aligned}$$

$$\begin{aligned} 3(-2) - 4y &= -30 \\ -4y &= -24 \\ y &= 6 \end{aligned}$$

The solution is $(-2, 6)$.

Mid-Chapter Quiz for Chapter 4

1. (a) $(1, -2)$

$$5(1) - 12(-2) \stackrel{?}{=} 2$$
$$5 + 24 \neq 2$$

This is not a solution.

(b) $(10, 4)$

$$5(10) - 12(4) \stackrel{?}{=} 2$$
$$50 - 48 = 2$$
$$2 = 2$$

$$2(10) + 1.5(4) \stackrel{?}{=} 26$$
$$20 + 6 = 26$$
$$26 = 26$$

This is a solution.

2.

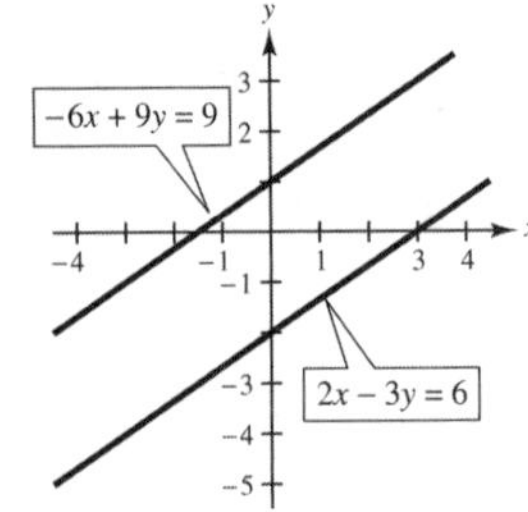

No solution

3.

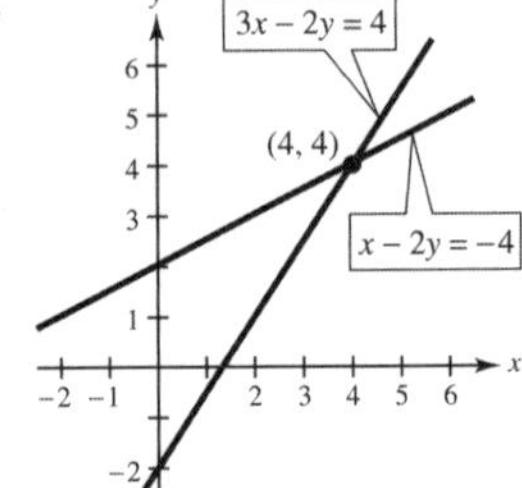

One solution

4.

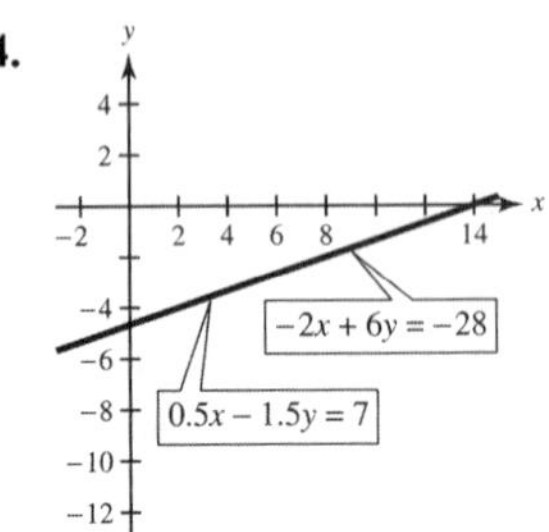

Infinitely many solutions

5. $x - y = 0 \qquad 2x = 8$

$y = x \qquad x = 4$

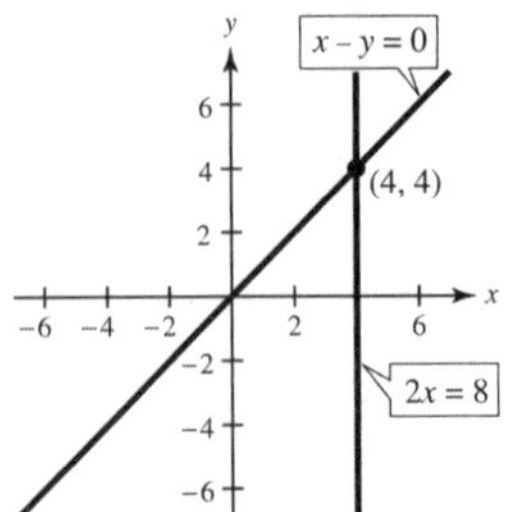

The solution is $(4, 4)$.

6.

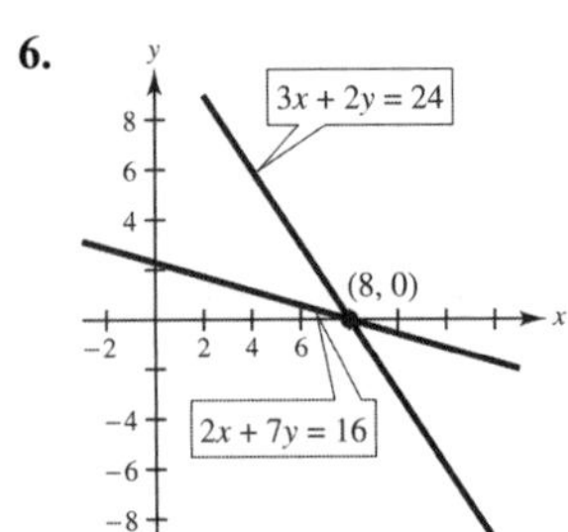

The solution is $(8, 0)$.

7.

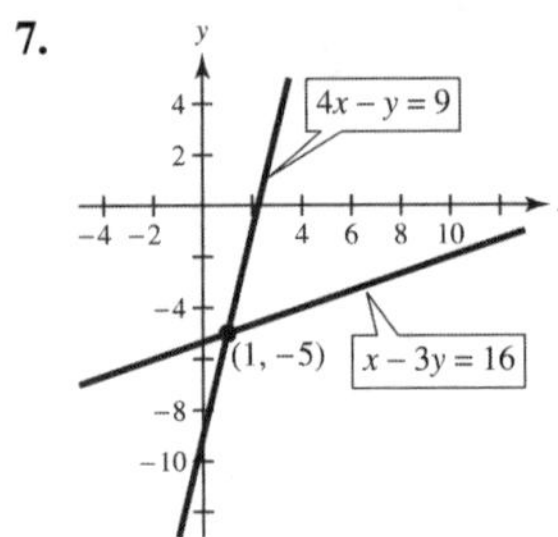

The solution is $(1, -5)$.

8. $2x - 3y = 4$

$$y = 2$$

$$2x - 3(2) = 4$$
$$2x - 6 = 4$$
$$2x = 10$$
$$x = 5$$

The solution is $(5, 2)$.

9. $5x - y = 32 \Rightarrow -y = -5x + 32 \Rightarrow y = 5x - 32$

$6x - 9y = 18$

$6x - 9(5x - 32) = 18$

$6x - 45x + 288 = 18$

$-39x = -270$

$x = \frac{-270}{-39} = \frac{90}{13}$

$y = 5\left(\frac{90}{13}\right) - 32 = \frac{450}{13} - \frac{416}{13} = \frac{34}{13}$

The solution is $\left(\frac{90}{13}, \frac{34}{13}\right)$.

10. $6x - 2y = 2$

$9x - 3y = 1$

$-2y = -6x + 2$

$y = 3x - 1$

$9x - 3(3x - 1) = 1$

$9x - 9x + 3 = 1$

$3 \neq 1$

No solution

11. $x + 10y = 18$

$5x + 2y = 42$

$x + 10y = 18$

$-48y = -48$

$x + 10y = 18$

$y = 1$

$x = 8$

$y = 1$

The solution is $(8, 1)$.

12. $x - 3y = 6 \Rightarrow -3x + 9y = -18$

$3x + y = 8 \Rightarrow 3x + y = 8$

$10y = -10$

$y = -1$

$x - 3(-1) = 6$

$x = 3$

The solution is $(3, -1)$.

13. $\begin{cases} a + b + c = 1 \\ 4a + 2b + c = 2 \\ 9a + 3b + c = 4 \end{cases}$

$\begin{cases} a + b + c = 1 \\ -2b - 3c = -2 \\ -6b - 8c = -5 \end{cases}$

$\begin{cases} a + b + c = 1 \\ b + \frac{3}{2}c = 1 \\ -6b - 8c = -5 \end{cases}$

$\begin{cases} a - \frac{1}{2}c = 0 \\ b + \frac{3}{2}c = 1 \\ c = 1 \end{cases}$

$a = \frac{1}{2}$

$b = -\frac{1}{2}$

$c = 1$

The solution is $\left(\frac{1}{2}, -\frac{1}{2}, 1\right)$.

14. $\begin{cases} x + 4z = 17 \\ -3x + 2y - z = -20 \\ x - 5y + 3z = 19 \end{cases}$

$\begin{cases} x + 4z = 17 \\ 2y + 11z = 31 \\ -5y - z = 2 \end{cases}$

$\begin{cases} x + 4z = 17 \\ y + \frac{11}{2}z = \frac{31}{2} \\ \frac{53}{2}z = \frac{159}{2} \end{cases}$

$\begin{cases} x + 4z = 17 \\ y + \frac{11}{2}z = \frac{31}{2} \\ z = 3 \end{cases}$

$x = 5$

$y = -1$

$z = 3$

The solution is $(5, -1, 3)$.

15. Answers will vary. Write any two equations that are satisfied by $(10, -12)$.

$x + y = -2$

$2x - y = 32$

16. Answers will vary. Write any three equations that are satisfied by $(2, -5, 10)$.

$\begin{cases} x + y + z = 7 \\ 2x - y = 9 \\ -2x + y + 3z = 21 \end{cases}$

17. *Verbal Model:* $\boxed{\text{Amount Solution 1}} + \boxed{\text{Amount Solution 2}} = \boxed{\text{Amount mixture}}$

$$0.20\boxed{\text{Amount Solution 1}} + 0.50\boxed{\text{Amount Solution 2}} = 0.30 \cdot 20$$

Labels: Amount Solution 1 $= x$

Amount Solution 2 $= y$

System:

$$x + y = 20$$
$$0.20x + 0.50y = 0.30(20)$$
$$x + y = 20$$
$$20x + 50y = 600$$

By substitution:

$$y = 20 - x$$
$$20x + 50(20 - x) = 600$$
$$20x + 1000 - 50x = 600$$
$$-30x = -400$$
$$x = 13\tfrac{1}{3} \text{ gallons at 20\% solution}$$
$$20 - x = 6\tfrac{2}{3} \text{ gallons at 50\% solution}$$

18. Let x = measure of first angle, y = measure of second angle, and z = measure of third angle.

$$\begin{cases} x + y + z = 180 \\ x = 2y - 14 \\ z = y + 30 \end{cases}$$

$$\begin{cases} x + y + z = 180 \\ x - 2y = -14 \\ -y + z = 30 \end{cases}$$

$$\begin{cases} x + y + z = 180 \\ -3y - z = -194 \\ -y + z = 30 \end{cases}$$

$$\begin{cases} x + y + z = 180 \\ -3y - z = -194 \\ -4y = -164 \\ y = 41 \end{cases}$$

$$-41 + z = 30$$
$$z = 71$$

$$x + 41 + 71 = 180$$
$$x = 68$$

The measures of the three angles are $68°$, $41°$, and $71°$.

Section 4.4 Matrices and Linear Systems

1. 4×2

3. 4×1

5. 2×2

7. 1×1

9. 1×4

11. (a) $\begin{bmatrix} 4 & -5 \\ -1 & 8 \end{bmatrix}$ (b) $\left[\begin{array}{cc:c} 4 & -5 & -2 \\ -1 & 8 & 10 \end{array}\right]$

13. (a) $\begin{bmatrix} 1 & 1 & 0 \\ 5 & -2 & -2 \\ 2 & 4 & 1 \end{bmatrix}$ (b) $\left[\begin{array}{ccc:c} 1 & 1 & 0 & 0 \\ 5 & -2 & -2 & 12 \\ 2 & 4 & 1 & 5 \end{array}\right]$

15. $\begin{cases} 4x + 3y = 8 \\ x - 2y = 3 \end{cases}$

17. $\begin{cases} x + 2z = -10 \\ 3y - z = 5 \\ 4x + 2y = 3 \end{cases}$

19. $\begin{bmatrix} 1 & 1 & -4 & 2 \\ 0 & 0 & 8 & 3 \\ 0 & 4 & 5 & 5 \end{bmatrix} \begin{matrix} \\ R_3 \\ R_2 \end{matrix} \begin{bmatrix} 1 & 1 & -4 & 2 \\ 0 & 4 & 5 & 5 \\ 0 & 0 & 8 & 3 \end{bmatrix}$

21. $\begin{bmatrix} 9 & -18 & 27 \\ 3 & 4 & 5 \end{bmatrix} \frac{1}{9}R_1 \rightarrow \begin{bmatrix} 1 & -2 & 3 \\ 3 & 4 & 5 \end{bmatrix}$

23. $\begin{bmatrix} 1 & 4 & 3 \\ 2 & 8 & 6 \end{bmatrix} -2R_1 + R_2 \rightarrow \begin{bmatrix} 1 & 4 & 3 \\ 0 & 0 & 0 \end{bmatrix}$

25. $x - 2y - z = 6$
$y + 4z = 5$
$4x + 2y + 3z = 8$

$$\left[\begin{array}{ccc:c} 1 & -2 & -1 & 6 \\ 0 & 1 & 4 & 5 \\ 4 & 2 & 3 & 8 \end{array}\right]$$

$$\begin{matrix} \\ \\ -4R_1 + R_3 \end{matrix} \left[\begin{array}{ccc:c} 1 & -2 & -1 & 6 \\ 0 & 1 & 4 & 5 \\ 0 & 10 & 7 & -16 \end{array}\right]$$

$$\begin{matrix} \\ \\ -10R_2 + R_3 \end{matrix} \left[\begin{array}{ccc:c} 1 & -2 & -1 & 6 \\ 0 & 1 & 4 & 5 \\ 0 & 0 & -33 & -66 \end{array}\right]$$

$$\begin{matrix} \\ \\ \frac{1}{-33}R_3 \end{matrix} \left[\begin{array}{ccc:c} 1 & -2 & -1 & 6 \\ 0 & 1 & 4 & 5 \\ 0 & 0 & 1 & 2 \end{array}\right]$$

$z = 2$ $\quad y + 4(2) = 5 \quad x - 2(-3) - (2) = 6$
$y = -3 \quad x + 6 - 2 = 6$
$x = 2$

The solution is $(2, -3, 2)$.

27. $6x - 4y = 2$
$5x + 2y = 7$

$$\left[\begin{array}{cc:c} 6 & -4 & 2 \\ 5 & 2 & 7 \end{array}\right]$$

$$\frac{1}{6}R_1 \left[\begin{array}{cc:c} 1 & -\frac{2}{3} & \frac{1}{3} \\ 5 & 2 & 7 \end{array}\right]$$

$$\begin{matrix} \\ -5R_1 + R_2 \end{matrix} \left[\begin{array}{cc:c} 1 & -\frac{2}{3} & \frac{1}{3} \\ 0 & \frac{16}{3} & \frac{16}{3} \end{array}\right]$$

$$\begin{matrix} \\ -\frac{3}{16}R_2 \end{matrix} \left[\begin{array}{cc:c} 1 & -\frac{2}{3} & \frac{1}{3} \\ 0 & 1 & 1 \end{array}\right]$$

$y = 1 \quad x - \frac{2}{3}(1) = \frac{1}{3}$
$x = 1$

The solution is $(1, 1)$.

29. $12 + 10y = -14$
$4x - 3y = -11$

$$\left[\begin{array}{cc:c} 12 & 10 & -14 \\ 4 & -3 & -11 \end{array}\right]$$

$$\frac{1}{12}R_1 \left[\begin{array}{cc:c} 1 & \frac{5}{6} & -\frac{7}{6} \\ 4 & -3 & -11 \end{array}\right]$$

$$\begin{matrix} \\ -4R_1 + R_2 \end{matrix} \left[\begin{array}{cc:c} 1 & \frac{5}{6} & -\frac{7}{6} \\ 0 & -\frac{19}{3} & -\frac{19}{3} \end{array}\right]$$

$$\begin{matrix} \\ -\frac{3}{19} \cdot R_2 \end{matrix} \left[\begin{array}{cc:c} 1 & \frac{5}{6} & -\frac{7}{6} \\ 0 & 1 & 1 \end{array}\right]$$

$y = 1 \quad x + \frac{5}{6}(1) = -\frac{7}{6}$
$x = -2$

The solution is $(-2, 1)$.

31. $2x + 4y \quad\quad = 10$

$2x + 2y + 3z = 3$

$-3x + y + 2z = -3$

$$\begin{bmatrix} 2 & 4 & 0 & \vdots & 10 \\ 2 & 2 & 3 & \vdots & 3 \\ -3 & 1 & 2 & \vdots & -3 \end{bmatrix}$$

$$\begin{matrix} \\ -R_1 + R_2 \\ \\ \end{matrix} \begin{bmatrix} 2 & 4 & 0 & \vdots & 10 \\ 0 & -2 & 3 & \vdots & -7 \\ -3 & 1 & 2 & \vdots & -3 \end{bmatrix}$$

$$\begin{matrix} \frac{1}{2}R_1 \\ \\ \\ \end{matrix} \begin{bmatrix} 1 & 2 & 0 & \vdots & 5 \\ 0 & -2 & 3 & \vdots & -7 \\ -3 & 1 & 2 & \vdots & -3 \end{bmatrix}$$

$$\begin{matrix} \\ \\ 3R_1 + R_3 \end{matrix} \begin{bmatrix} 1 & 2 & 0 & \vdots & 5 \\ 0 & -2 & 3 & \vdots & -7 \\ 0 & 7 & 2 & \vdots & 12 \end{bmatrix}$$

$$\begin{matrix} \\ -\frac{1}{2}R_2 \\ \\ \end{matrix} \begin{bmatrix} 1 & 2 & 0 & \vdots & 5 \\ 0 & 1 & -\frac{3}{2} & \vdots & \frac{7}{2} \\ 0 & 7 & 2 & \vdots & 12 \end{bmatrix}$$

$$\begin{matrix} \\ \\ -7R_2 + R_3 \end{matrix} \begin{bmatrix} 1 & 2 & 0 & \vdots & 5 \\ 0 & 1 & -\frac{3}{2} & \vdots & \frac{7}{2} \\ 0 & 0 & \frac{25}{2} & \vdots & -\frac{25}{2} \end{bmatrix}$$

$$\begin{matrix} \\ \\ \frac{2}{25}R_3 \end{matrix} \begin{bmatrix} 1 & 2 & 0 & \vdots & 5 \\ 0 & 1 & -\frac{3}{2} & \vdots & \frac{7}{2} \\ 0 & 0 & 1 & \vdots & -1 \end{bmatrix}$$

$z = -1$

$y - \frac{3}{2}(-1) = \frac{7}{2}$

$y = \frac{4}{2}$

$y = 2$

$x + 2(2) = 5$

$x + 4 = 5$

$x = 1$

The solution is $(1, 2, -1)$.

33. $-2x - 2y - 15z = 0$

$x + 2y + 2z = 18$

$3x + 3y + 22z = 2$

$$\begin{bmatrix} -2 & -2 & -15 & \vdots & 0 \\ 1 & 2 & 2 & \vdots & 18 \\ 3 & 3 & 22 & \vdots & 2 \end{bmatrix}$$

$$\begin{matrix} R_2 \\ R_1 \\ \\ \end{matrix} \begin{bmatrix} 1 & 2 & 2 & \vdots & 18 \\ -2 & -2 & -15 & \vdots & 0 \\ 3 & 3 & 22 & \vdots & 2 \end{bmatrix}$$

$$\begin{matrix} \\ 2R_1 + R_2 \\ -3R_1 + R_3 \end{matrix} \begin{bmatrix} 1 & 2 & 2 & \vdots & 18 \\ 0 & 2 & -11 & \vdots & 36 \\ 0 & -3 & 16 & \vdots & -52 \end{bmatrix}$$

$$\begin{matrix} \\ \\ \frac{3}{2}R_2 + R_3 \end{matrix} \begin{bmatrix} 1 & 2 & 2 & \vdots & 18 \\ 0 & 2 & -11 & \vdots & 36 \\ 0 & 0 & -\frac{1}{2} & \vdots & 2 \end{bmatrix}$$

$$\begin{matrix} \\ \frac{1}{2}R_2 \\ -2R_3 \end{matrix} \begin{bmatrix} 1 & 2 & 2 & \vdots & 18 \\ 0 & 1 & -\frac{11}{2} & \vdots & 18 \\ 0 & 0 & 1 & \vdots & -4 \end{bmatrix}$$

$z = -4$

$y - \frac{11}{2}(-4) = 18$

$y + 22 = 18$

$y = -4$

$x + 2(-4) + 2(-4) = 18$

$x - 8 - 8 = 18$

$x - 16 = 18$

$x = 34$

The solution is $(34, -4, -4)$

35. $x + y - 5z = 3$

$x \quad - 2z = 1$

$2x - y - z = 0$

$$\begin{bmatrix} 1 & 1 & -5 & \vdots & 3 \\ 1 & 0 & -2 & \vdots & 1 \\ 2 & -1 & -1 & \vdots & 0 \end{bmatrix}$$

$$\begin{matrix} \\ -R_1 + R_2 \\ -2R_1 + R_3 \end{matrix} \begin{bmatrix} 1 & 1 & -5 & \vdots & 3 \\ 0 & -1 & 3 & \vdots & -2 \\ 0 & -3 & 9 & \vdots & -6 \end{bmatrix}$$

$$\begin{matrix} \\ -R_2 \\ \\ \end{matrix} \begin{bmatrix} 1 & 1 & -5 & \vdots & 3 \\ 0 & 1 & -3 & \vdots & 2 \\ 0 & -3 & 9 & \vdots & -6 \end{bmatrix}$$

$$\begin{matrix} \\ \\ 3R_2 + R_3 \end{matrix} \begin{bmatrix} 1 & 1 & -5 & \vdots & 3 \\ 0 & 1 & -3 & \vdots & 2 \\ 0 & 0 & 0 & \vdots & 0 \end{bmatrix}$$

$y - 3z = 2$

$y = 2 + 3z$

$x + (2 + 3z) - 5z = 3$

$x = 1 + 2z$

Let $a = z$. (a is any real number).

$(1 + 2a, 2 + 3a, a)$

37. $2x \quad + 4z = 1$

$x + y + 3z = 0$

$x + 3y + 5z = 0$

$$\begin{bmatrix} 2 & 0 & 4 & \vdots & 1 \\ 1 & 1 & 3 & \vdots & 0 \\ 1 & 3 & 5 & \vdots & 0 \end{bmatrix}$$

$$\begin{matrix} R_2 \\ R_1 \\ \\ \end{matrix} \begin{bmatrix} 1 & 1 & 3 & \vdots & 0 \\ 2 & 0 & 4 & \vdots & 1 \\ 1 & 3 & 5 & \vdots & 0 \end{bmatrix}$$

$$\begin{matrix} \\ -2R_1 + R_2 \\ -R_1 + R_3 \end{matrix} \begin{bmatrix} 1 & 1 & 3 & \vdots & 0 \\ 0 & -2 & -2 & \vdots & 1 \\ 0 & 2 & 2 & \vdots & 0 \end{bmatrix}$$

$$\begin{matrix} \\ -\frac{1}{2}R_2 \\ \\ \end{matrix} \begin{bmatrix} 1 & 1 & 3 & \vdots & 0 \\ 0 & 1 & 1 & \vdots & -\frac{1}{2} \\ 0 & 2 & 2 & \vdots & 0 \end{bmatrix}$$

$$\begin{matrix} \\ \\ -2R_2 + R_3 \end{matrix} \begin{bmatrix} 1 & 1 & 3 & \vdots & 0 \\ 0 & 1 & 1 & \vdots & -\frac{1}{2} \\ 0 & 0 & 0 & \vdots & 1 \end{bmatrix}$$

Inconsistent; no solution

39. *Verbal Model:* $\boxed{\text{Money 1}} + \boxed{\text{Money 2}} + \boxed{\text{Money 3}} = \boxed{1{,}500{,}000}$

$0.08 \cdot \boxed{\text{Money 1}} + 0.09 \cdot \boxed{\text{Money 2}} + 0.12 \cdot \boxed{\text{Money 3}} = 113{,}000$

$\boxed{\text{Money 1}} = 4 \cdot \boxed{\text{Money 3}}$

Labels: x = Money 1

y = Money 2

z = Money 3

System:

$$\begin{aligned} x + y + z &= 1{,}500{,}000 \\ 0.08x + 0.09y + 0.12z &= 133{,}000 \\ x &= 4z \end{aligned}$$

$$\begin{bmatrix} 1 & 1 & 1 & \vdots & 1{,}500{,}000 \\ 8 & 9 & 12 & \vdots & 13{,}300{,}000 \\ 1 & 0 & -4 & \vdots & 0 \end{bmatrix}$$

$$\begin{matrix} \\ -8R_1 + R_2 \\ -R_1 + R_3 \end{matrix} \begin{bmatrix} 1 & 1 & 1 & \vdots & 1{,}500{,}000 \\ 0 & 1 & 4 & \vdots & 1{,}300{,}000 \\ 0 & -1 & -5 & \vdots & -1{,}500{,}000 \end{bmatrix}$$

$$\begin{matrix} \\ \\ R_2 + R_3 \end{matrix} \begin{bmatrix} 1 & 1 & 1 & \vdots & 1{,}500{,}000 \\ 0 & 1 & 4 & \vdots & 1{,}300{,}000 \\ 0 & 0 & -1 & \vdots & -200{,}000 \end{bmatrix}$$

$$\begin{matrix} \\ \\ -R_3 \end{matrix} \begin{bmatrix} 1 & 1 & 1 & \vdots & 1{,}500{,}000 \\ 0 & 1 & 4 & \vdots & 1{,}300{,}000 \\ 0 & 0 & 1 & \vdots & 200{,}000 \end{bmatrix}$$

$z = 200{,}000$

$$\begin{aligned} y + 4(200{,}000) &= 1{,}300{,}000 \\ y &= 500{,}000 \end{aligned}$$

$$\begin{aligned} x + 500{,}000 + 200{,}000 &= 1{,}500{,}000 \\ x &= 800{,}000 \end{aligned}$$

\$800,000 at 8%, \$500,000 at 9%, \$200,000 at 12%

41. 1×4: one row and four columns

2×2: two rows and two columns

4×1: four rows and one column

43. Row operations are performed on systems of equations and elementary row operations are performed on matrices.

45. $4x + 3y = 10$
$2x - y = 10$
$-2x + z = -9$

$$\begin{bmatrix} 4 & 3 & 0 & \vdots & 10 \\ 2 & -1 & 0 & \vdots & 10 \\ -2 & 0 & 1 & \vdots & -9 \end{bmatrix}$$

$$\begin{matrix} \\ \frac{1}{2}R_2 \\ \\ \end{matrix} \begin{bmatrix} 4 & 3 & 0 & \vdots & 10 \\ 1 & -\frac{1}{2} & 0 & \vdots & 5 \\ -2 & 0 & 1 & \vdots & -9 \end{bmatrix}$$

$$\begin{matrix} R_1 \\ R_2 \\ \\ \end{matrix} \begin{bmatrix} 1 & -\frac{1}{2} & 0 & \vdots & 5 \\ 4 & 3 & 0 & \vdots & 10 \\ -2 & 0 & 1 & \vdots & -9 \end{bmatrix}$$

$$\begin{matrix} \\ -4R_1 + R_2 \\ 2R_1 + R_3 \end{matrix} \begin{bmatrix} 1 & -\frac{1}{2} & 0 & \vdots & 5 \\ 0 & 5 & 0 & \vdots & -10 \\ 0 & -1 & 1 & \vdots & 1 \end{bmatrix}$$

$$\begin{matrix} \\ \\ -R_3 \end{matrix} \begin{bmatrix} 1 & -\frac{1}{2} & 0 & \vdots & 5 \\ 0 & 5 & 0 & \vdots & -10 \\ 0 & 1 & -1 & \vdots & -1 \end{bmatrix}$$

$$\begin{matrix} \\ R_2 \\ R_3 \end{matrix} \begin{bmatrix} 1 & -\frac{1}{2} & 0 & \vdots & 5 \\ 0 & 1 & -1 & \vdots & -1 \\ 0 & 5 & 0 & \vdots & -10 \end{bmatrix}$$

$$\begin{matrix} \frac{1}{2}R_2 + R_1 \\ \\ -5R_2 + R_3 \end{matrix} \begin{bmatrix} 1 & 0 & -\frac{1}{2} & \vdots & \frac{9}{2} \\ 0 & 1 & -1 & \vdots & -1 \\ 0 & 0 & 5 & \vdots & -5 \end{bmatrix}$$

$$\begin{matrix} \\ \\ \frac{1}{5}R_3 \end{matrix} \begin{bmatrix} 1 & 0 & -\frac{1}{2} & \vdots & \frac{9}{2} \\ 0 & 1 & -1 & \vdots & -1 \\ 0 & 0 & 1 & \vdots & -1 \end{bmatrix}$$

$z = -1$ $\quad y - 1(-1) = -1$ $\quad x - \frac{1}{2}(-1) = \frac{9}{2}$

$y = -2$ $\quad x = 4$

The solution is $(4, -2, -1)$.

47. $2x + y - 2z = 4$
$3x - 2y + 4z = 6$
$-4x + y + 6z = 12$

$$\begin{bmatrix} 2 & 1 & -2 & \vdots & 4 \\ 3 & -2 & 4 & \vdots & 6 \\ -4 & 1 & 6 & \vdots & 12 \end{bmatrix}$$

$$\begin{matrix} \\ -\frac{3}{2}R_1 + R_2 \\ 2R_1 + R_3 \end{matrix} \begin{bmatrix} 2 & 1 & -2 & \vdots & 4 \\ 0 & -\frac{7}{2} & 7 & \vdots & 0 \\ 0 & 3 & 2 & \vdots & 20 \end{bmatrix}$$

$$\begin{matrix} \frac{1}{2}R_1 \\ -\frac{2}{7}R_2 \\ \\ \end{matrix} \begin{bmatrix} 1 & \frac{1}{2} & -1 & \vdots & 2 \\ 0 & 1 & -2 & \vdots & 0 \\ 0 & 3 & 2 & \vdots & 20 \end{bmatrix}$$

$$\begin{matrix} \\ \\ -3R_2 + R_3 \end{matrix} \begin{bmatrix} 1 & \frac{1}{2} & -1 & \vdots & 2 \\ 0 & 1 & -2 & \vdots & 0 \\ 0 & 0 & 8 & \vdots & 20 \end{bmatrix}$$

$$\begin{matrix} \\ \\ \frac{1}{8}R_3 \end{matrix} \begin{bmatrix} 1 & \frac{1}{2} & -1 & \vdots & 2 \\ 0 & 1 & -2 & \vdots & 0 \\ 0 & 0 & 1 & \vdots & \frac{5}{2} \end{bmatrix}$$

$z = \frac{5}{2}$ $\quad y - 2\left(\frac{5}{2}\right) = 0$ $\quad x + \frac{1}{2}(5) - \left(\frac{5}{2}\right) = 2$

$y - 5 = 0$ $\quad x + \frac{5}{2} - \frac{5}{2} = 2$

$y = 5$ $\quad x = 2$

The solution is $\left(2, 5, \frac{5}{2}\right)$.

49. *Verbal Model:* [Theater A tickets] + [Theater B tickets] + [Theater C tickets] = 1500

1.50 [Theater A tickets] + 7.50 [Theater B tickets] + 8.50 [Theater C tickets] = 10,050

[Theater B tickets] = 2 [Theater A tickets]

[Theater C tickets] = 2 [Theater A tickets]

Labels: x = Theater A tickets

y = Theater B tickets

z = Theater C tickets

System:

$$\begin{aligned} x + y + z &= 1500 \\ 1.5x + 7.5y + 8.5z &= 10{,}050 \\ -2x + y &= 0 \\ -2x + z &= 0 \end{aligned}$$

$$\begin{bmatrix} 1 & 1 & 1 & \vdots & 1500 \\ 1.5 & 7.5 & 8.5 & \vdots & 10{,}050 \\ -2 & 1 & 0 & \vdots & 0 \\ -2 & 0 & 1 & \vdots & 0 \end{bmatrix}$$

$$\begin{matrix} \\ -1.5R_1 + R_2 \\ -R_4 + R_3 \\ 2R_1 + R_4 \end{matrix} \begin{bmatrix} 1 & 1 & 1 & \vdots & 1500 \\ 0 & 6 & 7 & \vdots & 7800 \\ 0 & 1 & -1 & \vdots & 0 \\ 0 & 2 & 3 & \vdots & 3000 \end{bmatrix}$$

$$\begin{matrix} \\ \frac{1}{6}R_2 \\ \\ -2R_3 + R_4 \end{matrix} \begin{bmatrix} 1 & 1 & 1 & \vdots & 1500 \\ 0 & 1 & \frac{7}{6} & \vdots & 1300 \\ 0 & 1 & -1 & \vdots & 0 \\ 0 & 0 & 5 & \vdots & 3000 \end{bmatrix}$$

$$\begin{matrix} \\ \\ \\ \frac{1}{5}R_4 \end{matrix} \begin{bmatrix} 1 & 1 & 1 & \vdots & 1500 \\ 0 & 1 & \frac{7}{6} & \vdots & 1300 \\ 0 & 1 & -1 & \vdots & 0 \\ 0 & 0 & 1 & \vdots & 600 \end{bmatrix}$$

$z = 600$ $\qquad$ $y - 1(600) = 0$ $\qquad$ $x + 600 + 600 = 1500$

$y = 600$ $\qquad$ $x = 300$

Theater A: 300 tickets

Theater B: 600 tickets

Theater C: 600 tickets

51. *Verbal Model:* $\boxed{\text{Number 1}} + \boxed{\text{Number 2}} + \boxed{\text{Number 3}} = \boxed{33}$

$\boxed{\text{Number 2}} = 3 + \boxed{\text{Number 1}}$

$\boxed{\text{Number 3}} = 4 \cdot \boxed{\text{Number 1}}$

Labels: Number 1 $= x$

Number 2 $= y$

Number 3 $= z$

System:

$$\begin{aligned} x + y + z &= 33 \\ y &= 3 + x \\ z &= 4x \end{aligned}$$

$$\begin{bmatrix} 1 & 1 & 1 & \vdots & 33 \\ -1 & 1 & 0 & \vdots & 3 \\ -4 & 0 & 1 & \vdots & 0 \end{bmatrix}$$

$$\begin{matrix} \\ R_1 + R_2 \\ 4R_1 + R_3 \end{matrix} \begin{bmatrix} 1 & 1 & 1 & \vdots & 33 \\ 0 & 2 & 1 & \vdots & 36 \\ 0 & 4 & 5 & \vdots & 132 \end{bmatrix}$$

$$\begin{matrix} \\ \frac{1}{2}R_2 \\ \\ \end{matrix} \begin{bmatrix} 1 & 1 & 1 & \vdots & 33 \\ 0 & 1 & \frac{1}{2} & \vdots & 18 \\ 0 & 4 & 5 & \vdots & 132 \end{bmatrix}$$

$$\begin{matrix} \\ \\ -4R_2 + R_3 \end{matrix} \begin{bmatrix} 1 & 1 & 1 & \vdots & 33 \\ 0 & 1 & \frac{1}{2} & \vdots & 18 \\ 0 & 0 & 3 & \vdots & 60 \end{bmatrix}$$

$$\begin{matrix} \\ \\ \frac{1}{3}R_3 \end{matrix} \begin{bmatrix} 1 & 1 & 1 & \vdots & 33 \\ 0 & 1 & \frac{1}{2} & \vdots & 18 \\ 0 & 0 & 1 & \vdots & 20 \end{bmatrix}$$

$z = 20$ $\qquad$ $y + \frac{1}{2}(20) = 18$ $\qquad$ $x + 8 + 20 = 33$

$\qquad\qquad y = 8$ $\qquad\qquad x = 5$

The three numbers are 5, 8, and 20.

53. *Verbal Model:*

	Computer chips	Resistors	Transistors
Copper	2	1	3
Zinc	2	3	2
Glass	1	2	2

Labels: Computer chips $= x$

Resistors $= y$

Transistors $= z$

System:

$$\begin{aligned} 2x + y + 3z &= 70 \\ 2x + 3y + 2z &= 80 \\ x + 2y + 2z &= 55 \end{aligned}$$

$$\begin{bmatrix} 2 & 1 & 3 & \vdots & 70 \\ 2 & 3 & 2 & \vdots & 80 \\ 1 & 2 & 2 & \vdots & 55 \end{bmatrix}$$

$$\begin{matrix} R_1 \\ \\ R_3 \end{matrix} \begin{bmatrix} 1 & 2 & 2 & \vdots & 55 \\ 2 & 3 & 2 & \vdots & 80 \\ 2 & 1 & 3 & \vdots & 70 \end{bmatrix}$$

$$\begin{matrix} \\ -2R_1 + R_2 \\ -2R_1 + R_3 \end{matrix} \begin{bmatrix} 1 & 2 & 2 & \vdots & 55 \\ 0 & -1 & -2 & \vdots & -30 \\ 0 & -3 & -1 & \vdots & -40 \end{bmatrix}$$

$$\begin{matrix} -R_2 \\ \\ \\ \end{matrix} \begin{bmatrix} 1 & 2 & 2 & \vdots & 55 \\ 0 & 1 & 2 & \vdots & 30 \\ 0 & -3 & -1 & \vdots & -40 \end{bmatrix}$$

$$\begin{matrix} -2R_2 + R_1 \\ 3R_2 + R_3 \\ \\ \end{matrix} \begin{bmatrix} 1 & 0 & -2 & \vdots & -5 \\ 0 & 1 & 2 & \vdots & 30 \\ 0 & 0 & 5 & \vdots & 50 \end{bmatrix}$$

$$\begin{matrix} \\ \\ \frac{1}{5}R_3 \end{matrix} \begin{bmatrix} 1 & 0 & -2 & \vdots & -5 \\ 0 & 1 & 2 & \vdots & 30 \\ 0 & 0 & 1 & \vdots & 10 \end{bmatrix}$$

$$\begin{matrix} 2R_3 + R_1 \\ -2R_3 + R_2 \\ \\ \end{matrix} \begin{bmatrix} 1 & 0 & 0 & \vdots & 15 \\ 0 & 1 & 0 & \vdots & 10 \\ 0 & 0 & 1 & \vdots & 10 \end{bmatrix}$$

15 computer chips, 10 resistors, 10 transistors

55. The order of the matrix is 3×5. There are 15 entries in the matrix, so the order is 3×5, 5×3, or 15×1. Because there are more columns than rows, the second number in the order must be larger than the first.

57. The row-echelon form of an augmented matrix that corresponds to a system of linear equations that is inconsistent occurs when there is a row in the matrix with all zero entries except in the last column.

59. The first entry in the first column is one and the other two entries are zero. In the second column, the first entry is a nonzero real number, the second entry is one, and the third entry is zero. In the third column, the first two entries are nonzero real numbers and the third entry is one.

61. $6(-7) = -42$

63. $5(4) - 3(-2) = 20 + 6 = 26$

65. $\begin{cases} x & & = 4 \\ & 3y + 2z & = -4 \\ x + & y + z & = 3 \end{cases}$

$\begin{cases} x & & = 4 \\ -3x & - z & = -13 \\ x + y + z & & = 3 \end{cases}$

$x = 4$

$-3(4) - z = -13$

$-z = -1$

$z = 1$

$4 + y + 1 = 3$

$y = -2$

The solution is $(4, -2, 1)$.

Section 4.5 Determinants and Linear Systems

1. $\det(A) = \begin{vmatrix} 2 & 1 \\ 3 & 4 \end{vmatrix} = 2(4) - 3(1) = 8 - 3 = 5$

3. $\det(A) = \begin{vmatrix} 5 & 2 \\ -6 & 3 \end{vmatrix} = 5(3) - (-6)(2) = 15 + 12 = 27$

5. $\det(A) = \begin{vmatrix} -4 & 0 \\ 9 & 0 \end{vmatrix} = (-4)(0) - 9(0) = 0$

7. $\det(A) = \begin{vmatrix} 3 & -3 \\ -6 & 6 \end{vmatrix} = 3(6) - (-6)(-3)$

$= 18 - 18 = 0$

9. $\det(A) = \begin{vmatrix} -7 & 6 \\ \frac{1}{2} & 3 \end{vmatrix} = (-7)(3) - \left(\frac{1}{2}\right)(6)$

$= -21 - 3 = -24$

11. $\det(A) = \begin{vmatrix} 0.4 & 0.7 \\ 0.7 & 0.4 \end{vmatrix} = 0.4(0.4) - 0.7(0.7)$

$= 0.16 - 0.49 = -0.33$

13. $\det(A) = \begin{vmatrix} 2 & 3 & -1 \\ 6 & 0 & 0 \\ 4 & 1 & 1 \end{vmatrix}$

$= -(6)\begin{vmatrix} 3 & -1 \\ 1 & 1 \end{vmatrix} + 0 + 0 \text{ (second row)}$

$= (-6)(4)$

$= -24$

15. $\det(A) = \begin{vmatrix} 1 & 1 & 2 \\ 3 & 1 & 0 \\ -2 & 0 & 3 \end{vmatrix} = (2)\begin{vmatrix} 3 & 1 \\ -2 & 0 \end{vmatrix} - (0)\begin{vmatrix} 1 & 1 \\ -2 & 0 \end{vmatrix} + (3)\begin{vmatrix} 1 & 1 \\ 3 & 1 \end{vmatrix} \text{ (third column)}$

$= (2)(2) - 0 + (3)(-2) = 4 - 6 = -2$

17. $\det(A) = \begin{vmatrix} 2 & 4 & 6 \\ 0 & 3 & 1 \\ 0 & 0 & -5 \end{vmatrix}$

$= (2)\begin{vmatrix} 3 & 1 \\ 0 & -5 \end{vmatrix} - 0 + 0 \text{ (first column)}$

$= (2)(-15) = -30$

19. $\det(A) = \begin{vmatrix} -2 & 2 & 3 \\ 1 & -1 & 0 \\ 0 & 1 & 4 \end{vmatrix}$

$= -(1)\begin{vmatrix} 2 & 3 \\ 1 & 4 \end{vmatrix} + (-1)\begin{vmatrix} -2 & 3 \\ 0 & 4 \end{vmatrix} - 0 \text{ (second row)}$

$= (-1)(5) + (-1)(-8) = -5 + 8 = 3$

21. $\begin{bmatrix} 1 & 2 & \vdots & 5 \\ -1 & 1 & \vdots & 1 \end{bmatrix}$

$$D = \begin{vmatrix} 1 & 2 \\ -1 & 1 \end{vmatrix} = 1 - (-2) = 3$$

$$x = \frac{D_x}{D} = \frac{\begin{vmatrix} 5 & 2 \\ 1 & 1 \end{vmatrix}}{3} = \frac{5 - 2}{3} = \frac{3}{3} = 1$$

$$y = \frac{D_y}{D} = \frac{\begin{vmatrix} 1 & 5 \\ -1 & 1 \end{vmatrix}}{3} = \frac{1 - (-5)}{3} = \frac{6}{3} = 2$$

The solution is $(1, 2)$.

23. $\begin{bmatrix} 3 & 4 & \vdots & -2 \\ 5 & 3 & \vdots & 4 \end{bmatrix}$

$$D = \begin{vmatrix} 3 & 4 \\ 5 & 3 \end{vmatrix} = 9 - 20 = -11$$

$$x = \frac{D_x}{D} = \frac{\begin{vmatrix} -2 & 4 \\ 4 & 3 \end{vmatrix}}{-11} = \frac{-6 - 16}{-11} = \frac{-22}{-11} = 2$$

$$y = \frac{D_y}{D} = \frac{\begin{vmatrix} 3 & -2 \\ 5 & 4 \end{vmatrix}}{-11} = \frac{12 - (-10)}{-11} = \frac{22}{-11} = -2$$

The solution is $(2, -2)$.

25. $\begin{bmatrix} 13 & -6 & \vdots & 17 \\ 26 & -12 & \vdots & 8 \end{bmatrix}$

$$D = \begin{vmatrix} 13 & -6 \\ 26 & -12 \end{vmatrix} = -156 + 156 = 0$$

Cannot be solved by Cramer's Rule because $D = 0$.

27. $\begin{bmatrix} 4 & -1 & 1 & \vdots & -5 \\ 2 & 2 & 3 & \vdots & 10 \\ 5 & -2 & 6 & \vdots & 1 \end{bmatrix}$

$$D = \begin{vmatrix} 4 & -1 & 1 \\ 2 & 2 & 3 \\ 5 & -2 & 6 \end{vmatrix} = (1)\begin{vmatrix} 2 & 2 \\ 5 & -2 \end{vmatrix} - (3)\begin{vmatrix} 4 & -1 \\ 5 & -2 \end{vmatrix} + (6)\begin{vmatrix} 4 & -1 \\ 2 & 2 \end{vmatrix} = (1)(-14) + (-3)(-3) + (6)(10) = -14 + 9 + 60 = 55$$

$$x = \frac{\begin{vmatrix} -5 & -1 & 1 \\ 10 & 2 & 3 \\ 1 & -2 & 6 \end{vmatrix}}{55} = \frac{(1)\begin{vmatrix} 10 & 2 \\ 1 & -2 \end{vmatrix} - (3)\begin{vmatrix} -5 & -1 \\ 1 & -2 \end{vmatrix} + (6)\begin{vmatrix} -5 & -1 \\ 10 & 2 \end{vmatrix}}{55} = \frac{(1)(-22) + (-3)(11) + (6)(0)}{55} = \frac{-22 - 33}{55} = \frac{-55}{55} = -1$$

$$y = \frac{\begin{vmatrix} 4 & -5 & 1 \\ 2 & 10 & 3 \\ 5 & 1 & 6 \end{vmatrix}}{55} = \frac{(1)\begin{vmatrix} 2 & 10 \\ 5 & 1 \end{vmatrix} - (3)\begin{vmatrix} 4 & -5 \\ 5 & 1 \end{vmatrix} + (6)\begin{vmatrix} 4 & -5 \\ 2 & 10 \end{vmatrix}}{55} = \frac{(1)(-48) + (-3)(29) + (6)(50)}{55} = \frac{-48 - 87 + 300}{55} = \frac{165}{55} = 3$$

$$z = \frac{\begin{vmatrix} 4 & -1 & -5 \\ 2 & 2 & 10 \\ 5 & -2 & 1 \end{vmatrix}}{55} = \frac{(5)\begin{vmatrix} -1 & -5 \\ 2 & 10 \end{vmatrix} - (-2)\begin{vmatrix} 4 & -5 \\ 2 & 10 \end{vmatrix} + (1)\begin{vmatrix} 4 & -1 \\ 2 & 2 \end{vmatrix}}{55} = \frac{(5)(0) + (2)(50) + (1)(10)}{55} = \frac{0 + 100 + 10}{55} = \frac{110}{55} = 2$$

The solution is $(-1, 3, 2)$.

29. $\begin{bmatrix} 4 & 3 & 4 & \vdots & 1 \\ 4 & -6 & 8 & \vdots & 8 \\ 1 & 9 & -2 & \vdots & -7 \end{bmatrix}$

$$D = \begin{vmatrix} 4 & 3 & 4 \\ 4 & -6 & 8 \\ -1 & 9 & -2 \end{vmatrix} = 4\begin{vmatrix} -6 & 8 \\ 9 & -2 \end{vmatrix} - 3\begin{vmatrix} 4 & 8 \\ -1 & -2 \end{vmatrix} + 4\begin{vmatrix} 4 & -6 \\ -1 & 9 \end{vmatrix} = (4)(-60) - (3)(0) + (4)(30) = -120$$

$$x = \frac{\begin{vmatrix} 1 & 3 & 4 \\ 8 & -6 & 8 \\ -7 & 9 & -2 \end{vmatrix}}{-120} = \frac{1\begin{vmatrix} -6 & 8 \\ 9 & -2 \end{vmatrix} - 3\begin{vmatrix} 8 & 8 \\ -7 & -2 \end{vmatrix} + 4\begin{vmatrix} 8 & -6 \\ -7 & 9 \end{vmatrix}}{-120}$$

$$= \frac{(1)(-60) - (3)(40) + (4)(30)}{-120} = \frac{-60}{-120} = \frac{1}{2}$$

$$y = \frac{\begin{vmatrix} 4 & 1 & 4 \\ 4 & 8 & 8 \\ -1 & -7 & -2 \end{vmatrix}}{-120} = \frac{4\begin{vmatrix} 8 & 8 \\ -7 & -2 \end{vmatrix} - 1\begin{vmatrix} 4 & 8 \\ -1 & -2 \end{vmatrix} + 4\begin{vmatrix} 4 & 8 \\ -1 & -7 \end{vmatrix}}{-120} = \frac{(4)(40) - (1)(0) + (4)(-20)}{-120} = \frac{80}{-120} = -\frac{2}{3}$$

$$z = \frac{\begin{vmatrix} 4 & 3 & 1 \\ 4 & -6 & 8 \\ -1 & 9 & -7 \end{vmatrix}}{-120} = \frac{4\begin{vmatrix} -6 & 8 \\ 9 & -7 \end{vmatrix} - 3\begin{vmatrix} 4 & 8 \\ -1 & -7 \end{vmatrix} + 1\begin{vmatrix} 4 & -6 \\ -1 & 9 \end{vmatrix}}{-120} = \frac{(4)(-30) - (3)(-20) + (1)(30)}{-120} = \frac{-30}{-120} = \frac{1}{4}$$

The solution is $\left(\frac{1}{2}, -\frac{2}{3}, \frac{1}{4}\right)$.

31. $\begin{bmatrix} 2 & 3 & 5 & \vdots & 4 \\ 3 & 5 & 9 & \vdots & 7 \\ 5 & 9 & 17 & \vdots & 13 \end{bmatrix}$

$$D = \begin{vmatrix} 2 & 3 & 5 \\ 3 & 5 & 9 \\ 5 & 9 & 17 \end{vmatrix} = (2)\begin{vmatrix} 5 & 9 \\ 9 & 17 \end{vmatrix} - (3)\begin{vmatrix} 3 & 9 \\ 5 & 17 \end{vmatrix} + 5\begin{vmatrix} 3 & 5 \\ 5 & 9 \end{vmatrix} = (2)(4) - (3)(6) + 5(2) = 8 - 18 + 10 = 0$$

Cannot be solved by Cramer's Rule because $D = 0$.

33. $D = \begin{vmatrix} 3 & -2 & 3 \\ 1 & 3 & 6 \\ 1 & 2 & 9 \end{vmatrix} = 48$

$$x = \frac{D_x}{D} = \frac{\begin{vmatrix} 8 & -2 & 3 \\ -3 & 3 & 6 \\ -5 & 2 & 9 \end{vmatrix}}{48} = \frac{153}{48} = \frac{51}{16}$$

$$y = \frac{D_y}{D} = \frac{\begin{vmatrix} 3 & 8 & 3 \\ 1 & -3 & 6 \\ 1 & -5 & 9 \end{vmatrix}}{48} = \frac{-21}{48} = \frac{-7}{16}$$

$$z = \frac{D_z}{D} = \frac{\begin{vmatrix} 3 & -2 & 8 \\ 1 & 3 & -3 \\ 1 & 2 & -5 \end{vmatrix}}{48} = \frac{-39}{48} = -\frac{13}{16}$$

$\left(\frac{51}{16}, -\frac{7}{16}, -\frac{13}{16}\right)$

35. $(x_1, y_1) = (0, 3), (x_2, y_2) = (4, 0), (x_3, y_3) = (8, 5)$

$$\begin{vmatrix} x_1 & y_1 & 1 \\ x_2 & y_2 & 1 \\ x_3 & y_3 & 1 \end{vmatrix} = \begin{vmatrix} 0 & 3 & 1 \\ 4 & 0 & 1 \\ 8 & 5 & 1 \end{vmatrix} = 32$$

Area $= +\frac{1}{2}(32) = 16$

37. $(x_1, y_1) = (-3, 4), (x_2, y_2) = (1, -2), (x_3, y_3) = (6, 1)$

$$\begin{vmatrix} x_1 & y_1 & 1 \\ x_2 & y_2 & 1 \\ x_3 & y_3 & 1 \end{vmatrix} = \begin{vmatrix} -3 & 4 & 1 \\ 1 & -2 & 1 \\ 6 & 1 & 1 \end{vmatrix}$$

$$= -3\begin{vmatrix} -2 & 1 \\ 1 & 1 \end{vmatrix} - 1\begin{vmatrix} 4 & 1 \\ 1 & 1 \end{vmatrix} + 6\begin{vmatrix} 4 & 1 \\ -2 & 1 \end{vmatrix}$$

$$= -3(-3) - 1(3) + 6(6)$$

$$= 9 - 3 + 36 = 42$$

Area $= +\frac{1}{2}(42) = 21$

39. $(x_1, y_1) = (-2, 1), (x_2, y_2) = (3, -1), (x_3, y_3) = (1, 6)$

$$\begin{vmatrix} x_1 & y_1 & 1 \\ x_2 & y_2 & 1 \\ x_3 & y_3 & 1 \end{vmatrix} = \begin{vmatrix} -2 & 1 & 1 \\ 3 & -1 & 1 \\ 1 & 6 & 1 \end{vmatrix} = (1)\begin{vmatrix} 3 & -1 \\ 1 & 6 \end{vmatrix} - (1)\begin{vmatrix} -2 & 1 \\ 1 & 6 \end{vmatrix} + (1)\begin{vmatrix} -2 & 1 \\ 3 & -1 \end{vmatrix} = (1)(19) - (1)(-13) + (1)(-1) = 19 + 13 - 1 = 31$$

Area $= +\frac{1}{2}(31) = \frac{31}{2}$ or $15\frac{1}{2}$

41. $(x_1, y_1) = \left(0, \frac{1}{2}\right), (x_2, y_2) = \left(\frac{5}{2}, 0\right), (x_3, y_3) = (4, 3)$

$$\begin{vmatrix} x_1 & y_1 & 1 \\ x_2 & y_2 & 1 \\ x_3 & y_3 & 1 \end{vmatrix} = \begin{vmatrix} 0 & \frac{1}{2} & 1 \\ \frac{5}{2} & 0 & 1 \\ 4 & 3 & 1 \end{vmatrix} = 0\begin{vmatrix} 0 & 1 \\ 3 & 1 \end{vmatrix} - \frac{1}{2}\begin{vmatrix} \frac{5}{2} & 1 \\ 4 & 1 \end{vmatrix} + 1\begin{vmatrix} \frac{5}{2} & 0 \\ 4 & 3 \end{vmatrix}$$

$$= 0 - \frac{1}{2}\left(\frac{5}{2} - 4\right) + 1\left(\frac{15}{2} - 0\right)$$

$$= -\frac{1}{2}\left(-\frac{3}{2}\right) + 1\left(\frac{15}{2}\right)$$

$$= \frac{3}{4} + \frac{15}{2}$$

$$= \frac{3}{4} + \frac{30}{4}$$

$$= \frac{33}{4}$$

Area $= \frac{1}{2}\left(\frac{33}{4}\right) = \frac{33}{8}$

43. Let $(x_1, y_1) = (-4, 1), (x_2, y_2) = (1, 4)$, and $(x_3, y_3) = (3, -2)$.

$$\begin{vmatrix} x_1 & y_1 & 1 \\ x_2 & y_2 & 1 \\ x_3 & y_3 & 1 \end{vmatrix} = \begin{vmatrix} -4 & 1 & 1 \\ 1 & 4 & 1 \\ 3 & -2 & 1 \end{vmatrix} = -4\begin{vmatrix} 4 & 1 \\ -2 & 1 \end{vmatrix} - 1\begin{vmatrix} 1 & 1 \\ 3 & 1 \end{vmatrix} + 1\begin{vmatrix} 1 & 4 \\ 3 & -2 \end{vmatrix} = -4(6) - 1(-2) + 1(-14) = -24 + 2 - 14 = -36$$

Area of triangle $= -\frac{1}{2}(-36) = 18$

Area of rectangle $=$ length $\cdot$ width $= |6 - (-4)| \cdot |4 - (-2)| = 10 \cdot 6 = 60$

Verbal Model: [Area of Rectangle] $-$ [Area of Triangle] $=$ [Area of Shaded region]

Equation: $60 - 18 = 42$

The area of the shaded region is 42.

45. Let $(x_1, y_1) = (-1, 2), (x_2, y_2) = (4, 0), (x_3, y_3) = (3, 5)$.

$$\begin{vmatrix} x_1 & y_1 & 1 \\ x_2 & y_2 & 1 \\ x_3 & y_3 & 1 \end{vmatrix} = \begin{vmatrix} -1 & 2 & 1 \\ 4 & 0 & 1 \\ 3 & 5 & 1 \end{vmatrix} = -4\begin{vmatrix} 2 & 1 \\ 5 & 1 \end{vmatrix} + 0 - 1\begin{vmatrix} -1 & 2 \\ 3 & 5 \end{vmatrix} = -4(-3) - 1(-11) = 12 + 11 = 23$$

Area $= \frac{1}{2}(23) = 11.5$

Let $(x_1, y_1) = (3, 5), (x_2, y_2) = (4, 0), (x_3, y_3) = (5, 4)$.

$$\begin{vmatrix} x_1 & y_1 & 1 \\ x_2 & y_2 & 1 \\ x_3 & y_3 & 1 \end{vmatrix} = \begin{vmatrix} 3 & 5 & 1 \\ 4 & 0 & 1 \\ 5 & 4 & 1 \end{vmatrix} = -4\begin{vmatrix} 5 & 1 \\ 4 & 1 \end{vmatrix} + 0 - 1\begin{vmatrix} 3 & 5 \\ 5 & 4 \end{vmatrix} = -4(1) - 1(-13) = -4 + 13 = 9$$

Area $= \frac{1}{2}(9) = 4.5$

Verbal Model: [Area of Shaded Region] $=$ [Area of Triangle 1] $+$ [Area of Triangle 2]

Equation: $A = 11.5 + 4.5 = 16$

47. Let $(x_1, y_1) = (-1, 11), (x_2, y_2) = (0, 8), (x_3, y_3) = (2, 2)$.

$$\begin{vmatrix} x_1 & y_1 & 1 \\ x_2 & y_2 & 1 \\ x_3 & y_3 & 1 \end{vmatrix} = \begin{vmatrix} -1 & 11 & 1 \\ 0 & 8 & 1 \\ 2 & 2 & 1 \end{vmatrix} = (-1)\begin{vmatrix} 8 & 1 \\ 2 & 1 \end{vmatrix} + 0 + (2)\begin{vmatrix} 11 & 1 \\ 8 & 1 \end{vmatrix} = (-1)(6) + (2)(3) = -6 + 6 = 0$$

The three points are collinear.

49. Let $(x_1, y_1) = (2, -4), (x_2, y_2) = (5, 2)$, and $(x_3, y_3) = (10, 10)$.

$$\begin{vmatrix} x_1 & y_1 & 1 \\ x_2 & y_2 & 1 \\ x_3 & y_3 & 1 \end{vmatrix} = \begin{vmatrix} 2 & -4 & 1 \\ 5 & 2 & 1 \\ 10 & 10 & 1 \end{vmatrix} = 1\begin{vmatrix} 5 & 2 \\ 10 & 10 \end{vmatrix} - 1\begin{vmatrix} 2 & -4 \\ 10 & 10 \end{vmatrix} + 1\begin{vmatrix} 2 & -4 \\ 5 & 2 \end{vmatrix} = 1(30) - 1(60) + 1(24) = 30 - 60 + 24 = -6$$

The three points are not collinear.

51. Let $(x_1, y_1) = \left(-2, \frac{1}{3}\right), (x_2, y_2) = (2, 1), (x_3, y_3) = \left(3, \frac{1}{5}\right)$.

$$\begin{vmatrix} x_1 & y_1 & 1 \\ x_2 & y_2 & 1 \\ x_3 & y_3 & 1 \end{vmatrix} = \begin{vmatrix} -2 & \frac{1}{3} & 1 \\ 2 & 1 & 1 \\ 3 & \frac{1}{5} & 1 \end{vmatrix} = (1)\begin{vmatrix} 2 & 1 \\ 3 & \frac{1}{5} \end{vmatrix} - (1)\begin{vmatrix} -2 & \frac{1}{3} \\ 3 & \frac{1}{5} \end{vmatrix} + (1)\begin{vmatrix} -2 & \frac{1}{3} \\ 2 & 1 \end{vmatrix}$$

$$= (1)\left(-\frac{13}{5}\right) - (1)\left(-\frac{7}{5}\right) + (1)\left(-\frac{8}{3}\right) = -\frac{13}{5} + \frac{7}{5} - \frac{8}{3} = -\frac{18}{15} - \frac{40}{15} = -\frac{58}{15}$$

The three points are not collinear.

53. Let $(x_1, y_1) = (-2, -1)$ and $(x_2, y_2) = (4, 2)$.

$$\begin{vmatrix} x & y & 1 \\ -2 & -1 & 1 \\ 4 & 2 & 1 \end{vmatrix} = 0$$

$$1\begin{vmatrix} -2 & -1 \\ 4 & 2 \end{vmatrix} - 1\begin{vmatrix} x & y \\ 4 & 2 \end{vmatrix} + 1\begin{vmatrix} x & y \\ -2 & -1 \end{vmatrix} = 0$$

$$\begin{aligned} 1(0) - (2x - 4y) + (-x + 2y) &= 0 \\ -2x + 4y - x + 2y &= 0 \\ -3x + 6y &= 0 \\ x - 2y &= 0 \end{aligned}$$

55. $(x_1, y_1) = (10, 7), (x_2, y_2) = (-2, -7)$

$$\begin{vmatrix} x & y & 1 \\ 10 & 7 & 1 \\ -2 & -7 & 1 \end{vmatrix} = 0$$

$$(1)\begin{vmatrix} 10 & 7 \\ -2 & -7 \end{vmatrix} - (1)\begin{vmatrix} x & y \\ -2 & -7 \end{vmatrix} + (1)\begin{vmatrix} x & y \\ 10 & 7 \end{vmatrix} = 0$$

$$\begin{aligned} (1)(-56) - (-7x + 2y) + (1)(7x - 10y) &= 0 \\ -56 + 7x - 2y + 7x - 10y &= 0 \\ 14x - 12y - 56 &= 0 \\ 7x - 6y - 28 &= 0 \end{aligned}$$

57. $(x_1, y_1) = \left(-2, \frac{3}{2}\right), (x_2, y_2) = (3, -3)$

$$\begin{vmatrix} x & y & 1 \\ -2 & \frac{3}{2} & 1 \\ 3 & -3 & 1 \end{vmatrix} = 0$$

$$x\begin{vmatrix} \frac{3}{2} & 1 \\ -3 & 1 \end{vmatrix} - y\begin{vmatrix} -2 & 1 \\ 3 & 1 \end{vmatrix} + 1\begin{vmatrix} -2 & \frac{3}{2} \\ 3 & -3 \end{vmatrix} = 0$$

$$\begin{aligned} \frac{9}{2}x + 5y + \frac{3}{2} &= 0 \\ 9x + 10y + 3 &= 0 \end{aligned}$$

59. $(x_1, y_1) = (2, 3.6), (x_2, y_2) = (8, 10)$

$$\begin{vmatrix} x & y & 1 \\ 2 & 3.6 & 1 \\ 8 & 10 & 1 \end{vmatrix} = 0$$

$$x\begin{vmatrix} 3.6 & 1 \\ 10 & 1 \end{vmatrix} - y\begin{vmatrix} 2 & 1 \\ 8 & 1 \end{vmatrix} + 1\begin{vmatrix} 2 & 3.6 \\ 8 & 10 \end{vmatrix} = 0$$

$$\begin{aligned} x(3.6 - 10) - y(2 - 8) + 1(20 - 28.8) &= 0 \\ -6.4x + 6y - 8.8 &= 0 \\ -3.2x + 3y - 4.4 &= 0 \\ 32x - 30y + 44 &= 0 \\ 16x - 15y + 22 &= 0 \end{aligned}$$

61. To find the determinant of a 2×2 matrix, find the difference of the products of the two diagonals of the matrix.

63. No. The matrix must be square.

65. Solution is 248.

67. Solution is 105.625.

69. Solution is 4.32.

71. From the diagram the coordinates of A, B, and C are determined to be $A(0, 20)$, $B(10, -5)$ and $C(28, 0)$.

$$\begin{vmatrix} x_1 & y_1 & 1 \\ x_2 & y_2 & 1 \\ x_3 & y_3 & 1 \end{vmatrix} = \begin{vmatrix} 0 & 20 & 1 \\ 10 & -5 & 1 \\ 28 & 0 & 1 \end{vmatrix} = -500$$

$$\text{Area} = -\tfrac{1}{2}(-500) = 250 \text{ mi}^2$$

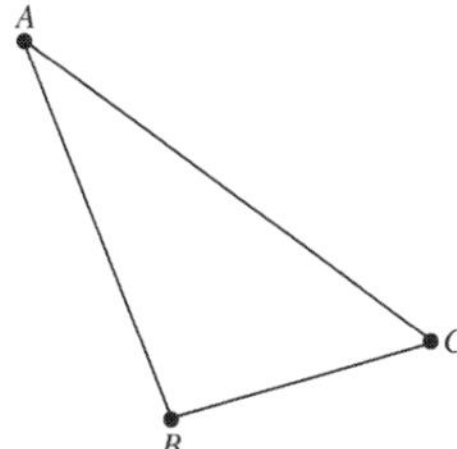

73. $\begin{bmatrix} 1 & 1 & -1 & \vdots & 0 \\ 1 & 0 & 2 & \vdots & 12 \\ 1 & -2 & 0 & \vdots & -4 \end{bmatrix}$

$$D = \begin{vmatrix} 1 & 1 & -1 \\ 1 & 0 & 2 \\ 1 & -2 & 0 \end{vmatrix} = 1\begin{vmatrix} 0 & 2 \\ -2 & 0 \end{vmatrix} - 1\begin{vmatrix} 1 & -1 \\ -2 & 0 \end{vmatrix} + 1\begin{vmatrix} 1 & -1 \\ 0 & 2 \end{vmatrix} = (1)(4) - (1)(-2) + (1)(2) = 4 + 2 + 2 = 8$$

$$I_1 = \frac{\begin{vmatrix} 0 & 1 & -1 \\ 12 & 0 & 2 \\ -4 & -2 & 0 \end{vmatrix}}{8} = \frac{0\begin{vmatrix} 0 & 2 \\ -2 & 0 \end{vmatrix} - 12\begin{vmatrix} 1 & -1 \\ -2 & 0 \end{vmatrix} + (-4)\begin{vmatrix} 1 & -1 \\ 0 & 2 \end{vmatrix}}{8} = \frac{0 - (12)(-2) - (4)(2)}{8} = \frac{24 - 8}{8} = \frac{16}{8} = 2$$

$$I_2 = \frac{\begin{vmatrix} 1 & 0 & -1 \\ 1 & 12 & 2 \\ 1 & -4 & 0 \end{vmatrix}}{8} = \frac{1\begin{vmatrix} 12 & 2 \\ -4 & 0 \end{vmatrix} - 1\begin{vmatrix} 0 & -1 \\ -4 & 0 \end{vmatrix} + 1\begin{vmatrix} 0 & -1 \\ 12 & 2 \end{vmatrix}}{8} = \frac{(1)(8) - (1)(-4) + 1(12)}{8} = \frac{8 + 4 + 12}{8} = \frac{24}{8} = 3$$

$$I_3 = \frac{\begin{vmatrix} 1 & 1 & 0 \\ 1 & 0 & 12 \\ 1 & -2 & -4 \end{vmatrix}}{8} = \frac{1\begin{vmatrix} 0 & 12 \\ -2 & -4 \end{vmatrix} - 1\begin{vmatrix} 1 & 0 \\ -2 & -4 \end{vmatrix} + 1\begin{vmatrix} 1 & 0 \\ 0 & 12 \end{vmatrix}}{8} = \frac{(1)(24) - (1)(-4) + (1)(12)}{8} = \frac{24 + 4 + 12}{8} = \frac{40}{8} = 5$$

75. (a) $\begin{bmatrix} k & 3k & \vdots & 2 \\ 2+k & k & \vdots & 5 \end{bmatrix}$

$$D = \begin{vmatrix} k & 3k \\ 2+k & k \end{vmatrix} = k^2 - 3k(2+k)$$

$$= k^2 - 6k - 3k^2 = -2k^2 - 6k$$

$$x = \frac{D_x}{D} = \frac{\begin{vmatrix} 2 & 3k \\ 5 & k \end{vmatrix}}{-2k^2 - 6k}$$

$$= \frac{2k - 15k}{-2k^2 - 6k} = \frac{-13k}{-2k(k+3)}$$

$$= \frac{13}{2(k+3)} = \frac{13}{2k+6}$$

$$y = \frac{D_y}{D} = \frac{\begin{vmatrix} k & 2 \\ 2+k & 5 \end{vmatrix}}{-2k^2 - 6k}$$

$$= \frac{5k - 2(2+k)}{-2k^2 - 6k} = \frac{5k - 4 - 2k}{-2k^2 - 6k}$$

$$= \frac{3k - 4}{-2k^2 - 6k} = \frac{-1(4 - 3k)}{-1(2k^2 + 6k)}$$

$$= \frac{4 - 3k}{2k^2 + 6k}$$

(b) $-2k^2 - 6k = 0$

$-2k(k+3) = 0$

$-2k = 0 \qquad k + 3 = 0$

$k = 0 \qquad k = -3$

77. A determinant is a real number associated with a square matrix.

79. The determinant is zero. Because two rows are identical, each term is zero when expanding by minors along the other row. Therefore, the sum is zero.

81. 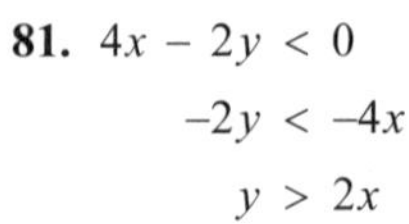

$4x - 2y < 0$

$-2y < -4x$

$y > 2x$

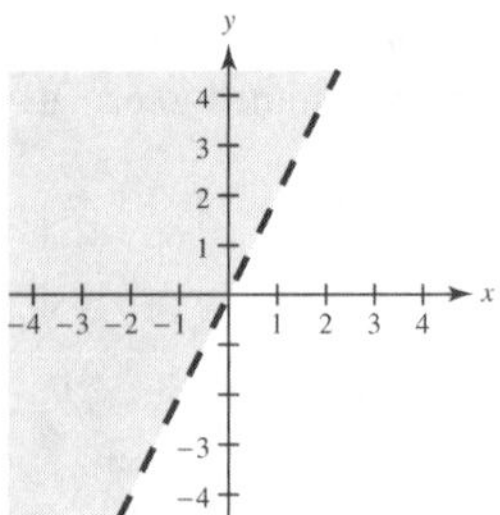

83. $-x + 3y > 12$

$3y > x + 12$

$y > \frac{1}{3}x + 4$

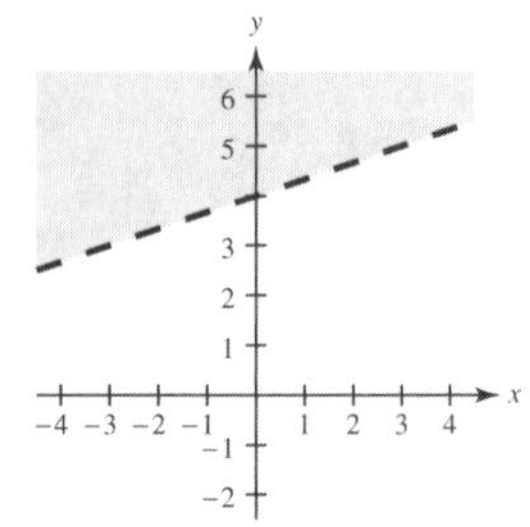

85. The graph of $h(x)$ has a vertical shift c units upward.

87. $f(x) - 2$

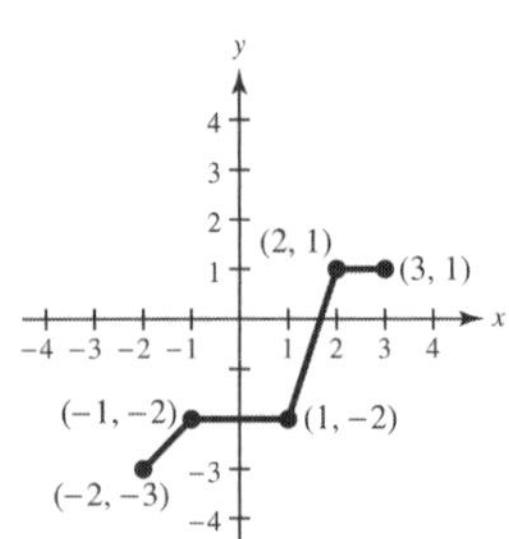

89. $f(-x)$

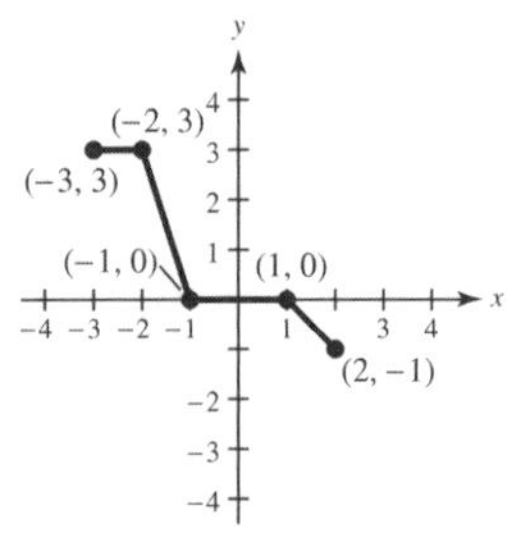

Section 4.6 Systems of Linear Inequalities

1. $\begin{cases} x + y \le 3 \\ x - y \le 1 \end{cases} \Rightarrow \begin{matrix} y \le -x + 3 \\ y \ge x - 1 \end{matrix}$

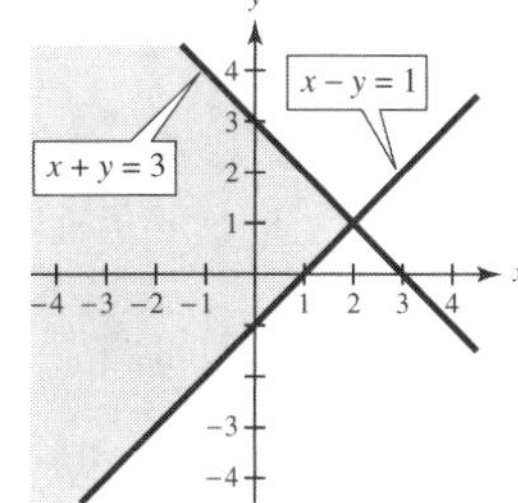

Solid lines at $y = -x + 3$ and $y = x - 1a$. Shade below $y = -x + 3$ and above $y = x - 1$.

3. $\begin{cases} 2x - 4y \le 6 \\ x + y \ge 2 \end{cases} \Rightarrow \begin{matrix} y \ge \frac{1}{2}x - \frac{3}{2} \\ y \ge -x + 2 \end{matrix}$

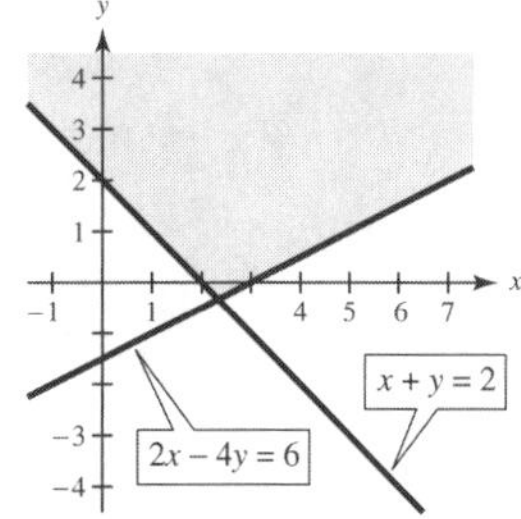

Solid lines at $y = \frac{1}{2}x - \frac{3}{2}$ and $y = -x + 2$. Shade above each line.

5. $\begin{cases} x + 2y \le 6 \\ x - 2y \le 0 \end{cases} \Rightarrow \begin{matrix} y \le -\frac{1}{2}x + 3 \\ y \ge \frac{1}{2}x \end{matrix}$

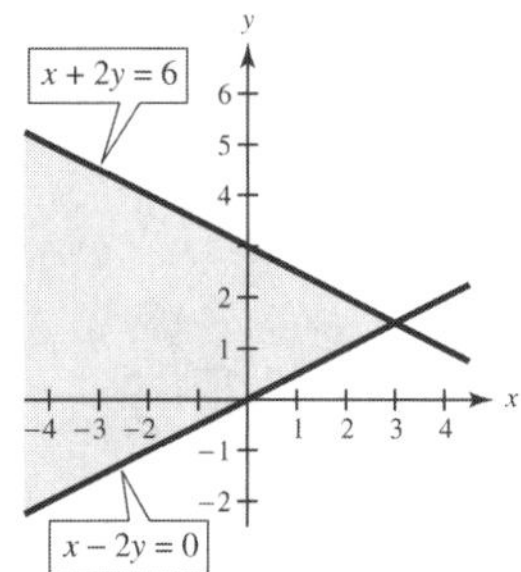

Solid lines at $y = -\frac{1}{2}x + 3$ and $y = \frac{1}{2}x$. Shade below $y = -\frac{1}{2}x + 3$ and above $y = \frac{1}{2}x$.

7. $\begin{cases} x - 2y > 4 \\ 2x + y > 6 \end{cases} \Rightarrow \begin{matrix} y < \frac{1}{2}x - 2 \\ y > -2x + 6 \end{matrix}$

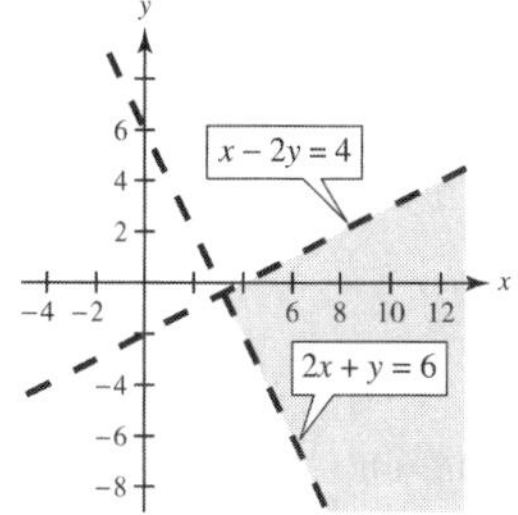

Dotted lines at $y = \frac{1}{2}x - 2$ and $y = -2x + 6$. Shade below $y = \frac{1}{2}x - 2$ and above $y = -2x + 6$.

9. $\begin{cases} x + y > -1 \\ x + y < 3 \end{cases} \Rightarrow \begin{matrix} y > -x - 1 \\ y < -x + 3 \end{matrix}$

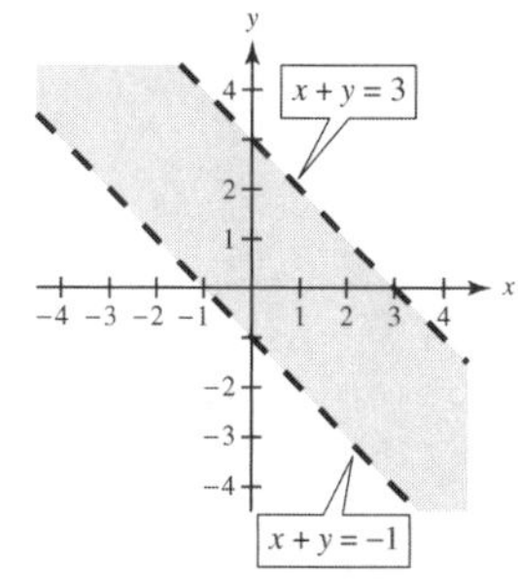

Dotted lines at $y = -x - 1$ and $y = -x + 3$. Shade above $y = -x - 1$ and below $y = -x + 3$.

11. $\begin{cases} y \ge \frac{4}{3}x + 1 \\ y \le 5x - 2 \end{cases}$

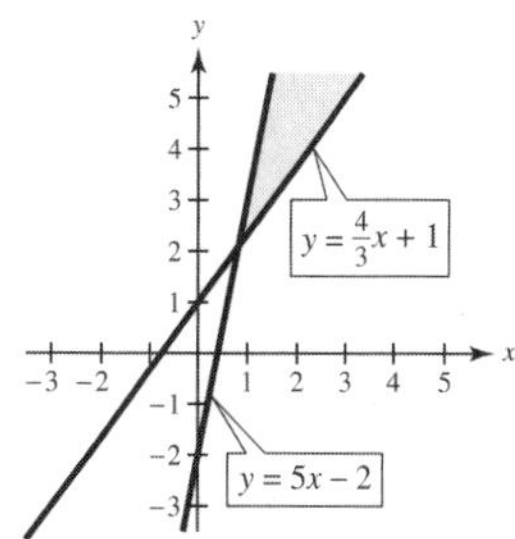

Solid lines at $y = \frac{4}{3}x + 1$ and $y = 5x - 2$. Shade above $y = \frac{4}{3}x + 1$ and below $5x - 2$.

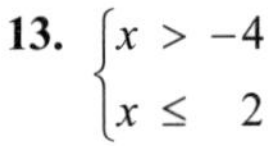

13. $\begin{cases} x > -4 \\ x \leq 2 \end{cases}$

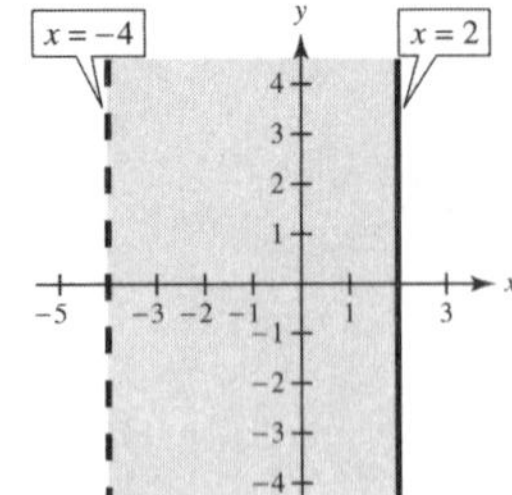

Dotted line at $x = -4$. Solid line at $x = 2$. Shade to the right of $x = -4$ and to the left of $x = 2$.

15. $\begin{cases} x < 3 \\ x > -2 \end{cases}$

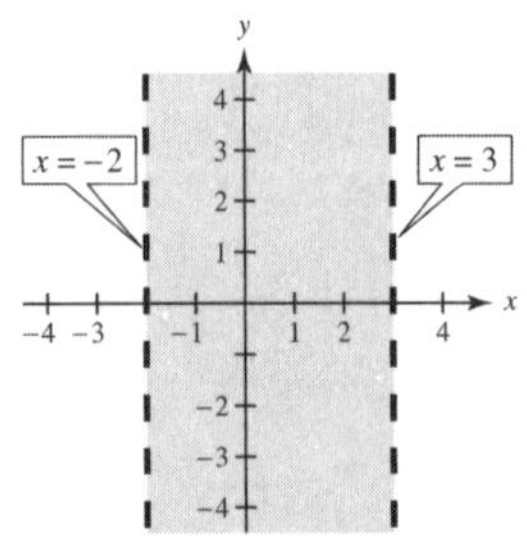

Dotted lines at $x = 3$ and $x = -2$. Shade to the right of $x = -2$ and to the left of $x = 3$.

17. $\begin{cases} x \leq 5 \\ x > -6 \end{cases}$

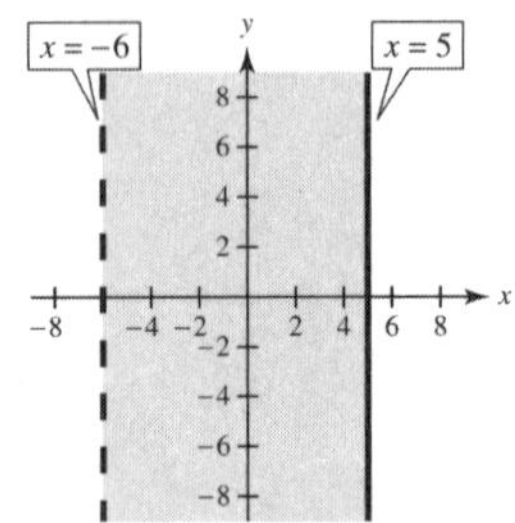

Solid line at $x = 5$. Dotted line at $x = -6$. Shade to the left of $x = 5$ and to the right of $x = -6$.

19. $\begin{cases} x < 3 \\ x > -3 \end{cases}$

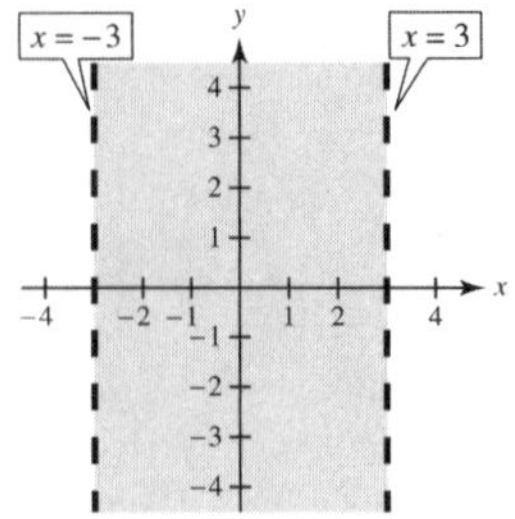

Dotted line at $x = 3$. Dotted line at $x = -3$. Shade to the left of $x = 3$ and to the right of $x = -3$.

21. $\begin{cases} x + y \leq 4 \\ x \geq 0 \\ y \geq 0 \end{cases} \Rightarrow \begin{array}{l} y \leq -x + 4 \\ x \geq 0 \\ y \geq 0 \end{array}$

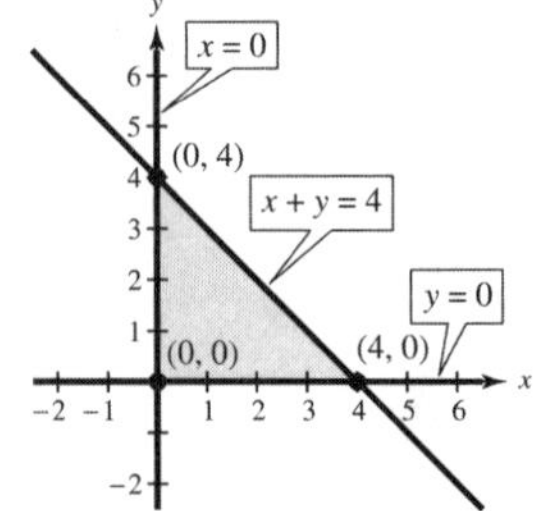

Solid line at $y = -x + 4$. $x = 0$ is the y-axis and $y = 0$ is the x-axis. Shade below $y = -x + 4$ in the first quadrant.

23. $\begin{cases} 4x - 2y > 8 \\ x \geq 0 \\ y \leq 0 \end{cases} \Rightarrow \begin{array}{l} y < 2x - 4 \\ x \geq 0 \\ y \leq 0 \end{array}$

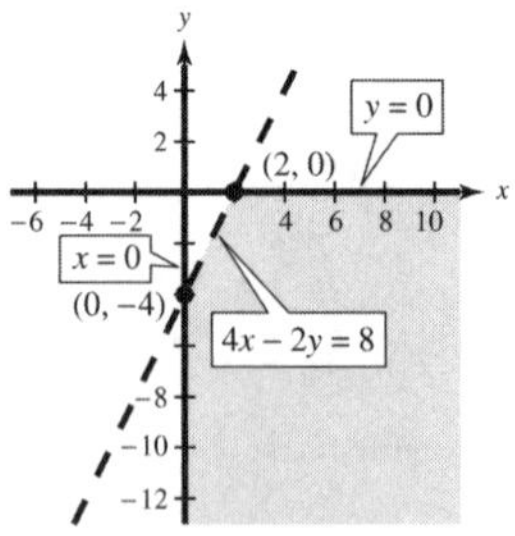

Dotted line at $y = 2x - 4$. $x = 0$ is the y-axis and $y = 0$ is the x-axis. Shade below $y = 2x - 4$ in the fourth quadrant.

25. $\begin{cases} y > -5 \\ x \le 2 \\ y \le x + 2 \end{cases}$

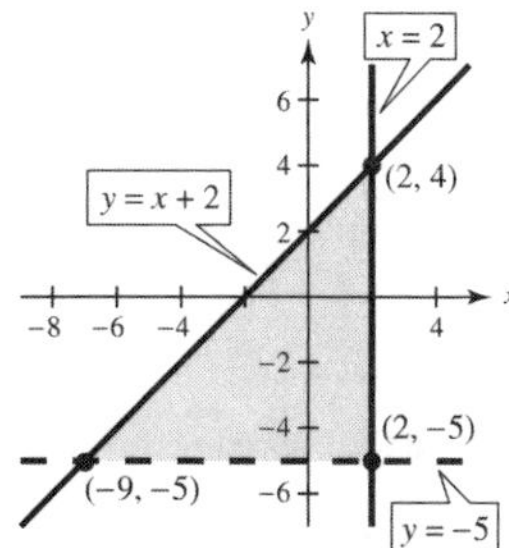

Dotted line at $y = -5$. Solid lines at $x = 2$ and $y = x + 2$. Shade above $y = -5$, to the left of $x = 2$ and below $y = x + 2$.

27. $\begin{cases} x \ge 1 \\ x - 2y \le 3 \\ 3x + 2y \ge 9 \\ x + y \le 6 \end{cases} \Rightarrow \begin{array}{l} x \ge 1 \\ y \ge \frac{1}{2}x - \frac{3}{2} \\ y \ge -\frac{3}{2}x + \frac{9}{2} \\ y \le -x + 6 \end{array}$

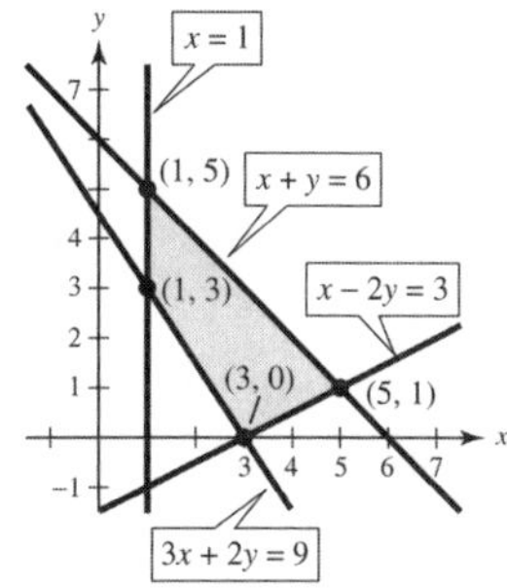

Solid lines at $x = 1$, $y = \frac{1}{2}x - \frac{3}{2}$, $y = -\frac{3}{2}x + \frac{9}{2}$, and $y = -x + 6$. Shade to the right of $x = 1$, above $y = \frac{1}{2}x - \frac{3}{2}$ and $y = -\frac{3}{2}x + \frac{9}{2}$ and below $y = -x + 6$.

29. $\begin{cases} x - y \le 8 \\ 2x + 5y \le 25 \\ x \ge 0 \\ y \ge 0 \end{cases}$

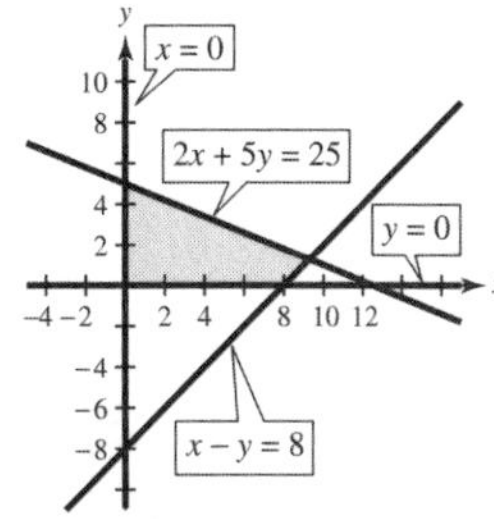

Solid lines at all lines. Shade below $2x + 5y = 25$, above $y = 0$ and $x - y = 8$, and to the right of $x = 0$.

31. Line 1: vertical $x = 1$

Line 2: points $(1, -2)$ and $(3, 0)$

$$m = \frac{0 + 2}{3 - 1} = \frac{2}{2} = 1$$

$$y - 0 = 1(x - 3)$$

$$y = x - 3$$

Line 3: points $(1, 4)$ and $(3, 0)$

$$m = \frac{0 - 4}{3 - 1} = \frac{-4}{2} = -2$$

$$y - 0 = -2(x - 3)$$

$$y = -2x + 6$$

System of linear inequalities:

$x \ge 1$	Region on and to the right of $x = 1$
$y \ge x - 3$	Region on and above line $y = x - 3$.
$y \le -2x + 6$	Region on and below line $y = -2x + 6$.

33. Line 1: $x = -2$

Line 2: $x = 2$

Line 3: points $(-2, 3)$ and $(2, 1)$

$$m = \frac{1 - 3}{2 - (-2)} = \frac{-2}{4} = -\frac{1}{2}$$

$$y - 1 = -\frac{1}{2}(x - 2)$$

$$y - 1 = -\frac{1}{2}x + 1$$

$$y = -\frac{1}{2} + 2$$

Line 4: points $(-2, -3)$ and $(2, -5)$

$$m = \frac{-5 - (-3)}{2 - (-2)} = \frac{-2}{4} = -\frac{1}{2}$$

$$y - (-3) = -\frac{1}{2}(x - (-2))$$

$$y + 3 = -\frac{1}{2}x - 1$$

$$y = -\frac{1}{2}x - 4$$

System of linear inequalities:

$x \ge -2$:	Region on and to the right of $x = -2$
$x \le 2$:	Region on and to the left of $x = 2$
$y \le -\frac{1}{2}x + 2$:	Region on and below $y = -\frac{1}{2}x + 2$
$y \ge -\frac{1}{2}x - 4$:	Region on and above $y = -\frac{1}{2}x - 4$

35. *Verbal Model:* [Number of hours in assembly] · [Number of tables] + [Number of hours in assembly] · [Number of chairs] ≤ 12

[Number of hours in finishing] · [Number of tables] + [Number of hours in finishing] · [Number of chairs] ≤ 16

Labels: Number of tables $= x$

Number of chairs $= y$

System:

$$1x + \tfrac{3}{2}y \le 12$$
$$\tfrac{4}{3}x + \tfrac{3}{4}y \le 16$$
$$x \ge 0$$
$$y \ge 0$$

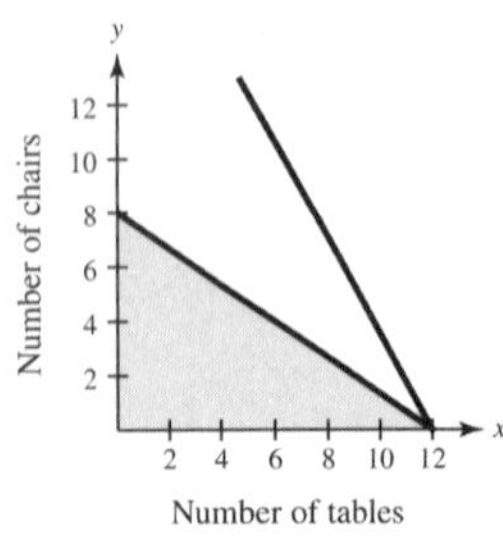

37. A system of linear inequalities in two variables consists of two or more linear inequalities in two variables.

39. Not necessarily. Two boundary lines can sometimes intersect at a point that is not a vertex of the region.

41. $\begin{cases} x + y \le 1 \\ -x + y \le 1 \\ y \ge 0 \end{cases} \Rightarrow \begin{matrix} y \le -x + 1 \\ y \le x + 1 \\ y \ge 0 \end{matrix}$

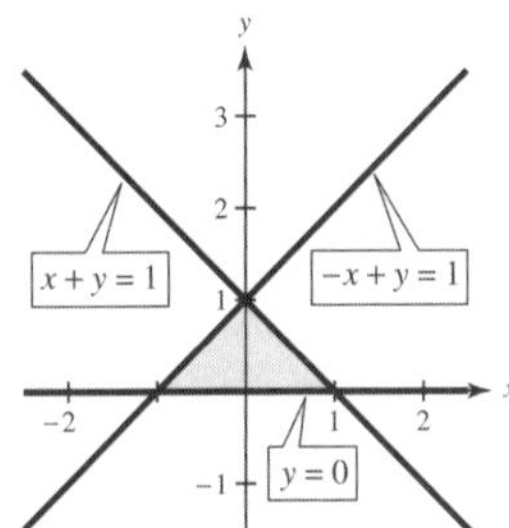

Solid lines at $y = -x + 1$ and $y = x + 1$. $y = 0$ is the x-axis. Shade below $y = -x + 1$ and $y = x + 1$ and above the x-axis.

43. $\begin{cases} x + y \le 5 \\ x - 2y \ge 2 \\ y \ge 3 \end{cases} \Rightarrow \begin{matrix} y \le -x + 5 \\ y \le \frac{1}{2}x - 1 \\ y \ge 3 \end{matrix}$

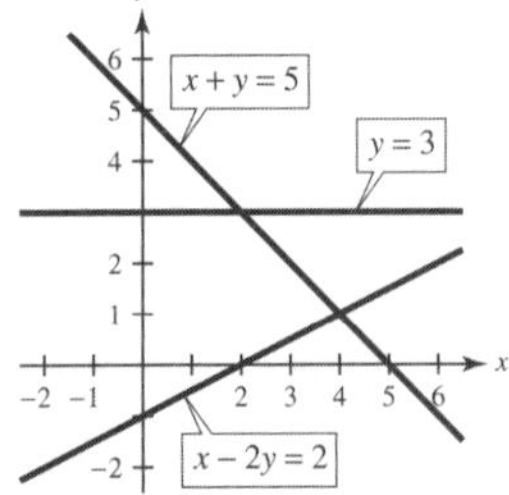

Solid lines at $y = -x + 5$, $y = \frac{1}{2}x - 1$, and $y = 3$.

Shade below $y = -x + 5$ and $y = \frac{1}{2}x - 1$ and above $y = 3$. The half-planes do not all intersect, so there is no solution.

45. $\begin{cases} -3x + 2y < 6 \\ x - 4y > -2 \\ 2x + y < 3 \end{cases} \Rightarrow \begin{matrix} y < -\frac{3}{2}x + 3 \\ y < \frac{1}{4}x + \frac{1}{2} \\ y < -2x + 3 \end{matrix}$

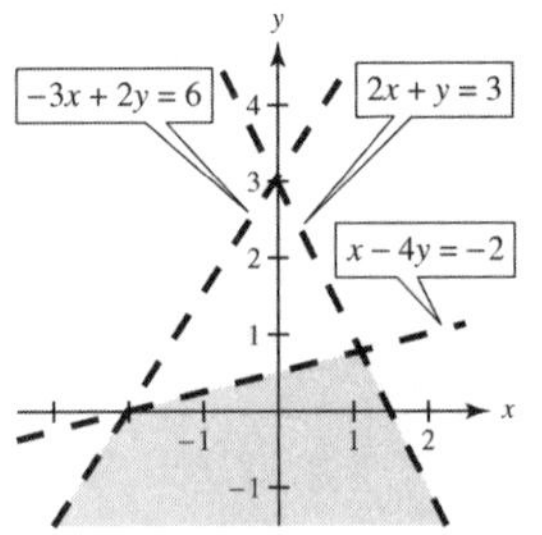

Dotted lines at $y = -\frac{3}{2}x + 3$, $y = \frac{1}{4}x + \frac{1}{2}$, and $y = -2x + 3$. Shade below all three lines.

47. *Verbal Model:* [Amount in account X] + [Amount in account Y] ≤ 25,000

[Amount in account Y] ≥ 4000

[Amount in account X] ≥ 3 · [Amount in account Y]

Labels: Amount in account X $= x$

Amount in account Y $= y$

System:

$$\begin{aligned} x + y &\le 25{,}000 \\ y &\ge 4000 \\ x &\ge 3y \end{aligned}$$

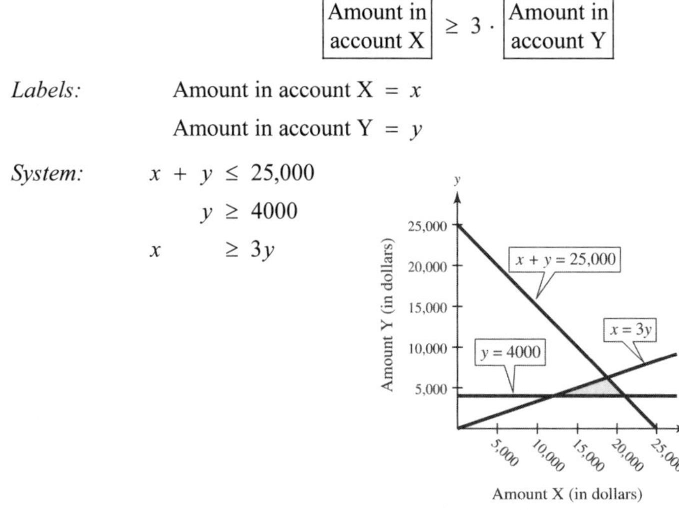

49. *Verbal Model:* 30 · [Reserved seat tickets] + 20 · [General admission tickets] ≥ 75,000

[Reserved seat tickets] + [General admission tickets] ≤ 3000

[Reserved seat tickets] ≤ 2000

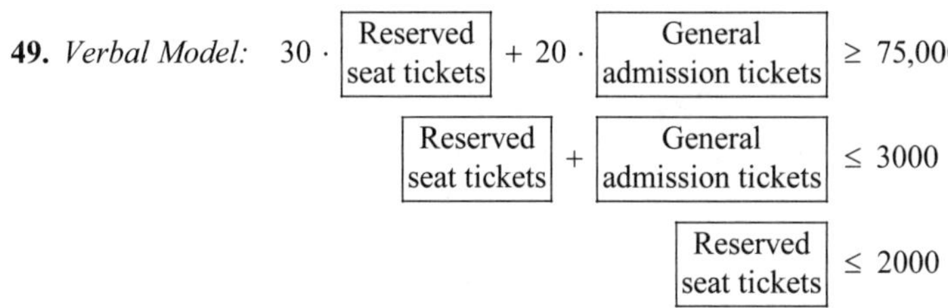

Labels: Reserved seat tickets $= x$

General admission tickets $= y$

System:

$$\begin{aligned} 30x + 20y &\ge 75{,}000 \\ x + y &\le 3000 \\ x &\le 2000 \end{aligned}$$

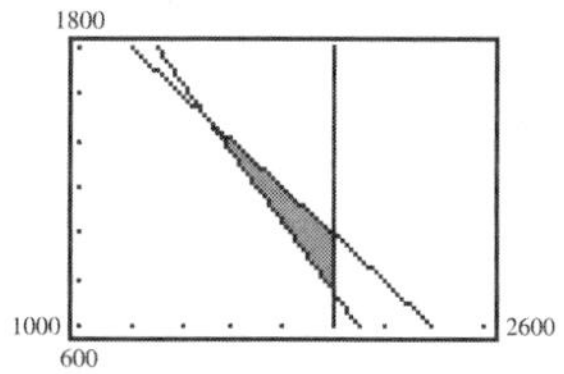

51. Line 1: horizontal, $y = 22$

Line 2: horizontal, $y = 10$

Line 3: points $(-12, 0)$ and $(-8, -8)$

$$m = \frac{-8 - 0}{-8 + 12} = -\frac{8}{4} = -2$$

$$y - 0 = -2(x + 12)$$
$$y = -2x - 24$$

Line 4: points $(12, 0)$ and $(8, -8)$

$$m = \frac{-8 - 0}{8 - 12} = \frac{-8}{-4} = 2$$

$$y - 0 = 2(x - 12)$$
$$y = 2x - 24$$

System of linear inequalities:

$y \le 22$: Region on and below $y = 22$

$y \ge 10$: Region on and above $y = 10$

$y \ge -2x - 24$: Region on and above the line $y = -2x - 24$

$y \ge 2x - 24$: Region on and above the line $y = 2x - 24$

53. Check to see if (x_1, y_1) satisfies each inequality in the system.

55. The solution set of a system of linear equations is usually finite whereas the solution set of a system of linear inequalities is often infinite.

57. Let $y = 0$:

$$y = 8 - 3x$$
$$0 = 8 - 3x$$
$$-8 = -3x$$
$$\tfrac{8}{3} = x$$

The x-intercept is $\left(\tfrac{8}{3}, 0\right)$.

Let $x = 0$:

$$y = 8 - 3x$$
$$y = 8 - 3(0)$$
$$y = 8$$

The y-intercept is $(0, 8)$.

59. Let $y = 0$:

$$3x - 6y = 12$$
$$3x - 6(0) = 12$$
$$3x = 12$$
$$x = 4$$

The x-intercept is $(4, 0)$.

Let $x = 0$:

$$3x - 6y = 12$$
$$3(0) - 6y = 12$$
$$-6y = 12$$
$$y = -2$$

The y-intercept is $(0, -2)$.

61. Let $y = 0$:

$$y = |x - 1| - 2$$
$$0 = |x - 1| - 2$$
$$2 = |x - 1|$$
$$2 = x - 1 \quad \text{or} \quad -2 = x - 1$$
$$3 = x \qquad\qquad -1 = x$$

The x-intercepts are $(-1, 0)$ and $(3, 0)$.

Let $x = 0$:

$$y = |x - 1| - 2$$
$$y = |0 - 1| - 2$$
$$y = |-1| - 2$$
$$y = 1 - 2$$
$$y = -1$$

The y-intercept is $(0, -1)$.

63. (a) $f(x) = x^2 + x$

$$f(3) = 3^2 + 3 = 9 + 3 = 12$$

(b) $f(x) = x^2 + x$

$$f(-2) = (-2)^2 + (-2) = 4 - 2 = 2$$

65. (a) $f(x) = \dfrac{x + 3}{x - 1}$

$$f(8) = \frac{8 + 3}{8 - 1} = \frac{11}{7}$$

(b) $f(x) = \dfrac{x + 3}{x - 1}$

$$f(k - 2) = \frac{(k - 2) + 3}{(k - 2) - 1} = \frac{k + 1}{k - 3}$$

Review Exercises for Chapter 4

1. (a) $(3, 4)$

$$3(3) + 7(4) \stackrel{?}{=} 2$$
$$37 \neq 2$$

Not a solution

(b) $(3, -1)$

$$3(3) + 7(-1) \stackrel{?}{=} 2 \qquad 5(3) + 6(-1) \stackrel{?}{=} 9$$
$$2 = 2 \qquad 9 = 9$$

Solution

3. Solve each equation for y.

$$x + y = 2 \qquad x - y = 2$$
$$y = -x + 2 \qquad -y = -x + 2$$
$$y = x - 2$$

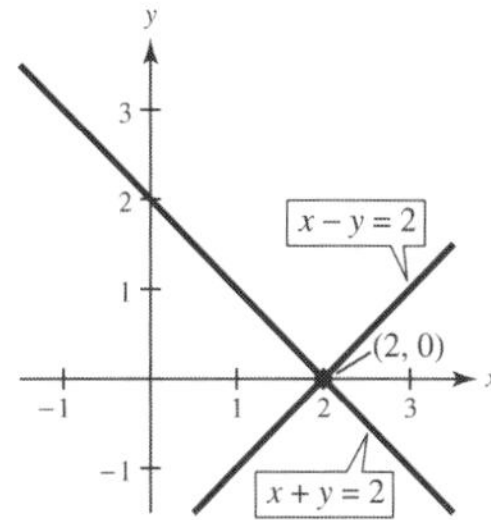

Point of intersection is $(2, 0)$.

5. $x + y = -1$

$3x + 2y = 0$

Solve each equation for y.

$$x + y = -1 \qquad 3x + 2y = 0$$
$$y = -x - 1 \qquad 2y = -3x$$
$$y = -\frac{3}{2}x$$

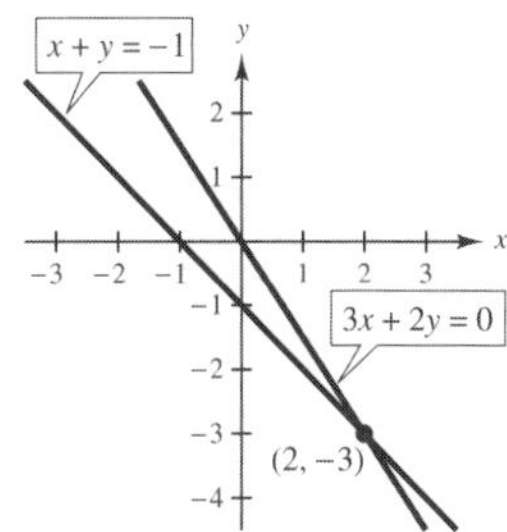

Point of intersection is $(2, -3)$.

7. Solve each equation for y.

$$2x + y = 4 \qquad -4x - 2y = -8$$
$$y = -2x + 4 \qquad -2y = 4x - 8$$
$$y = -2x + 4$$

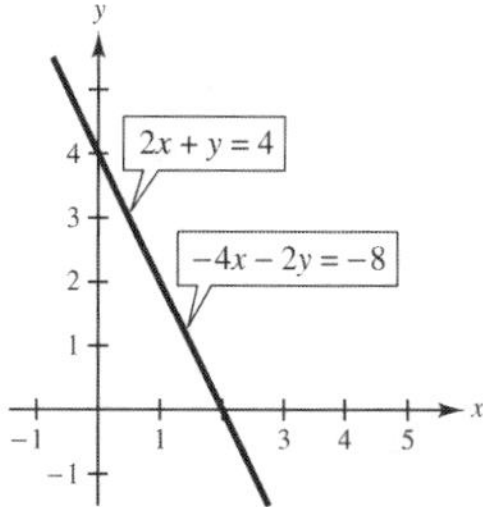

Infinitely many solutions

9. $2x - 3y = -1$

$x + 4y = 16$

$$x + 4y = 16$$
$$x = -4y + 16$$

$$2(-4y + 16) - 3y = -1$$
$$-8y + 32 - 3y = -1$$
$$-11y = -33$$
$$y = 3$$

$$x + 4(3) = 16$$
$$x = 4$$

The solution is $(4, 3)$.

11. $-x + y = 6$

$15x + y = -10$

Solve for y and substitute into second equation.

$$y = x + 6$$

$$15x + (x + 6) = -10$$
$$15x + x + 6 = -10$$
$$16x = -16$$
$$x = -1$$

$$y = (-1) + 6$$
$$y = 5$$

The solution is $(-1, 5)$.

13. $-3x - 3y = 3$

$x + y = -1$

Solve for y and substitute into first equation.

$y = -x - 1$

$-3x - 3(-x - 1) = 3$

$-3x + 3x + 3 = 3$

$3 = 3$

Infinitely many solutions

15. *Verbal Model:*

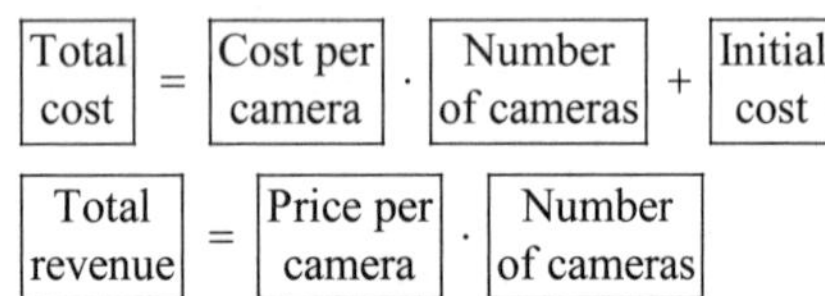

Total revenue = Price per camera · Number of cameras

Labels: Total cost $= C$

Cost per camera $= 4.45$

Number of cameras $= x$

Initial cost $= 25{,}000$

Total revenue $= R$

Price per camera $= 8.95$

System: $C = 4.45x + 25{,}000$

$R = 8.95x$

Break-even point occurs when $R = C$.

$8.95x = 4.45x + 25{,}000$

$4.5x = 25{,}000$

$x \approx 5555.56$

5556 one-time-use cameras must be sold before the business breaks even.

17. $x + y = 0 \Rightarrow x + y = 0$

$2x + y = 0 \Rightarrow -2x - y = 0$

$-x = 0$

$x = 0$

$0 + y = 0$

$y = 0$

19. $2x - y = 2 \Rightarrow 16x - 8y = 16$

$6x + 8y = 39 \Rightarrow 6x + 8y = 39$

$22x = 55$

$x = \frac{55}{22} = \frac{5}{2}$

$2\left(\frac{5}{2}\right) - y = 2$

$5 - y = 2$

$-y = -3$

$y = 3$

The solution is $\left(\frac{5}{2}, 3\right)$.

21. $4x + y = -3$

$-4x + 3y = 23$

$4y = 20$

$y = 5$

$4x + 5 = -3$

$4x = -8$

$x = -2$

The solution is $(-2, 5)$.

23. *Verbal Model:* Number of adult tickets + Number of children tickets = 500

Value of adult tickets + Value of children tickets = 3400

Labels: Number of adult tickets $= x$

Number of children tickets $= y$

System: $x + y = 500$

$7.50x + 4.00y = 3400$

$y = 500 - x$

$7.50x + 4.00(500 - x) = 3400$

$7.50x + 2000 - 4.00x = 3400$

$3.5x = 1400$

$x = 400$

$y = 500 - 400 = 100$

There were 400 adult tickets and 100 children tickets sold.

25. *Verbal Model:* $\boxed{\text{Gallons of solution 1}} + \boxed{\text{Gallons of solution 2}} = 50$

$\boxed{\text{Value of solution 1}} + \boxed{\text{Value of solution 2}} = 0.90(50)$

Labels: Gallons solution 1 $= x$

Gallons solution 2 $= y$

System:

$$\begin{aligned} x + \quad y &= 50 \\ 1.00x + 0.75y &= 0.90(50) \end{aligned}$$

$$\begin{aligned} y &= 50 - x \\ 1.00x + 0.75(50 - x) &= 0.90(50) \\ 1.00x + 37.5 - 0.75x &= 45 \\ 0.25x &= 7.5 \\ x &= 30 \text{ gallons at } 100\% \\ y &= 50 - 30 = 20 \text{ gallons at } 75\% \end{aligned}$$

27. $\begin{cases} x = 3 \\ x + 2y = 7 \\ -3x - y + 4z = 9 \end{cases}$

$$\begin{aligned} 3 + 2y &= 7 \\ 2y &= 4 \\ y &= 2 \end{aligned}$$

$$\begin{aligned} -3(3) - 2 + 4z &= 9 \\ -9 - 2 + 4z &= 9 \\ 4z &= 20 \\ z &= 5 \end{aligned}$$

The solution is $(3, 2, 5)$.

29. $\begin{cases} x + 2y = 6 \\ 3y = 9 \\ x + 2z = 12 \end{cases}$

$$\begin{aligned} 3y &= 9 \\ y &= 3 \end{aligned}$$

$$\begin{aligned} x + 2(3) &= 6 \\ x &= 0 \end{aligned}$$

$$\begin{aligned} 0 + 2z &= 12 \\ z &= 6 \end{aligned}$$

The solution is $(0, 3, 6)$.

31. $\begin{cases} x - y - 2z = -1 \\ 2x + 3y + z = -2 \\ 5x + 4y + 2z = 4 \end{cases}$

$\begin{cases} x - y - 2z = -1 \\ 5y + 5z = 0 \\ 9y + 12z = 9 \end{cases}$

$\begin{cases} x - y - 2z = -1 \\ y + z = 0 \\ 9y + 12z = 9 \end{cases}$

$\begin{cases} x - y - 2z = -1 \\ y + z = 0 \\ 3z = 9 \end{cases}$

$\begin{cases} x - y - 2z = -1 \\ y + z = 0 \\ z = 3 \end{cases}$

$$\begin{aligned} y + 3 &= 0 \\ y &= -3 \end{aligned}$$

$$\begin{aligned} x - (-3) - 2(3) &= -1 \\ x + 3 - 6 &= -1 \\ x &= 2 \end{aligned}$$

The solution is $(2, -3, 3)$.

33. $\begin{cases} x - y - 2z = 1 \\ -2x + y + 3z = -5 \\ 3x + 4y - z = 6 \end{cases}$

$\begin{cases} x - y - z = 1 \\ -y + z = -3 \\ 7y + 2z = 3 \end{cases}$

$$\begin{aligned} x - y - z &= 1 \\ -y + z &= -3 \\ 9z &= -18 \\ z &= -2 \\ -y + (-2) &= -3 \\ -y &= -1 \\ y &= 1 \\ x - 1 - (-2) &= 1 \\ x &= 0 \end{aligned}$$

The solution is $(0, 1, -2)$.

35. *Verbal Model:* [Amount in 7% investment] + [Amount in 9% investment] + [Amount in 11% investment] = [20,000]

[Interest from 7% investment] + [Interest from 9% investment] + [Interest from 11% investment] = [1780]

[Amount in 9% investment] − [Amount in 11% investment] = [−2000]

System:
$$\begin{aligned} x + y + z &= 20{,}000 \\ 0.07x + 0.09y + 0.11z &= 1780 \\ y - z &= -2000 \end{aligned}$$

$$\begin{aligned} x + y + z &= 20{,}000 \\ 7x + 9y + 11z &= 178{,}000 \\ y - z &= -2000 \end{aligned}$$

$$\begin{aligned} x + y + z &= 20{,}000 \\ 2y + 4z &= 38{,}000 \\ y - z &= -2000 \end{aligned}$$

$$\begin{aligned} x + y + z &= 20{,}000 \\ y + 2z &= 19{,}000 \\ y - z &= -2000 \end{aligned}$$

$$\begin{aligned} x + y + z &= 20{,}000 \\ y + 2z &= 19{,}000 \\ -3z &= -21{,}000 \end{aligned}$$

$$\begin{aligned} x + y + z &= 20{,}000 \\ y + 2z &= 19{,}000 \\ z &= 7000 \end{aligned}$$

$$\begin{aligned} y + 2(7000) &= 19{,}000 \\ y &= 5000 \end{aligned} \qquad \begin{aligned} x + 5000 + 7000 &= 20{,}000 \\ x &= 8000 \end{aligned}$$

7% investment: \$8000; 9% investment: \$5000; 11% investment: \$7000

37. (a) $\begin{bmatrix} 7 & -5 \\ 1 & -1 \end{bmatrix}$

(b) $\left[\begin{array}{cc:c} 7 & -5 & 11 \\ 1 & -1 & -5 \end{array}\right]$

39.
$$\begin{aligned} 4x - y &= 2 \\ 6x + 3y + 2z &= 1 \\ y + 4z &= 0 \end{aligned}$$

41.
$$\left[\begin{array}{cc:c} 5 & 4 & 2 \\ -1 & 1 & -22 \end{array}\right]$$

$$\begin{matrix} R_1 \\ R_2 \end{matrix} \left[\begin{array}{cc:c} -1 & 1 & -22 \\ 5 & 4 & 2 \end{array}\right]$$

$$-R_1 \left[\begin{array}{cc:c} 1 & -1 & 22 \\ 5 & 4 & 2 \end{array}\right]$$

$$-5R_1 + R_2 \left[\begin{array}{cc:c} 1 & -1 & 22 \\ 0 & 9 & -108 \end{array}\right]$$

$$\tfrac{1}{9}R_2 \left[\begin{array}{cc:c} 1 & -1 & 22 \\ 0 & 1 & -12 \end{array}\right]$$

$y = -12 \qquad x - (-12) = 22$

$x = 10$

The solution is $(10, -12)$.

43.
$$\left[\begin{array}{cc:c} 0.2 & -0.1 & 0.07 \\ 0.4 & -0.5 & -0.01 \end{array}\right]$$

$$\begin{matrix} 10R_1 \\ 10R_2 \end{matrix} \left[\begin{array}{cc:c} 2 & -1 & 0.7 \\ 4 & -5 & -0.1 \end{array}\right]$$

$$\tfrac{1}{2}R_1 \left[\begin{array}{cc:c} 1 & -\frac{1}{2} & 0.35 \\ 4 & -5 & -0.1 \end{array}\right]$$

$$-4R_1 + R_2 \left[\begin{array}{cc:c} 1 & -\frac{1}{2} & 0.35 \\ 0 & -3 & -1.5 \end{array}\right]$$

$$-\tfrac{1}{3}R_2 \left[\begin{array}{cc:c} 1 & -\frac{1}{2} & 0.35 \\ 0 & 1 & 0.5 \end{array}\right]$$

$y = 0.5 \qquad x - \frac{1}{2}(0.5) = 0.35$

$x = 0.6$

The solution is $(0.6, 0.5)$.

45.
$$\begin{aligned} x + 4y + 4z &= 7 \\ -3x + 2y + 3z &= 0 \\ 4x - 2z &= -2 \end{aligned}$$

$$\left[\begin{array}{ccc:c} 1 & 4 & 4 & 7 \\ -3 & 2 & 3 & 0 \\ 4 & 0 & -2 & -2 \end{array}\right]$$

$$\begin{matrix} \\ 3R_1 + R_2 \\ -4R_1 + R_3 \end{matrix} \left[\begin{array}{ccc:c} 1 & 4 & 4 & 7 \\ 0 & 14 & 15 & 21 \\ 0 & -16 & -18 & -30 \end{array}\right]$$

$$\tfrac{1}{14}R_2 \left[\begin{array}{ccc:c} 1 & 4 & 4 & 7 \\ 0 & 1 & \frac{15}{14} & \frac{3}{2} \\ 0 & -16 & -18 & -30 \end{array}\right]$$

$$16R_2 + R_3 \left[\begin{array}{ccc:c} 1 & 4 & 4 & 7 \\ 0 & 1 & \frac{15}{14} & \frac{3}{2} \\ 0 & 0 & -\frac{6}{7} & -6 \end{array}\right]$$

$$-\tfrac{7}{6}R_3 \left[\begin{array}{ccc:c} 1 & 4 & 4 & 7 \\ 0 & 1 & \frac{15}{14} & \frac{3}{2} \\ 0 & 0 & 1 & 7 \end{array}\right]$$

$z = 7$

$y + \frac{15}{14}(7) = \frac{3}{2}$

$y + \frac{15}{2} = \frac{3}{2}$

$y = -6$

$x + 4(-6) + 4(7) = 7$

$x + 4 = 7$

$x = 3$

The solution is $(3, -6, 7)$.

47. $\begin{bmatrix} 2 & 3 & 3 & \vdots & 3 \\ 6 & 6 & 12 & \vdots & 13 \\ 12 & 9 & -1 & \vdots & 2 \end{bmatrix}$

$\frac{1}{2}R_1$ $\begin{bmatrix} 1 & \frac{3}{2} & \frac{3}{2} & \vdots & \frac{3}{2} \\ 6 & 6 & 12 & \vdots & 13 \\ 12 & 9 & -1 & \vdots & 2 \end{bmatrix}$

$\begin{matrix} \\ -6R_1 + R_2 \\ -12R_1 + R_3 \end{matrix} \begin{bmatrix} 1 & \frac{3}{2} & \frac{3}{2} & \vdots & \frac{3}{2} \\ 0 & -3 & 3 & \vdots & 4 \\ 0 & -9 & -19 & \vdots & -16 \end{bmatrix}$

$\begin{matrix} \\ -\frac{1}{3}R_2 \\ \\ \end{matrix} \begin{bmatrix} 1 & \frac{3}{2} & \frac{3}{2} & \vdots & \frac{3}{2} \\ 0 & 1 & -1 & \vdots & -\frac{4}{3} \\ 0 & -9 & -19 & \vdots & -16 \end{bmatrix}$

$\begin{matrix} \\ \\ 9R_2 + R_3 \end{matrix} \begin{bmatrix} 1 & \frac{3}{2} & \frac{3}{2} & \vdots & \frac{3}{2} \\ 0 & 1 & -1 & \vdots & -\frac{4}{3} \\ 0 & 0 & -28 & \vdots & -28 \end{bmatrix}$

$\begin{matrix} \\ \\ -\frac{1}{28}R_3 \end{matrix} \begin{bmatrix} 1 & \frac{3}{2} & \frac{3}{2} & \vdots & \frac{3}{2} \\ 0 & 1 & -1 & \vdots & -\frac{4}{3} \\ 0 & 0 & 1 & \vdots & 1 \end{bmatrix}$

$z = 1$

$y - 1 = -\frac{4}{3}$

$y = -\frac{1}{3}$

$x + \frac{3}{2}\left(-\frac{1}{3}\right) + \frac{3}{2}(1) = \frac{3}{2}$

$x - \frac{1}{2} + \frac{3}{2} = \frac{3}{2}$

$x + 1 = \frac{3}{2}$

$x = \frac{1}{2}$

The solution is $\left(\frac{1}{2}, -\frac{1}{3}, 1\right)$.

49. $\begin{vmatrix} 9 & 8 \\ 10 & 10 \end{vmatrix} = 9(10) - 10(8) = 90 - 80 = 10$

51. $\begin{vmatrix} 8 & 6 & 3 \\ 6 & 3 & 0 \\ 3 & 0 & 2 \end{vmatrix} = (3)\begin{vmatrix} 6 & 3 \\ 3 & 0 \end{vmatrix} - 0\begin{vmatrix} 8 & 3 \\ 6 & 0 \end{vmatrix} + 2\begin{vmatrix} 8 & 6 \\ 6 & 3 \end{vmatrix}$ (third row)

$= (3)(-9) - 0 + (2)(-12)$

$= -27 - 24$

$= -51$

53. $\begin{vmatrix} 8 & 3 & 2 \\ 1 & -2 & 4 \\ 6 & 0 & 5 \end{vmatrix} = 6\begin{vmatrix} 3 & 2 \\ -2 & 4 \end{vmatrix} - 0 + 5\begin{vmatrix} 8 & 3 \\ 1 & -2 \end{vmatrix}$ (third row)

$= (6)(16) + (5)(-19) = 1$

55. $\begin{bmatrix} 7 & 12 & \vdots & 63 \\ 2 & 3 & \vdots & 15 \end{bmatrix}$

$D = \begin{vmatrix} 7 & 12 \\ 2 & 3 \end{vmatrix} = 21 - 24 = -3$

$x = \frac{D_x}{D} = \frac{\begin{vmatrix} 63 & 12 \\ 15 & 3 \end{vmatrix}}{-3} = \frac{189 - 180}{-3} = \frac{9}{-3} = -3$

$y = \frac{D_y}{D} = \frac{\begin{vmatrix} 7 & 63 \\ 2 & 15 \end{vmatrix}}{-3} = \frac{105 - 126}{-3} = \frac{-21}{-3} = 7$

The solution is $(-3, 7)$.

57. $\begin{bmatrix} -1 & 1 & 2 & \vdots & 1 \\ 2 & 3 & 1 & \vdots & -2 \\ 5 & 4 & 2 & \vdots & 4 \end{bmatrix}$

$$D = \begin{vmatrix} -1 & 1 & 2 \\ 2 & 3 & 1 \\ 5 & 4 & 2 \end{vmatrix} = (-1)\begin{vmatrix} 3 & 1 \\ 4 & 2 \end{vmatrix} - (1)\begin{vmatrix} 2 & 1 \\ 5 & 2 \end{vmatrix} + (2)\begin{vmatrix} 2 & 3 \\ 5 & 4 \end{vmatrix} = (-1)(2) - (1)(-1) + (2)(-7) = -2 + 1 - 14 = -15$$

$$x = \frac{\begin{vmatrix} 1 & 1 & 2 \\ -2 & 3 & 1 \\ 4 & 4 & 2 \end{vmatrix}}{-15} = \frac{(1)\begin{vmatrix} 3 & 1 \\ 4 & 2 \end{vmatrix} - (1)\begin{vmatrix} -2 & 1 \\ 4 & 2 \end{vmatrix} + (2)\begin{vmatrix} -2 & 3 \\ 4 & 4 \end{vmatrix}}{-15} = \frac{(1)(2) - (1)(-8) + (2)(-20)}{-15} = \frac{2 + 8 - 40}{-15} = \frac{-30}{-15} = 2$$

$$y = \frac{\begin{vmatrix} -1 & 1 & 2 \\ 2 & -2 & 1 \\ 5 & 4 & 2 \end{vmatrix}}{-15} = \frac{(-1)\begin{vmatrix} -2 & 1 \\ 4 & 2 \end{vmatrix} - (1)\begin{vmatrix} 2 & 1 \\ 5 & 2 \end{vmatrix} + (2)\begin{vmatrix} 2 & -2 \\ 5 & 4 \end{vmatrix}}{-15} = \frac{(-1)(-8) - (1)(-1) + (2)(18)}{-15} = \frac{8 + 1 + 36}{-15} = \frac{45}{-15} = -3$$

$$z = \frac{\begin{vmatrix} -1 & 1 & 1 \\ 2 & 3 & -2 \\ 5 & 4 & 4 \end{vmatrix}}{-15} = \frac{(-1)\begin{vmatrix} 3 & -2 \\ 4 & 4 \end{vmatrix} - (1)\begin{vmatrix} 2 & -2 \\ 5 & 4 \end{vmatrix} + (1)\begin{vmatrix} 2 & 3 \\ 5 & 4 \end{vmatrix}}{-15} = \frac{(-1)(20) - (1)(18) + (1)(-7)}{-15} = \frac{-20 - 18 - 7}{-15} = \frac{-45}{-15} = 3$$

The solution is $(2, -3, 3)$.

59. $(x_1, y_1) = (1, 0), (x_2, y_2) = (5, 0), (x_3, y_3) = (5, 8)$

$$\begin{vmatrix} x_1 & y_1 & 1 \\ x_2 & y_2 & 1 \\ x_3 & y_3 & 1 \end{vmatrix} = \begin{vmatrix} 1 & 0 & 1 \\ 5 & 0 & 1 \\ 5 & 8 & 1 \end{vmatrix} = -0 + 0 - (8)\begin{vmatrix} 1 & 1 \\ 5 & 1 \end{vmatrix} = (-8)(-4) = 32$$

Area $= +\frac{1}{2}(32) = 16$

61. $(x_1, y_1) = (1, 2), (x_2, y_2) = (4, -5), (x_3, y_3) = (3, 2)$

$$\begin{vmatrix} x_1 & y_1 & 1 \\ x_2 & y_2 & 1 \\ x_3 & y_3 & 1 \end{vmatrix} = \begin{vmatrix} 1 & 2 & 1 \\ 4 & -5 & 1 \\ 3 & 2 & 1 \end{vmatrix} = (1)\begin{vmatrix} 4 & -5 \\ 3 & 2 \end{vmatrix} - (1)\begin{vmatrix} 1 & 2 \\ 3 & 2 \end{vmatrix} + (1)\begin{vmatrix} 1 & 2 \\ 4 & -5 \end{vmatrix} = (1)(23) - (1)(-4) + (1)(-13) = 23 + 4 - 13 = 14$$

Area $= +\frac{1}{2}(14) = 7$

63. $(x_1, y_1) = (1, 2), (x_2, y_2) = (5, 0), (x_3, y_3) = (10, -2)$

$$\begin{vmatrix} x_1 & y_1 & 1 \\ x_2 & y_2 & 1 \\ x_3 & y_3 & 1 \end{vmatrix} = \begin{vmatrix} 1 & 2 & 1 \\ 5 & 0 & 1 \\ 10 & -2 & 1 \end{vmatrix} = 1\begin{vmatrix} 0 & 1 \\ -2 & 1 \end{vmatrix} - 2\begin{vmatrix} 5 & 1 \\ 10 & 1 \end{vmatrix} + 1\begin{vmatrix} 5 & 0 \\ 10 & -2 \end{vmatrix} = 1(2) - 2(-5) + 1(-10) = 2 + 10 - 10 = 2$$

The points are not collinear.

65. $\begin{vmatrix} x & y & 1 \\ -4 & 0 & 1 \\ 4 & 4 & 1 \end{vmatrix} = 0$

$$-(-4)\begin{vmatrix} y & 1 \\ 4 & 1 \end{vmatrix} + 0 - (1)\begin{vmatrix} x & y \\ 4 & 4 \end{vmatrix} = 0$$

$$\begin{aligned} (4)(y - 4) - (1)(4x - 4y) &= 0 \\ 4y - 16 - 4x + 4y &= 0 \\ -4x + 8y - 16 &= 0 \\ x - 2y + 4 &= 0 \end{aligned}$$

67. $\begin{cases} x + y < 5 \\ x > 2 \\ y \geq 0 \end{cases} \Rightarrow \begin{array}{l} y < -x + 5 \\ x > 2 \\ y \geq 0 \end{array}$

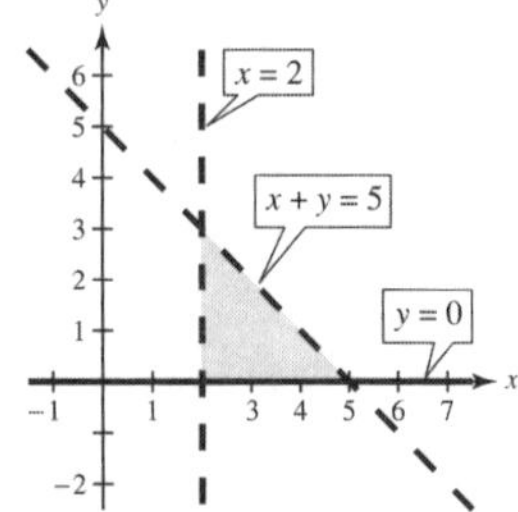

Dotted lines at $y = -x + 5$ and $x = 2$. $y = 0$ is the x-axis. Shade below $y = -x + 5$ and above $y = 0$. Shade to the right of $x = 2$.

69.

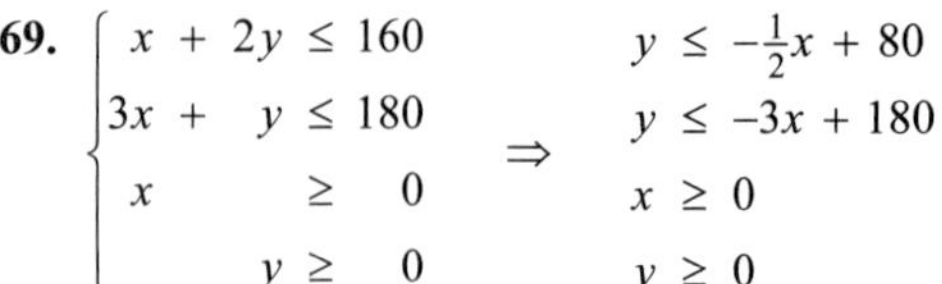

$\begin{cases} x + 2y \leq 160 \\ 3x + y \leq 180 \\ x \geq 0 \\ y \geq 0 \end{cases} \Rightarrow \begin{array}{l} y \leq -\frac{1}{2}x + 80 \\ y \leq -3x + 180 \\ x \geq 0 \\ y \geq 0 \end{array}$

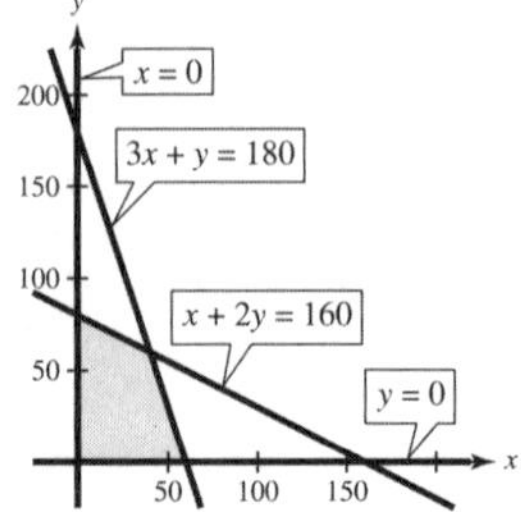

Solid lines at $y = -\frac{1}{2}x + 80$ and $y = -3x + 180$. $x = 0$ is the y-axis and $y = 0$ is the x-axis. Shade below $y = -\frac{1}{2}x + 80$ and $y = -3x + 180$. Shade to the right of $x = 0$ and above $y = 0$.

71. *Verbal Model:* [Cartons of soup for soup kitchen] + [Cartons of soup for homeless shelter] ≤ 500

[Cartons of soup for soup kitchen] ≥ 150

[Cartons of soup for homeless shelter] ≥ 220

Labels: Cartons of soup for soup kitchen $= x$

Cartons of soup for homeless shelter $= y$

System: $\begin{array}{l} x + y \leq 500 \\ x \geq 150 \\ y \geq 220 \end{array}$

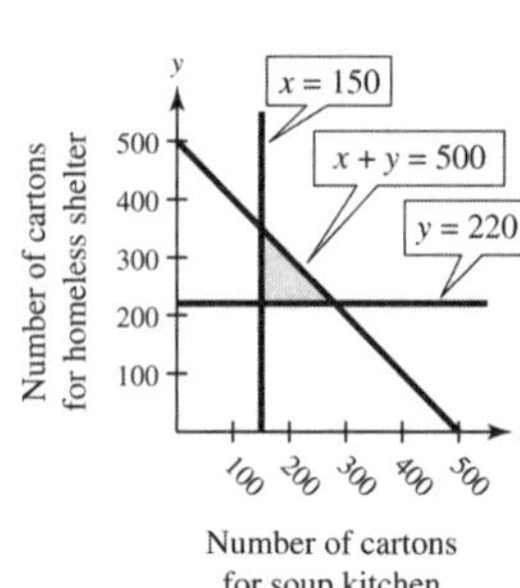

Chapter Test for Chapter 4

1. (a) $(2, 1)$

$$\begin{aligned} 2x - 2y &= 2 & -x + 2y &= 0 \\ 2(2) - 2(1) &\stackrel{?}{=} 2 & -2 + 2(1) &\stackrel{?}{=} 0 \\ 4 - 2 &\stackrel{?}{=} 2 & -2 + 2 &\stackrel{?}{=} 0 \\ 2 &= 2 & 0 &= 0 \end{aligned}$$

$(2, 1)$ is a solution.

(b) $(4, 3)$

$$\begin{aligned} 2x - 2y &= 2 & -x + 2y &= 0 \\ 2(4) - 2(3) &\stackrel{?}{=} 2 & -4 + 2(3) &\stackrel{?}{=} 0 \\ 8 - 6 &\stackrel{?}{=} 2 & -4 + 6 &\stackrel{?}{=} 0 \\ 2 &= 2 & 2 &\neq 0 \end{aligned}$$

$(4, 3)$ is not a solution.

2.

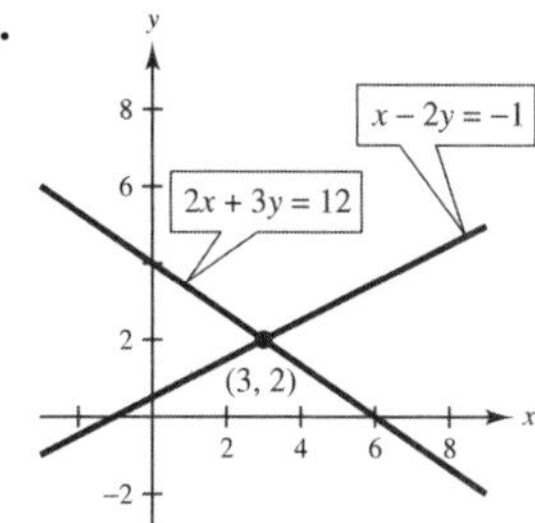

The solution is $(3, 2)$.

3. $$\begin{aligned} 4x - y &= 1 \\ 4x - 3y &= -5 \end{aligned}$$

$$\begin{aligned} 4x - y &= 1 \\ -y &= -4x + 1 \\ y &= 4x - 1 \end{aligned}$$

$$\begin{aligned} 4x - 3(4x - 1) &= -5 \\ 4x - 12x + 3 &= -5 \\ -8x + 3 &= -5 \\ -8x &= -8 \\ x &= 1 \end{aligned}$$

$$\begin{aligned} 4(1) - y &= 1 \\ -y &= -3 \\ y &= 3 \end{aligned}$$

The solution is $(1, 3)$.

4. $$\begin{aligned} 2x - 2y &= -2 \\ 3x + y &= 9 \end{aligned}$$

Solve for y.

$y = -3x + 9$

Substitute into first equation.

$$\begin{aligned} 2x - 2(-3x + 9) &= -2 \\ 2x + 6x - 18 &= -2 \\ 8x &= 16 \\ x &= 2 \end{aligned}$$

$y = -3(2) + 9 = 3$

The solution is $(2, 3)$.

5. $$\begin{aligned} 3x - 4y &= -14 \\ -3x + y &= 8 \\ \hline -3y &= -6 \\ y &= 2 \end{aligned} \qquad \begin{aligned} 3x - 4(2) &= -14 \\ 3x &= -6 \\ x &= -2 \end{aligned}$$

The solution is $(-2, 2)$.

6. $$\begin{cases} x + 2y - 4z = 0 \\ 3x + y - 2z = 5 \\ 3x - y + 2z = 7 \end{cases}$$

$$\begin{cases} x + 2y - 4z = 0 \\ -5y + 10z = 5 \\ -7y + 14z = 7 \end{cases}$$

$$\begin{cases} x + 2y - 4z = 0 \\ y - 2z = -1 \\ -7y + 14z = 7 \end{cases}$$

$$\begin{cases} x + 2y - 4z = 0 \\ y - 2z = -1 \\ 0 = 0 \end{cases}$$

Let $a = z$. (a is any real number.)

$$\begin{aligned} y &= 2z - 1 \\ x &= -2y + 4z \\ &= -2(2z - 1) + 4z \\ &= -4z + 2 + 4z \\ x &= 2 \end{aligned}$$

A solution is $(2, 2a - 1, a)$.

7. $$\begin{bmatrix}1&0&-3&\vdots&-10\\0&-2&2&\vdots&0\\1&-2&0&\vdots&-7\end{bmatrix}$$

$$\begin{array}{r}\\-\frac{1}{2}R_2\\-R_1+R_3\end{array}\begin{bmatrix}1&0&-3&\vdots&-10\\0&1&-1&\vdots&0\\0&-2&3&\vdots&3\end{bmatrix}$$

$$\begin{array}{r}\\\\2R_2+R_3\end{array}\begin{bmatrix}1&0&-3&\vdots&-10\\0&1&-1&\vdots&0\\0&0&1&\vdots&3\end{bmatrix}$$

$z = 3 \qquad y - 3 = 0 \qquad x - 3(3) = -10$

$\qquad\qquad y = 3 \qquad\qquad x = -1$

The solution is $(-1, 3, 3)$.

8. $$\begin{bmatrix}1&-3&1&\vdots&-3\\3&2&-5&\vdots&18\\0&1&1&\vdots&-1\end{bmatrix}$$

$$\begin{array}{r}\\-3R_1+R_2\\\\\end{array}\begin{bmatrix}1&-3&1&\vdots&-3\\0&11&-8&\vdots&27\\0&1&1&\vdots&-1\end{bmatrix}$$

$$\begin{array}{r}\\R_2\\R_3\end{array}\begin{bmatrix}1&-3&1&\vdots&-3\\0&1&1&\vdots&-1\\0&11&-8&\vdots&27\end{bmatrix}$$

$$\begin{array}{r}3R_2+R_1\\\\-11R_2+R_3\end{array}\begin{bmatrix}1&0&4&\vdots&-6\\0&1&1&\vdots&-1\\0&0&-19&\vdots&38\end{bmatrix}$$

$$\begin{array}{r}\\\\-\frac{1}{19}R_3\end{array}\begin{bmatrix}1&0&4&\vdots&-6\\0&1&1&\vdots&-1\\0&0&1&\vdots&-2\end{bmatrix}$$

$z = -2 \qquad y + (-2) = -1 \qquad x + 4(-2) = -6$

$\qquad\qquad y = 1 \qquad\qquad x = 2$

The solution is $(2, 1, -2)$.

9. $$\begin{bmatrix}2&-7&\vdots&7\\3&7&\vdots&13\end{bmatrix}$$

$$D = \begin{vmatrix}2&-7\\3&7\end{vmatrix} = 14 + 21 = 35$$

$$x = \frac{D_x}{D} = \frac{\begin{vmatrix}7&-7\\13&7\end{vmatrix}}{35} = \frac{49+91}{35} = \frac{140}{35} = 4$$

$$y = \frac{D_y}{D} = \frac{\begin{vmatrix}2&7\\3&13\end{vmatrix}}{35} = \frac{26-21}{35} = \frac{5}{35} = \frac{1}{7}$$

The solution is $\left(4, \frac{1}{7}\right)$.

10. $$\begin{bmatrix}3&-2&1&\vdots&12\\1&-3&0&\vdots&2\\-3&0&-9&\vdots&-6\end{bmatrix}$$

$$\begin{array}{r}R_1\\\\-\frac{1}{3}R_3\end{array}\begin{bmatrix}1&0&3&\vdots&2\\1&-3&0&\vdots&2\\3&-2&1&\vdots&12\end{bmatrix}$$

$$\begin{array}{r}\\-1R_1+R_2\\-3R_1+R_3\end{array}\begin{bmatrix}1&0&3&\vdots&2\\0&-3&-3&\vdots&0\\0&-2&-8&\vdots&6\end{bmatrix}$$

$$\begin{array}{r}\\-\frac{1}{3}R_2\\-\frac{1}{2}R_3\end{array}\begin{bmatrix}1&0&3&\vdots&2\\0&1&1&\vdots&0\\0&1&4&\vdots&-3\end{bmatrix}$$

$$\begin{array}{r}\\\\-R_2+R_3\end{array}\begin{bmatrix}1&0&3&\vdots&2\\0&1&1&\vdots&0\\0&0&3&\vdots&-3\end{bmatrix}$$

$$\begin{array}{r}\\\\\frac{1}{3}R_3\end{array}\begin{bmatrix}1&0&3&\vdots&2\\0&1&1&\vdots&0\\0&0&1&\vdots&-1\end{bmatrix}$$

$$\begin{array}{r}-3R_3+R_1\\-R_3+R_2\\\\\end{array}\begin{bmatrix}1&0&0&\vdots&5\\0&1&0&\vdots&1\\0&0&1&\vdots&-1\end{bmatrix}$$

The solution is $(5, 1, -1)$.

11. $$\begin{bmatrix}4&1&2&\vdots&-4\\0&3&1&\vdots&8\\-3&1&-3&\vdots&5\end{bmatrix}$$

$$\begin{array}{r}\frac{1}{4}R_1\\\\\\\end{array}\begin{bmatrix}1&\frac{1}{4}&\frac{1}{2}&\vdots&-1\\0&3&1&\vdots&8\\-3&1&-3&\vdots&5\end{bmatrix}$$

$$\begin{array}{r}\\\frac{1}{3}R_2\\3R_1+R_3\end{array}\begin{bmatrix}1&\frac{1}{4}&\frac{1}{2}&\vdots&-1\\0&1&\frac{1}{3}&\vdots&\frac{8}{3}\\0&\frac{7}{4}&-\frac{3}{2}&\vdots&2\end{bmatrix}$$

$$\begin{array}{r}-\frac{1}{4}R_2+R_1\\\\-\frac{7}{4}R_2+R_3\end{array}\begin{bmatrix}1&0&\frac{5}{12}&\vdots&-\frac{5}{3}\\0&1&\frac{1}{3}&\vdots&\frac{8}{3}\\0&0&-\frac{25}{12}&\vdots&-\frac{8}{3}\end{bmatrix}$$

$$\begin{array}{r}\\\\-\frac{12}{25}R_3\end{array}\begin{bmatrix}1&0&\frac{5}{12}&\vdots&-\frac{5}{3}\\0&1&\frac{1}{3}&\vdots&\frac{8}{3}\\0&0&1&\vdots&\frac{32}{25}\end{bmatrix}$$

$$\begin{array}{r}-\frac{5}{12}R_3+R_1\\-\frac{1}{3}R_3+R_2\\\\\end{array}\begin{bmatrix}1&0&0&\vdots&-\frac{11}{5}\\0&1&0&\vdots&\frac{56}{25}\\0&0&1&\vdots&\frac{32}{25}\end{bmatrix}$$

The solution is $\left(-\frac{11}{5}, \frac{56}{25}, \frac{32}{25}\right)$.

12. $\begin{vmatrix} 2 & -2 & 0 \\ -1 & 3 & 1 \\ 2 & 8 & 1 \end{vmatrix} = 0\begin{vmatrix} -1 & 3 \\ 2 & 8 \end{vmatrix} - 1\begin{vmatrix} 2 & -2 \\ 2 & 8 \end{vmatrix} + 1\begin{vmatrix} 2 & -2 \\ -1 & 3 \end{vmatrix} = 0(-14) - 1(20) + 1(4) = 0 - 20 + 4 = -16$

13. $(x_1, y_1) = (0, 0), (x_2, y_2) = (5, 4), (x_3, y_3) = (6, 0)$

$$\begin{vmatrix} x_1 & y_1 & 1 \\ x_2 & y_2 & 1 \\ x_3 & y_3 & 1 \end{vmatrix} = \begin{vmatrix} 0 & 0 & 1 \\ 5 & 4 & 1 \\ 6 & 0 & 1 \end{vmatrix} = (1)\begin{vmatrix} 5 & 4 \\ 6 & 0 \end{vmatrix} = (1)(-24) = -24$$

Area $= -\frac{1}{2}(-24) = 12$

14. $\begin{cases} x - 2y > -3 \\ 2x + 3y \le 22 \\ y \ge 0 \end{cases} \Rightarrow \begin{array}{l} y < \frac{1}{2}x + \frac{3}{2} \\ y \le -\frac{2}{3}x + \frac{22}{3} \\ y \ge 0 \end{array}$

Dotted line at $y = \frac{1}{2}x + \frac{3}{2}$ and solid line at $y = -\frac{2}{3}x + \frac{22}{3}$. $y = 0$ is the x-axis.

Shade below $y = \frac{1}{2}x + \frac{3}{2}$ and $y = -\frac{2}{3}x + \frac{22}{3}$ and above $y = 0$.

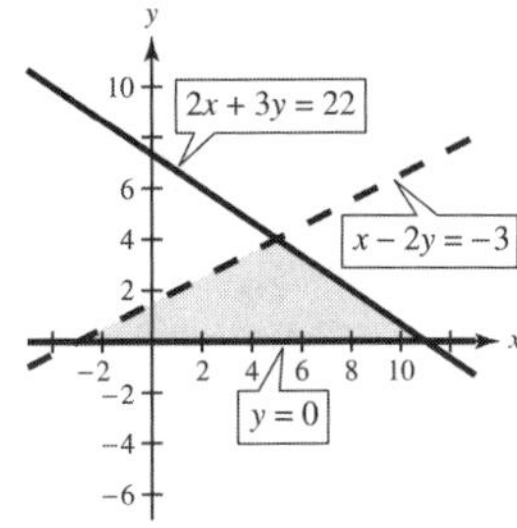

15. *Verbal Model:* $2 \cdot$ [Length] $+ 2 \cdot$ [Width] $= 68$

[Width] $= \frac{8}{9} \cdot$ [Length]

Labels: Length $= x$

Width $= y$

System: $2x + 2y = 68$

$y = \frac{8}{9}x$

Solve by substitution.

$$\begin{aligned} 2x + 2\left(\tfrac{8}{9}x\right) &= 68 \\ 18x + 16x &= 612 \\ 34x &= 612 \\ x &= 18 \\ y &= \tfrac{8}{9}(18) \\ y &= 16 \end{aligned}$$

16 feet × 18 feet

16. *Verbal Model:* $\boxed{\text{Investment 1}} + \boxed{\text{Investment 2}} + \boxed{\text{Investment 3}} = \$25{,}000$

$4.5\% \cdot \boxed{\text{Investment 1}} + 5\% \cdot \boxed{\text{Investment 2}} + 8\% \cdot \boxed{\text{Investment 3}} = \$1275$

$\boxed{\text{Investment 1}} + 4000 = \boxed{\text{Investment 3}} + 10{,}000$

Labels: Investment 1 $= x$

Investment 2 $= y$

Investment 3 $= z$

System:

$$\begin{aligned} x + y + z &= 25{,}000 \\ 0.045x + 0.05y + 0.08z &= 1275 \end{aligned}$$

$y + 4000 = z + 10{,}000 \Rightarrow y - z = 6000$

Using determinants and Cramer's Rule:

$$\begin{bmatrix} 1 & 1 & 1 & 25{,}000 \\ 0.045 & 0.05 & 0.08 & 1275 \\ 0 & 1 & -1 & 6000 \end{bmatrix} \qquad D = \begin{vmatrix} 1 & 1 & 1 \\ 0.045 & 0.05 & 0.08 \\ 0 & 1 & -1 \end{vmatrix} = -0.04$$

$$x = \frac{D_x}{D} = \frac{\begin{vmatrix} 25{,}000 & 1 & 1 \\ 1275 & 0.05 & 0.08 \\ 6000 & 1 & -1 \end{vmatrix}}{-0.04} = \frac{-520}{-0.04} = \$13{,}000$$

$$y = \frac{D_y}{D} = \frac{\begin{vmatrix} 1 & 25{,}000 & 1 \\ 0.045 & 1275 & 0.08 \\ 0 & 6000 & -1 \end{vmatrix}}{-0.04} = \frac{-360}{-0.04} = \$9000$$

$$z = \frac{D_z}{D} = \frac{\begin{vmatrix} 1 & 1 & 25{,}000 \\ 0.045 & 0.05 & 1275 \\ 0 & 1 & 6000 \end{vmatrix}}{-0.04} = \frac{-120}{-0.04} = \$3000$$

Use your graphing calculator to find each determinant.

\$13,000 at 4.5%, \$9,000 at 5%, \$3,000 at 8%

17. *Verbal Model:* $30 \cdot \boxed{\text{Reserved seat tickets}} + 40 \cdot \boxed{\text{Floor seat tickets}} \ge 300{,}000$

$\boxed{\text{Reserved seat tickets}} \le 9000$

$\boxed{\text{Floor seat tickets}} \le 4000$

Labels: Reserved seat tickets $= x$

Floor seat tickets $= y$

System:

$$\begin{aligned} 30x + 40y &\ge 300{,}000 \\ x &\le 9000 \\ y &\le 4000 \end{aligned}$$

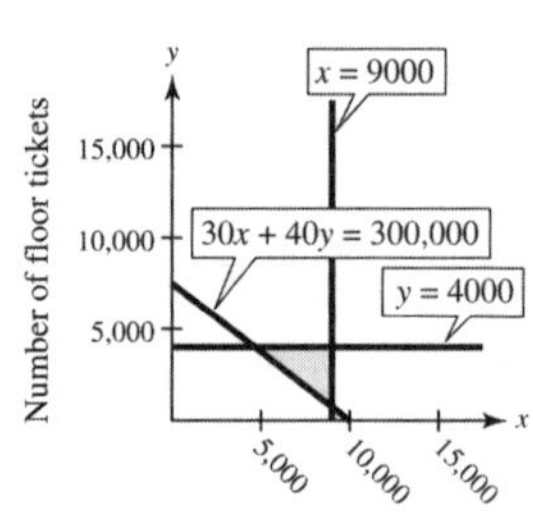

Cumulative Test for Chapters 1–4

1. (a) $-2 > -4$

(b) $\frac{2}{3} > \frac{1}{2}$

(c) $-4.5 = -|-4.5|$

2. "The number n is tripled and the product is decreased by 8," is expressed by $3n - 8$.

3. $t(3t - 1) - 2(t + 4) = 3t^2 - t - 2t - 8$

$= 3t^2 - 3t - 8$

4. $4x(x + x^2) - 6(x^2 + 4) = 4x^2 + 4x^3 - 6x^2 - 24$

$= 4x^3 + 4x^2 - 6x^2 - 24$

$= 4x^3 - 2x^2 - 24$

5. $12 - 5(3 - x) = x + 3$

$12 - 15 + 5x = x + 3$

$-3 + 5x = x + 3$

$-3 + 5x - x = x + 3 - x$

$3 - 3 + 4x = 3 + 3$

$4x = 6$

$\dfrac{4x}{4} = \dfrac{6}{4}$

$x = \dfrac{3}{2}$

6. $1 - \dfrac{x + 2}{4} = \dfrac{7}{8}$

$8\left[1 - \dfrac{x + 2}{4}\right] = \left[\dfrac{7}{8}\right]8$

$8 - 2(x + 2) = 7$

$8 - 2x - 4 = 7$

$4 - 2x = 7$

$4 - 4 - 2x = 7 - 4$

$-2x = 3$

$\dfrac{-2x}{-2} = -\dfrac{3}{2}$

$x = -\dfrac{3}{2}$

7. $|x - 2| \geq 3$

$x - 2 \leq -3$ or $x - 2 \geq 3$

$x \leq -1$ $\qquad$ $x \geq 5$

8. $-12 \leq 4x - 6 < 10$

$-6 \leq 4x < 16$

$-\frac{6}{4} \leq x < 4$

$-\frac{3}{2} \leq x < 4$

9. $1150 + 0.20(1150) = 1150 + 230 = 1380$

Your new premium is $1380.

10. $\dfrac{9}{4.5} = \dfrac{13}{x}$

$9x = 13(4.5)$

$x = \dfrac{13(4.5)}{9}$

$x = 6.5$

11. *Verbal Model:* [Weekly cost for Company B] > [Weekly cost for Company A]

Labels:

Numbers of miles driven in one week $= m$	(miles)
Weekly cost for Company A $= 240$	(dollars)
Weekly cost for Company B $= 100 + 0.25m$	(dollars)

Inequality: $100 + 0.25m > 240$

$0.25m > 140$

$m > 560$

So, the car from Company B is more expensive if you drive more than 560 miles in a week.

12. No, $x - y^2 = 0$ does not represent y as a function of x.

13. $f(x) = \sqrt{x - 2}$

$D = x \geq 2$

$x - 2 \geq 0$

$2 \leq x < \infty$

14. (a) $f(x) = x^2 - 2x$

$f(3) = 3^2 - 2(3) = 9 - 6 = 3$

(b) $f(x) = x^2 - 2x$

$f(-3c) = (-3c)^2 - 2(-3c) = 9c^2 + 6c$

15. $m = \dfrac{6 - 0}{4 + 4} = \dfrac{6}{8} = \dfrac{3}{4}$

$d = \sqrt{(-4 - 4)^2 + (0 - 6)^2}$

$= \sqrt{64 + 36}$

$= \sqrt{100}$

$= 10$

Midpoint $= \left(\dfrac{-4 + 4}{2}, \dfrac{0 + 6}{2}\right) = (0, 3)$

16. (a) $2x - y = 1$

$$-y = -2x + 1$$
$$y = 2x - 1$$
$$m = 2$$
$$y - 1 = 2(x + 2)$$
$$y - 1 = 2x + 4$$
$$2x - y + 5 = 0$$

(b) $3x + 2y = 5$

$$2y = -3x + 5$$
$$y = -\tfrac{3}{2}x + 5$$
$$m = \tfrac{2}{3}$$
$$y - 1 = \tfrac{2}{3}(x + 2)$$
$$y - 1 = \tfrac{2}{3}x + \tfrac{4}{3}$$
$$y = \tfrac{2}{3}x + \tfrac{7}{3}$$
$$3y = 2x + 7$$
$$2x - 3y + 7 = 0$$

17. $4x + 3y - 12 = 0$

$$4(0) + 3y - 12 = 0$$
$$3y = 12$$
$$y = 4$$

$(0, 4)$

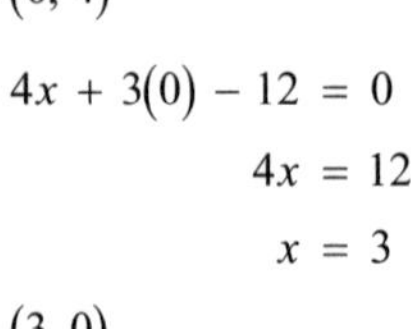

$$4x + 3(0) - 12 = 0$$
$$4x = 12$$
$$x = 3$$

$(3, 0)$

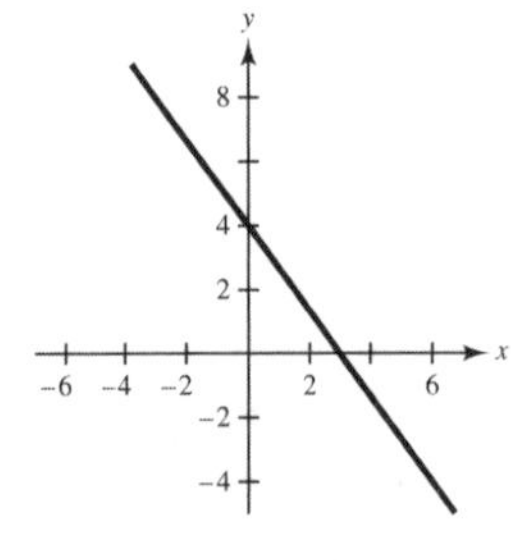

18. $y = 2 - (x - 3)^2$

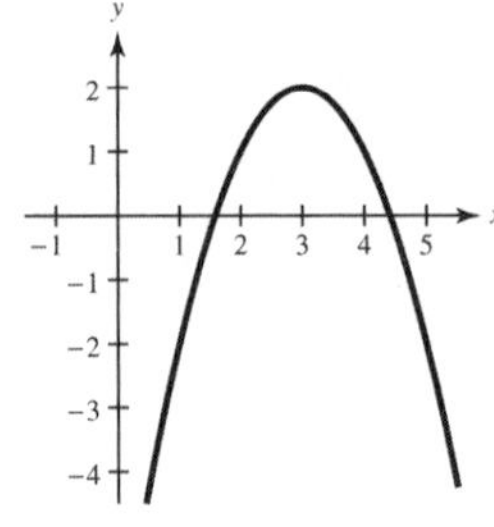

19. $\begin{cases} x + y = 6 \\ 2x - y = 3 \end{cases}$

Solve for y.

$y = 6 - x$

Substitute into second equation.

$$2x - (6 - x) = 3$$
$$2x - 6 + x = 3$$
$$3x = 9$$
$$x = 3$$
$$y = 6 - 3$$
$$= 3$$

The solution is $(3, 3)$.

20. $\begin{cases} 2x + y = 6 \\ 3x - 2y = 16 \end{cases}$

$$4x + 2y = 12$$
$$3x - 2y = 16$$
$$7x = 28$$
$$x = 4$$
$$2(4) + y = 6$$
$$8 + y = 6$$
$$y = -2$$

The solution is $(4, -2)$.

21. $\begin{cases} 2x + y - 2z = 1 \\ x - z = 1 \\ 3x + 3y + z = 12 \end{cases}$

$$\begin{bmatrix} 2 & 1 & -2 & \vdots & 1 \\ 1 & 0 & -1 & \vdots & 1 \\ 3 & 3 & 1 & \vdots & 12 \end{bmatrix}$$

$R_1 \leftrightarrow R_2$ $\begin{bmatrix} 1 & 0 & -1 & \vdots & 1 \\ 2 & 1 & -2 & \vdots & 1 \\ 3 & 3 & 1 & \vdots & 12 \end{bmatrix}$

$-2R_1 + R_2$, $-3R_1 + R_3$ $\begin{bmatrix} 1 & 0 & -1 & \vdots & 1 \\ 0 & 1 & 0 & \vdots & -1 \\ 0 & 3 & 4 & \vdots & 9 \end{bmatrix}$

$-3R_2 + R_3$ $\begin{bmatrix} 1 & 0 & -1 & \vdots & 1 \\ 0 & 1 & 0 & \vdots & -1 \\ 0 & 0 & 4 & \vdots & 12 \end{bmatrix}$

$\tfrac{1}{4}R_3$ $\begin{bmatrix} 1 & 0 & -1 & \vdots & 1 \\ 0 & 1 & 0 & \vdots & -1 \\ 0 & 0 & 1 & \vdots & 3 \end{bmatrix}$

$R_3 + R_1$ $\begin{bmatrix} 1 & 0 & 0 & \vdots & 4 \\ 0 & 1 & 0 & \vdots & -1 \\ 0 & 0 & 1 & \vdots & 3 \end{bmatrix}$

The solution is $(4, -1, 3)$.

CHAPTER 5
Polynomials and Factoring

Section 5.1 Integer Exponents and Scientific Notation..**146**

Section 5.2 Adding and Subtracting Polynomials..**148**

Section 5.3 Multiplying Polynomials..**151**

Mid-Chapter Quiz..**155**

Section 5.4 Factoring by Grouping and Special Forms **156**

Section 5.5 Factoring Trinomials **158**

Section 5.6 Solving Polynomial Equations by Factoring....................................**161**

Review Exercises **165**

Chapter Test **169**

CHAPTER 5
Polynomials and Factoring

Section 5.1 Integer Exponents and Scientific Notation

1. $(u^3v)(2v^2) = 2 \cdot u^3 \cdot v^{1+2} = 2u^3v^3$

3. $-3(x^3)^2 = -3(x^{3 \cdot 2}) = -3x^6$

5. $(-5z^2)^3 = (-5)^3 \cdot (z^2)^3 = -125z^{2 \cdot 3} = -125z^6$

7. $(2u)^4(4u) = 2^4u^4 \cdot 4u = 16 \cdot 4 \cdot u^{4+1} = 64u^5$

9. $\dfrac{27m^5n^6}{9mn^3} = \dfrac{27}{9} \cdot \dfrac{m^5}{m} \cdot \dfrac{n^6}{n^3}$

$= 3 \cdot m^{5-1} \cdot n^{6-3} = 3m^4n^3$

11. $-\left(\dfrac{2a}{3y}\right)^2 = -\dfrac{(2)^2a^2}{(3)^2y^2} = -\dfrac{4a^2}{9y^2}$

13. $\dfrac{x^ny^{2n}}{x^3y^2} = (x^{n-3})(y^{2n-2}) = x^{n-3}y^{2n-2}$

15. $\dfrac{(-2x^2y)^3}{9x^2y^2} = \dfrac{(-2)^3(x^2)^3y^3}{9x^2y^2}$

$= \dfrac{(-8)x^6y^3}{9x^2y^2} = -\dfrac{8x^{6-2}y^{3-2}}{9} = -\dfrac{8x^4y}{9}$

17. $(-3)^0 = 1$

19. $5^{-2} = \dfrac{1}{5^2} = \dfrac{1}{25}$

21. $\left(\dfrac{2}{3}\right)^{-1} = \dfrac{3}{2}$

23. $-10^{-3} = -\dfrac{1}{10^3} = -\dfrac{1}{1000}$

25. $7x^{-4} = \dfrac{7}{x^4}$

27. $(8x)^{-1} = \dfrac{1}{8x}$

29. $(4x)^{-3} = \dfrac{1}{(4x)^3} = \dfrac{1}{64x^3}$

31. $y^4 \cdot y^{-2} = y^{4+(-2)} = y^2$

33. $\dfrac{1}{x^{-6}} = x^6$

35. $\dfrac{(4t)^0}{t^{-2}} = \dfrac{1}{t^{-2}} = t^2$

37. $\dfrac{1}{(5y)^{-2}} = (5y)^2 = 5^2y^2 = 25y^2$

39. $\dfrac{(5u)^{-4}}{(5u)^0} = (5u)^{-4-0} = (5u)^{-4} = \dfrac{1}{(5u)^4} = \dfrac{1}{625u^4}$

41. $\left(\dfrac{x}{10}\right)^{-1} = \dfrac{10}{x}$

43. $(2x^2)^{-2} = \dfrac{1}{(2x^2)^2} = \dfrac{1}{4x^4}$

45. $-\left(\dfrac{5x}{y^3}\right)^{-3} = -\left(\dfrac{y^3}{5x}\right)^3 = -\dfrac{(y^3)^3}{(5x)^3} = -\dfrac{y^{3 \cdot 3}}{5^3 \cdot x^3} = -\dfrac{y^9}{125x^3}$

47. $\dfrac{6x^3y^{-3}}{12x^{-2}y} = \dfrac{6x^{3-(-2)}y^{-3-1}}{6 \cdot 2} = \dfrac{x^5y^{-4}}{2} = \dfrac{x^5}{2y^4}$

49. $\left(\dfrac{3u^2v^{-1}}{3^3u^{-1}v^3}\right)^{-2} = \left(\dfrac{3u^{2-(-1)}v^{-1-3}}{3^3}\right)^{-2}$

$= \left(\dfrac{u^3v^{-4}}{3^2}\right)^{-2}$

$= \left(\dfrac{3^2}{u^3v^{-4}}\right)^2$

$= \dfrac{3^4}{u^6v^{-8}}$

$= \dfrac{81v^8}{u^6}$

51. $\dfrac{5x^0y}{(6x)^0y^3} = \dfrac{5y}{y^3} = 5y^{1-3} = 5y^{-2} = \dfrac{5}{y^2}$

53. $0.00031 = 3.1 \times 10^{-4}$

55. $3{,}600{,}000 = 3.6 \times 10^6$

57. $9{,}460{,}800{,}000{,}000 = 9.4608 \times 10^{12}$

59. $7.2 \times 10^8 = 7.2 \times 100{,}000{,}000 = 720{,}000{,}000$

61. $1.359 \times 10^{-7} = 0.0000001359$

63. $1.5 \times 10^7 = 15{,}000{,}000$

65. $$(4{,}500{,}000)(2{,}000{,}000{,}000) = (4.5 \times 10^6)(2 \times 10^9) = (4.5)(2) \times 10^{15} = 9 \times 10^{15}$$

67. $$\frac{64{,}000{,}000}{0.00004} = \frac{6.4 \times 10^7}{4.0 \times 10^{-5}} = 1.6 \times 10^{7-(-5)} = 1.6 \times 10^{12}$$

69. $$(7900)(5{,}700{,}000{,}000) = (7.9 \times 10^3)(5.7 \times 10^9) = 45.03 \times 10^{3+9} = 45.03 \times 10^{12} = 4.503 \times 10^{13} \approx 4.50 \times 10^{13}$$

71. $$\frac{(0.0000565)(2{,}850{,}000{,}000{,}000)}{0.00465} = \frac{(5.65 \times 10^{-5})(2.85 \times 10^{12})}{4.65 \times 10^{-3}} = \frac{(5.65)(2.85)}{4.65} \times 10^{10} \approx 3.4629032 \times 10^{10} \approx 3.46 \times 10^{10}$$

73. $$\frac{1.99 \times 10^{30}}{5.98 \times 10^{24}} = \frac{1.99}{5.98} \times 10^6 \approx 0.3327759 \times 10^6 \approx 3.33 \times 10^5$$

75. The value of c is 2.4 since the number has 24 to the left of the zeros.

77. To write a small number in scientific notation, write the number in the form $c \times 10^{-n}$, where $1 \le c < 10$, by moving the decimal point n places to the right.

79. $$\left[\frac{(-5u^3v)^2}{10u^2v}\right]^2 = \left[\frac{(-5)^2 \cdot (u^3)^2 \cdot (v)^2}{10u^2v}\right]^2 = \left[\frac{25u^6v^2}{10u^2v}\right]^2 = \left[\frac{25}{10} \cdot \frac{u^6}{u^2} \cdot \frac{v^2}{v}\right]^2 = \left[\frac{5}{2} \cdot u^{6-2} \cdot v^{2-1}\right]^2 = \left[\frac{5}{2}u^4v\right]^2 = \frac{25}{4}u^8v^2$$

81. $$\frac{x^{2n+4}y^{4n}}{x^5y^{2n+1}} = x^{2n+4-5}y^{4n-(2n+1)} = x^{2n-1}y^{4n-2n-1} = x^{2n-1}y^{2n-1}$$

83. $$(2x^3y^{-1})^{-3}(4xy^{-6}) = (2^{-3}x^{-9}y^3)(4xy^{-6}) = \frac{4x^{-9+1}y^{3+(-6)}}{2^3} = \frac{4x^{-8}y^{-3}}{8} = \frac{1}{2x^8y^3}$$

85. $u^4(6u^{-3}v^0)(7v)^0 = u^4(6u^{-3})(1) = 6u^{4+(-3)} = 6u$

87. $\left[(x^{-4}y^{-6})^{-1}\right]^2 = (x^4y^6)^2 = x^8y^{12}$

89. $$\frac{(2a^{-2}b^4)^3b}{(10a^3b)^2} = \frac{2^3a^{-6}b^{12} \cdot b}{10^2a^6b^2} = \frac{8a^{-6-6}b^{12+1-2}}{100} = \frac{2a^{-12}b^{11}}{25} = \frac{2b^{11}}{25a^{12}}$$

91. $(2 \times 10^9)(3.4 \times 10^{-4}) = (2)(3.4)(10^5) = 6.8 \times 10^5$

93. $(5 \times 10^4)^2 = 5^2 \times 10^8 = 25 \times 10^8 = 2.5 \times 10^9$

95. $$\frac{3.6 \times 10^{12}}{6 \times 10^5} = \frac{3.6}{6} \times 10^{12-5} = 0.6 \times 10^7 = 6.0 \times 10^6$$

97. $70{,}000{,}000{,}000{,}000{,}000{,}000{,}000 = 7 \times 10^{22}$ stars

99. $(43{,}783)(309{,}000{,}000) = (4.3783 \times 10^4)(3.09 \times 10^8)$
$= 13.528947 \times 10^{12}$
$\approx 1.35 \times 10^{13}$
$= \$13{,}500{,}000{,}000{,}000$

101. Scientific notation makes it easier to multiply or divide very large or very small numbers because the properties of exponents make it more efficient.

103. False. $\dfrac{1}{3^{-3}} = \dfrac{1}{\left(\frac{1}{3}\right)^3} = \dfrac{1}{\frac{1}{27}} = 27$, which is greater than 1.

105. $a^m \cdot b^n = ab^{m+n}$ is false because the product rule can be applied only to exponential expressions with the same base.

107. $(a^m)^n = a^{m+n}$ is false because the power-to-power rule applied to this expression raises the base to the product of the exponents.

109. $3x + 4x - x = (3 + 4 - 1)x = 6x$

111. $a^2 + 2ab - b^2 + ab + 4b^2 = a^2 + (2 + 1)ab + (-1 + 4)b^2 = a^2 + 3ab + 3b^2$

113.

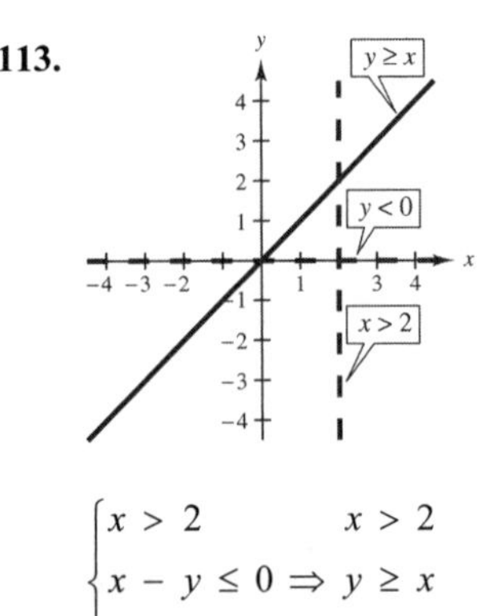

$$\begin{cases} x > 2 \\ x - y \le 0 \\ y < 0 \end{cases} \Rightarrow \begin{matrix} x > 2 \\ y \ge x \\ y < 0 \end{matrix}$$

Section 5.2 Adding and Subtracting Polynomials

1. Standard form: $4y + 16$
Degree: 1
Leading coefficient: 4

3. Standard form: $x^2 + 2x - 6$
Degree: 2
Leading coefficient: 1

5. Standard form: $-42x^3 - 10x^2 + 3x + 5$
Degree: 3
Leading coefficient: -42

7. Standard form: $t^5 - 14t^4 - 20t + 4$
Degree: 5
Leading coefficient: 1

9. Standard form: -4
Degree: 0
Leading coefficient: -4

11. $5 + (2 + 3x) = (5 + 2) + 3x = 3x + 7$

13. $(2x^2 - 3) + (5x^2 + 6) = (2x^2 + 5x^2) + (-3 + 6)$
$= 7x^2 + 3$

15. $(5y + 6) + (4y^2 - 6y - 3) = 4y^2 + (5y - 6y) + (6 - 3) = 4y^2 - y + 3$

17. $(2 - 8y) + (-2y^4 + 3y + 2) = (-2y^4) + (-8y + 3y) + (2 + 2) = -2y^4 - 5y + 4$

19. $(x^3 + 9) + (2x^2 + 5) + (x^3 - 14) = (x^3 + x^3) + 2x^2 + (9 + 5 - 14) = 2x^3 + 2x^2$

21.
$$\begin{array}{r} 5x^2 - 3x + 4 \\ -3x^2 \qquad\quad - 4 \\ \hline 2x^2 - 3x \qquad \end{array}$$

23.
$$\begin{array}{r} 4x^3 - 2x^2 + 8x \qquad \\ 4x^2 + x - 6 \\ \hline 4x^3 + 2x^2 + 9x - 6 \end{array}$$

25.
$$\begin{array}{r} 5p^2 - 4p + 2 \\ -3p^2 + 2p - 7 \\ \hline 2p^2 - 2p - 5 \end{array}$$

27.
$$\begin{array}{l} -3.6b^2 + 2.5b \\ -2.4b^2 - 3.1b + 7.1 \\ 6.6b^2 \\ \hline 0.6b^2 - 0.6b + 7.1 \end{array}$$

29.
$$\begin{aligned} (4 - y^3) - (4 + y^3) &= (4 - y^3) + (-4 - y^3) \\ &= (4 - 4) + (-y^3 - y^3) \\ &= -2y^3 \end{aligned}$$

31.
$$\begin{aligned} (3x^2 - 2x + 1) - (2x^2 + x - 1) &= (3x^2 - 2x + 1) + (-2x^2 - x + 1) \\ &= (3x^2 - 2x^2) + (-2x - x) + (1 + 1) \\ &= x^2 - 3x + 2 \end{aligned}$$

33.
$$\begin{aligned} (6t^3 - 12) - (-t^3 + t - 2) &= (6t^3 - 12) + (t^3 - t + 2) \\ &= (6t^3 + t^3) - t + (-12 + 2) \\ &= 7t^3 - t - 10 \end{aligned}$$

35.
$$\begin{array}{rcr} x^2 - x + 3 & \Rightarrow & x^2 - x + 3 \\ -\quad (x - 2) & \Rightarrow & -x + 2 \\ \hline & & x^2 - 2x + 5 \end{array}$$

37.
$$\begin{array}{rcr} 2x^2 - 4x + 5 & \Rightarrow & 2x^2 - 4x + 5 \\ -(-4x^2 + 5x - 6) & \Rightarrow & -4x^2 - 5x + 6 \\ \hline & & -2x^2 - 9x + 11 \end{array}$$

39.
$$\begin{array}{rcr} -3x^7 \qquad + 6x^4 + 4 & \Rightarrow & -3x^7 \qquad + 6x^4 + 4 \\ -(8x^7 + 10x^5 - 2x^4 - 12) & \Rightarrow & -8x^7 - 10x^5 + 2x^4 + 12 \\ \hline & & -11x^7 - 10x^5 + 8x^4 + 16 \end{array}$$

41.
$$\begin{aligned} -(2x^3 - 3) + (4x^3 - 2x) &= -2x^3 + 3 + 4x^3 - 2x \\ &= (-2x^3 + 4x^3) + (-2x) + (3) \\ &= 2x^3 - 2x + 3 \end{aligned}$$

43.
$$\begin{aligned} (4x^5 - 10x^3 + 6x) - (8x^5 - 3x^3 + 11) + (4x^5 + 5x^3 - x^2) &= (4x^5 - 10x^3 + 6x) + (-8x^5 + 3x^3 - 11) + (4x^5 + 5x^3 - x^2) \\ &= (4x^5 - 8x^5 + 4x^5) + (-10x^3 + 3x^3 + 5x^3) - x^2 + 6x - 11 \\ &= -2x^3 - x^2 + 6x - 11 \end{aligned}$$

45.
$$\begin{aligned} (5n^2 + 6) + [(2n - 3n^2) - (2n^2 + 2n + 6)] &= (5n^2 + 6) + [2n - 3n^2 - 2n^2 - 2n - 6] \\ &= (5n^2 + 6) + [(-3n^2 - 2n^2) + (2n - 2n) - 6] \\ &= (5n^2 + 6) + (-5n^2 + 0 - 6) \\ &= (5n^2 - 5n^2) + (6 - 6) = 0 \end{aligned}$$

47.
$$\begin{aligned} (8x^3 - 4x^2 + 3x) - [(x^3 - 4x^2 + 5) + (x - 5)] &= (8x^3 - 4x^2 + 3x) - [x^3 - 4x^2 + x] \\ &= (8x^3 - 4x^2 + 3x) + (-x^3 + 4x^2 - x) \\ &= (8x^3 - x^3) + (-4x^2 + 4x^2) + (3x - x) \\ &= 7x^3 + 2x \end{aligned}$$

49. $3(4x^2 - 1) + (3x^3 - 7x^2 + 5) = 12x^2 - 3 + 3x^3 - 7x^2 + 5$

$= 3x^3 + 5x^2 + 2$

51. $2(t^2 + 12) - 5(t^2 + 5) + 6(t^2 + 5) = 2t^2 + 24 - 5t^2 - 25 + 6t^2 + 30$

$= (2t^2 - 5t^2 + 6t^2) + (24 - 25 + 30)$

$= 3t^2 + 29$

53. $15v - 3(3v - v^2) + 9(8v + 3) = 15v - 9v + 3v^2 + 72v + 27$

$= (3v^2) + (15v - 9v + 72v) + 27$

$= 3v^2 + 78v + 27$

55. $5s - [6s - (30s + 8)] = 5s - [6s - 30s - 8]$

$= (5s - 6s + 30s) + (8)$

$= 29s + 8$

57. $h(t) = -16t^2 + 40t + 500$

$h(1) = -16(1)^2 + 40(1) + 500 = 524$ feet

$h(5) = -16(5)^2 + 40(5) + 500 = 300$ feet

$h(6) = -16(6)^2 + 40(6) + 500 = 164$ feet

59. $P = (4x + 1) + (2x + 8) + (6x - 8)$

$= (4x + 2x + 6x) + (1 + 8 - 8)$

$= 12x + 1$

When $x = 6$:

$P = 12(6) + 1$

$= 73$ units

61. $P = (3x + 1) + (x) + (x) + (2x + 4) + (2x + 4) + (x) + (x) + (3x + 1)$

$= (3x + x + x + 2x + 2x + x + x + 3x) + (1 + 4 + 4 + 1)$

$= 14x + 10$

When $x = 4$:

$P = 14(4) + 10$

$= 66$ units

63. The like terms are $3x^2$ and x^2, and $2x$ and $-4x$, since they have the same variables with the same exponents.

65. The step in adding polynomials vertically that corresponds to grouping like terms when adding polynomials horizontally is aligning the terms of the polynomials by degrees.

67. $\left(\frac{2}{3}x^3 - 4x + 1\right) + \left(-\frac{3}{5} + 7x - \frac{1}{2}x^3\right) = \left(\frac{2}{3}x^3 - \frac{1}{2}x^3\right) + (-4x + 7x) + \left(1 - \frac{3}{5}\right)$

$= \left(\frac{4}{6}x^3 - \frac{3}{6}x^3\right) + 3x + \left(\frac{5}{5} - \frac{3}{5}\right)$

$= \frac{1}{6}x^3 + 3x + \frac{2}{5}$

69. $\left(\frac{1}{4}y^2 - 5y\right) - \left(12 + 4y - \frac{3}{2}y^2\right) = \left(\frac{1}{4}y^2 - 5y\right) + \left(-12 - 4y + \frac{3}{2}y^2\right)$

$= \left(\frac{1}{4}y^2 + \frac{3}{2}y^2\right) + (-5y - 4y) - 12$

$= \frac{7}{4}y^2 - 9y - 12$

71. $(2x^{2r} - 6x^r - 3) + (3x^{2r} - 2x^r + 6) = (2x^{2r} + 3x^{2r}) + (-6x^r - 2x^r) + (-3 + 6) = 5x^{2r} - 8x^r + 3$

73. Area $= 12(x + 6) - 7x = 12x + 72 - 7x = 5x + 72$

75. Area of region $= \left(6 \cdot \frac{3}{2}x\right) + \left(6 \cdot \frac{9}{2}x\right)$ or $6 \cdot \left[\frac{3}{2}x + \frac{9}{2}x\right] = 9x + 27x$ or $6\left[\frac{12}{2}x\right] = 36x$

77. The free-falling object was dropped.

$-16(0)^2 + 100 = 100$ feet

79. The free-falling object was thrown upward.

$h(0) = -16(0)^2 + 40(0) + 12 = 12$ feet

81. *Verbal Model:* $\boxed{\text{Profit}} = \boxed{\text{Revenue}} - \boxed{\text{Cost}}$

Equation:
$$P = R - C$$
$$P = 14x - (8x + 15{,}000)$$
$$P = 6x - 15{,}000$$
$$P = 6(5000) - 15{,}000$$
$$P = \$15{,}000$$

83. The degree of the term ax^k is k. The degree of a polynomial is the degree of its highest-degree term.

85. (a) A polynomial is a trinomial is sometimes true. A polynomial is a trinomial when it has three terms.

(b) A trinomial is a polynomial is always true. Every trinomial is a polynomial.

87. No, not every trinomial is a second-degree polynomial. For example, $x^3 + 2x + 3$ is a trinomial of third-degree.

89. B has 2 rows and 3 columns so the order is 2×3.

91. A is a square matrix because it has an equal number of rows and columns.

93.
$$\begin{bmatrix} 6 & -3 & \vdots & 0 \\ 2 & 1 & \vdots & 4 \end{bmatrix}$$
$$\tfrac{1}{6}R_1 \quad \begin{bmatrix} 1 & -\tfrac{1}{2} & \vdots & 0 \\ 2 & 1 & \vdots & 4 \end{bmatrix}$$
$$-2R_1 + R_2 \quad \begin{bmatrix} 1 & -\tfrac{1}{2} & \vdots & 0 \\ 0 & 2 & \vdots & 4 \end{bmatrix}$$
$$\tfrac{1}{2}R_2 \quad \begin{bmatrix} 1 & -\tfrac{1}{2} & \vdots & 0 \\ 0 & 1 & \vdots & 2 \end{bmatrix}$$

95. $B = \begin{bmatrix} 6 & -3 & \vdots & 0 \\ 2 & 1 & \vdots & 4 \end{bmatrix} \Rightarrow \begin{array}{l} 6x - 3y = 0 \\ 2x + y = 4 \end{array}$

97. $|A| = (3)(4) - (1)(2) = 12 - 2 = 10$

Section 5.3 Multiplying Polynomials

1. $(3x - 4)(6x) = (3x)(6x) - (4)(6x) = 18x^2 - 24x$

3. $2y(5 - y) = (2y)(5) - (2y)(y) = -2y^2 + 10y$

5. $4x(2x^2 - 3x + 5) = (4x)(2x^2) - (4x)(3x) + (4x)(5)$
$$= 8x^3 - 12x^2 + 20x$$

7. $(-3x)(2 + 5x) = (-3x)(2) + (-3x)(5x)$
$$= -6x - 15x^2$$
$$= -15x^2 - 6x$$

9. $-2m^2(7 - 4m + 2m^2) = -2m^2(7) - 2m^2(-4m) - 2m^2(2m^2) = -14m^2 + 8m^3 - 4m^4 = -4m^4 + 8m^3 - 14m^2$

11. $-x^3(x^4 - 2x^3 + 5x - 6) = -x^3(x^4) - x^3(-2x^3) - x^3(5x) - x^3(-6) = -x^7 + 2x^6 - 5x^4 + 6x^3$

13. $(x + 2)(x + 4) = x^2 + 4x + 2x + 8 = x^2 + 6x + 8$

15. $(x - 3)(x - 3) = x^2 - 3x - 3x + 9 = x^2 - 6x + 9$

17. $(2x - 3)(x + 5) = 2x^2 + 10x - 3x - 15$
$$= 2x^2 + 7x - 15$$

19. $(5x - 2)(2x - 6) = 10x^2 - 30x - 4x + 12$
$$= 10x^2 - 34x + 12$$

21. $(2x^2 - 1)(x + 2) = 2x^3 + 4x^2 - x - 2$

23.
$$\begin{array}{r} 2x + 1 \\ \times\ 7x^2 - 14x + 9 \\ \hline 18x + 9 \\ -28x^2 - 14x \phantom{{}+9} \\ 14x^3 + 7x^2 \phantom{{}-14x+9} \\ \hline 14x^3 - 21x^2 + 4x + 9 \end{array}$$

25.
$$\begin{array}{r} -x^2 + 2x - 1 \\ \times \qquad 2x + 1 \\ \hline -x^2 + 2x - 1 \\ -2x^3 + 4x^2 - 2x \phantom{{}-1} \\ \hline -2x^3 + 3x^2 \phantom{{}+2x} - 1 \end{array}$$

27.
$$\begin{array}{rrrrr} & & t^2 & +\ t & -\ 2 \\ \times & & t^2 & -\ t & +\ 2 \\ \hline & & 2t^2 & +\ 2t & -\ 4 \\ & -t^3 & -\ t^2 & +\ 2t & \\ t^4 & +\ t^3 & -\ 2t^2 & & \\ \hline t^4 & & -\ t^2 & +\ 4t & -\ 4 \end{array}$$

29.

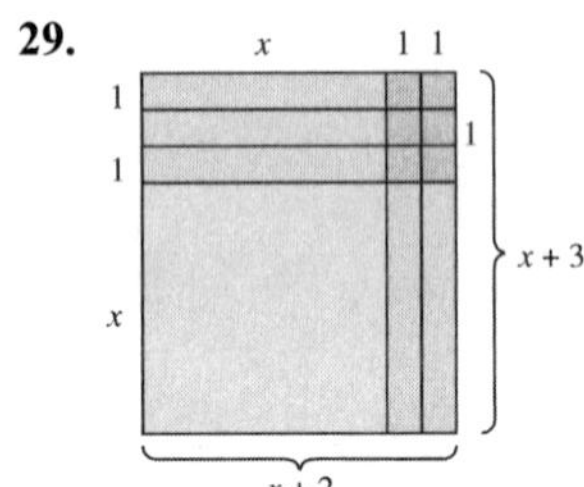

31.

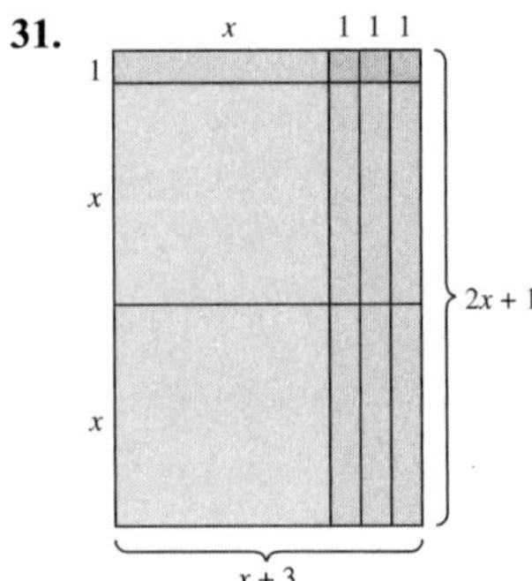

33. $(x - 8)(x + 8) = (x)^2 - (8)^2 = x^2 - 64$

35. $(2 + 7y)(2 - 7y) = (2)^2 - (7y)^2 = 4 - 49y^2$

37. $(6x - 9y)(6x + 9y) = (6x)^2 - (9y)^2 = 36x^2 - 81y^2$

39. $(x + 5)^2 = (x^2) + 2(x)(5) + (5)^2 = x^2 + 10x + 25$

41. $(x - 10)^2 = (x)^2 - 2(x)(10) + 10^2 = x^2 - 20x + 100$

43. $(2x + 5)^2 = (2x)^2 + 2(2x)(5) + (5)^2$
$= 4x^2 + 20x + 25$

45. $(k + 5)^3 = (k + 5)^2(k + 5)$
$= (k^2 + 10k + 25)(k + 5)$
$= k^2(k + 5) + 10k(k + 5) + 25(k + 5)$
$= k^3 + 5k^2 + 10k^2 + 50k + 25k + 125$
$= k^3 + 15k^2 + 75k + 125$

47. $(u + v)^3 = (u + v)(u + v)(u + v)$
$= (u^2 + uv + uv + v^2)(u + v)$
$= (u^2 + 2uv + v^2)(u + v)$

$$\begin{array}{rrrr} & u^2 & +\ 2uv & +\ v^2 \\ & & u & +\ v \\ \hline & u^2v & +\ 2uv^2 & +\ v^3 \\ u^3 +\ 2u^2v & +\ uv^2 & & \\ \hline u^3 +\ 3u^2v & +\ 3uv^2 & +\ v^3 & \end{array}$$

49. (a) *Verbal Model:* [Volume] = [Length] · [Width] · [Height]

Function: $V(n) = n \cdot (n + 2) \cdot (n + 4) = n(n^2 + 6n + 8) = n^3 + 6n^2 + 8n$

(b) $V(3) = 3^3 + 6(3)^2 + 8(3) = 27 + 54 + 24 = 105$ cubic inches

(c) *Verbal Model:* [Area] = [Length] · [Width]

Function: $A(n) = n \cdot (n + 2) = n^2 + 2n$

(d) *Function:* $A(n + 5) = (n + 5)(n + 5 + 2)$
$= (n + 5)(n + 7)$
$= n^2 + 7n + 5n + 35$
$= n^2 + 12n + 35$

$A(n + 5) = (n + 5)^2 + 2(n + 5)$
$= n^2 + 10n + 25 + 2n + 10$
$= n^2 + 12n + 35$

51. (a) *Verbal Model:* [Revenue] = [Number of units sold] · [Price per unit]

Equation: $R = x(175 - 0.02x) = 175x - 0.02x^2 = -0.02x^2 + 175x$

(b) $R = -0.02(3000)^2 + 175(3000) = -180{,}000 + 525{,}000 = 345{,}000$ cents $= \$3450$

53. The binomials are $x - y$ and $x + 2$.

55. To multiply $(x + y)$ and $(x - y)$ without using the FOIL Method, use the special product formula for the sum and difference of two terms.

57. $(x - 1)(x^2 - 4x + 6) = (x - 1)(x^2) + (x - 1)(-4x) + (x - 1)(6) = x^3 - x^2 - 4x^2 + 4x + 6x - 6 = x^3 - 5x^2 + 10x - 6$

59. $$\begin{aligned}(3a + 2)(a^2 + 3a + 1) &= (3a + 2)(a^2) + (3a + 2)(3a) + (3a + 2)(1)\\ &= 3a^3 + 2a^2 + 9a^2 + 6a + 3a + 2\\ &= 3a^3 + 11a^2 + 9a + 2\end{aligned}$$

61. $$\begin{aligned}(2u^2 + 3u - 4)(4u + 5) &= (4u + 5)(2u^2) + (4u + 5)(3u) + (4u + 5)(-4)\\ &= 8u^3 + 10u^2 + 12u^2 + 15u - 16u - 20\\ &= 8u^3 + 22u^2 - u - 20\end{aligned}$$

63. $u^2v(3u^4 - 5u^2 + 6uv^3) = u^2v(3u^4) + u^2v(-5u^2) + u^2v(6uv^3) = 3u^6v - 5u^4v + 6u^3v^4$

65. $(2t - 1)(t + 1) + 1(2t - 5)(t - 1) = 2t^2 + 2t - t - 1 + 2t^2 - 2t - 5t + 5 = 4t^2 - 6t + 4$

67. $$\begin{aligned}[u - (v - 3)][u + (v - 3)] &= (u)^2 - (v - 3)^2\\ &= u^2 - \left[v^2 - 2(v)(3) + (3)^2\right]\\ &= u^2 - (v^2 - 6v + 9)\\ &= u^2 - v^2 + 6v - 9\end{aligned}$$

69. $(6x^m - 5)(2x^{2m} - 3) = 6x^m(2x^{2m} - 3) + (-5)(2x^{2m} - 3) = 12x^{m+2m} - 18x^m - 10x^{2m} + 15 = 12x^{3m} - 10x^{2m} - 18x^m + 15$

71. *Verbal Model:* [Area of shaded region] = [Area of outside rectangle] − [Area of inside rectangle]

Function: $A(x) = 3x(3x + 10) - x(x + 4) = 9x^2 + 30x - x^2 - 4x = 8x^2 + 26x$

73. *Verbal Model:* [Area of shaded region] = [Area of larger triangle] − [Area of smaller triangle]

Function: $A(x) = \frac{1}{2}(2x)(1.6x) = \frac{1}{2}x(0.8x) = 1.6x^2 - 0.4x^2 = 1.2x^2$

75. $A = \text{length} \cdot \text{width} = (x + a)(x + a) = (x + a)^2$

$A = \text{sum of area of pieces} = x^2 + ax + ax + a^2$

Sum of the parts is equal to the whole, so

$(x + a)^2 = x^2 + 2ax + a^2$.

This special product is the square of a binomial.

77. (a) *Verbal Model:*

[Perimeter] = 2 [Length] + 2 [Width]

$P = 2\left[\frac{5}{2}(2x)\right] + 2(2x) = 10x + 4x = 14x$

(b) *Verbal Model:* [Area] = [Length] · [Width]

$A = \frac{5}{2}(2x) \cdot 2x = 10x^2$

79. Interest $= 5000(1 + r)^2$
$$= 5000(1 + r)(1 + r)$$
$$= 5000(1 + 2r + r^2)$$
$$= 5000 + 10{,}000r + 5{,}000r^2$$
$$= 5000r^2 + 10{,}000r + 5000$$

81. (a) $(x - 1)(x + 1) = x^2 - 1$

(b) $(x - 1)(x^2 + x + 1) = x^3 + x^2 + x - x^2 - x - 1 = x^3 - 1$

(c) $(x - 1)(x^3 + x^2 + x + 1) = x^4 + x^3 + x^2 + x - x^3 - x^2 - x - 1 = x^4 - 1$

$(x - 1)(x^4 + x^3 + x^2 + x + 1) = x^5 - 1$

83. When two polynomials are multiplied together, an understanding of the Distributive Property is essential because the Distributive Property is used to multiply each term of the first polynomial by each term of the second polynomial.

85. The degree of the product of two polynomials of degrees m and n is $m + n$.

87. $y = 5 - \frac{1}{2}x$

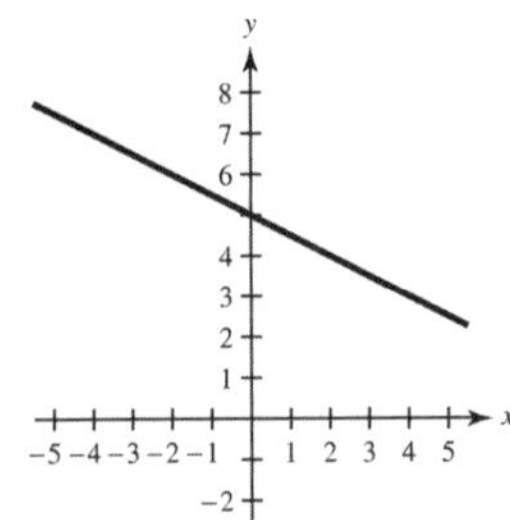

Function

89. $y - 4x + 1 = 0$
$$y = 4x - 1$$

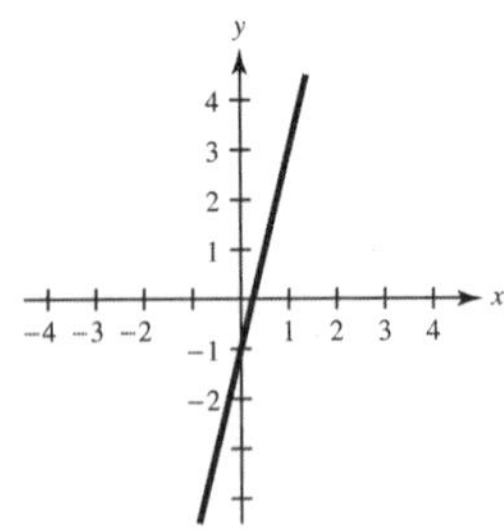

Function

91. To graph the equation $|y| + 2x = 0$, plot a few points and draw the graph. Use the Vertical Line Test to then determine that this equation is not a function.

x	$\lvert y\rvert + 2x$	point
0	0	(0, 0)
−1	2	(−1, 2)
−1	−2	(−1, −2)
−2	4	(−2, 4)
−2	−4	(−2, −4)

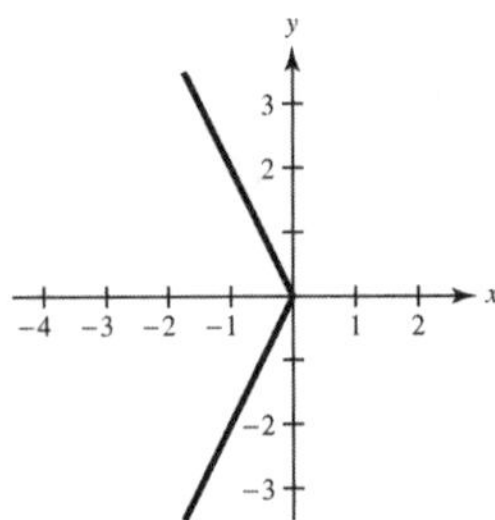

93. $2^{-5} = \dfrac{1}{2^5} = \dfrac{1}{32}$

95. $\dfrac{4^2}{4^{-1}} = 4^{2-(-1)} = 4^3 = 64$

97. $(6^3 + 3^{-6})^0 = 1$

Mid-Chapter Quiz for Chapter 5

1. Standard form: $-2x^4 + 4x^3 - 2x + 3$

Degree: 4

Leading coefficient: -2

2. (a) $0.00000054 = 5.4 \times 10^{-7}$

(b) $664{,}000{,}000 = 6.64 \times 10^8$

3. $(5y^2)(-y^4)(2y^3) = (5)(-1)(2)y^{2+4+3} = -10y^9$

4. $(-6x)(-3x^2)^2 = (-6x)(-3)^2(x^2)^2$

$= (-6x)(9)x^4$

$= (-6)(9)x^{1+4}$

$= -54x^5$

5. $(-5n^2)(-2n^3) = 10n^5$

6. $(3m^3)^2(-2m^4) = (9m^6)(-2m^4) = -18m^{10}$

7. $\dfrac{6x^{-7}}{(-2x^2)^{-3}} = \dfrac{6(-2x^2)^3}{x^7}$

$= \dfrac{6(-2)^3(x^2)^3}{x^7}$

$= \dfrac{6(-8)x^6}{x^7}$

$= -\dfrac{48}{x}$

8. $\left(\dfrac{4y^2}{5x}\right)^{-2} = \left(\dfrac{5x}{4y^2}\right)^2 = \left(\dfrac{5x}{4y^2}\right)\left(\dfrac{5x}{4y^2}\right) = \dfrac{25x^{1+1}}{16y^{2+2}} = \dfrac{25x^2}{16y^4}$

9. $\left(\dfrac{3a^{-2}b^5}{9a^{-4}b^0}\right)^{-2} = \left(\dfrac{9a^{-4}b^0}{3a^{-2}b^5}\right)^2$

$= \left(\dfrac{9a^{-4}b^0}{3a^{-2}b^5}\right)\left(\dfrac{9a^{-4}b^0}{3a^{-2}b^5}\right)$

$= \dfrac{81a^{-4+(-4)}b^{0+0}}{9a^{-2+(-2)}b^{5+5}}$

$= \dfrac{9a^{-8}b^0}{a^{-4}b^{10}}$

$= 9a^{-8-(-4)}b^{0-10}$

$= 9a^{-4}b^{-10}$

$= \dfrac{9}{a^4b^{10}}$

10. $\left(\dfrac{5x^0y^{-7}}{2x^{-2}y^4}\right)^{-3} = \left(\dfrac{2x^{-2}y^4}{5x^0y^{-7}}\right)^3$

$= \left(\dfrac{2x^{-2}y^4}{5x^0y^{-7}}\right)\left(\dfrac{2x^{-2}y^4}{5x^0y^{-7}}\right)\left(\dfrac{2x^{-2}y^4}{5x^0y^{-7}}\right)$

$= \dfrac{8x^{-2+(-2)+(-2)}y^{4+4+4}}{125x^{0+0+0}y^{-7+(-7)+(-7)}}$

$= \dfrac{8x^{-6}y^{12}}{125x^0y^{-21}}$

$= \dfrac{8x^{-6-0}y^{12-(-21)}}{125}$

$= \dfrac{8x^{-6}y^{33}}{125}$

$= \dfrac{8y^{33}}{125x^6}$

11. $(2t^3 + 3t^2 - 2) + (t^3 + 9) = 3t^3 + 3t^2 + 7$

12. $(3 - 7y) + (7y^2 + 2y - 3) = 7y^2 - 5y$

13. $(7x^3 - 3x^2 + 1) - (x^2 - 2x^3) = 7x^3 - 3x^2 + 1 - x^2 + 2x^3 = 9x^3 - 4x^2 + 1$

14. $(5 - u) - 2[3 - (u^2 + 1)] = (5 - u) - 2[3 - u^2 - 1] = (5 - u) - 2[2 - u^2] = 5 - u - 4 + 2u^2 = 2u^2 - u + 1$

15. $7y(4 - 3y) = 28y - 21y^2$

16. $(k + 8)(k + 5) = k^2 + 5k + 8k + 40 = k^2 + 13k + 40$

17. $(4x - y)(6x - 5y) = 24x^2 - 20xy - 6xy + 5y^2 = 24x^2 - 26xy + 5y^2$

18. $2z(z + 5) - 7(z + 5) = 2z^2 + 10z - 7z - 35 = 2z^2 + 3z - 35$

19. $(6r + 5)(6r - 5) = 36r^2 - 25$

20. $(2x - 3)^2 = (2x - 3)(2x - 3) = 4x^2 - 12x + 9$

21. $(x + 1)(x^2 - x + 1) = x^3 - x^2 + x + x^2 - x + 1 = x^3 + 1$

22. $(x^2 - 3x + 2)(x^2 + 5x - 10) = x^2(x^2 + 5x - 10) - 3x(x^2 + 5x - 10) + 2(x^2 + 5x - 10)$

$= x^4 + 5x^3 - 10x^2 - 3x^3 - 15x^2 + 30x + 2x^2 + 10x - 20$

$= x^4 + 2x^3 - 23x^2 + 40x - 20$

23. *Verbal Model:* [Area of shaded region] = [Area of large triangle] − [Area of small triangle]

Equation: $A = \frac{1}{2}(x + 2)^2 - \frac{1}{2}x^2$

$= \frac{1}{2}(x^2 + 4x + 4) - \frac{1}{2}x^2$

$= \frac{1}{2}x^2 + 2x + 2 - \frac{1}{2}x^2$

$= 2x + 2$

24. $h\left(\frac{3}{2}\right) = -16\left(\frac{3}{2}\right)^2 + 32\left(\frac{3}{2}\right) + 100 = -16\left(\frac{9}{4}\right) + 16(3) + 100 = -4(9) + 48 + 100 = -36 + 148 = 112$ feet

$h(3) = -16(3)^2 + 32(3) + 100 = -16(9) + 96 + 100 = -144 + 196 = 52$ feet

25. $P = R - C = 24x - (5x + 2250) = 24x - 5x - 2250 = 19x - 2250$

$P(1500) = 19(1500) - 2250 = 28500 - 2250 = \$26,250$

Section 5.4 Factoring by Grouping and Special Forms

1. $8t^2 + 8t = 8t(t + 1)$

3. $28x^2 + 16x - 8 = 4(7x^2 + 4x - 2)$

5. $7 - 14x = -7(-1 + 2x) = -7(2x - 1)$

7. $2y - 2 - 6y^2 = -2(-y + 1 + 3y^2) = -2(3y^2 - y + 1)$

9. (a) $45l - l^2 = l(45 - l)$

So, the width is $45 - l$.

(b) When $l = 26$:

Width $= 45 - (26) = 19$ feet

11. $2y(y - 4) + 5(y - 4) = (y - 4)(2y + 5)$

13. $5x(3x + 2) - 3(3x + 2) = (3x + 2)(5x - 3)$

15. $2(7a + 6) - 3a^2(7a + 6) = (7a + 6)(2 - 3a^2)$

17. $x^2 + 25x + x + 25 = (x^2 + 25x) + (x + 25)$

$= x(x + 25) + 1(x + 25)$

$= (x + 25)(x + 1)$

19. $y^2 - 6y + 2y - 12 = (y^2 - 6y) + (2y - 12)$

$= y(y - 6) + 2(y - 6)$

$= (y - 6)(y + 2)$

21. $x^3 + 2x^2 + x + 2 = (x^3 + 2x^2) + (x + 2)$

$= x^2(x + 2) + 1(x + 2)$

$= (x + 2)(x^2 + 1)$

23. $3a^3 - 12a^2 - 2a + 8 = (3a^3 - 12a^2) + (-2a + 8)$

$= 3a^2(a - 4) - 2(a - 4)$

$= (a - 4)(3a^2 - 2)$

25. $x^2 - 9 = x^2 - 3^2 = (x + 3)(x - 3)$

27. $1 - a^2 = 1^2 - a^2 = (1 - a)(1 + a)$

29. $4z^2 - y^2 = (2z - y)(2z + y)$

31. $36x^2 - 25y^2 = (6x)^2 - (5y)^2 = (6x - 5y)(6x + 5y)$

33. $(x - 1)^2 - 16 = [(x - 1) - 4][(x - 1) + 4] = (x - 5)(x + 3)$

35. $81 - (z + 5)^2 = 9^2 - (z + 5)^2 = [9 - (z + 5)][9 + (z + 5)] = [9 - z - 5][9 + z + 5] = (4 - z)(14 + z)$

37. $(2x + 5)^2 - (x - 4)^2 = [(2x + 5) - (x - 4)][(2x + 5) + (x - 4)]$
$= [2x + 5 - x + 4][2x + 5 + x - 4] = (x + 9)(3x + 1)$

39. $x^3 - 8 = x^3 - 2^3 = (x - 2)(x^2 + 2x + 4)$

41. $y^3 + 64 = y^3 + 4^3 = (y + 4)(y^2 - 4y + 16)$

43. $8t^3 - 27 = (2t)^3 - 3^3 = (2t - 3)(4t^2 + 6t + 9)$

45. $27u^3 + 1 = (3u)^3 + 1^3$
$= (3u + 1)(9u^2 - 3u + 1)$

47. $x^3 + 27y^3 = x^3 + (3y)^3$
$= (x + 3y)(x^2 - 3xy + 9y^2)$

49. $8 - 50x^2 = 2(4 - 25x^2)$
$= 2[2^2 - (5x)^2]$
$= 2[2 - 5x][2 + 5x]$

51. $8x^3 + 64 = 8(x^3 + 8)$
$= 8(x^3 + 2^3)$
$= 8(x + 2)(x^2 - 2x + 4)$

53. $y^4 - 81 = (y^2)^2 - 9^2$
$= (y^2 - 9)(y^2 + 9)$
$= (y - 3)(y + 3)(y^2 + 9)$

55. $3x^4 - 300x^2 = 3x^2(x^2 - 100)$
$= 3x^2(x - 10)(x + 10)$

57. $6x^6 - 48y^6 = 6(x^6 - 8y^6)$
$= 6[(x^2)^3 - (2y^2)^3]$
$= 6(x^2 - 2y^2)(x^4 + 2x^2y^2 + 4y^4)$

59. The greatest common factor of $2x^3 - 18x$ is $2x$.

61. The completely factored form of $2x^3 - 18x$ is $2x(x + 3)(x - 3)$.

63. $4x^{2n} - 25 = (2x^n)^2 - 5^2$
$= (2x^n - 5)(2x^n + 5)$

65. $81 - 16y^{4n} = 9^2 - (4y^{2n})^2$
$= (9 - 4y^{2n})(9 + 4y^{2n})$
$= [3^2 - (2y^n)^2](9 + 4y^{2n})$
$= (3 - 2y^n)(3 + 2y^n)(9 + 4y^{2n})$

67. $2x^{3r} + 8x^r + 4x^{2r} = 2x^r(x^{2r} + 4 + 2x^r)$
$= 2x^r(x^{2r} + 2x^r + 4)$

69. $3x^3 + 4x^2 - 3x - 4 = (3x^3 + 4x^2) + (-3x - 4)$
$= x^2(3x + 4) - 1(3x + 4)$
$= (x^2 - 1)(3x + 4)$
$= (x - 1)(x + 1)(3x + 4)$

or

$= (3x^3 - 3x) + (4x^2 - 4)$
$= 3x(x^2 - 1) + 4(x^2 - 1)$
$= (x^2 - 1)(3x + 4)$
$= (x - 1)(x + 1)(3x + 4)$

71. $S = 2\pi r^2 + 2\pi rh = 2\pi r(r + 2h)$

73. $R = 800x - 0.25x^2 = x(800 - 0.25x)$
$R = xp$
$p = 800 - 0.25x$

75. $P + Prt = P(1 + rt)$

77. (a) Total volume $= a^3$

Volume of I $= a^2(a - b)$

Volume of II $= ab(a - b)$

Volume of III $= b^2(a - b)$

Volume of IV $= b^3$

(b) Volume of I + Volume of II + Volume of III $= a^2(a - b) + ab(a - b) + b^2(a - b)$

$= (a - b)(a^2 + ab + b^2)$

(c) Subtract Volume of IV from the total volume. This is equal to the sum of the remaining volumes:

$a^3 - b^3 = (a - b)(a^2 + ab + b^2)$

The equation provides the formula for a difference of two cubes.

79. A polynomial is in factored form when the polynomial is written as a product of polynomials.

81. Check a result after factoring by multiplying the factors to see if the product is the original polynomial.

83. The polynomial (a) $x^2 + y^2$ cannot be factored at all because it is prime.

85. $\begin{vmatrix} 3 & 4 \\ 2 & 1 \end{vmatrix} = (3)(1) - (2)(4) = 3 - 8 = -5$

87. $\begin{vmatrix} -1 & 3 & 0 \\ -2 & 0 & 6 \\ 0 & 4 & 2 \end{vmatrix} = -1\begin{vmatrix} 0 & 6 \\ 4 & 2 \end{vmatrix} - 3\begin{vmatrix} -2 & 6 \\ 0 & 2 \end{vmatrix} + 0$

$= (-1)[0 - 24] - (3)[-4 - 0]$

$= 24 + 12 = 36$

89. $(x + 7)(x - 7) = x^2 - 7^2 = x^2 - 49$

91. $(2x - 3)^2 = (2x)^2 - 2(2x)(3) + 3^2 = 4x^2 - 12x + 9$

Section 5.5 Factoring Trinomials

1. $x^2 + 4x + 4 = x^2 + 2(2x) + 2^2 = (x + 2)^2$

3. $25y^2 - 10y + 1 = (5y)^2 - 2(5y) + 1 = (5y - 1)^2$

5. $9b^2 + 12b + 4 = (3b)^2 + 2(3b)(2) + 2^2 = (3b + 2)^2$

7. $36x^2 - 60xy + 25y^2 = (6x)^2 - 2(6x)(5y) + (5y)^2$

$= (6x - 5y)^2$

9. $3m^3 - 18m^2 + 27m = 3m(m^2 - 6m + 9)$

$= 3m(m^2 - 2(3m) + 3^2)$

$= 3m(m - 3)^2$

11. $20v^4 - 60v^3 + 45v^2 = 5v^2(4v^2 - 12v + 9)$

$= 5v^2\left[(2v)^2 - 2(2v)(3) + 3^2\right]$

$= 5v^2(2v - 3)^2$

13. $a^2 + 6a + 8 = (a + 4)(a + 2)$

15. $y^2 - y - 20 = (y + 4)(y - 5)$

17. $x^2 + 6x + 5 = (x + 1)(x + 5)$

19. $x^2 - 5x + 6 = (x - 3)(x - 2)$

21. $y^2 + 7y - 30 = (y + 10)(y - 3)$

23. $t^2 - 6t - 16 = (t - 8)(t + 2)$

25. $x^2 + 10x + 24 = (x + 4)(x + 6)$

27. $z^2 - 6z + 8 = (z - 4)(z - 2)$

29. Area $= x^2 + x - 12 = (x - 3)(x + 4)$

The length is $x + 4$.

31. $x^2 - 15x - 16 = (x + 1)(x - 16)$

33. $x^2 + x - 72 = (x - 8)(x + 9)$

35. $x^2 - 20x + 96 = (x - 12)(x - 8)$

37. $5x^2 + 18x + 9 = (x + 3)(5x + 3)$

39. $2y^2 - 3y - 27 = (y + 3)(2y - 9)$

41. $5z^2 + 2z - 3 = (5z - 3)(z + 1)$

43. $2t^2 - 7t - 4 = (2t + 1)(t - 4)$

45. $6x^2 - 5x - 25 = (3x + 5)(2x - 5)$

47. $3x^2 + 10x + 8 = 3x^2 + 6x + 4x + 8$

$= (3x^2 + 6x) + (4x + 8)$

$= 3x(x + 2) + 4(x + 2)$

$= (3x + 4)(x + 2)$

49. $5x^2 - 12x - 9 = 5x^2 - 15x + 3x - 9$

$= 5x(x - 3) + 3(x - 3)$

$= (5x + 3)(x - 3)$

51. $15x^2 - 11x + 2 = 15x^2 - 6x - 5x + 2$

$= (15x^2 - 6x) + (-5x + 2)$

$= 3x(5x - 2) - 1(5x - 2)$

$= (3x - 1)(5x - 2)$

53. $10y^2 - 7y - 12 = 10y^2 - 15y + 8y - 12$

$= 5y(2y - 3) + 4(2y - 3)$

$= (5y + 4)(2y - 3)$

55. $2 + 5x - 12x^2 = -(12x^2 - 5x - 2)$

$= -(12x^2 + 3x - 8x - 2)$

$= -[3x(4x + 1) - 2(4x + 1)]$

$= -(3x - 2)(4x + 1)$

57. $3x^3 - 3x = 3x(x^2 - 1) = 3x(x^2 - 1^2) = 3x(x + 1)(x - 1)$

59. $3x^3 - 18x^2 + 27x = 3x(x^2 - 6x + 9)$

$= 3x(x - 3)(x - 3)$

$= 3x(x - 3)^2$

61. $10t^3 + 2t^2 - 36t = 2t(5t^2 + t - 18)$

$= 2t(5t - 9)(t + 2)$

63. $x^3 + 2x^2 - 16x - 32 = (x^3 + 2x^2) + (-16x - 32)$

$= x^2(x + 2) - 16(x + 2)$

$= (x + 2)(x^2 - 16)$

$= (x + 2)(x - 4)(x + 4)$

65. *Verbal Model:* [Area of shaded region] = [Area of rectangle] − [Area of squares]

Equation: $\text{Area} = (8 \cdot 18) - 4 \cdot x^2$

$= 144 - 4x^2$

$= 4(36 - x^2)$

$= 4(6 + x)(6 - x)$

67. When factoring $x^2 + 5x + 6$, you know that the signs of the factors of 6 are positive because 6 is positive, so the factors must have like signs and match the sign of 5.

69. To factor $x^2 + 6x + 9$, use the method for factoring a perfect square trinomial.

71. $a^2 + 2ab + b^2 = (a + b)^2$

73. $x^2 + bx + 81 = x^2 + bx + 9^2$

(a) $b = 18$

$x^2 + 18x + 9^2 = x^2 + 2(9x) + 9^2 = (x + 9)^2$

or

(b) $b = -18$

$x^2 - 18x + 9^2 = x^2 - 2(9x) + 9^2 = (x - 9)^2$

75. $4x^2 + bx + 9 = (2x)^2 + bx + 3^2$

(a) $b = 12$

$(2x)^2 + 12x + 3^2 = (2x)^2 + 2(2x)(3) + 3^2$

$= (2x + 3)^2$

or

(b) $b = -12$

$(2x)^2 - 12x + 3^2 = (2x)^2 - 2(2x)(3) + 3^2$

$= (2x - 3)^2$

77. $c = 16$

$x^2 + 8x + c = x^2 + 2(4x) + c$

$= x^2 + 2(4x) + 4^2$

$= (x + 4)^2$

79. $c = 9$

$$y^2 - 6y + c = y^2 - 2(3y) + c$$
$$= y^2 - 2(3y) + 3^2 = (y - 3)^2$$

81. $x^2 + bx + 8$

$b = 6 \qquad x^2 + 6x + 8 = (x + 4)(x + 2)$

$b = -6 \qquad x^2 - 6x + 8 = (x - 4)(x - 2)$

$b = 9 \qquad x^2 + 9x + 8 = (x + 8)(x + 1)$

$b = -9 \qquad x^2 - 9x + 8 = (x - 8)(x - 1)$

83. $b = 20: \quad x^2 + 20x - 21 = (x + 21)(x - 1)$

$b = -20: \quad x^2 - 20x - 21 = (x - 21)(x + 1)$

$b = 4: \quad x^2 + 4x - 21 = (x + 7)(x - 3)$

$b = -4: \quad x^2 - 4x - 21 = (x - 7)(x + 3)$

85. $4w^2 - 3w + 8$ is prime.

87. $60y^3 + 35y^2 - 50y = 5y(12y^2 + 7y - 10)$

$$= 5y(3y - 2)(4y + 5)$$

89. $54x^3 - 2 = 2(27x^3 - 1) = 2(3x - 1)(9x^2 + 3x + 1)$

91. $49 - (r - 2)^2 = -1\left[-49 + (r - 2)^2\right]$

$$= -1\left[(r - 2)^2 - 49\right]$$
$$= -1\left[(r - 2)^2 - 7^2\right]$$
$$= -\left[(r - 2) + 7\right]\left[(r - 2) - 7\right]$$
$$= -(r + 5)(r - 9)$$

93. $x^8 - 1 = (x^4)^2 - 1^2 = (x^4 - 1)(x^4 + 1)$

$$= \left[(x^2)^2 - 1^2\right](x^4 + 1)$$
$$= (x^2 - 1)(x^2 + 1)(x^4 + 1)$$
$$= (x - 1)(x + 1)(x^2 + 1)(x^4 + 1)$$

95. (a) $8n^3 + 24n^2 + 16n = 8n(n^2 + 3n + 2)$

$$= 8n(n + 1)(n + 2)$$

(b) Factor 8 and then distribute a factor of 2 into each binomial factor.

$$8n(n + 1)(n + 2) = (2 \cdot 2 \cdot 2)n(n + 1)(n + 2)$$
$$= 2n(2n + 2)(2n + 4)$$

(c) If $n = 10, \qquad 2n = 2(10) = 20$

$$2n + 2 = 2(10) + 2 = 22$$
$$2n + 4 = 2(10) + 4 = 24$$

97. To factor $x^2 - 5x + 6$ begin by finding the factors of 6 whose sum is -5. They are -2 and -3. The factorization is $(x - 2)(x - 3)$.

99. An example of a prime trinomial is $x^2 + x + 1$.

101. No, $x(x + 2) - 2(x + 2)$ is not in factored form. It is not yet a product.

$$x(x + 2) - 2(x + 2) = (x + 2)(x - 2)$$

103. For each pair of factors of 12, the signs must be the same to yield a positive product. So, it would be better to try 3, 4 and $-3, -4$ instead of -3, 4 and 3, -4.

105. *Verbal Model:* [Compared number] = [Percent] · [Base number]

Labels: Compared number $= a$

Percent $= p$

Base number $= b$

Equation: $a = p \cdot b$

$a = 1.25 \cdot 340$

$a = 425$

107. *Verbal Model:* [Compared number] = [Percent] · [Base number]

Labels: Compared number $= a$

Percent $= p$

Base number $= b$

Equation:

$$a = p \cdot b$$
$$725 = p \cdot 2000$$
$$\frac{725}{2000} = p$$
$$36.25\% = p$$

109. *Verbal Model:* $\boxed{\text{Amount of solution 1}} + \boxed{\text{Amount of solution 2}} = \boxed{\text{Amount of final solution}}$

Labels:
Percent of solution 1 $= 20\%$
Liters of solution 1 $= x$
Percent of solution 2 $= 60\%$
Liters of solution 2 $= 10 - x$
Percent of final solution $= 30\%$
Liters of final solution $= 10$ L

Equation:
$$0.20x + 0.60(10 - x) = 0.30(10)$$
$$0.20x + 6 - 0.60x = 3$$
$$-0.40x = -3$$
$$x = 7.5 \text{ L}$$
$$10 - x = 2.5 \text{ L}$$

111. *Verbal Model:* $\boxed{\text{Amount of solution 1}} + \boxed{\text{Amount of solution 2}} = \boxed{\text{Amount of final solution}}$

Labels:
Percent of solution 1 $= 60\%$
Gallons of solution 1 $= x$
Percent of solution 2 $= 90\%$
Gallons of solution 2 $= 120 - x$
Percent of final solution $= 85\%$
Gallons of final solution $= 120$ gal

Equation:
$$0.60x + 0.90(120 - x) = 0.85(120)$$
$$0.60x + 108 - 0.90x = 102$$
$$-0.30x = -6$$
$$x = 20 \text{ gal}$$
$$120 - x = 100 \text{ gal}$$

Section 5.6 Solving Polynomial Equations by Factoring

1. No, $x(x - 1) = 2$ *is not* in the correct form to apply the Zero-Factor Property.

3. No, $(x + 2) + (x - 1) = 0$ *is not* in the correct form to apply the Zero-Factor Property.

5. Yes, $(x - 1)(x + 2) = 0$ *is* in the correct form to apply the Zero-Factor Property.

7. $x(x - 4) = 0$

$x = 0$

$x - 4 = 0$

$x = 4$

9. $(y - 3)(y + 10) = 0$

$y - 3 = 0 \qquad y + 10 = 0$

$y = 3 \qquad y = -10$

11. $25(a + 4)(a - 2) = 0$

$a + 4 = 0 \qquad a - 2 = 0$

$a = -4 \qquad a = 2$

13. $(2t + 5)(3t + 1) = 0$

$2t + 5 = 0 \qquad 3t + 1 = 0$

$t = -\frac{5}{2} \qquad t = -\frac{1}{3}$

15. $(x - 3)(2x + 1)(x + 4) = 0$

$x - 3 = 0 \qquad 2x + 1 = 0 \qquad x + 4 = 0$

$x = 3 \qquad x = -\frac{1}{2} \qquad x = -4$

17. $x^2 + 2x = 0$
$x(x + 2) = 0$
$x = 0 \qquad x + 2 = 0$
$x = -2$

19. $x^2 - 8x = 0$
$x(x - 8) = 0$
$x = 0 \qquad x - 8 = 0$
$x = 8$

21. $x^2 - 25 = 0$
$(x + 5)(x - 5) = 0$
$x + 5 = 0 \qquad x - 5 = 0$
$x = -5 \qquad x = 5$

23. $x^2 - 3x - 10 = 0$
$(x - 5)(x + 2) = 0$
$x - 5 = 0 \qquad x + 2 = 0$
$x = 5 \qquad x = -2$

25. $x^2 - 10x + 24 = 0$
$(x - 6)(x - 4) = 0$
$x - 6 = 0 \qquad x - 4 = 0$
$x = 6 \qquad x = 4$

27. $4x^2 + 15x - 25 = 0$
$(4x - 5)(x + 5) = 0$
$4x - 5 = 0 \qquad x + 5 = 0$
$4x = 5 \qquad x = -5$
$x = \frac{5}{4}$

29. $7 + 13x - 2x^2 = 0$
$(7 - x)(1 + 2x) = 0$
$7 - x = 0 \qquad 1 + 2x = 0$
$7 = x \qquad -\frac{1}{2} = x$

31. $3y^2 - 2 = -y$
$3y^2 + y - 2 = 0$
$(3y - 2)(y + 1) = 0$
$3y - 2 = 0 \qquad y + 1 = 0$
$y = \frac{2}{3} \qquad y = -1$

33. $-13x + 36 = -x^2$
$x^2 - 13x + 36 = 0$
$(x - 9)(x - 4) = 0$
$x - 9 = 0 \qquad x - 4 = 0$
$x = 9 \qquad x = 4$

35. $m^2 - 8m + 18 = 2$
$m^2 - 8m + 16 = 0$
$(m - 4)^2 = 0$
$m - 4 = 0$
$m = 4$

37. $x^2 + 16x + 57 = -7$
$x^2 + 16x + 64 = 0$
$(x + 8)^2 = 0$
$x + 8 = 0$
$x = -8$

39. $4z^2 - 12z + 15 = 6$
$4z^2 - 12z + 9 = 0$
$(2z - 3)^2 = 0$
$2z - 3 = 0$
$z = \frac{3}{2}$

41. $x^3 - 19x^2 + 84x = 0$
$x(x^2 - 19x + 84) = 0$
$x(x - 12)(x - 7) = 0$
$x = 0 \qquad x - 12 = 0 \qquad x - 7 = 0$
$x = 0 \qquad x = 12 \qquad x = 7$

43. $6t^3 = t^2 + t$
$6t^3 - t^2 - t = 0$
$t(6t^2 - t - 1) = 0$
$t(3t + 1)(2t - 1) = 0$
$t = 0 \qquad 3t + 1 = 0 \qquad 2t - 1 = 0$
$t = 0 \qquad t = -\frac{1}{3} \qquad t = \frac{1}{2}$

45. $z^2(z + 2) - 4(z + 2) = 0$
$(z + 2)(z^2 - 4) = 0$
$(z + 2)(z - 2)(z + 2) = 0$
$z + 2 = 0 \qquad z - 2 = 0 \qquad z + 2 = 0$
$z = -2 \qquad z = 2 \qquad z = -2$

49. $c^3 - 3c^2 - 9c + 27 = 0$

$c^2(c - 3) - 9(c - 3) = 0$

$(c - 3)(c^2 - 9) = 0$

$(c - 3)(c - 3)(c + 3) = 0$

$c - 3 = 0 \qquad c - 3 = 0 \qquad c + 3 = 0$

$c = 3 \qquad c = 3 \qquad c = -3$

51. $x^4 - 3x^3 - x^2 + 3x = 0$

$x^3(x - 3) - x(x - 3) = 0$

$(x - 3)(x^3 - x) = 0$

$(x - 3)x(x^2 - 1) = 0$

$(x - 3)x(x - 1)(x + 1) = 0$

$x - 3 = 0 \qquad x = 0 \qquad x - 1 = 0 \qquad x + 1 = 0$

$x = 3 \qquad x = 1 \qquad x = -1$

53. $8x^4 + 12x^3 - 32x^2 - 48x = 0$

$4x^3(2x + 3) - 16x(2x + 3) = 0$

$(2x + 3)(4x^3 - 16x) = 0$

$(2x + 3)\,4x(x^2 - 4) = 0$

$(2x + 3)(4x)(x - 2)(x + 2) = 0$

$2x + 3 = 0 \qquad 4x = 0 \qquad x - 2 = 0 \qquad x + 2 = 0$

$x = -\frac{3}{2} \qquad x = 0 \qquad x = 2 \qquad x = -2$

55. *Verbal Model:* [Length] · [Width] = [Area]

Labels: Length $= w + 7$

Width $= w$

Equation: $(w + 7) \cdot w = 540$

$w^2 + 7w - 540 = 0$

$(w + 27)(w - 20) = 0$

$w + 27 = 0 \qquad w - 20 = 0$

$w = -27 \qquad w = 20$ feet

Reject $\qquad w + 7 = 27$ feet

20 feet × 27 feet

57. *Verbal Model:* [$\frac{1}{2}$] · [Base] · [Height] = [Area]

Labels: Base $= x$

Height $= \frac{3}{2}x$

Equation: $\frac{1}{2} \cdot x \cdot \frac{3}{2}x = 27$

$\frac{3}{4}x^2 - 27 = 0$

$3x^2 - 108 = 0$

$3(x^2 - 36) = 0$

$3(x + 6)(x - 6) = 0$

$x + 6 = 0 \qquad x - 6 = 0$

$x = -6 \qquad x = 6$ inches

Reject $\qquad \frac{3}{2}x = 9$ inches

The base is 6 inches. The height is 9 inches.

59. $h = -16t^2 + 400$

$0 = -16t^2 + 400$

$0 = -16(t^2 - 25)$

$0 = -16(t + 5)(t - 5)$

$t + 5 = 0 \qquad t - 5 = 0$

$t = -5 \qquad t = 5$ seconds

Reject

The tool reaches the ground after 5 seconds.

61. $h = -16t^2 + 30t$

$0 = -16t\left(t - \frac{30}{16}\right)$

$-16t = 0 \qquad t - \frac{30}{16} = 0$

$t = 0 \qquad t = \frac{30}{16}$

$t \approx 1.9$

About 1.9 seconds

63. Yes. It has the form $ax^2 + bx + c = 0$, with $a \neq 0$.

65. The factors $2x$ and $x + 9$ are set equal to zero to form two linear equations.

67. $8x^2 = 5x$

$8x^2 - 5x = 0$

$x(8x - 5) = 0$

$x = 0 \qquad 8x - 5 = 0$

$8x = 5$

$x = \frac{5}{8}$

69.
$$x(x-5)=36$$
$$x^2-5x=36$$
$$x^2-5x-36=0$$
$$(x-9)(x+4)=0$$
$$x-9=0 \qquad x+4=0$$
$$x=9 \qquad x=-4$$

71. $x(x+2)-10(x+2)=0$
$$(x+2)(x-10)=0$$
$$x+2=0 \qquad x-10=0$$
$$x=-2 \qquad x=10$$

73.
$$(x-4)(x+5)=10$$
$$x^2+x-20-10=0$$
$$x^2+x-30=0$$
$$(x+6)(x-5)=0$$
$$x+6=0 \qquad x-5=0$$
$$x=-6 \qquad x=5$$

75.
$$81-(x+4)^2=0$$
$$[9-(x+4)][9+(x+4)]=0$$
$$(5-x)(13+x)=0$$
$$5-x=0 \qquad 13+x=0$$
$$-x=-5 \qquad x=-13$$
$$x=5$$

77.
$$(t-2)^2=16$$
$$(t-2)^2-16=0$$
$$(t-2+4)(t-2-4)=0$$
$$(t+2)(t-6)=0$$
$$t+2=0 \qquad t-6=0$$
$$t=-2 \qquad t=6$$

79. From the graph, the x-intercepts are $(-3, 0)$ and $(3, 0)$. The solutions of the equation $0 = x^2 - 9$ are 3 and -3.

$$0=(x-3)(x+3)$$
$$0=x-3 \qquad 0=x+3$$
$$3=x \qquad -3=x$$

81. From the graph, the x-intercepts are $(0, 0)$ and $(3, 0)$.

The solutions of the equation $0 = x^3 - 6x^2 + 9x$ are 0 and 3.

$$x^3-6x^2+9x=0$$
$$x(x^2-6x+9)=0$$
$$x(x-3)(x-3)=0$$
$$x=0 \qquad x-3=0 \qquad x-3=0$$
$$x=3 \qquad x=3$$

83. $x=-2,\ x=6$
$$[x-(-2)][x-6]=0$$
$$(x+2)(x-6)=0$$
$$x^2-4x-12=0$$

85. *Verbal Model:* [Number] + [Its square] = [240]

Labels: Number $= x$

Its square $= x^2$

Equation:
$$x+x^2=240$$
$$x^2+x-240=0$$
$$(x+16)(x-15)=0$$
$$x+16=0 \qquad x-15=0$$
$$x=-16 \qquad x=15$$

Reject

87. $h=-16t^2+80$
$$16=-16t^2+80$$
$$0=-16t^2+64$$
$$0=-16(t^2-4)$$
$$0=-16(t+2)(t-2)$$
$$t+2=0 \qquad t-2=0$$
$$t=-2 \qquad t=2$$

Reject

The object reaches the balcony after 2 seconds.

89. *Verbal Model:* [Revenue] = [Cost]

Equation:
$$R=C$$
$$140x-x^2=2000+50x$$
$$0=x^2-90x+2000$$
$$0=(x-40)(x-50)$$
$$x-40=0 \qquad x-50=0$$
$$x=40 \qquad x=50$$

Units Units

40 or 50 systems can be produced and sold.

91. The maximum number of solutions of an nth degree polynomial equation is n. The third-degree equation $(x + 1)^3 = 0$ has only one solution, $x = -1$.

93. When a quadratic equation has a repeated solution, the graph of the equation has one x-intercept, which is the vertex of the graph.

95. A solution to a polynomial expression is the value of x when y is zero. If a polynomial is not factorable, the equation can still have real number solutions for x when y is zero.

97. Unit price $= \dfrac{\$0.75}{12} = \$0.0625$ per ounce

99. Unit price $= \dfrac{\$2.13}{30} = \$0.071$ per ounce

101. The domain of $f(x) = \dfrac{x + 3}{x + 1}$ is all real numbers such that $x \neq -1$ because $x + 1 \neq 0$ means $x \neq -1$.

103. The domain of $g(x) = \sqrt{3 - x}$ is all real numbers such that $x \leq 3$ because $3 - x \geq 0$ means $-x \geq -3$ and $x \leq 3$.

Review Exercises for Chapter 5

1. $x^4 \cdot x^5 = x^{4+5} = x^9$

3. $(u^2)^3 = u^{2\cdot 3} = u^6$

5. $(-2z)^3 = (-2)^3 z^3 = -8z^3$

7. $-(u^2v)^2(-4u^3v) = -(u^4v^2)(-4u^3v)$
$= 4u^{4+3}v^{2+1}$
$= 4u^7v^3$

9. $\dfrac{12z^5}{6z^2} = \left(\dfrac{12}{6}\right) \cdot z^{5-2} = 2z^3$

11. $\dfrac{25g^4d^2}{80g^2d^2} = \dfrac{25}{80} \cdot \dfrac{g^4}{g^2} \cdot \dfrac{d^2}{d^2} = \dfrac{5}{16} \cdot g^{4-2}d^{2-2} = \dfrac{5}{16}g^2$

13. $\left(\dfrac{72x^4}{6x^2}\right)^2 = (12x^{4-2})^2 = (12x^2)^2 = 144x^4$

15. $(2^3 \cdot 3^2)^{-1} = (8 \cdot 9)^{-1} = 72^{-1} = \frac{1}{72}$

17. $\left(\dfrac{3}{4}\right)^{-3} = \left(\dfrac{4}{3}\right)^3 = \dfrac{4^3}{3^3} = \frac{64}{27}$

19. $(6y^4)(2y^{-3}) = 12y^{4+(-3)} = 12y^1 = 12y$

21. $\dfrac{4x^{-2}}{2x} = 2x^{-2-1} = 2x^{-3} = \dfrac{2}{x^3}$

23. $(x^3y^{-4})^0 = 1$

25. $\dfrac{7a^6b^{-2}}{14a^{-1}b^4} = \dfrac{7}{14} \cdot \dfrac{a^6}{a^{-1}} \cdot \dfrac{b^{-2}}{b^4}$
$= \dfrac{1}{2} \cdot a^{6-(-1)} \cdot b^{-2-4}$
$= \dfrac{1}{2}a^7b^{-6}$
$= \dfrac{a^7}{2b^6}$

27. $\left(\dfrac{3x^{-1}y^2}{12x^5y^{-3}}\right)^{-1} = \dfrac{12x^5y^{-3}}{3x^{-1}y^2}$
$= 4x^{5-(-1)}y^{(-3)-2}$
$= 4x^6y^{-5}$
$= \dfrac{4x^6}{y^5}$

29. $u^3(5u^0v^{-1})(9u)^2 = u^3(5v^{-1})(81u^2)$
$= (5)(81)u^{3+2}v^{-1}$
$= \dfrac{405u^5}{v}$

31. $0.0000319 = 3.19 \times 10^{-5}$

33. $17{,}350{,}000 = 1.735 \times 10^7$

35. $1.95 \times 10^6 = 1{,}950{,}000$

37. $2.05 \times 10^{-5} = 0.0000205$

39. $(6 \times 10^3)^2 = 6^2 \times 10^6 = 36 \times 10^6 = 36{,}000{,}000$

41. $\dfrac{3.5 \times 10^7}{7 \times 10^4} = \dfrac{3.5}{7} \times 10^{7-4}$
$= 0.5 \times 10^3$
$= 5 \times 10^2$
$= 500$

43. Standard form: $-x^4 + 6x^3 + 5x^2 - 4x$

Leading coefficient: –1

Degree: 4

45. $x^4 + 3x^5 - 4 - 6x$

Standard form: $3x^5 + x^4 - 6x - 4$

Degree: 5

Leading coefficient: 3

47. $(10x + 8) + (x^2 + 3x) = x^2 + (10x + 3x) + 8$
$= x^2 + 13x + 8$

49. $(5x^3 - 6x + 11) + (5 + 6x - x^2 - 8x^3) = (5x^3 - 8x^3) - x^2 + (-6x + 6x) + (11 + 5) = -3x^3 - x^2 + 16$

51. $(3y - 4) - (2y^2 + 1) = (3y - 4) + (-2y^2 - 1) = -2y^2 + 3y + (-4 - 1) = -2y^2 + 3y - 5$

53. $(-x^3 - 3x) - 4(2x^3 - 3x + 1) = -x^3 - 3x - 8x^3 + 12x - 4 = (-x^3 - 8x^3) + (-3x + 12x) + (-4) = -9x^3 + 9x - 4$

55. $3y^2 - [2y + 3(y^2 + 5)] = 3y^2 - [2y + 3y^2 + 15] = 3y^2 - 2y - 3y^2 - 15 = (3y^2 - 3y^2) - 2y - 15 = -2y - 15$

57. $(3x^5 + 4x^2 - 8x + 12) - (2x^5 + x) + (3x^2 - 4x^3 - 9) = (3x^5 - 2x^5) - 4x^3 + (4x^2 + 3x^2) + (-8x - x) + (12 - 9)$
$= x^5 - 4x^3 + 7x^2 - 9x + 3$

59.
$$\begin{array}{r} 3x^2 + 5x \phantom{{}+6} \\ \underline{-4x^2 - x + 6} \\ -x^2 + 4x + 6 \end{array}$$

61.
$$\begin{array}{r} 3t - 5 \\ \underline{-(t^2 - t - 5)} \end{array} \Rightarrow \begin{array}{r} 3t - 5 \\ \underline{-t^2 + t + 5} \\ -t^2 + 4t \phantom{{}+5} \end{array}$$

63. $P = (6x) + (3x + 5) + (3x) + (x) + (3x) + (2x + 5)$
$= (6x + 3x + 3x + x + 3x + 2x) + (5 + 5)$
$= 18 + 10$

When $x = \frac{2}{3}$: $P = 18\left(\frac{2}{3}\right) + 10 = 22$ units

65. *Verbal Model:* Profit = Revenue – Cost

Equation: $P(x) = 35x - (16x + 3000)$
$= 35x - 16x - 3000$
$= 19x - 3000$

$P(1200) = 19(1200) - 3000$
$= 22{,}800 - 3000$
$= \$19{,}800$

67. $-2x^3(x + 4) = -2x^4 - 8x^3$

69. $3x(2x^2 - 5x + 3) = 6x^3 - 15x^2 + 9x$

71. $(x - 2)(x + 7) = x^2 + 7x - 2x - 14$
$= x^2 + 5x - 14$

73. $(5x + 3)(3x - 4) = 15x^2 - 20x + 9x - 12$
$= 15x^2 - 11x - 12$

75. $(4x^2 + 3)(6x^2 + 1) = 24x^4 + 4x^2 + 18x^2 + 3$
$= 24x^4 + 22x^2 + 3$

77. $(2x^2 - 3x + 2)(2x + 3) = 2x^2(2x + 3) - 3x(2x + 3) + 2(2x + 3)$
$= 4x^3 + 6x^2 - 6x^2 - 9x + 4x + 6$
$= 4x^3 + (6x^2 - 6x^2) + (-9x + 4x) + 6$
$= 4x^3 - 5x + 6$

79. $2u(u - 7) - (u + 1)(u - 7) = 2u(u - 7) - u(u - 7) - 1(u - 7)$
$= 2u^2 - 14u - u^2 + 7u - u + 7$
$= (2u^2 - u^2) + (-14u + 7u - u) + 7$
$= u^2 - 8u + 7$

81. $(4x - 7)^2 = (4x)^2 - 2(4x)(7) + (-7)^2$
$= 16x^2 - 56x + 49$

83. $(6v + 9)(6v - 9) = (6v)^2 - 9^2$
$= 36v^2 - 81$

85. $[(u - 3) + v][(u - 3) - v] = (u - 3)^2 - v^2$
$= u^2 - 2(u)(3) + (-3)^2 - v^2$
$= u^2 - 6u + 9 - v^2$

87. *Verbal Model:* [Area of shaded region] = [Area of larger rectangle] − [Area of smaller rectangle]

Labels: Width of larger rectangle $= 2x$
Length of larger rectangle $= 2x + 5$
Width of smaller rectangle $= 2x - 3$
Length of smaller rectangle $= 2x + 1$

Equation: Area $= 2x(2x + 5) - (2x + 1)(2x - 3)$
$= 4x^2 + 10x - (4x^2 - 6x + 2x - 3)$
$= 4x^2 + 10x - 4x^2 + 4x + 3$
$= 14x + 3$

89. Interest $= 1000(1 + 0.06)^2 = \$1123.60$

91. $24x^2 - 18 = 6(4x^2 - 3)$

93. $-3b^2 + b = -b(3b - 1)$

95. $6x^2 + 15x^3 - 3x = 3x(2x + 5x^2 - 1)$

97. $28(x + 5) - 70(x + 5) = (28 - 70)(x + 5)$
$= -42(x + 5)$

99. $v^3 - 2v^2 - v + 2 = v^2(v - 2) - 1(v - 2)$
$= (v - 2)(v^2 - 1)$
$= (v - 2)(v - 1)(v + 1)$

101. $t^3 + 3t^2 + 3t + 9 = t^2(t + 3) + 3(t + 3)$
$= (t^2 + 3)(t + 3)$

103. $x^2 - 36 = x^2 - 6^2 = (x - 6)(x + 6)$

105. $(u + 6)^2 - 81 = (u + 6 - 9)(u + 6 + 9)$
$= (u - 3)(u + 15)$

107. $u^3 - 1 = (u - 1)(u^2 + u + 1)$

109. $8x^3 + 27 = (2x)^3 + (3)^3$
$= (2x + 3)(4x^2 - 6x + 9)$

111. $x^3 - x = x(x^2 - 1)$
$= x(x - 1)(x + 1)$

113. $24 + 3u^3 = 3u^3 + 24$
$= 3(u^3 + 8)$
$= 3(u^3 + 2^3)$
$= 3(u + 2)(u^2 - 2u + 4)$

115. $x^2 - 18x + 81 = x^2 - 2(9)x + 9^2$
$= (x - 9)^2$

117. $4s^2 + 40st + 100t^2 = (2s)^2 + 2(2s)(10) + (10t)^2$
$= (2s + 10t)^2$

119. $x^2 + 2x - 35 = (x + 7)(x - 5)$

121. $2x^2 - 7x + 6 = (2x - 3)(x - 2)$

123. $18x^2 + 27x + 10 = (3x + 2)(6x + 5)$

125. $4x^2 - 3x - 1 = 4x^2 - 4x + x - 1$
$= 4x(x - 1) + 1(x - 1)$
$= (4x + 1)(x - 1)$

127. $5x^2 - 12x + 7 = 5x^2 - 5x - 7x + 7$
$= 5x(x - 1) - 7(x - 1)$
$= (5x - 7)(x - 1)$

129. $7s^2 + 10s - 8 = 7s^2 + 14s - 4s - 8$
$= 7s(s + 2) - 4(s + 2)$
$= (7s - 4)(s + 2)$

131. $4a - 64a^3 = 4a(1 - 16a^2)$
$= 4a(1 - 4a)(1 + 4a)$

133. $z^3 + z^2 + 3z + 3 = z^2(z + 1) + 3(z + 1)$
$= (z^2 + 3)(z + 1)$

135. $\frac{1}{4}x^2 + xy + y^2 = \left(\frac{1}{2}x\right)^2 + 2\left(\frac{1}{2}\right)xy + y^2$
$= \left(\frac{1}{2}x + y\right)^2$

137. $x^2 - 10x + 25 - y^2 = (x - 5)^2 - y^2$
$= [(x - 5) - y][(x - 5) + y]$
$= (x - 5 - y)(x - 5 + y)$
$= (x - y - 5)(x + y - 5)$

139. $4x(x - 2) = 0$
$4x = 0 \qquad x - 2 = 0$
$x = 0 \qquad x = 2$

141. $(2x + 1)(x - 3) = 0$
$2x + 1 = 0 \qquad x - 3 = 0$
$x = -\frac{1}{2} \qquad x = 3$

143. $(x + 10)(4x - 1)(5x + 9) = 0$
$x + 10 = 0 \qquad 4x - 1 = 0 \qquad 5x + 9 = 0$
$x = -10 \qquad x = \frac{1}{4} \qquad x = -\frac{9}{5}$

145. $3s^2 - 2s - 8 = 0$
$(3s + 4)(s - 2) = 0$
$3s + 4 = 0 \qquad s - 2 = 0$
$s = -\frac{4}{3} \qquad s = 2$

147. $m(2m - 1) + 3(2m - 1) = 0$
$(m + 3)(2m - 1) = 0$
$m + 3 = 0 \qquad 2m - 1 = 0$
$m = -3 \qquad 2m = 1$
$m = \frac{1}{2}$

149. $z(5 - z) + 36 = 0$
$5z - z^2 + 36 = 0$
$z^2 - 5z - 36 = 0$
$(z - 9)(z + 4) = 0$
$z - 9 = 0 \qquad z + 4 = 0$
$z = 9 \qquad z = -4$

151. $v^2 - 100 = 0$
$(v - 10)(v + 10) = 0$
$v - 10 = 0 \qquad v + 10 = 0$
$v = 10 \qquad v = -10$

153. $2y^4 + 2y^3 - 24y^2 = 0$
$2y^2(y^2 + y - 12) = 0$
$2y^2(y + 4)(y - 3) = 0$
$2y^2 = 0 \qquad y + 4 = 0 \qquad y - 3 = 0$
$y = 0 \qquad y = -4 \qquad y = 3$

155. $x^3 - 11x^2 + 18x = 0$
$x(x^2 - 11x + 18) = 0$
$x(x - 9)(x - 2) = 0$
$x = 0 \qquad x - 9 = 0 \qquad x - 2 = 0$
$x = 9 \qquad x = 2$

157. $b^3 - 6b^2 - b + 6 = 0$
$b^2(b - 6) - (b - 6) = 0$
$(b - 6)(b^2 - 1) = 0$
$(b - 6)(b - 1)(b + 1) = 0$
$b - 6 = 0 \qquad b - 1 = 0 \qquad b + 1 = 0$
$b = 6 \qquad b = 1 \qquad b = -1$

159. $x^4 - 5x^3 - 9x^2 + 45x = 0$

$x^3(x - 5) - 9x(x - 5) = 0$

$(x^3 - 9x)(x - 5) = 0$

$x(x^2 - 9)(x - 5) = 0$

$x(x - 3)(x + 3)(x - 5) = 0$

$x = 0 \qquad x - 3 = 0 \qquad x + 3 = 0 \qquad x - 5 = 0$

$\qquad\qquad x = 3 \qquad x = -3 \qquad x = 5$

161. *Verbal Model:* $\boxed{\text{First odd integer}} \cdot \boxed{\text{Second odd integer}} = 99$

Labels: First odd integer $= 2n + 1$

Second odd integer $= 2n + 3$

Equation: $(2n + 1)(2n + 3) = 99$

$4n^2 + 8n + 3 = 99$

$4n^2 + 8n - 96 = 0$

$4(n^2 + 2n - 24) = 0$

$4(n + 6)(n - 4) = 0$

$n + 6 = 0 \qquad n - 4 = 0 \qquad 2n + 1 = 9$

$n = -6 \qquad n = 4 \qquad 2n + 3 = 11$

Reject

163. *Verbal Model:* Area $= \boxed{\text{Length}} \cdot \boxed{\text{Width}}$

Labels: Length $= 2\frac{1}{4}x$

Width $= x$

Equation: $900 = 2\frac{1}{4}x \cdot x$

$900 = \frac{9}{4}x^2$

$3600 = 9x^2$

$400 = x^2$

$20 = x$

$45 = 2\frac{1}{4}x$

45 inches × 20 inches

165. $h = -16t^2 + 3600$

$0 = -16t^2 + 3600$

$0 = -16(t^2 - 225)$

$0 = -16(t + 15)(t - 15)$

$t + 15 = 0 \qquad t - 15 = 0$

$t = -15 \qquad t = 15$

Reject

The object reaches the ground after 15 seconds.

Chapter Test for Chapter 5

1. $3 - 4.5x + 8.2x^3 = 8.2x^3 - 4.5x + 3$

Degree: 3

Leading coefficient: 8.2

2. (a) $690{,}000{,}000 = 6.9 \times 10^8$

(b) $4.72 \times 10^{-5} = 0.0000472$

3. (a) $\dfrac{2^{-1}x^5y^{-3}}{4x^{-2}y^2} = \dfrac{1}{2 \cdot 4} \cdot x^{5-(-2)} \cdot y^{-3-2}$

$= \dfrac{1}{8} \cdot x^7y^{-5} = \dfrac{x^7}{8y^5}$

(b) $\left(\dfrac{-2x^2y}{z^{-3}}\right)^{-2} = \left(\dfrac{z^{-3}}{-2x^2y}\right)^2 = \dfrac{z^{-6}}{4x^4y^2} = \dfrac{1}{4x^4y^2z^6}$

4. (a) $\left(\dfrac{-2u^2}{v^{-1}}\right)^3\left(\dfrac{3v^2}{u^{-3}}\right) = \left(-\dfrac{8u^6}{v^{-3}}\right)\left(\dfrac{3v^2}{u^{-3}}\right) = -24u^{6-(-3)}v^{2-(-3)} = -24u^9v^5$

(b) $\dfrac{(-3x^2y^{-1})^4}{6x^2y^0} = \dfrac{81x^8y^{-4}}{6x^2} = \dfrac{27x^{8-2}}{2y^4} = \dfrac{27x^6}{2y^4}$

5. (a) $(5a^2 - 3a + 4) + (a^2 - 4) = 6a^2 - 3a$

(b) $(16 - y^2) - (16 + 2y + y^2) = 16 - y^2 - 16 - 2y - y^2 = -2y^2 - 2y$

6. (a) $-2(2x^4 - 5) + 4x(x^3 + 2x - 1) = -4x^4 + 10 + 4x^4 + 8x^2 - 4x = 8x^2 - 4x + 10$

(b) $4t - [3t - (10t + 7)] = 4t - [3t - 10t - 7] = 4t - 3t + 10t + 7 = 11t + 7$

7. (a) $-3x(x - 4) = -3x^2 + 12x$

(b) $(2x - 3y)(x + 5y) = 2x^2 + 7xy - 15y^2$

8. (a) $(x-1)[2x+(x-3)] = (x-1)(3x-3) = 3x^2 - 6x + 3$

(b) $(2s-3)(3s^2-4s+7) = 6s^3 - 8s^2 + 14s - 9s^2 + 12s - 21 = 6s^3 - 17s^2 + 26s - 21$

9. (a) $(2w-7)^2 = (2w)^2 - 2(2w)(7) + 7^2 = 4w^2 - 28w + 49$

(b) $(4-(a+b))(4+(a+b)) = 4^2 - (a+b)^2 = 16 - (a^2+2ab+b^2) = 16 - a^2 - 2ab - b^2$

10. $18y^2 - 12y = 6y(3y-2)$

11. $v^2 - \frac{16}{9} = \left(v - \frac{4}{3}\right)\left(v + \frac{4}{3}\right)$

12. $$\begin{aligned} x^3 - 3x^2 - 4x + 12 &= x^2(x-3) - 4(x-3) \\ &= (x-3)(x^2-4) \\ &= (x-3)(x-2)(x+2) \end{aligned}$$

13. $9u^2 - 6u + 1 = (3u-1)(3u-1)$ or $(3u-1)^2$

14. $$\begin{aligned} 6x^2 - 26x - 20 &= 2(3x^2 - 13x - 10) \\ &= 2(3x+2)(x-5) \end{aligned}$$

15. $x^3 + 27 = (x+3)(x^2 - 3x + 9)$

16. $$\begin{aligned} (x-3)(x+2) &= 14 \\ x^2 + 2x - 3x - 6 &= 14 \\ x^2 - x - 20 &= 0 \\ (x+4)(x-5) &= 0 \end{aligned}$$

$x + 4 = 0 \qquad x - 5 = 0$

$x = -4 \qquad x = 5$

17. $$\begin{aligned} (y+2)^2 - 9 &= 0 \\ [(y+2)-3][(y+2)+3] &= 0 \end{aligned}$$

$y - 1 = 0 \qquad y + 5 = 0$

$y = 1 \qquad y = -5$

18. $$\begin{aligned} 12 + 5y - 3y^2 &= 0 \\ (3-y)(4+3y) &= 0 \end{aligned}$$

$3 - y = 0 \qquad 4 + 3y = 0$

$3 = y \qquad -\frac{4}{3} = y$

19. $$\begin{aligned} 2x^3 + 10x^2 + 8x &= 0 \\ 2x(x^2 + 5x + 4) &= 0 \\ 2x(x+4)(x+1) &= 0 \end{aligned}$$

$2x = 0 \qquad x + 4 = 0 \qquad x + 1 = 0$

$x = 0 \qquad x = -4 \qquad x = -1$

20. Area $= 2x(x+15) - x(x+4)$

Shaded region $= 2x^2 + 30x - x^2 - 4x$

$= x^2 + 26x$

21. *Verbal Model:* [Area rectangle] = [Length] · [Width]

Labels: Length $= \frac{3}{2}w$

Width $= w$

Equation: $$\begin{aligned} 54 &= \tfrac{3}{2}w \cdot w \\ 108 &= 3w^2 \\ 36 &= w^2 \end{aligned}$$

6 centimeters = width

9 centimeters = length

22. *Verbal Model:* [Area of a triangle] = $\frac{1}{2}$ · [Base] · [Height]

Labels: Base $= x$

Height $= 2x + 2$

Equation: $$\begin{aligned} 20 &= \tfrac{1}{2} \cdot x \cdot (2x+2) \\ 20 &= x^2 + x \\ 0 &= x^2 + x - 20 \\ 0 &= (x+5)(x-4) \end{aligned}$$

$x + 5 = 0 \qquad x - 4 = 0$

$x = -5 \qquad x = 4$ feet

Reject $\qquad 2x + 2 = 10$ feet

The base is 4 feet and the height is 10 feet.

23. *Verbal Model:* [Revenue] = [Cost]

Labels: Revenue $= R$

Cost $= C$

Equation: $$\begin{aligned} R &= C \\ x^2 - 35x &= 150 + 12x \\ x^2 - 47x - 150 &= 0 \\ (x-50)(x+3) &= 0 \end{aligned}$$

$x - 50 = 0 \qquad x + 3 = 0$

$x = 50 \qquad x = -3$

50 computer desks

CHAPTER 6
Rational Expressions, Equations, and Functions

Section 6.1 Rational Expressions and Functions **172**

Section 6.2 Multiplying and Dividing Rational Expressions **173**

Section 6.3 Adding and Subtracting Rational Expressions **176**

Section 6.4 Complex Fractions **180**

Mid-Chapter Quiz **184**

Section 6.5 Dividing Polynomials and Synthetic Division **185**

Section 6.6 Solving Rational Equations **189**

Section 6.7 Variation **197**

Review Exercises **199**

Chapter Test **205**

CHAPTER 6
Rational Expressions, Equations, and Functions

Section 6.1 Rational Expressions and Functions

1. $x - 3 \neq 0$

$x \neq 3$

$D = (-\infty, 3) \cup (3, \infty)$

3. $z(z - 4) \neq 0$

$z \neq 0 \quad z - 4 \neq 0$

$z \neq 4$

$D = (-\infty, 0) \cup (0, 4) \cup (4, \infty)$

5. $t^2 - 16 \neq 0$

$(t - 4)(t + 4) \neq 0$

$t \neq 4 \quad t \neq -4$

$D = (-\infty, -4) \cup (-4, 4) \cup (4, \infty)$

7. Since length must be positive, $x \geq 0$. Since $\dfrac{500}{x}$ must be defined, $x \neq 0$. Therefore, the domain is $x > 0$ or $(0, \infty)$.

9. $x =$ units of a product

$D = \{1, 2, 3, 4, \ldots\}$

11. $$\frac{3x^2 - 9x}{12x^2} = \frac{3x(x - 3)}{12x^2} = \frac{(x - 3)}{4x}$$

13. $$\frac{x^2(x - 8)}{x(x - 8)} = \frac{x \cdot x(x - 8)}{x(x - 8)}$$
$$= x, \; x \neq 0, x \neq 8$$

15. $$\frac{u^2 - 12u + 36}{u - 6} = \frac{(u - 6)(u - 6)}{u - 6}$$
$$= u - 6, \quad u \neq 6$$

17. $$\frac{y^3 - 4y}{y^2 + 4y - 12} = \frac{y(y^2 - 4)}{(y + 6)(y - 2)}$$
$$= \frac{y(y - 2)(y + 2)}{(y + 6)(y - 2)}$$
$$= \frac{y(y + 2)}{y + 6}, \quad y \neq 2$$

19. $$\frac{3x^2 - 7x - 20}{12 + x - x^2} = \frac{(3x + 5)(x - 4)}{-1(x^2 - x - 12)}$$
$$= \frac{(3x + 5)(x - 4)}{-1(x - 4)(x + 3)}$$
$$= -\frac{3x + 5}{x + 3}, \quad x \neq 4$$

21. $$\frac{2x^2 + 19x + 24}{2x^2 - 3x - 9} = \frac{(2x + 3)(x + 8)}{(2x + 3)(x - 3)}$$
$$= \frac{x + 8}{x - 3}, \quad x \neq -\frac{3}{2}$$

23. $$\frac{3xy^2}{xy^2 + x} = \frac{3xy^2}{x(y^2 + 1)} = \frac{3y^2}{y^2 + 1}, \quad x \neq 0$$

25. $$\frac{y^2 - 64x^2}{5(3y + 24x)} = \frac{(y - 8x)(y + 8x)}{15(y + 8x)}$$
$$= \frac{y - 8x}{15}, \quad y \neq -8x$$

27. $$\frac{5xy + 3x^2y^2}{xy^3} = \frac{xy(5 + 3xy)}{xy \cdot y^2} = \frac{5 + 3xy}{y^2}, \quad x \neq 0$$

29. $$\frac{u^2 - 4v^2}{u^2 + uv - 2v^2} = \frac{(u - 2v)(u + 2v)}{(u - v)(u + 2v)}$$
$$= \frac{u - 2v}{u - v}, \quad u \neq -2v$$

31. $$\frac{3m^2 - 12n^2}{m^2 + 4mn + 4n^2} = \frac{3(m^2 - 4n^2)}{(m + 2n)(m + 2n)}$$
$$= \frac{3(m - 2n)(m + 2n)}{(m + 2n)(m + 2n)}$$
$$= \frac{3(m - 2n)}{m + 2n}$$

33. $$\frac{\text{Area of shaded portion}}{\text{Area of total figure}} = \frac{x(x + 1)}{(x + 1)(x + 3)}$$
$$= \frac{x}{x + 3}, \quad x > 0$$

35. $$\frac{\text{Area of shaded portion}}{\text{Area of total figure}} = \frac{\frac{1}{2}x(x + 1)}{\frac{1}{2}(2x)(2x + 2)}$$
$$= \frac{1}{4}, \quad x > 0$$

37. The rational expression is in simplified form if the numerator and denominator have no factors in common (other than ± 1).

39. A change in signs is one additional step sometimes needed in order to find common factors.

41. $\dfrac{x+5}{3x} = \dfrac{(x+5)(x(x-2))}{3x^2(x-2)}, \quad x \neq 2$

43. $\dfrac{8x(\ )}{x^2-2x-15} = \dfrac{8x(\ )}{(x-5)(x+3)}$

$= \dfrac{8x(x+3)}{(x-5)(x+3)}$

$= \dfrac{8x}{x-5}, \quad x \neq -3$

45. (a) *Verbal Model:* $\boxed{\text{Total cost}} = \boxed{\text{Number of units}} \cdot \boxed{\text{Cost per unit}} + \boxed{\text{Initial cost}}$

Labels: Total cost $= C$

Number of units $= x$

Equation: $C = 6.50x + 60{,}000$

(b) *Verbal Model:* $\boxed{\text{Average cost}} = \boxed{\text{Total cost}} \div \boxed{\text{Number of units}}$

Label: Average cost $= \overline{C}$

Equation: $\overline{C} = \dfrac{C}{x}; \overline{C} = \dfrac{6.50x + 60{,}000}{x}$

(c) $D = \{1, 2, 3, 4, \ldots\}$

(d) $\overline{C}(11{,}000) = \dfrac{6.50(11{,}000) + 60{,}000}{11{,}000} \approx \$11.95$

47. $\dfrac{\text{Circular pool volume}}{\text{Rectangular pool volume}} = \dfrac{\pi(3d)^2(d+2)}{d(3d)(3d+6)}$

$= \dfrac{\pi(3d)^2(d+2)}{3d^2 \cdot 3(d+2)}$

$= \dfrac{\pi(3d)^2(d+2)}{(3d)^2(d+2)}$

$= \pi, \quad d > 0$

49. A rational expression is undefined if the denominator is equal to zero. To find the implied domain restriction of a rational function, find the values that make the denominator equal to zero (by setting the denominator equal to zero), and exclude those values from the domain.

51. $\dfrac{1}{x^2+1}$

There are many correct answers.

53. The student incorrectly divided out; the denominator may not be split up.

Correct solution:

$\dfrac{x^2+7x}{x+7} = \dfrac{x(x+7)}{x+7} = x, \quad x \neq -7$

55. To write the polynomial $g(x)$, multiply $f(x)$ by $(x-2)$ and divide by $(x-2)$.

$g(x) = \dfrac{f(x)(x-2)}{(x-2)}, \quad x \neq 2$

57. $\left(\dfrac{1}{4}\right)\left(\dfrac{3}{4}\right) = \dfrac{1 \cdot 3}{4 \cdot 4} = \dfrac{3}{16}$

59. $\dfrac{1}{3}\left(\dfrac{3}{5}\right)(5) = \dfrac{1 \cdot 3 \cdot 5}{3 \cdot 5 \cdot 1} = 1$

61. $(-2a^3)(-2a) = -2 \cdot -2 \cdot a^3 \cdot a = 4a^4$

63. $(-3b)(b^2 - 3b + 5) = -3b(b^2) - 3b(-3b) - 3b(5)$

$= -3b^3 + 9b^2 - 15b$

Section 6.2 Multiplying and Dividing Rational Expressions

1. $\dfrac{7}{3y} = \dfrac{7x^2}{3y(x^2)}, \quad x \neq 0$

3. $\dfrac{3x}{x-4} = \dfrac{3x(x+2)^2}{(x-4)(x+2)^2}, \quad x \neq -2$

5. $4x \cdot \dfrac{7}{12x} = \dfrac{4x(7)}{12x} = \dfrac{7}{3}, \quad x \neq 0$

7. $\dfrac{8s^3}{9s} \cdot \dfrac{6s^2}{32s} = \dfrac{8s^3 \cdot 3 \cdot 2s \cdot s}{3 \cdot 3 \cdot s \cdot 8 \cdot 2 \cdot 2 \cdot s} = \dfrac{s^3}{6}, \quad s \neq 0$

9. $16u^4 \cdot \dfrac{12}{8u^2} = \dfrac{8 \cdot 2 \cdot u^2 \cdot u^2 \cdot 12}{8 \cdot u^2} = 24u^2, \quad u \neq 0$

11. $\dfrac{8}{3 + 4x} \cdot (9 + 12x) = \dfrac{8 \cdot 3(3 + 4x)}{3 + 4x} = 24, \quad x \neq -\dfrac{3}{4}$

13. $\dfrac{8u^2v}{3u + v} \cdot \dfrac{u + v}{12u} = \dfrac{4 \cdot 2 \cdot u \cdot u \cdot v(u + v)}{(3u + v) \cdot 4 \cdot 3 \cdot u}$

$= \dfrac{2uv(u + v)}{3(3u + v)}, \quad u \neq 0$

15. $\dfrac{12 - r}{3} \cdot \dfrac{3}{r - 12} = \dfrac{-1(r - 12) \cdot 3}{3(r - 12)} = -1, \quad r \neq 12$

17. $\dfrac{(2x - 3)(x + 8)}{x^3} \cdot \dfrac{x}{3 - 2x} = \dfrac{(2x - 3)(x + 8)x}{x \cdot x^2 \cdot -1(2x - 3)}$

$= -\dfrac{x + 8}{x^2}, \quad x \neq \dfrac{3}{2}$

19. $\dfrac{4r - 12}{r - 2} \cdot \dfrac{r^2 - 4}{r - 3} = \dfrac{4(r - 3)(r - 2)(r + 2)}{(r - 2) \cdot (r - 3)}$

$= 4(r + 2), \quad r \neq 3, r \neq 2$

21. $\dfrac{2t^2 - t - 15}{t + 2} \cdot \dfrac{t^2 - t - 6}{t^2 - 6t + 9} = \dfrac{(2t + 5)(t - 3)(t - 3)(t + 2)}{(t + 2)(t - 3)(t - 3)} = 2t + 5, \quad t \neq 3, t \neq -2$

23. $(4y^2 - x^2) \cdot \dfrac{xy}{(x - 2y)^2} = \dfrac{-1(x^2 - 4y^2)(xy)}{(x - 2y)^2} = -\dfrac{-(x - 2y)(x + 2y)(xy)}{(x - 2y)(x - 2y)} = -\dfrac{xy(x + 2y)}{(x - 2y)}$

25. $\dfrac{x^2 + 2xy - 3y^2}{(x + y)^2} \cdot \dfrac{x^2 - y^2}{x + 3y} = \dfrac{(x + 3y)(x - y)}{(x + y)^2} \cdot \dfrac{(x - y)(x + y)}{x + 3y} = \dfrac{(x - y)^2}{x + y}, \quad x \neq -3y$

27. $\dfrac{x + 5}{x - 5} \cdot \dfrac{2x^2 - 9x - 5}{3x^2 + x - 2} \cdot \dfrac{x^2 - 1}{x^2 + 7x + 10} = \dfrac{x + 5}{x - 5} \cdot \dfrac{(2x + 1)(x - 5)}{(3x - 2)(x + 1)} \cdot \dfrac{(x - 1)(x + 1)}{(x + 5)(x + 2)}$

$= \dfrac{(x + 5)(2x + 1)(x - 5)(x - 1)(x + 1)}{(x - 5)(3x - 2)(x + 1)(x + 5)(x + 2)}$

$= \dfrac{(2x + 1)(x - 1)}{(3x - 2)(x + 2)}, \quad x \neq \pm 5, -1$

29. $\dfrac{9 - x^2}{2x + 3} \cdot \dfrac{4x^2 + 8x - 5}{4x^2 - 8x + 3} \cdot \dfrac{6x^4 - 2x^3}{8x^2 + 4x} = \dfrac{(3 - x)(3 + x)}{2x + 3} \cdot \dfrac{(2x + 5)(2x - 1)}{(2x - 3)(2x - 1)} \cdot \dfrac{2x^3(3x - 1)}{4x(2x + 1)}$

$= \dfrac{-1(x - 3)(x + 3)(2x + 5)x^2(3x - 1)}{(2x + 3)(2x - 3)2(2x + 1)}$

$= \dfrac{x^2(x^2 - 9)(2x + 5)(3x - 1)}{2(2x + 3)(3 - 2x)(2x + 1)}, \quad x \neq 0, \dfrac{1}{2}$

31. $\dfrac{x}{x + 2} \div \dfrac{3}{x + 1} = \dfrac{x}{x + 2} \cdot \dfrac{x + 1}{3} = \dfrac{x(x + 1)}{3(x + 2)}, \quad x \neq -1$

33. $x^2 \div \dfrac{3x}{4} = x^2 \cdot \dfrac{4}{3x} = \dfrac{4x}{3}, \quad x \neq 0$

35. $\dfrac{2x}{5} \div \dfrac{x^2}{15} = \dfrac{2x}{5} \cdot \dfrac{15}{x^2} = \dfrac{2(3)(5)x}{5x^2} = \dfrac{6}{x}$

37. $\dfrac{4x}{3x - 3} \div \dfrac{x^2 + 2x}{x^2 + x - 2} = \dfrac{4x}{3(x - 1)} \cdot \dfrac{(x + 2)(x - 1)}{x(x + 2)} = \dfrac{4x}{3x} = \dfrac{4}{3}, \quad x \neq -2, 0, 1$

39. $\dfrac{7xy^2}{10x^2y} \div \dfrac{21x^3}{45xy} = \dfrac{7xy^2}{10x^2y} \cdot \dfrac{45xy}{21x^3} = \dfrac{7 \cdot x \cdot y^2 \cdot 3 \cdot 3 \cdot 5 \cdot x \cdot y}{2 \cdot 5 \cdot x^2 \cdot y \cdot 3 \cdot 7 \cdot x^3} = \dfrac{3y^2}{2x^3}, \quad y \neq 0$

41. $\dfrac{3(a + b)}{4} \div \dfrac{(a + b)^2}{2} = \dfrac{3(a + b)}{4} \cdot \dfrac{2}{(a + b)^2} = \dfrac{3(a + b) \cdot 2}{2 \cdot 2 \cdot (a + b)(a + b)} = \dfrac{3}{2(a + b)}$

43. $$\frac{2x+2y}{3} \div \frac{x^2-y^2}{x-y} = \frac{2(x+y)}{3} \cdot \frac{x-y}{(x-y)(x+y)}$$
$$= \frac{2(x+y)(x-y)}{3(x-y)(x+y)}$$
$$= \frac{2}{3}, \quad x \neq y, \quad x \neq -y$$

45. $$\frac{(x^3y)^2}{(x+2y)^2} \div \frac{x^2y}{(x+2y)^3} = \frac{(x^3y)^2}{(x+2y)^2} \cdot \frac{(x+2y)^3}{x^2y}$$
$$= \frac{(x^3y)(x^3y)(x+2y)^2(x+2y)}{(x+2y)^2x^2y}$$
$$= \frac{(x^3y)(x^2 \cdot xy)(x+2y)}{x^2y} = x^4y(x+2y), \quad x \neq 0, y \neq 0, x \neq -2y$$

47. $$\frac{x^2+2x-15}{x^2+11x+30} \div \frac{x^2-8x+15}{x^2+2x-24} = \frac{(x+5)(x-3)}{(x+5)(x+6)} \cdot \frac{(x+6)(x-4)}{(x-3)(x-5)}$$
$$= \frac{(x+5)(x-3)(x+6)(x-4)}{(x+5)(x+6)(x-3)(x-5)}$$
$$= \frac{x-4}{x-5}, \quad x \neq -6, x \neq -5, x \neq 3$$

49. To find a model Y for the annual per capita income, divide the total personal income by the population.

$$Y = (6591.43t + 102{,}139.0) \div \frac{0.184t + 4.3}{0.029t + 1}$$
$$= (6591.43t + 102{,}139.0) \cdot \frac{0.029t + 1}{0.184t + 4.3}$$
$$= \frac{(6591.43t + 102{,}139.0)(0.029t + 1)}{0.184t + 4.3}, \quad 4 \leq t \leq 9$$

51. To multiply two rational expressions, let a, b, c, and d represent real numbers, variables, or algebraic expressions such that $b \neq 0$ and $d \neq 0$. Then the product of $\frac{a}{b}$ and $\frac{c}{d}$ is $\frac{a}{b} \cdot \frac{c}{d} = \frac{ac}{bd}$.

53. To divide rational expressions, multiply the first expression by the reciprocal of the second expression.

55. $$\left[\frac{x^2}{9} \cdot \frac{3(x+4)}{x^2+2x}\right] \div \frac{x}{x+2} = \frac{x^2}{9} \cdot \frac{3(x+4)}{x(x+2)} \cdot \frac{x+2}{x}$$
$$= \frac{x+4}{3}, \quad x \neq -2, 0$$

57. $$\left[\frac{xy+y}{4x} \div (3x+3)\right] \div \frac{y}{3x} = \frac{y(x+1)}{4x} \cdot \frac{1}{3(x+1)} \cdot \frac{3x}{y}$$
$$= \frac{1}{4}, \quad x \neq -1, 0, y \neq 0$$

59. $$\frac{2x^2+5x-25}{3x^2+5x+2} \cdot \frac{3x^2+2x}{x+5} \div \left(\frac{x}{x+1}\right)^2 = \frac{(2x-5)(x+5)}{(3x+2)(x+1)} \cdot \frac{x(3x+2)}{x+5} \cdot \left(\frac{x+1}{x}\right)^2$$
$$= \frac{(2x-5)(x+5)x(3x+2)(x+1)(x+1)}{(3x+2)(x+1)(x+5)x \cdot x}$$
$$= \frac{(2x-5)(x+1)}{x}, \quad x \neq -1, -5, -\frac{2}{3}$$

61. $x^3 \cdot \dfrac{x^{2n} - 9}{x^{2n} + 4x^n + 3} \div \dfrac{x^{2n} - 2x^n - 3}{x} = x^3 \cdot \dfrac{(x^n - 3)(x^n + 3)}{(x^n + 3)(x^n + 1)} \cdot \dfrac{x}{(x^n - 3)(x^n + 1)}$

$= \dfrac{x^4}{(x^n + 1)^2}, \quad x^n \neq -3, 3, 0$

63. $\dfrac{\text{Unshaded Area}}{\text{Total Area}} = \dfrac{x \cdot x}{2x(4x + 2)} = \dfrac{x}{2(2)(2x + 1)} = \dfrac{x}{4(2x + 1)}, \quad x > 0$

65. $\dfrac{\text{Unshaded Area}}{\text{Total Area}} = \dfrac{\pi x^2}{2x(4x + 2)} = \dfrac{\pi x}{2(2)(2x + 1)} = \dfrac{\pi x}{4(2x + 1)}, \quad x > 0$

67. In simplifying a product of national expressions, you divide the common factors out of the numerator and denominator.

69. The domain needs to be restricted, $x \neq a, x \neq b$.

71. The first expression needs to be multiplied by the reciprocal of the second expression (not the second by the reciprocal of the first), and the domain needs to be restricted.

$\dfrac{x^2 - 4}{5x} \div \dfrac{x + 2}{x - 2} = \dfrac{x^2 - 4}{5x} \cdot \dfrac{x - 2}{x + 2} = \dfrac{(x - 2)^2(x + 2)}{5x(x + 2)} = \dfrac{(x - 2)^2}{5x}, \quad x \neq \pm 2$

73. $\dfrac{1}{8} + \dfrac{3}{8} + \dfrac{5}{8} = \dfrac{1 + 3 + 5}{8} = \dfrac{9}{8}$

75. $\dfrac{3}{5} + \dfrac{4}{15} = \dfrac{9}{15} + \dfrac{4}{15} = \dfrac{9 + 4}{15} = \dfrac{13}{15}$

77. $x^2 + 3x = 0$

$x(x + 3) = 0$

$x = 0 \qquad x + 3 = 0$

$x = -3$

Section 6.3 Adding and Subtracting Rational Expressions

1. $\dfrac{5x}{6} + \dfrac{4x}{6} = \dfrac{5x + 4x}{6} = \dfrac{9x}{6} = \dfrac{3x}{2}$

3. $\dfrac{2}{3a} - \dfrac{11}{3a} = \dfrac{2 - 11}{3a} = \dfrac{-9}{3a} = -\dfrac{3}{a}$

5. $\dfrac{x}{9} - \dfrac{x + 2}{9} = \dfrac{x - (x + 2)}{9} = \dfrac{x - x - 2}{9} = -\dfrac{2}{9}$

7. $\dfrac{2x - 1}{x(x - 3)} + \dfrac{1 - x}{x(x - 3)} = \dfrac{2x - 1 + 1 - x}{x(x - 3)}$

$= \dfrac{x}{x(x - 3)}$

$= \dfrac{1}{x - 3}, \quad x \neq 0$

9. $\dfrac{c}{c^2 + 3c - 4} - \dfrac{1}{c^2 + 3c - 4} = \dfrac{c - 1}{(c + 4)(c - 1)}$

$= \dfrac{1}{c + 4}, \quad c \neq 1$

11. $5x^2 = 5 \cdot x \cdot x$

$20x^3 = 5 \cdot 2 \cdot 2 \cdot x \cdot x \cdot x$

$\text{LCM} = 20x^3$

13. $9y^3 = 3 \cdot 3 \cdot y \cdot y \cdot y$

$12y = 2 \cdot 2 \cdot 3 \cdot y$

$\text{LCM} = 3 \cdot 3 \cdot 2 \cdot 2 \cdot y \cdot y \cdot y = 36y^3$

15. $15x^2 = 5 \cdot 3 \cdot x \cdot x$

$3(x + 5) = 3 \cdot (x + 5)$

$\text{LCM} = 15x^2(x + 5)$

17. $63z^2(z + 1) = 7 \cdot 9 \cdot z \cdot z(z + 1)$

$14(z + 1)^4 = 7 \cdot 2 \cdot (z + 1)^4$

$\text{LCM} = 126z^2(z + 1)^4$

19. $8t(t + 2) = 2 \cdot 2 \cdot 2 \cdot t \cdot (t + 2)$

$14(t^2 - 4) = 2 \cdot 7 \cdot (t + 2)(t - 2)$

$\text{LCM} = 2 \cdot 2 \cdot 2 \cdot 7 \cdot t \cdot (t + 2)(t - 2) = 56t(t^2 - 4)$

21. $2y^2 + y - 1 = (2y - 1)(y + 1)$

$4y^2 - 2y = 2y(2y - 1)$

$\text{LCM} = 2y(2y - 1)(y + 1)$

23. $\dfrac{5}{4x} - \dfrac{3}{5} = \dfrac{5(5)}{4x(5)} - \dfrac{3(4x)}{5(4x)} = \dfrac{25}{20x} - \dfrac{12x}{20x} = \dfrac{25 - 12x}{20x}$

25. $\dfrac{7}{a} + \dfrac{14}{a^2} = \dfrac{7(a)}{a(a)} + \dfrac{14(1)}{a^2(1)} = \dfrac{7a}{a^2} + \dfrac{14}{a^2} = \dfrac{7a + 14}{a^2}$

27.
$$\begin{aligned} 25 + \frac{10}{x + 4} &= \frac{25(x + 4)}{1(x + 4)} + \frac{10(1)}{(x + 4)(1)} \\ &= \frac{25(x + 4)}{x + 4} + \frac{10}{x + 4} \\ &= \frac{25x + 100 + 10}{x + 4} = \frac{25x + 110}{x + 4} \end{aligned}$$

29.
$$\begin{aligned} \frac{x}{x + 3} - \frac{5}{x - 2} &= \frac{x(x - 2)}{(x + 3)(x - 2)} - \frac{5(x + 3)}{(x - 2)(x + 3)} \\ &= \frac{x^2 - 2x - 5x - 15}{(x + 3)(x - 2)} \\ &= \frac{x^2 - 7x - 15}{(x + 3)(x - 2)} \end{aligned}$$

31.
$$\begin{aligned} \frac{12}{x^2 - 9} - \frac{2}{x - 3} &= \frac{12}{(x - 3)(x + 3)} - \frac{2}{x - 3} \\ &= \frac{12}{(x - 3)(x + 3)} - \frac{2(x + 3)}{(x - 3)(x + 3)} \\ &= \frac{12 - 2(x + 3)}{(x - 3)(x + 3)} \\ &= \frac{12 - 2x - 6}{(x - 3)(x + 3)} \\ &= \frac{6 - 2x}{(x - 3)(x + 3)} \\ &= \frac{-2(-3 + x)}{(x - 3)(x + 3)} \\ &= \frac{-2}{x + 3} \\ &= -\frac{2}{x + 3}, \quad x \neq 3 \end{aligned}$$

33.
$$\begin{aligned} \frac{20}{x - 4} + \frac{20}{4 - x} &= \frac{20(1)}{(x - 4)(1)} + \frac{20(-1)}{(4 - x)(-1)} \\ &= \frac{20}{(x - 4)} - \frac{20}{x - 4} \\ &= \frac{20 - 20}{x - 4} = 0, \quad x \neq 4 \end{aligned}$$

35.
$$\begin{aligned} \frac{3x}{x - 8} - \frac{6}{8 - x} &= \frac{3x(1)}{(x - 8)(1)} - \frac{6(-1)}{(8 - x)(-1)} \\ &= \frac{3x}{x - 8} + \frac{6}{x - 8} \\ &= \frac{3x + 6}{x - 8} \end{aligned}$$

37.
$$\begin{aligned} \frac{3x}{3x - 2} + \frac{2}{2 - 3x} &= \frac{3x(1)}{3x - 2(1)} + \frac{2(-1)}{(2 - 3x)(-1)} \\ &= \frac{3x}{3x - 2} + \frac{-2}{3x - 2} \\ &= \frac{3x - 2}{3x - 2} = 1, \quad x \neq \frac{2}{3} \end{aligned}$$

39.
$$\begin{aligned} \frac{3}{x - 5} + \frac{2}{x + 5} &= \frac{3(x + 5)}{(x - 5)(x + 5)} + \frac{2(x - 5)}{(x + 5)(x - 5)} \\ &= \frac{3x + 15 + 2x - 10}{(x - 5)(x + 5)} \\ &= \frac{5x + 5}{(x - 5)(x + 5)} \end{aligned}$$

41.
$$\begin{aligned} \frac{9}{5x} + \frac{3}{x - 1} &= \frac{9(x - 1)}{5x(x - 1)} + \frac{3(5x)}{5x(x - 1)} \\ &= \frac{9x - 9 + 15x}{5x(x - 1)} \\ &= \frac{24x - 9}{5x(x - 1)} \\ &= \frac{3(8x - 3)}{5x(x - 1)} \end{aligned}$$

43. $\dfrac{x}{x^2 - x - 30} - \dfrac{1}{x + 5} = \dfrac{x}{(x + 5)(x - 6)} - \dfrac{(x - 6)}{(x + 5)(x - 6)} = \dfrac{x - (x - 6)}{(x + 5)(x - 6)} = \dfrac{x - x + 6}{(x + 5)(x - 6)} = \dfrac{6}{(x + 5)(x - 6)}$

45. $\dfrac{4}{x - 4} + \dfrac{16}{(x - 4)^2} = \dfrac{4(x - 4)}{(x - 4)(x - 4)} + \dfrac{16(1)}{(x - 4)^2(1)} = \dfrac{4x - 16}{(x - 4)^2} + \dfrac{16}{(x - 4)^2} = \dfrac{4x - 16 + 16}{(x - 4)^2} = \dfrac{4x}{(x - 4)^2}$

47. $$\frac{y}{x^2 + xy} - \frac{x}{xy + y^2} = \frac{y}{x(x + y)} - \frac{x}{y(x + y)}$$
$$= \frac{y(y)}{x(x + y)(y)} - \frac{x(x)}{y(x + y)(x)}$$
$$= \frac{y^2}{xy(x + y)} - \frac{x^2}{xy(x + y)}$$
$$= \frac{y^2 - x^2}{xy(x + y)}$$
$$= \frac{(y - x)(y + x)}{xy(x + y)}$$
$$= \frac{y - x}{xy}, \quad x \neq -y$$

49. $$\frac{4}{x} - \frac{2}{x^2} + \frac{4}{x + 3} = \frac{4x(x + 3)}{x(x)(x + 3)} - \frac{2(x + 3)}{x^2(x + 3)} + \frac{4(x^2)}{(x + 3)x^2}$$
$$= \frac{4x^2 + 12x}{x^2(x + 3)} - \frac{2x + 6}{x^2(x + 3)} + \frac{4x^2}{x^2(x + 3)}$$
$$= \frac{4x^2 + 12x - 2x - 6 + 4x^2}{x^2(x + 3)}$$
$$= \frac{8x^2 + 10x - 6}{x^2(x + 3)}$$
$$= \frac{2(4x^2 + 5x - 3)}{x^2(x + 3)}$$

51. $$\frac{3u}{u^2 - 2uv + v^2} + \frac{2}{u - v} - \frac{u}{u - v} = \frac{3u}{(u - v)^2} + \frac{2 - u}{u - v}$$
$$= \frac{3u(1)}{(u - v)^2(1)} + \frac{(2 - u)(u - v)}{(u - v)(u - v)}$$
$$= \frac{3u}{(u - v)^2} + \frac{2u - 2v - u^2 + uv}{(u - v)^2}$$
$$= \frac{3u + 2u - 2v - u^2 + uv}{(u - v)^2}$$
$$= \frac{5u - 2v - u^2 + uv}{(u - v)^2}$$
$$= -\frac{u^2 - uv - 5u + 2v}{(u - v)^2}$$

53. To find the model T, find the sum of M and F.

$$T = \frac{0.150t + 5.53}{-0.049t + 1} + \frac{-0.313t + 7.71}{-0.055t + 1}$$
$$= \frac{(-0.150t + 5.53)(-0.055t + 1) + (-0.313t + 7.71)(-0.049t + 1)}{(-0.049t + 1)(-0.055t + 1)}$$
$$= \frac{0.023587t^2 - 1.14494t + 13.24}{(-0.049t + 1)(-0.055t + 1)}, \quad 5 \leq t \leq 10$$

55. Two rational expressions with like denominators have a common denominator is a true statement.

57. To add or subtract rational expressions with like denominators, simply add (or subtract) the terms in the numerator and keep the common denominator.

59. $$\frac{4}{x^2} - \frac{4}{x^2 + 1} = \frac{4(x^2 + 1)}{x^2(x^2 + 1)} - \frac{4x^2}{(x^2 + 1)x^2}$$
$$= \frac{4(x^2 + 1)}{x^2(x^2 + 1)} - \frac{4x^2}{x^2(x^2 + 1)}$$
$$= \frac{4x^2 + 4 - 4x^2}{x^2(x^2 + 1)}$$
$$= \frac{4}{x^2(x^2 + 1)}$$

61. $$\frac{x+2}{x-1}-\frac{2}{x+6}-\frac{14}{x^2+5x-6}=\frac{(x+2)(x+6)}{(x-1)(x+6)}-\frac{2(x-1)}{(x+6)(x-1)}-\frac{14(1)}{(x+6)(x-1)(1)}$$
$$=\frac{x^2+8x+12}{(x-1)(x+6)}-\frac{2x-2}{(x+6)(x-1)}-\frac{14}{(x+6)(x-1)}$$
$$=\frac{x^2+8x+12-2x+2-14}{(x-1)(x+6)}$$
$$=\frac{x^2+6x}{(x-1)(x+6)}$$
$$=\frac{x(x+6)}{(x-1)(x+6)}$$
$$=\frac{x}{x-1},\quad x\neq -6$$

63. $\frac{t}{4}+\frac{t}{6}=\frac{t(3)}{4(3)}+\frac{t(2)}{6(2)}=\frac{3t}{12}+\frac{2t}{12}=\frac{5t}{12}$

65. $A+B+C=0$

$-B+C=0$

$-A=4$

$A=-4$

Substitute into first equation.

$-4+B+C=0$

$B+C=4$

Solve first and second equations.

$B+C=4$

$-B+C=0$

$2C=4$

$C=2$

$B+2=4$

$B=2$

$$\frac{4}{x^3-x}=\frac{-4}{x}+\frac{2}{x+1}+\frac{2}{x-1}$$
$$=\frac{-4(x+1)(x-1)}{x(x+1)(x-1)}+\frac{2(x)(x-1)}{(x+1)(x)(x-1)}+\frac{2(x)(x+1)}{(x-1)(x)(x+1)}$$
$$=\frac{-4(x^2-1)+2x(x-1)+2x(x+1)}{x(x^2-1)}$$
$$=\frac{-4x^2+4+2x^2-2x+2x^2+2x}{x^3-x}$$
$$=\frac{4}{x^3-x}$$

67. $\frac{x-1}{x+4}-\frac{4x-11}{x+4}=\frac{(x-1)-(4x-11)}{x+4}=\frac{x-1-4x+11}{x+4}=\frac{-3x+10}{x+4}$

The subtraction must be distributed to both terms of the numerator of the second fraction.

69. Yes. $\frac{3}{2}(x+2)+\frac{x}{x+2}$

71. $5v+(4-3v)=(5v-3v)+4=2v+4$

73. $(x^2-4x+3)-(6-2x)=x^2-4x+3-6+2x$
$=x^2-2x-3$

75. $x^2-7x+12=(x-3)(x-4)$

77. $2a^2-9a-18=(a-6)(2a+3)$

Section 6.4 Complex Fractions

1. $\dfrac{\left(\dfrac{3}{16}\right)}{\left(\dfrac{9}{12}\right)} = \dfrac{3}{16} \div \dfrac{9}{12} = \dfrac{3}{16} \cdot \dfrac{12}{9} = \dfrac{3 \cdot 2 \cdot 2 \cdot 3}{2 \cdot 2 \cdot 2 \cdot 2 \cdot 3 \cdot 3} = \dfrac{1}{4}$

3. $\dfrac{\left(\dfrac{8x^2y}{3z^2}\right)}{\left(\dfrac{4xy}{9z^5}\right)} = \dfrac{8x^2y}{3z^2} \div \dfrac{4xy}{9z^5}$

$= \dfrac{8x^2y}{3z^2} \cdot \dfrac{9z^5}{4xy}$

$= \dfrac{4 \cdot 2 \cdot 3 \cdot 3x^2 \cdot y \cdot z^5}{3 \cdot 4 \cdot x \cdot y \cdot z^2}$

$= 6xz^3, x, y, z \neq 0$

5. $\dfrac{\left(\dfrac{6x^3}{(5y)^2}\right)}{\left(\dfrac{(3x)^2}{15y^4}\right)} = \dfrac{6x^3}{25y^2} \div \dfrac{9x^2}{15y^4}$

$= \dfrac{6x^3}{25y^2} \cdot \dfrac{15y^4}{9x^2}$

$= \dfrac{3 \cdot 2 \cdot 5 \cdot 3 \cdot x^3 \cdot y^4}{5 \cdot 5 \cdot 3 \cdot 3 \cdot x^2 \cdot y^2}$

$= \dfrac{2xy^2}{5}, \quad x \neq 0, y \neq 0$

7. $\dfrac{\left(\dfrac{y}{3-y}\right)}{\left(\dfrac{y^2}{y-3}\right)} = \dfrac{y}{3-y} \div \dfrac{y^2}{y-3}$

$= \dfrac{y}{-1(y-3)} \cdot \dfrac{y-3}{y^2}$

$= \dfrac{y(y-3)}{-1y^2(y-3)}$

$= -\dfrac{1}{y}, \quad y \neq 3$

9. $\dfrac{\left(\dfrac{25x^2}{x-5}\right)}{\left(\dfrac{10x}{5+4x-x^2}\right)} = \dfrac{25x^2}{x-5} \div \dfrac{10x}{5+4x-x^2}$

$= \dfrac{25x^2}{x-5} \cdot \dfrac{5+4x-x^2}{10x}$

$= \dfrac{5 \cdot 5 \cdot x \cdot x \cdot (-1)(x^2-4x-5)}{(x-5) \cdot 5 \cdot 2 \cdot x}$

$= \dfrac{5 \cdot x \cdot (-1)(x-5)(x+1)}{(x-5)2}$

$= -\dfrac{5x(x+1)}{2}, \quad x \neq 0, 5, -1$

11. $\dfrac{\left(\dfrac{x^2+3x-10}{x+4}\right)}{3x-6} = \dfrac{x^2+3x-10}{x+4} \div \dfrac{3x-6}{1}$

$= \dfrac{(x+5)(x-2)}{x+4} \cdot \dfrac{1}{3(x-2)}$

$= \dfrac{(x+5)(x-2)}{(x+4)3(x-2)}$

$= \dfrac{x+5}{3(x+4)}, \quad x \neq 2$

13. $\dfrac{2x-14}{\left(\dfrac{x^2-9x+14}{x+3}\right)} = \dfrac{2x-14}{1} \div \dfrac{x^2-9x+14}{x+3}$

$= \dfrac{2(x-7)}{1} \cdot \dfrac{x+3}{(x-7)(x-2)}$

$= \dfrac{2(x-7)(x+3)}{(x-7)(x-2)}$

$= \dfrac{2(x+3)}{x-2}, \quad x \neq -3, 7$

15. $\dfrac{\left(\dfrac{6x^2-17x+5}{3x^2+3x}\right)}{\dfrac{3x-1}{3x+1}} = \dfrac{6x^2-17x+5}{3x^2+3x} \div \dfrac{3x-1}{3x+1}$

$= \dfrac{(3x-1)(2x-5)}{3x(x+1)} \cdot \dfrac{3x+1}{3x-1}$

$= \dfrac{(3x-1)(2x-5)(3x+1)}{3x(x+1)(3x-1)}$

$= \dfrac{(2x-5)(3x+1)}{3x(x+1)}, \quad x \neq \pm\dfrac{1}{3}$

17. $$\frac{16x^2+8x+1}{3x^2+8x-3} \div \frac{4x^2-3x-1}{x^2+6x+9} = \frac{16x^2+8x+1}{3x^2+8x-3} \cdot \frac{x^2+6x+9}{4x^2-3x-1}$$
$$= \frac{(4x+1)(4x+1)}{(3x-1)(x+3)} \cdot \frac{(x+3)(x+3)}{(4x+1)(x-1)}$$
$$= \frac{(4x+1)(4x+1)(x+3)(x+3)}{(3x-1)(x+3)(4x+1)(x-1)}$$
$$= \frac{(4x+1)(x+3)}{(3x-1)(x-1)}, \quad x \neq -3, -\frac{1}{4}$$

19. $$\frac{x^2+3x-2x-6}{x^2-4} \div \frac{x+3}{x^2+4x+4} = \frac{x(x+3)-2(x+3)}{x^2-4} \cdot \frac{x^2+4x+4}{x+3}$$
$$= \frac{(x+3)(x-2)}{(x-2)(x+2)} \cdot \frac{(x+2)(x+2)}{x+3}$$
$$= \frac{(x+3)(x-2)(x+2)(x+2)}{(x-2)(x+2)(x+3)}$$
$$= (x+2), \quad x \neq \pm 2, -3$$

21. $$\frac{\left(\dfrac{x^2-3x-10}{x^2-4x+4}\right)}{\left(\dfrac{21+4x-x^2}{x^2-5x-14}\right)} = \frac{x^2-3x-10}{x^2-4x+4} \div \frac{21+4x-x^2}{x^2-5x-14}$$
$$= \frac{x^2-3x-10}{x^2-4x+4} \cdot \frac{x^2-5x-14}{-1(x^2-4x-21)}$$
$$= \frac{(x-5)(x+2)}{(x-2)(x-2)} \cdot \frac{(x-7)(x+2)}{-1(x-7)(x+3)}$$
$$= -\frac{(x^2-3x-10)(x+2)}{(x^2-4x+4)(x+3)}$$
$$= -\frac{(x-5)(x+2)(x+2)}{(x-2)^2(x+3)}$$
$$= -\frac{(x+2)^2(x-5)}{(x-2)^2(x+3)}, \quad x \neq -2, 7$$

23. $$\frac{\left(1+\dfrac{4}{y}\right)}{y} = \frac{\left(1+\dfrac{4}{y}\right)}{y} \cdot \frac{y}{y} = \frac{y+4}{y^2}$$

25. $$\frac{\left(\dfrac{4}{x}+3\right)}{\left(\dfrac{4}{x}-3\right)} = \frac{\left(\dfrac{4}{x}+3\right)}{\left(\dfrac{4}{x}-3\right)} \cdot \frac{x}{x} = \frac{4+3x}{4-3x}, \quad x \neq 0$$

27. $$\frac{\left(\dfrac{x}{2}\right)}{\left(2+\dfrac{3}{x}\right)} = \frac{\dfrac{x}{2}}{2+\dfrac{3}{x}} \cdot \frac{2x}{2x}$$
$$= \frac{x^2}{4x+6} = \frac{x^2}{2(2x+3)}, \quad x \neq 0$$

29. $$\frac{\left(3+\dfrac{9}{x-3}\right)}{\left(4+\dfrac{12}{x-3}\right)} = \frac{\left(3+\dfrac{9}{x-3}\right)}{\left(4+\dfrac{12}{x-3}\right)} \cdot \frac{x-3}{x-3}$$
$$= \frac{3(x-3)+\dfrac{9}{x-3}(x-3)}{4(x-3)+\dfrac{12}{x-3}(x-3)}$$
$$= \frac{3x-9+9}{4x-12+12}$$
$$= \frac{3x}{4x} = \frac{3}{4}, \quad x \neq 0, 3$$

31. $$\frac{\left(\frac{1}{x}-\frac{1}{x+1}\right)}{\left(\frac{1}{x+1}\right)} = \frac{\frac{1}{x}-\frac{1}{x+1}}{\frac{1}{x+1}} \cdot \frac{x(x+1)}{x(x+1)} = \frac{x+1-x}{x} = \frac{1}{x},\quad x \neq -1$$

33. $$\frac{2y-y^{-1}}{10-y^{-2}} = \frac{2y-\frac{1}{y}}{10-\frac{1}{y^2}} \cdot \frac{y^2}{y^2} = \frac{2y^3-y}{10y^2-1} = \frac{y(2y^2-1)}{10y^2-1},\quad y \neq 0$$

35. $$\frac{7x^2+2x^{-1}}{5x^{-3}+x} = \frac{7x^2+\frac{2}{x}}{\frac{5}{x^3}+x} \cdot \frac{x^3}{x^3} = \frac{7x^5+2x^2}{5+x^4} = \frac{x^2(7x^3+2)}{x^4+5},\quad x \neq 0$$

37. $$\frac{1}{\left(\frac{1}{R_1}+\frac{1}{R_2}\right)} = \frac{1}{\left(\frac{R_2}{R_1R_2}+\frac{R_1}{R_1R_2}\right)} = \frac{1}{\left(\frac{R_2+R_1}{R_1R_2}\right)} = \frac{1}{1}\cdot\frac{R_1R_2}{R_1+R_2} = \frac{R_1R_2}{R_1+R_2}$$

39. A complex fraction is a fraction that has a fraction in its numerator or denominator, or both. Examples of complex fractions:

$$\frac{\left(\frac{x-3}{x}\right)}{\left(\frac{3x-9}{4}\right)},\quad \frac{x+5}{\left(\frac{6}{3x+15}\right)},\quad \frac{\left(\frac{2x}{3}-\frac{x}{4}\right)}{\left(\frac{x+1}{6}\right)}$$

41. To simplify a complex fraction, multiply the numerator and denominator by the least common denominator of all of the fractions in the numerator and denominator.

43. $$\frac{\left(\frac{y}{x}-\frac{x}{y}\right)}{\left(\frac{x+y}{xy}\right)} = \frac{\left(\frac{y}{x}-\frac{x}{y}\right)}{\left(\frac{x+y}{xy}\right)} \cdot \frac{xy}{xy} = \frac{\frac{y}{x}(xy)-\frac{x}{y}(xy)}{\left(\frac{x+y}{xy}\right)xy} = \frac{y^2-x^2}{x+y} = \frac{(y-x)(y+x)}{x+y} = y-x,\quad x \neq 0, y \neq 0, x \neq -y$$

45. $$\frac{\left(\frac{x}{x-3}-\frac{2}{3}\right)}{\left(\frac{10}{3x}+\frac{x^2}{x-3}\right)} = \frac{\left(\frac{x}{x-3}-\frac{2}{3}\right)}{\left(\frac{10}{3x}+\frac{x^2}{x-3}\right)} \cdot \frac{3x(x-3)}{3x(x-3)} = \frac{3x^2-2x(x-3)}{10(x-3)+3x^3} = \frac{3x^2-2x^2+6x}{10x-30+3x^3} = \frac{x^2+6x}{3x^3+10x-30},\quad x \neq 0, x \neq 3$$

47. $$\frac{\left(\frac{10}{(x+1)}\right)}{\left(\frac{1}{2x+2}+\frac{3}{x+1}\right)} = \frac{\left(\frac{10}{x+1}\right)}{\left(\frac{1}{2x+2}+\frac{3}{x+1}\right)} \cdot \frac{2(x+1)}{2(x+1)} = \frac{10(2)}{1+3(2)} = \frac{20}{7},\quad x \neq -1$$

49. $$\frac{x^{-1}+y^{-1}}{x^{-1}-y^{-1}} = \frac{\frac{1}{x}+\frac{1}{y}}{\frac{1}{x}-\frac{1}{y}} \cdot \frac{xy}{xy} = \frac{y+x}{y-x},\quad x \neq 0, y \neq 0$$

51. $$\frac{x^{-2}-y^{-2}}{(x+y)^2} = \frac{\dfrac{\dfrac{1}{x^2}-\dfrac{1}{y^2}}{(x+y)^2}}{1}\cdot\frac{x^2y^2}{x^2y^2}$$
$$= \frac{y^2-x^2}{x^2y^2(x+y)(x+y)}$$
$$= \frac{(y-x)(y+x)}{x^2y^2(x+y)(x+y)}$$
$$= \frac{y-x}{x^2y^2(x+y)}$$

53. $$\frac{f(2+h)-f(2)}{h} = \frac{\dfrac{1}{2+h}-\dfrac{1}{2}}{h}$$
$$= \frac{\dfrac{1}{2+h}-\dfrac{1}{2}}{h}\cdot\frac{2(2+h)}{2(2+h)}$$
$$= \frac{2-(2+h)}{2h(2+h)}$$
$$= \frac{2-2-h}{2h(2+h)}$$
$$= \frac{-h}{2h(2+h)}$$
$$= \frac{-1}{2(2+h)},\quad h \neq 0$$

55. $$\frac{\dfrac{x}{5}+\dfrac{x}{6}}{2} = \frac{\dfrac{x}{5}+\dfrac{x}{6}}{2}\cdot\frac{30}{30} = \frac{6x+5x}{60} = \frac{11x}{60}$$

57. $$\frac{\dfrac{2x}{3}+\dfrac{x}{4}}{2} = \frac{\dfrac{2x}{3}+\dfrac{x}{4}}{2}\cdot\frac{12}{12}$$
$$= \frac{\dfrac{2x}{3}(12)+\dfrac{x}{4}(12)}{2(12)}$$
$$= \frac{8x+3x}{24}$$
$$= \frac{11x}{24}$$

59. $$\frac{\dfrac{b+5}{4}+\dfrac{2}{b}}{2} = \frac{\dfrac{b+5}{4}+\dfrac{2}{b}}{2}\cdot\frac{4b}{4b}$$
$$= \frac{\dfrac{b+5}{4}(4b)+\dfrac{2}{b}(4b)}{2(4b)}$$
$$= \frac{(b+5)b+8}{8b}$$
$$= \frac{b^2+5b+8}{8b}$$

61. $$\frac{\dfrac{x}{6}-\dfrac{x}{9}}{4} = \frac{\dfrac{x}{6}-\dfrac{x}{9}}{\dfrac{4}{1}}\cdot\frac{18}{18} = \frac{3x-2x}{72} = \frac{x}{72}$$

$$x_1 = \frac{x}{9}+\frac{x}{72} = \frac{8x}{72}+\frac{x}{72} = \frac{9x}{72} = \frac{x}{8}$$
$$x_2 = \frac{x}{8}+\frac{x}{72} = \frac{9x}{72}+\frac{x}{72} = \frac{10x}{72} = \frac{5x}{36}$$
$$x_3 = \frac{5x}{36}+\frac{x}{72} = \frac{10x}{72}+\frac{x}{72} = \frac{11x}{72}$$

63. $$\frac{1}{\left(\dfrac{1}{R_1}+\dfrac{1}{R_2}+\dfrac{1}{R_3}\right)} = \frac{1}{\left(\dfrac{R_2R_3}{R_1R_2R_3}+\dfrac{R_1R_3}{R_1R_2R_3}+\dfrac{R_1R_2}{R_1R_2R_3}\right)} = \frac{1}{\left(\dfrac{R_1R_2+R_1R_3+R_2R_3}{R_1R_2R_3}\right)} = \frac{R_1R_2R_3}{R_1R_2+R_1R_3+R_2R_3}$$

65. No. A complex fraction can be written as the division of two rational expressions, so the simplified form will be a rational expression.

67. In the second step, the set of parentheses cannot be moved because division is not associative.

$$\frac{\left(\dfrac{a}{b}\right)}{b} = \frac{a}{b}\cdot\frac{1}{b} = \frac{a}{b^2}$$

69. $(2y)^3(3y)^2 = 8y^3\cdot 9y^2 = 72y^5$

71. $3x^2+5x-2 = (3x-1)(x+2)$

73. $\dfrac{x^2}{2}\div 4x = \dfrac{x^2}{2}\cdot\dfrac{1}{4x} = \dfrac{x^2}{8x} = \dfrac{x}{8},\quad x \neq 0$

75. $$\frac{(x+1)^2}{x+2}\div\frac{x+1}{(x+2)^3} = \frac{(x+1)^2(x+2)^3}{(x+2)(x+1)} = (x+1)(x+2)^2,\quad x \neq -2,\quad x \neq -1$$

Mid-Chapter Quiz for Chapter 6

1. Domain: All real numbers x such that $x \neq -1$ and $x \neq 0$.

2. Domain: All real numbers x such that $x \neq -2$ and $x \neq 2$.

3. $\dfrac{9y^2}{6y} = \dfrac{3y}{2}, \quad y \neq 0$

4. $\dfrac{6u^4v^3}{15uv^3} = \dfrac{3 \cdot 2 \cdot u^4 \cdot v^3}{3 \cdot 5 \cdot u \cdot v^3} = \dfrac{2u^3}{5}, \quad u \neq 0, v \neq 0$

5. $$\frac{4x^2 - 1}{x - 2x^2} = \frac{(2x-1)(2x+1)}{x(1-2x)} = \frac{(2x-1)(2x+1)}{-x(2x-1)} = \frac{2x+1}{-x}, \quad x \neq \frac{1}{2}$$

6. $$\frac{(z+3)^2}{2z^2 + 5z - 3} = \frac{(z+3)(z+3)}{(2z-1)(z+3)} = \frac{z+3}{2z-1}, \quad z \neq -3$$

7. $$\frac{5a^2b + 3ab^3}{a^2b^2} = \frac{ab(5a + 3b^2)}{a^2b^2} = \frac{5a + 3b^2}{ab}$$

8. $$\frac{2mn^2 - n^3}{2m^2 + mn - n^2} = \frac{n^2(2m - n)}{(2m - n)(m + n)} = \frac{n^2}{m+n}, \quad 2m - n \neq 0$$

9. $$\frac{11t^2}{6} \cdot \frac{9}{33t} = \frac{11t^2(9)}{6(33t)} = \frac{t}{2}, \quad t \neq 0$$

10. $$(x^2 + 2x) \cdot \frac{5}{x^2 - 4} = \frac{x(x+2)5}{(x-2)(x+2)} = \frac{5x}{x-2}, \quad x \neq -2$$

11. $$\frac{4}{3(x-1)} \cdot \frac{12x}{6(x^2 + 2x - 3)} = \frac{4(12x)}{3(x-1)6(x+3)(x-1)} = \frac{8x}{3(x-1)^2(x+3)} = \frac{8x}{3(x-1)(x^2+2x-3)} = \frac{8x}{3(x-1)(x+3)(x-1)}$$

12. $$\frac{32z^4}{5x^5y^5} \div \frac{80z^5}{25x^8y^6} = \frac{32z^4 \cdot 25x^8y^6}{5x^5y^5 \cdot 80z^5} = \frac{16 \cdot 2 \cdot 5 \cdot 5 \cdot x^8 \cdot y^6 \cdot z^4}{16 \cdot 5 \cdot 5 \cdot x^5 \cdot y^5 \cdot z^5} = \frac{2x^3y}{z}, \quad x \neq 0, y \neq 0$$

13. $$\frac{a-b}{9a+9b} \div \frac{a^2 - b^2}{a^2 + 2a + 1} = \frac{a-b}{9(a+b)} \cdot \frac{(a+1)(a+1)}{(a-b)(a+b)} = \frac{(a-b)(a+1)^2}{9(a+b)^2(a-b)} = \frac{(a+1)^2}{9(a+b)^2}, \quad a \neq b$$

14. $$\frac{5u}{3(u+v)} \cdot \frac{2(u^2 - v^2)}{3v} \div \frac{25u^2}{18(u-v)} = \frac{5u \cdot 2(u-v)(u+v) \cdot 18(u-v)}{3(u+v)(3v)(25u^2)} = \frac{4(u-v)^2}{5uv}, \quad u \neq \pm v$$

15. $$\frac{5x-6}{x-2} + \frac{2x-5}{x-2} = \frac{5x - 6 + 2x - 5}{x-2} = \frac{7x - 11}{x-2}$$

16. $$\frac{x}{x^2 - 9} - \frac{4(x-3)}{x+3} = \frac{x}{(x-3)(x+3)} - \frac{4(x-3)^2}{(x-3)(x+3)}$$
$$= \frac{x - 4(x-3)^2}{(x-3)(x+3)} = \frac{x - 4(x^2 - 6x + 9)}{(x-3)(x+3)} = \frac{x - 4x^2 + 24x - 36}{(x-3)(x+3)} = \frac{-4x^2 + 25x - 36}{(x-3)(x+3)} = -\frac{4x^2 - 25x + 36}{(x-3)(x+3)}$$

17. $$\frac{x^2 + 2}{x^2 - x - 2} + \frac{1}{x+1} - \frac{x}{x-2} = \frac{x^2 + 2}{(x-2)(x+1)} + \frac{1(x-2)}{(x-2)(x+1)} - \frac{x(x+1)}{(x-2)(x+1)}$$
$$= \frac{x^2 + 2 + x - 2 - x^2 - x}{(x-2)(x+1)} = \frac{0}{(x-2)(x+1)} = 0, \quad x \neq 2, x \neq -1$$

18. $\dfrac{\frac{9t^2}{3-t}}{\frac{6t}{t-3}} \cdot \dfrac{t-3}{t-3} = \dfrac{-9t^2}{6t} = -\dfrac{3t}{2}, \quad t \neq 3$

19. $\dfrac{\frac{10}{x^2+2x}}{\frac{15}{x^2+3x+2}} = \dfrac{\frac{10}{x(x+2)}}{\frac{15}{(x+2)(x+1)}} \cdot \dfrac{x(x+2)(x+1)}{x(x+2)(x+1)}$

$= \dfrac{10(x+1)}{15x}$

$= \dfrac{2(x+1)}{3x}, \quad x \neq -2, x \neq -1$

20. $\dfrac{3x^{-1} - y^{-1}}{(x-y)^{-1}} = \dfrac{\left(\frac{3}{x} - \frac{1}{y}\right)}{\left(\frac{1}{x-y}\right)} = \dfrac{\left(\frac{3}{x} - \frac{1}{y}\right)}{\left(\frac{1}{x-y}\right)} \cdot \dfrac{xy(x-y)}{xy(x-y)}$

$= \dfrac{3y(x-y) - x(x-y)}{xy}$

$= \dfrac{3xy - 3y^2 - x^2 + xy}{xy}$

$= \dfrac{-3y^2 + 4xy - x^2}{xy}$

$= \dfrac{-1(3y^2 - 4xy + x^2)}{xy}$

$= \dfrac{-1(3y-x)(y-x)}{xy}$

$= \dfrac{(3y-x)(x-y)}{xy}, \quad x \neq y$

21. (a) *Verbal Model:* [Total cost] = [Set up cost] + 144 · [Number of arrangements]

Labels: Total cost $= C$

Number of arrangements $= x$

Equation: $C = 25{,}000 + 144x$

(b) *Verbal Model:* [Average cost] = [Total cost] ÷ [Number of arrangements]

Labels: Average cost $= \overline{C}$

Total cost $= C$

Number of arrangements $= x$

Equation: $\overline{C} = \dfrac{25{,}000 + 144x}{x}$

(c) $\overline{C}(500) = \dfrac{25{,}000 + 144(500)}{500} = \dfrac{97000}{500} = \$194$

22. $\dfrac{x + \frac{x}{2} + \frac{2x}{3}}{3} = \dfrac{x + \frac{x}{2} + \frac{2x}{3}}{3} \cdot \dfrac{6}{6} = \dfrac{6x + 3x + 4x}{18} = \dfrac{13x}{18}$

Section 6.5 Dividing Polynomials and Synthetic Division

1. $(7x^3 - 2x^2) \div x = \dfrac{7x^3 - 2x^2}{x}$

$= \dfrac{7x^3}{x} - \dfrac{2x^2}{x}$

$= 7x^2 - 2x, \quad x \neq 0$

3. $\dfrac{m^4 + 2m^2 - 7}{m} = \dfrac{m^4}{m} + \dfrac{2m^2}{m} - \dfrac{7}{m}$

$= m^3 + 2m - \dfrac{7}{m}$

5. $(4x^2 - 2x) \div (-x) = \dfrac{4x^2 - 2x}{-x}$

$= \dfrac{4x^2}{-x} - \dfrac{2x}{-x}$

$= -4x + 2, \quad x \neq 0$

7. $\dfrac{50z^3 + 30z}{-5z} = \dfrac{50z^3}{-5z} + \dfrac{30z}{-5z}$

$= -10z^2 - 6, \quad z \neq 0$

9. $\dfrac{4v^4 + 10v^3 - 8v^2}{4v^2} = \dfrac{4v^4}{4v^2} + \dfrac{10v^3}{4v^2} - \dfrac{8v^2}{4v^2} = v^2 + \dfrac{5}{2}v - 2, \quad v \neq 0$

11. $\left(5x^2y - 8xy + 7xy^2\right) \div 2xy = \dfrac{5x^2y - 8xy + 7xy^2}{2xy} = \dfrac{5x^2y}{2xy} - \dfrac{8xy}{2xy} + \dfrac{7xy^2}{2xy} = \dfrac{5x}{2} - 4 + \dfrac{7}{2}y, \quad x \neq 0, y \neq 0$

13.
$$\begin{array}{r} 112 \\ 9\overline{)1013} \\ \underline{9} \\ 11 \\ \underline{9} \\ 23 \\ \underline{18} \\ 5 \end{array}$$

So $1013 \div 9 = 112 + \frac{5}{9}$

15.
$$\begin{array}{r} 215 \\ 15\overline{)3235} \\ \underline{30} \\ 23 \\ \underline{15} \\ 85 \\ \underline{75} \\ 10 \end{array}$$

So $3235 \div 15 = 215 + \frac{10}{15} = 215 + \frac{2}{3}$

17.
$$\begin{array}{r} 242 \\ 25\overline{)6055} \\ \underline{50} \\ 105 \\ \underline{100} \\ 55 \\ \underline{50} \\ 5 \end{array}$$

So, $\frac{6055}{25} = 242 + \frac{5}{25}$

$= 242 + \frac{1}{5}$

19.
$$\dfrac{x^2 - 8x + 15}{x - 3} = \begin{array}{r} x - 5, \quad x \neq 3 \\ x - 3\overline{)x^2 - 8x + 15} \\ \underline{x^2 - 3x\phantom{{}+15}} \\ -5x + 15 \\ \underline{-5x + 15} \end{array}$$

21.
$$\begin{array}{r} x + 7, \quad x \neq 3 \\ -x + 3\overline{)-x^2 - 4x + 21} \\ \underline{-x^2 + 3x\phantom{{}+21}} \\ -7x + 21 \\ \underline{-7x + 21} \end{array}$$

23.
$$\begin{array}{r} 3t - 4, \quad t \neq \frac{3}{2} \\ 2t - 3\overline{)6t^2 - 17t + 12} \\ \underline{6t^2 - 9t\phantom{{}+12}} \\ -8t + 12 \\ \underline{-8t + 12} \end{array}$$

25.
$$\begin{array}{r} 3x^2 - 3x + 1 + \dfrac{2}{3x + 2} \\ 3x + 2\overline{)9x^3 - 3x^2 - 3x + 4} \\ \underline{9x^3 + 6x^2\phantom{{}-3x+4}} \\ -9x^2 - 3x\phantom{{}+4} \\ \underline{-9x^2 - 6x\phantom{{}+4}} \\ 3x + 4 \\ \underline{3x + 2} \\ 2 \end{array}$$

27.
$$\begin{array}{r} x - 4 + \dfrac{32}{x + 4} \\ x + 4\overline{)x^2 + 0x + 16} \\ \underline{x^2 + 4x\phantom{{}+16}} \\ -4x + 16 \\ \underline{-4x - 16} \\ 32 \end{array}$$

29.
$$\begin{array}{r} x^2 - 5x + 25, \quad x \neq -5 \\ x + 5\overline{)x^3 + 0x^2 + 0x + 125} \\ \underline{x^3 + 5x^2\phantom{{}+0x+125}} \\ -5x^2 + 0x\phantom{{}+125} \\ \underline{-5x^2 - 25x\phantom{{}+125}} \\ 25x + 125 \\ \underline{25x + 125} \end{array}$$

31.
$$\begin{array}{r} x + 2 + \dfrac{1}{x^2 + 2x + 3} \\ x^2 + 2x + 3\overline{)x^3 + 4x^2 + 7x + 7} \\ \underline{x^3 + 2x^2 + 3x\phantom{{}+7}} \\ 2x^2 + 4x + 7 \\ \underline{2x^2 + 4x + 6} \\ 1 \end{array}$$

33.

$$\begin{array}{r} 4x^2 + 12x + 25 + \dfrac{52x - 55}{x^2 - 3x + 2} \\ x^2 - 3x + 2\overline{)4x^4 + 0x^3 - 3x^2 + x - 5} \\ \underline{4x^4 - 12x^3 + 8x^2} \\ 12x^3 - 11x^2 + x \\ \underline{12x^3 - 36x^2 + 24x} \\ 25x^2 - 23x - 5 \\ \underline{25x^2 - 75x + 50} \\ 52x - 55 \end{array}$$

35. $(x^2 + x - 6) \div (x - 2)$

$$\begin{array}{r|rrr} 2 & 1 & 1 & -6 \\ & & 2 & 6 \\ \hline & 1 & 3 & 0 \end{array}$$

$(x^2 + x - 6) \div (x - 2) = x + 3,\ x \neq 2$

37. $\dfrac{x^3 + 3x^2 - 1}{x + 4}$

$$\begin{array}{r|rrrr} -4 & 1 & 3 & 0 & -1 \\ & & -4 & 4 & -16 \\ \hline & 1 & -1 & 4 & -17 \end{array}$$

$$\frac{x^3 + 3x^2 - 1}{x + 4} = x^2 - x + 4 + \frac{-17}{x + 4}$$

39. $\dfrac{5x^3 - 6x^2 + 8}{x - 4}$

$$\begin{array}{r|rrrr} 4 & 5 & -6 & 0 & 8 \\ & & 20 & 56 & 224 \\ \hline & 5 & 14 & 56 & 232 \end{array}$$

$$\frac{5x^3 - 6x^2 + 8}{x - 4} = 5x^2 + 14x + 56 + \frac{232}{x - 4}$$

41. $\dfrac{0.1x^2 + 0.8x + 1}{x - 0.2}$

$$\begin{array}{r|rrr} 0.2 & 0.1 & 0.8 & 1 \\ & & 0.02 & 0.164 \\ \hline & 0.1 & 0.82 & 1.164 \end{array}$$

$$\frac{0.1x^2 + 0.8x + 1}{x - 0.2} = 0.1x + 0.82 + \frac{1.164}{x - 0.2}$$

43.

$$\begin{array}{r|rrrr} 3 & 1 & -1 & -14 & 24 \\ & & 3 & 6 & -24 \\ \hline & 1 & 2 & -8 & 0 \end{array}$$

$$x^2 + 2x - 8 = (x + 4)(x - 2)$$

$$x^3 - x^2 - 14x + 24 = (x - 3)(x + 4)(x - 2)$$

45.

$$\begin{array}{r|rrrr} 1 & 4 & 0 & -3 & -1 \\ & & 4 & 4 & 1 \\ \hline & 4 & 4 & 1 & 0 \end{array}$$

$$4x^2 + 4x + 1 = (2x + 1)(2x + 1)$$

$$4x^3 - 3x^2 - 1 = (x - 1)(2x + 1)^2$$

47.

$$\begin{array}{r|rrrrr} -4 & 1 & 7 & 3 & -63 & -108 \\ & & -4 & -12 & 36 & 108 \\ \hline & 1 & 3 & -9 & -27 & 0 \end{array}$$

$$\begin{aligned} x^3 + 3x^2 - 9x - 27 &= x^2(x + 3) - 9(x + 3) \\ &= (x^2 - 9)(x + 3) \\ &= (x + 3)(x - 3)(x + 3) \end{aligned}$$

$$x^4 + 7x^3 + 3x^2 - 63x - 108 = (x + 3)^2(x - 3)(x + 4)$$

49. $\dfrac{15x^2 - 2x - 8}{x - \frac{4}{5}}$

$$\begin{array}{r|rrr} \frac{4}{5} & 15 & -2 & -8 \\ & & 12 & 8 \\ \hline & 15 & 10 & 0 \end{array}$$

$$\begin{aligned} 15x^2 - 2x - 8 &= (15x + 10)\left(x - \frac{4}{5}\right) \\ &= 5(3x + 2)\left(x - \frac{4}{5}\right) \end{aligned}$$

51. $\dfrac{x^3 + 2x^2 - 4x + c}{x - 2}$

$$\begin{array}{r|rrrr} 2 & 1 & 2 & -4 & c \\ & & 2 & 8 & 8 \\ \hline & 1 & 4 & 4 & 0 \end{array}$$

$$\begin{aligned} c + 8 &= 0 \\ c &= -8 \end{aligned}$$

53. The equation $1253 \div 12 = 104 + \frac{5}{12}$ has a dividend of 1253, a divisor of 12, a quotient of 104 and a remainder of 5.

55. A divisor divides evenly into a dividend when the remainder is zero.

57.
$$\begin{array}{r} x^2 + 2x - 5 - \dfrac{10}{x-2} \\ x - 2 \overline{\big) x^3 + 0x^2 - 9x + 0} \\ \underline{x^3 - 2x^2} \qquad\qquad \\ 2x^2 - 9x \qquad \\ \underline{2x^2 - 4x} \qquad \\ -5x + 0 \\ \underline{-5x + 10} \\ -10 \end{array}$$

59. $\dfrac{8u^2v}{2u} + \dfrac{3(uv)^2}{uv} = 4uv + \dfrac{3u^2v^2}{uv}$

$= 4uv + 3uv$

$= 7uv, \quad u \neq 0, v \neq 0$

61. $\dfrac{x^2 + 3x + 2}{x + 2} + (2x + 3) = \dfrac{(x + 2)(x + 1)}{x + 2} + (2x + 3)$

$= (x + 1) + (2x + 3)$

$= (x + 2x) + (1 + 3)$

$= 3x + 4, \qquad x \neq -2$

63.
$$\begin{array}{r} x^{2n} + x^n + 4, \; x^n \neq -2 \\ x^n + 2 \overline{\big) x^{3n} + 3x^{2n} + 6x^n + 8} \\ \underline{x^{3n} + 2x^{2n}} \qquad\qquad\quad \\ x^{2n} + 6x^n \qquad \\ \underline{x^{2n} + 2x^n} \qquad \\ 4x^n + 8 \\ \underline{4x^n + 8} \end{array}$$

65. Dividend = Divisor · Quotient + Remainder

$= (x - 6) \cdot (x^2 - x + 1) - 4$

$= x^3 + x^2 + x - 6x^2 - 6x - 6 - 4$

$= x^3 - 5x^2 - 5x - 10$

67. *Verbal Model:*

$\boxed{\text{Area of base}} = \boxed{\text{Volume}} \div \boxed{\text{Height}}$

$= (x^3 + 3x^2 + 3x + 1) \div (x + 1)$

$$\begin{array}{r|rrrr} -1 & 1 & 3 & 3 & 1 \\ & & -1 & -2 & -1 \\ \hline & 1 & 2 & 1 & 0 \end{array}$$

Area of base $= x^2 + 2x + 1$

69. Volume = Area of triangle · Height (of prism)

Area of triangle $= \dfrac{\text{Volume}}{\text{Height (of prism)}}$

$= \dfrac{x^3 + 18x^2 + 80x + 96}{x + 12}$

$= x^2 + 6x + 8$

Area of triangle $= \dfrac{1}{2} \cdot \text{Base} \cdot \text{Height}$

Height $= \dfrac{2(\text{Area of triangle})}{\text{Base}}$

$= \dfrac{2(x^2 + 6x + 8)}{x + 2}$

$= 2x + 8$ or $2(x + 4)$

71. x is not a factor of the numerator.

73. $\dfrac{x^2 + 4}{x + 1} = x - 1 + \dfrac{5}{x + 1}$

Divisor: $x + 1$

Dividend: $x^2 + 4$

Quotient: $x - 1$

Remainder: 5

75. $7 - 3x > 4 - x$

$-3x + x > 4 - 7$

$-2x > -3$

$x < \frac{3}{2}$

77. $|x - 3| < 2$

$-2 < x - 3 < 2$

$3 - 2 < x < 2 + 3$

$1 < x < 5$

79. $\left|\frac{1}{4}x - 1\right| \geq 3$

$\frac{1}{4}x - 1 \leq -3$ or $\frac{1}{4}x - 1 \geq 3$

$\frac{1}{4}x \leq -2$ $\qquad \frac{1}{4}x \geq 4$

$x \leq -8$ or $\qquad x \geq 16$

81. $(-3, y)$, y is a real number

This point is in Quadrant II or III because the x-coordinate is a negative number which means the point is to the left of the y-axis.

83. The set of points whose x-coordinates are 0 are located on the y-axis.

Section 6.6 Solving Rational Equations

1. (a) $x = 0$

$$\frac{0}{3} - \frac{0}{5} \stackrel{?}{=} \frac{4}{3}$$
$$0 \neq \frac{4}{3}$$

Not a solution

(b) $x = -2$

$$\frac{-2}{3} - \frac{-2}{5} \stackrel{?}{=} \frac{4}{3}$$
$$\frac{-10}{15} + \frac{6}{15} \stackrel{?}{=} \frac{20}{15}$$
$$\frac{-4}{15} \neq \frac{20}{15}$$

Not a solution

(c) $x = \frac{1}{8}$

$$\frac{\frac{1}{8}}{3} - \frac{\frac{1}{8}}{5} \stackrel{?}{=} \frac{4}{3}$$
$$\frac{1}{24} - \frac{1}{40} \stackrel{?}{=} \frac{4}{3}$$
$$\frac{5}{120} - \frac{3}{120} \stackrel{?}{=} \frac{160}{120}$$
$$\frac{2}{120} \neq \frac{160}{120}$$

Not a solution

(d) $x = 10$

$$\frac{10}{3} - \frac{10}{5} \stackrel{?}{=} \frac{4}{3}$$
$$\frac{50}{15} - \frac{30}{15} \stackrel{?}{=} \frac{20}{15}$$
$$\frac{20}{15} = \frac{20}{15}$$

Solution

3.
$$\frac{x}{6} - 1 = \frac{2}{3}$$
$$6\left(\frac{x}{6} - 1\right) = \left(\frac{2}{3}\right)6$$
$$x - 6 = 4$$
$$x = 10$$

Check:
$$\frac{10}{6} - 1 \stackrel{?}{=} \frac{2}{3}$$
$$\frac{5}{3} - \frac{3}{3} \stackrel{?}{=} \frac{2}{3}$$
$$\frac{2}{3} = \frac{2}{3}$$

5.
$$\frac{1}{4} = \frac{z+1}{8}$$
$$4(z+1) = 1(8)$$
$$4z + 4 = 8$$
$$4z = 4$$
$$z = 1$$

Check:
$$\frac{1}{4} \stackrel{?}{=} \frac{1+1}{8}$$
$$\frac{1}{4} \stackrel{?}{=} \frac{2}{8}$$
$$\frac{1}{4} = \frac{1}{4}$$

7.
$$\frac{x}{4} + \frac{x}{2} = \frac{2x}{3}$$
$$12\left(\frac{x}{4} + \frac{x}{2}\right) = \left(\frac{2x}{3}\right)12$$
$$3x + 6x = 8x$$
$$9x = 8x$$
$$x = 0$$

Check:
$$\frac{0}{4} + \frac{0}{2} \stackrel{?}{=} \frac{2(0)}{3}$$
$$0 = 0$$

9.
$$\frac{z+2}{3} = 4 - \frac{z}{12}$$
$$12\left(\frac{z+2}{3}\right) = \left(4 - \frac{z}{12}\right)12$$
$$4(z+2) = 48 - z$$
$$4z + 8 = 48 - z$$
$$5z = 40$$
$$z = 8$$

Check:
$$\frac{8+2}{3} \stackrel{?}{=} 4 - \frac{8}{12}$$
$$\frac{10}{3} \stackrel{?}{=} \frac{12}{3} - \frac{2}{3}$$
$$\frac{10}{3} = \frac{10}{3}$$

11.
$$\frac{x-5}{5} + 3 = -\frac{x}{4}$$
$$20\left(\frac{x-5}{5} + 3\right) = \left(-\frac{x}{4}\right)20$$
$$4(x-5) + 60 = -5x$$
$$4x - 20 + 60 = -5x$$
$$4x + 40 = -5x$$
$$40 = -9x$$
$$-\frac{40}{9} = x$$

Check:
$$\frac{-\frac{40}{9} - 5}{5} + 3 \stackrel{?}{=} -\frac{-\frac{40}{9}}{4}$$
$$\frac{-\frac{40}{9} - \frac{45}{9}}{5} + 3 \stackrel{?}{=} \frac{40}{9} \cdot \frac{1}{4}$$
$$\frac{1}{5}\left(-\frac{85}{9}\right) + 3 \stackrel{?}{=} \frac{10}{9}$$
$$-\frac{17}{9} + \frac{27}{9} \stackrel{?}{=} \frac{10}{9}$$
$$\frac{10}{9} = \frac{10}{9}$$

13. $\frac{x^2}{2} - \frac{3x}{5} = -\frac{1}{10}$

$10\left(\frac{x^2}{2} - \frac{3x}{5}\right) = \left(-\frac{1}{10}\right)10$

$5x^2 - 6x = -1$

$5x^2 - 6x + 1 = 0$

$(5x - 1)(x - 1) = 0$

$5x - 1 = 0 \qquad x - 1 = 0$

$x = \frac{1}{5} \qquad x = 1$

Check: $\frac{\left(\frac{1}{5}\right)^2}{2} - \frac{3\left(\frac{1}{5}\right)}{5} \stackrel{?}{=} \frac{-1}{10}$

$\frac{\frac{1}{25}}{2} - \frac{\frac{3}{5}}{5} \stackrel{?}{=} \frac{-1}{10}$

$\frac{1}{50} - \frac{3}{25} \stackrel{?}{=} \frac{-1}{10}$

$\frac{1}{50} - \frac{6}{50} \stackrel{?}{=} \frac{-1}{10}$

$-\frac{5}{50} \stackrel{?}{=} \frac{-1}{10}$

$\frac{-1}{10} = \frac{-1}{10}$

Check: $\frac{1^2}{2} - \frac{3(1)}{5} \stackrel{?}{=} \frac{-1}{10}$

$\frac{1}{2} - \frac{3}{5} \stackrel{?}{=} \frac{-1}{10}$

$\frac{5}{10} - \frac{6}{10} \stackrel{?}{=} \frac{-1}{10}$

$-\frac{1}{10} = \frac{-1}{10}$

15. $\frac{t}{2} = 12 - \frac{3t^2}{2}$

$2\left(\frac{t}{2}\right) = \left(12 - \frac{3t^2}{2}\right)2$

$t = 24 - 3t^2$

$3t^2 + t - 24 = 0$

$(3t - 8)(t + 3) = 0$

$3t - 8 = 0 \qquad t + 3 = 0$

$t = \frac{8}{3} \qquad t = -3$

Check: $\frac{\frac{8}{3}}{2} \stackrel{?}{=} 12 - \frac{3\left(\frac{8}{3}\right)^2}{2}$

$\frac{8}{6} \stackrel{?}{=} 12 - \frac{3\left(\frac{64}{9}\right)}{2}$

$\frac{4}{3} \stackrel{?}{=} 12 - \frac{\frac{64}{3}}{2}$

$\frac{4}{3} \stackrel{?}{=} 12 - \frac{64}{6}$

$\frac{4}{3} \stackrel{?}{=} \frac{36}{3} - \frac{32}{3}$

$\frac{4}{3} = \frac{4}{3}$

Check: $\frac{-3}{2} \stackrel{?}{=} 12 - \frac{3(-3)^2}{2}$

$\frac{-3}{2} \stackrel{?}{=} 12 - \frac{27}{2}$

$\frac{-3}{2} \stackrel{?}{=} \frac{24}{2} - \frac{27}{2}$

$\frac{-3}{2} = \frac{-3}{2}$

17. $\frac{9}{25 - y} = -\frac{1}{4}$

$4(25 - y)\left(\frac{9}{25 - y}\right) = \left(-\frac{1}{4}\right)4(25 - y)$

$36 = -(25 - y)$

$36 = -25 + y$

$61 = y$

Check: $\frac{9}{25 - 61} \stackrel{?}{=} -\frac{1}{4}$

$-\frac{9}{36} = -\frac{1}{4}$

$-\frac{1}{4} = -\frac{1}{4}$

19.

$$5 - \frac{12}{a} = \frac{5}{3}$$
$$3a\left(5 - \frac{12}{a}\right) = \left(\frac{5}{3}\right)3a$$
$$15a - 36 = 5a$$
$$10a = 36$$
$$a = \frac{36}{10}$$
$$a = \frac{18}{5}$$

Check:
$$5 - \frac{12}{\frac{18}{5}} \stackrel{?}{=} \frac{5}{3}$$
$$5 - \frac{60}{18} = \frac{5}{3}$$
$$\frac{15}{3} - \frac{10}{3} = \frac{5}{3}$$
$$\frac{5}{3} = \frac{5}{3}$$

21.

$$\frac{4}{x} - \frac{7}{5x} = -\frac{1}{2}$$
$$10x\left(\frac{4}{x} - \frac{7}{5x}\right) = \left(-\frac{1}{2}\right)10x$$
$$40 - 14 = -5x$$
$$26 = -5x$$
$$-\frac{26}{5} = x$$

Check:
$$\frac{4}{-\frac{26}{5}} - \frac{7}{5\left(-\frac{26}{5}\right)} \stackrel{?}{=} -\frac{1}{2}$$
$$-\frac{20}{26} + \frac{7}{26} \stackrel{?}{=} -\frac{1}{2}$$
$$-\frac{13}{26} \stackrel{?}{=} -\frac{1}{2}$$
$$-\frac{1}{2} = -\frac{1}{2}$$

23.

$$\frac{12}{y+5} + \frac{1}{2} = 2$$
$$2(y+5)\left(\frac{12}{y+5} + \frac{1}{2}\right) = (2)2(y+5)$$
$$24 + y + 5 = 4(y+5)$$
$$y + 29 = 4y + 20$$
$$9 = 3y$$
$$3 = y$$

Check:
$$\frac{12}{3+5} + \frac{1}{2} \stackrel{?}{=} 2$$
$$\frac{3}{2} + \frac{1}{2} = 2$$
$$\frac{4}{2} = 2$$
$$2 = 2$$

25.

$$\frac{5}{x} = \frac{25}{3(x+2)}$$
$$3x(x+2)\left(\frac{5}{x}\right) = \left(\frac{25}{3(x+2)}\right)3x(x+2)$$
$$15(x+2) = 25x$$
$$15x + 30 = 25x$$
$$30 = 10x$$
$$3 = x$$

Check:
$$\frac{5}{3} \stackrel{?}{=} \frac{25}{3(3+2)}$$
$$\frac{5}{3} = \frac{25}{15}$$
$$\frac{5}{3} = \frac{5}{3}$$

27.

$$\frac{8}{3x+5} = \frac{1}{x+2}$$
$$(3x+5)(x+2)\left(\frac{8}{3x+5}\right) = \left(\frac{1}{x+2}\right)(3x+5)(x+2)$$
$$8(x+2) = 3x + 5$$
$$8x + 16 = 3x + 5$$
$$5x = -11$$
$$x = -\frac{11}{5}$$

Check:
$$\frac{8}{3\left(-\frac{11}{5}\right) + 5} \stackrel{?}{=} \frac{1}{-\frac{11}{5} + 2}$$
$$\frac{8}{-\frac{33}{5} + \frac{25}{5}} = \frac{1}{-\frac{11}{5} + \frac{10}{5}}$$
$$\frac{8}{-\frac{8}{5}} = \frac{1}{-\frac{1}{5}}$$
$$-5 = -5$$

29.

$$\begin{aligned}
\frac{3}{x+2} - \frac{1}{x} &= \frac{1}{5x} \\
5x(x+2)\left(\frac{3}{x+2} - \frac{1}{x}\right) &= \left(\frac{1}{5x}\right)5x(x+2) \\
15x - 5(x+2) &= x + 2 \\
15x - 5x - 10 &= x + 2 \\
10x - 10 &= x + 2 \\
9x &= 12 \\
x &= \frac{12}{9} \\
x &= \frac{4}{3}
\end{aligned}$$

Check:

$$\begin{aligned}
\frac{1}{\frac{4}{3}+2} - \frac{1}{\frac{4}{3}} &\stackrel{?}{=} \frac{1}{5\left(\frac{4}{3}\right)} \\
\frac{1}{\frac{10}{3}} - \frac{1}{\frac{4}{3}} &= \frac{1}{\frac{20}{3}} \\
\frac{3}{10} - \frac{3}{4} &= \frac{3}{20} \\
\frac{18}{20} - \frac{15}{20} &= \frac{3}{20} \\
\frac{3}{20} &= \frac{3}{20}
\end{aligned}$$

31.

$$\frac{1}{x-4} + 2 = \frac{2x}{x-4}$$ The least common denominator is $x - 4$.

$$\left(x-4\right)\left(\frac{1}{x-4} + 2\right) = \left(\frac{2x}{x-4}\right)\left(x-4\right)$$ Multiply both sides by $x - 4$.

$$\begin{aligned}
1 + 2(x-4) &= 2x, \quad x \neq 4 \\
1 + 2x - 8 &= 2x \\
2x - 7 &= 2x \\
2x - 7 - 2x &= 2x - 2x \\
-7 &\neq 0
\end{aligned}$$

The original equation has no solution.

33.

$$\begin{aligned}
\frac{4}{x(x-1)} + \frac{3}{x} &= \frac{4}{x-1} \\
x(x-1)\left(\frac{4}{x(x-1)} + \frac{3}{x}\right) &= \left(\frac{4}{x-1}\right)x(x-1) \\
4 + 3(x-1) &= 4x \\
4 + 3x - 3 &= 4x \\
1 &= x
\end{aligned}$$

Check:

$$\begin{aligned}
\frac{4}{1(1-1)} + \frac{3}{1} &\stackrel{?}{=} \frac{4}{1-1} \\
\frac{4}{0} + 3 &= \frac{4}{0}
\end{aligned}$$

Division by zero is undefined. Solution $x = 1$ extraneous. No solution.

35.

$$\begin{aligned}
\frac{2}{x-10} - \frac{3}{x-2} &= \frac{6}{x^2 - 12x + 20} \\
\frac{2}{x-10} - \frac{3}{x-2} &= \frac{6}{(x-10)(x-2)} \\
(x-10)(x-2)\left(\frac{2}{x-10} - \frac{3}{x-2}\right) &= \left(\frac{6}{(x-10)(x-2)}\right)(x-10)(x-2) \\
2(x-2) - 3(x-10) &= 6 \\
2x - 4 - 3x + 30 &= 6 \\
-x + 26 &= 6 \\
-x &= -20 \\
x &= 20
\end{aligned}$$

Check:

$$\begin{aligned}
\frac{2}{20-10} - \frac{3}{20-2} &\stackrel{?}{=} \frac{6}{(20)^2 - 12(20) + 20} \\
\frac{2}{10} - \frac{3}{18} &\stackrel{?}{=} \frac{6}{400 - 240 + 20} \\
\frac{1}{5} - \frac{1}{6} &\stackrel{?}{=} \frac{6}{180} \\
\frac{6}{30} - \frac{5}{30} &= \frac{1}{30} \\
\frac{1}{30} &= \frac{1}{30}
\end{aligned}$$

37.

$$\begin{aligned}
1 - \frac{6}{4 - x} &= \frac{x + 2}{x^2 - 16} \\
1 + \frac{6}{x - 4} &= \frac{x + 2}{(x - 4)(x + 4)} \\
(x - 4)(x + 4)\left(1 + \frac{6}{x - 4}\right) &= \left(\frac{x + 2}{(x - 4)(x + 4)}\right)(x - 4)(x + 4) \\
(x^2 - 16) + 6(x + 4) &= x + 2 \\
x^2 - 16 + 6x + 24 &= x + 2 \\
x^2 + 5x + 6 &= 0 \\
(x + 3)(x + 2) &= 0 \\
x + 3 &= 0 \qquad x + 2 = 0 \\
x &= -3 \qquad x = -2
\end{aligned}$$

Check: $1 - \frac{6}{4 - (-3)} \stackrel{?}{=} \frac{-3 + 2}{(-3)^2 - 16}$

$1 - \frac{6}{7} \stackrel{?}{=} \frac{-1}{9 - 16}$

$\frac{1}{7} = \frac{1}{7}$

Check: $1 - \frac{6}{4 - (-2)} \stackrel{?}{=} \frac{-2 + 2}{(-2)^2 - 16}$

$1 - \frac{6}{6} \stackrel{?}{=} \frac{0}{4 - 16}$

$0 = 0$

39.

$$\begin{aligned}
\frac{2x}{5} &= \frac{x^2 - 5x}{5x} \\
5x\left(\frac{2x}{5}\right) &= \left(\frac{x^2 - 5x}{5x}\right)5x \\
2x^2 &= x^2 - 5x \\
x^2 + 5x &= 0 \\
x(x + 5) &= 0 \\
x = 0 \quad x + 5 &= 0 \\
x &= -5
\end{aligned}$$

Check:

$x = 0$

$\frac{2(0)}{5} \stackrel{?}{=} \frac{0^2 - 5(0)}{5(0)}$

$0 \neq$ undefined

so $x = 0$ is extraneous.

$x = -5$

$\frac{2(-5)}{5} \stackrel{?}{=} \frac{(-5)^2 - 5(-5)}{5(-5)}$

$\frac{-10}{5} = \frac{25 + 25}{-25}$

$-2 = -2$

41.

$$\begin{aligned}
\frac{y + 1}{y + 10} &= \frac{y - 2}{y + 4} \\
(y + 1)(y + 4) &= (y + 10)(y - 2) \\
y^2 + 5y + 4 &= y^2 + 8y - 20 \\
5y + 4 &= 8y - 20 \\
-3y &= -24 \\
y &= 8
\end{aligned}$$

Check: $\frac{8 + 1}{8 + 10} \stackrel{?}{=} \frac{8 - 2}{8 + 4}$

$\frac{9}{18} \stackrel{?}{=} \frac{6}{12}$

$\frac{1}{2} = \frac{1}{2}$

43.
$$\frac{x}{x-2}+\frac{3x}{x-4}=\frac{-2(x-6)}{x^2-6x+8}$$
$$(x-2)(x-4)\left(\frac{x}{x-2}+\frac{3x}{x-4}\right)=\left(\frac{-2(x-6)}{(x-4)(x-2)}\right)(x-2)(x-4)$$
$$x(x-4)+3x(x-2)=-2(x-6)$$
$$x^2-4x+3x^2-6x=-2x+12$$
$$4x^2-8x-12=0$$
$$x^2-2x-3=0$$
$$(x-3)(x+1)=0$$
$$x=3\quad x=-1$$

Check: $\frac{3}{3-2}+\frac{3(3)}{3-4}\stackrel{?}{=}\frac{-2(3-6)}{3^2-6(3)+8}$
$$3+\frac{9}{-1}\stackrel{?}{=}\frac{6}{-1}$$
$$-6=-6$$

Check: $\frac{-1}{-1-2}+\frac{3(-1)}{-1-4}\stackrel{?}{=}\frac{-2(-1-6)}{(-1)^2-6(-1)+8}$
$$\frac{1}{3}+\frac{3}{5}\stackrel{?}{=}\frac{14}{15}$$
$$\frac{5}{15}+\frac{9}{15}\stackrel{?}{=}\frac{14}{15}$$
$$\frac{14}{15}=\frac{14}{15}$$

45. $\frac{x}{3}=\frac{1+\frac{4}{x}}{1+\frac{2}{x}}$
$$\frac{x}{3}=\frac{1+\frac{4}{x}}{1+\frac{2}{x}}\cdot\frac{x}{x}$$
$$3(x+2)\left(\frac{x}{3}\right)=\left(\frac{x+4}{x+2}\right)3(x+2)$$
$$x(x+2)=3(x+4)$$
$$x^2+2x=3x+12$$
$$x^2-x-12=0$$
$$(x-4)(x+3)=0$$
$$x-4=0\qquad x+3=0$$
$$x=4\qquad x=-3$$

Check: $\frac{4}{3}\stackrel{?}{=}\frac{1+\frac{4}{4}}{1+\frac{2}{4}}$
$$\frac{4}{3}\stackrel{?}{=}\frac{1+1}{1+\frac{1}{2}}$$
$$\frac{4}{3}\stackrel{?}{=}\frac{2}{\frac{3}{2}}$$
$$\frac{4}{3}=\frac{4}{3}$$

Check: $\frac{-3}{3}\stackrel{?}{=}\frac{1+\frac{4}{-3}}{1+\frac{2}{-3}}$
$$-1\stackrel{?}{=}\frac{-\frac{1}{3}}{\frac{1}{3}}$$
$$-1=-1$$

47. *Verbal Model:* $\boxed{\frac{\text{Hits}}{\text{Total bats}}}=\boxed{\text{Batting average}}$

Label: x = additional safe hits

Equation:
$$\frac{8+x}{47+x}=0.250$$
$$(47+x)\left(\frac{8+x}{47+x}\right)=0.250(47+x)$$
$$8+x=11.75+0.250x$$
$$0.75x=3.75$$
$$x=5\text{ hits}$$

49. *Verbal Model:* Distance = Rate · Time

$$\frac{\text{Distance person 1}}{\text{Rate person 1}} = \frac{\text{Distance person 2}}{\text{Rate person 2}}$$

Label: Rate person 1 = $x + 1.5$

Rate person 2 = x

Equation:

$$\frac{4}{x + 1.5} = \frac{3}{x}$$

$$x(x + 1.5)\left(\frac{4}{x + 1.5}\right) = \left(\frac{3}{x}\right)x(x + 1.5)$$

$$4x = 3(x + 1.5)$$

$$4x = 3x + 4.5$$

$$x = 4.5 \text{ mph person 2}$$

$$x + 1.5 = 6 \text{ mph person 1}$$

51. A rational equation is an equation containing one or more rational expressions.

53. The domain of a rational equation must be restricted if any value of any of the variables makes any of the denominators zero.

55.

$$\frac{1}{2} = \frac{18}{x^2}$$

$$2x^2\left(\frac{1}{2}\right) = \left(\frac{18}{x^2}\right)2x^2$$

$$x^2 = 36$$

$$x^2 - 36 = 0$$

$$(x - 6)(x + 6) = 0$$

$$x = 6 \quad x = -6$$

Check: $\frac{1}{2} \stackrel{?}{=} \frac{18}{6^2}$ $\qquad$ $\frac{1}{2} \stackrel{?}{=} \frac{18}{(-6)^2}$

$\frac{1}{2} = \frac{18}{36}$ $\qquad$ $\frac{1}{2} = \frac{18}{36}$

$\frac{1}{2} = \frac{1}{2}$ $\qquad$ $\frac{1}{2} = \frac{1}{2}$

57.

$$\frac{x + 5}{4} - \frac{3x - 8}{3} = \frac{4 - x}{12}$$

$$12\left(\frac{x + 5}{4} - \frac{3x - 8}{3}\right) = \left(\frac{4 - x}{12}\right)12$$

$$3(x + 5) - 4(3x - 8) = 4 - x$$

$$3x + 15 - 12x + 32 = 4 - x$$

$$-9x + 47 = 4 - x$$

$$-8x = -43$$

$$x = \frac{43}{8}$$

Check:

$$\frac{\frac{43}{8} + 5}{4} - \frac{3\left(\frac{43}{8}\right) - 8}{3} \stackrel{?}{=} \frac{4 - \left(\frac{43}{8}\right)}{12}$$

$$\frac{1}{4}\left(\frac{43}{8} + \frac{40}{8}\right) - \frac{1}{3}\left(\frac{129}{8} - \frac{64}{8}\right) \stackrel{?}{=} \frac{1}{12}\left(\frac{32}{8} - \frac{43}{8}\right)$$

$$\frac{1}{4}\left(\frac{83}{8}\right) - \frac{1}{3}\left(\frac{65}{8}\right) \stackrel{?}{=} \frac{1}{12}\left(-\frac{11}{8}\right)$$

$$\frac{1}{8}\left(\frac{83}{4} - \frac{65}{3}\right) \stackrel{?}{=} \frac{1}{8}\left(-\frac{11}{12}\right)$$

$$\frac{1}{8}\left(\frac{249}{12} - \frac{260}{12}\right) \stackrel{?}{=} \frac{1}{8}\left(-\frac{11}{12}\right)$$

$$\frac{1}{8}\left(-\frac{11}{12}\right) = \frac{1}{8}\left(-\frac{11}{12}\right)$$

59.

$$\frac{16}{x^2 - 16} + \frac{x}{2x - 8} = \frac{1}{2} \rightarrow \text{equation}$$

$$2(x - 4)(x + 4)\left(\frac{16}{x^2 - 16} + \frac{x}{2(x - 4)}\right) = \left(\frac{1}{2}\right)2(x - 4)(x + 4)$$

$$32 + x(x + 4) = x^2 - 16$$

$$32 + x^2 + 4x = x^2 - 16$$

$$4x = -48$$

$$x = -12$$

61. $\dfrac{5}{x+3} + \dfrac{5}{3} + 3 \rightarrow$ expression

$$\frac{3(5)}{3(x+3)} + \frac{5(x+3)}{3(x+3)} + \frac{3(3)(x+3)}{3(x+3)} = \frac{15 + 5x + 15 + 9x + 27}{3(x+3)} = \frac{14x+57}{3(x+3)}$$

63. (a) x-intercept: $(-2, 0)$

(b)
$$0 = \frac{x+2}{x-2}$$
$$(x-2)(0) = \left(\frac{x+2}{x-2}\right)(x-2)$$
$$0 = x + 2$$
$$-2 = x$$

(a) and (b) $(-2, 0)$

65. (a) x-intercepts: $(-1, 0)$ and $(1, 0)$

(b)
$$0 = x - \frac{1}{x}$$
$$x(0) = \left(x - \frac{1}{x}\right)x$$
$$0 = x^2 - 1$$
$$0 = (x-1)(x+1)$$
$$x - 1 = 0 \qquad x + 1 = 0$$
$$x = 1 \qquad x = -1$$

(a) and (b) $(-1, 0), (1, 0)$

67. *Verbal Model:*

[Twice a number] + [3 times the reciprocal] = $\dfrac{203}{10}$

Labels: *Number* $= x$

Reciprocal $= \dfrac{1}{x}$

Equation:
$$2x + \frac{3}{x} = \frac{203}{10}$$
$$10x\left(2x + \frac{3}{x}\right) = \left(\frac{203}{10}\right)10x$$
$$20x^2 + 30 = 203x$$
$$20x^2 - 203x + 30 = 0$$
$$(20x - 3)(x - 10) = 0$$
$$20x - 3 = 0 \qquad x - 10 = 0$$
$$x = \frac{3}{20} \qquad x = 10$$

69. *Verbal Model:*

[Rate roofer 1] + [Rate roofer 2] = [Rate together]

Labels: Rate roofer 1 $= \dfrac{1}{15}$

Rate roofer 2 $= \dfrac{1}{21}$

Rate together $= \dfrac{1}{x}$

Equation:
$$\frac{1}{15} + \frac{1}{21} = \frac{1}{x}$$
$$105x\left(\frac{1}{15} + \frac{1}{21}\right) = \left(\frac{1}{x}\right)105x$$
$$7x + 5x = 105$$
$$12x = 105$$
$$x = \frac{105}{12} = 8.75 \text{ hours}$$
$$= 8 \text{ hours } 45 \text{ minutes}$$

71. An extraneous solution is a "trial solution" that does not satisfy the original equation.

73. When the equation is solved, the solution is $x = 0$. However, if $x = 0$, then there is division by zero, so the equation has no solution.

75. $x^2 - 81 = x^2 - (9)^2 = (x+9)(x-9)$

77. $4x^2 - \dfrac{1}{4} = (2x)^2 - \left(\dfrac{1}{2}\right)^2 = \left(2x + \dfrac{1}{2}\right)\left(2x - \dfrac{1}{2}\right)$

79. $f(x) = \dfrac{2x^2}{5}$

$5 \neq 0$

Domain: $(-\infty, \infty)$

Section 6.7 Variation

1.
$$R = kx$$
$$4825 = k(500)$$
$$\tfrac{4825}{500} = k$$
$$9.65 = k$$
$$R = 9.65x$$

3. (a)
$$d = kF \qquad d = \tfrac{1}{10}F$$
$$5 = k(50) \qquad d = \tfrac{1}{10}(20)$$
$$\tfrac{5}{50} = k \qquad d = 2 \text{ inches}$$
$$\tfrac{1}{10} = k$$

(b)
$$d = \tfrac{1}{10}F$$
$$1.5 = \tfrac{1}{10}F$$
$$15 \text{ pounds} = F$$

5.
$$d = kF$$
$$7 = k(10.5)$$
$$\tfrac{7}{10.5} = k$$
$$\tfrac{70}{105} = k$$
$$\tfrac{2}{3} = k$$
$$12 = \tfrac{2}{3}F$$
$$18 \text{ pounds} = F$$

7.
$$d = ks^2$$
$$75 = k(30)^2$$
$$75 = k(900)$$
$$\tfrac{75}{900} = k$$
$$\tfrac{1}{12} = k$$
$$d = \tfrac{1}{12}s^2$$
$$d = \tfrac{1}{12}(48)^2$$
$$d = \tfrac{1}{12}(2304)$$
$$d = 192 \text{ feet}$$

9.
$$x = \frac{k}{p}$$
$$800 = \frac{k}{5}$$
$$4000 = k$$
$$x = \frac{4000}{6}$$
$$x = 666.\overline{6} \approx 667 \text{ boxes}$$

11.
$$D = \frac{ka}{p}$$
$$2000 = \frac{k(500)}{5}$$
$$\frac{5(2000)}{500} = k$$
$$20 = k$$
$$2000 = \frac{20(600)}{p}$$
$$p = \frac{20(600)}{2000}$$
$$p = \$6$$

13.
$$i = kpt$$
$$10 = k(600)\left(\tfrac{1}{3}\right)$$
$$\tfrac{10}{200} = k$$
$$\tfrac{1}{20} = k$$
$$i = \tfrac{1}{20}(900)\left(\tfrac{1}{2}\right)$$
$$i = \tfrac{900}{40}$$
$$i = \$22.50$$

15. In a problem, y varies directly as x and the constant of proportionality is positive. If one of the variables increases, the other variable also increases because if one side of the equation increases, so must the other side.

17. The statements, y varies directly as x and y is directly proportional to the square of x, are not equivalent. The equation $y = kx$ is not equivalent to $y = kx^2$.

19. $I = kV$

21. $p = \dfrac{k}{d}$

23. $A = klw$

25.
$$h = kr$$
$$28 = k(12)$$
$$\tfrac{28}{12} = k$$
$$\tfrac{7}{3} = k$$
$$h = \tfrac{7}{3}r$$

27. $F = kxy$

$500 = k(15)(8)$

$\frac{500}{120} = k$

$\frac{25}{6} = k$

$F = \frac{25}{6}xy$

29. $d = k\left(\frac{x^2}{r}\right)$

$3000 = k\left(\frac{10^2}{4}\right)$

$3000 = k(25)$

$120 = k$

$d = \frac{120x^2}{r}$

31.

x	10	20	30	40	50
y	$\frac{2}{5}$	$\frac{1}{5}$	$\frac{2}{15}$	$\frac{1}{10}$	$\frac{2}{25}$

$\frac{2}{5} = \frac{k}{10}$ $\quad$ $\frac{1}{5} = \frac{k}{20}$

$4 = k$ $\quad$ $4 = k$

Using any two pairs of numbers, k is 4. So, $y = \frac{k}{x}$ with $k = 4$.

33. $P = kw^3$

$400 = k(20)^3$

$400 = k(8000)$

$\frac{400}{8000} = k$

$\frac{1}{20} = k$

$P = \frac{1}{20}w^3$

$= \frac{1}{20}(30)^3$

$= \frac{1}{20}(27{,}000)$

$= 1350$ watts

35. (a) $P = \frac{kWD^2}{L}$

(b) Unchanged

(c) Increases by a factor of 8.

(d) Increases by a factor of 4.

(e) Increases by a factor of $\frac{1}{4}$.

(f) $2000 = \frac{k(3)8^2}{120}$

$2000 = \frac{k(192)}{120}$

$1250 = k$

$L = \frac{1250(3)10^2}{120}$

$L = 3125$ pounds

37. In a joint variation problem where z varied jointly as x and y, if x increases, then z and y do not both necessarily increase.

39. $y = \frac{k}{x^2}$

$y = \frac{k}{(2x)^2}$

$y = \frac{k}{4x^2}$

y will be $\frac{1}{4}$ as great.

41. $(6)(6)(6)(6) = 6^4$

43. $\left(\frac{1}{5}\right)\left(\frac{1}{5}\right)\left(\frac{1}{5}\right)\left(\frac{1}{5}\right)\left(\frac{1}{5}\right) = \left(\frac{1}{5}\right)^5$

45.

$$\begin{array}{r|rrr} -2 & 1 & -5 & -14 \\ & & -2 & 14 \\ \hline & 1 & -7 & 0 \end{array}$$

$(x^2 - 5x - 14) \div (x + 2) = x - 7, x \neq -2$

47.

$$\begin{array}{r|rrr} 3 & 4 & -14 & 6 \\ & & 12 & -6 \\ \hline & 4 & -2 & 0 \end{array}$$

$\frac{4x^5 - 14x^4 + 6x^3}{x - 3} = 4x^4 - 2x^3, x \neq 3$

Review Exercises for Chapter 6

1. $y - 8 \neq 0$

$y \neq 8$

Domain: $(-\infty, 8) \cup (8, \infty)$

3. $f(x) = \dfrac{2x}{x^2 + 1}$

$x^2 + 1 \neq 0$

Domain: $(-\infty, \infty)$

5. $u^2 - 7u + 6 \neq 0$

$(u - 6)(u - 1) \neq 0$

$u \neq 6, \quad u \neq 1$

Domain: $(-\infty, 1) \cup (1, 6) \cup (6, \infty)$

7. Domain: $(0, \infty]$

$P = 2\left(w + \dfrac{36}{w}\right), \quad w \neq 0$

9. $\dfrac{6x^4y^2}{15xy^2} = \dfrac{2 \cdot 3x \cdot x^3 \cdot y^2}{5 \cdot 3x \cdot y^2}$

$= \dfrac{2x^3}{5}, \quad x \neq 0, y \neq 0$

11. $\dfrac{5b - 15}{30b - 120} = \dfrac{5(b - 3)}{30(b - 4)}$

$= \dfrac{5(b - 3)}{5 \cdot 6(b - 4)}$

$= \dfrac{b - 3}{6(b - 4)}$

13. $\dfrac{9x - 9y}{y - x} = \dfrac{9(x - y)}{-1(x - y)}$

$= -9, \quad x \neq y$

15. $\dfrac{x^2 - 5x}{2x^2 - 50} = \dfrac{x(x - 5)}{2(x^2 - 25)}$

$= \dfrac{x(x - 5)}{2(x - 5)(x + 5)}$

$= \dfrac{x}{2(x + 5)}, \quad x \neq 5$

17. $\dfrac{\text{Area of shaded region}}{\text{Area of whole figure}} = \dfrac{x(x + 2)}{(x + 2)2x}$

$= \dfrac{x}{2x} = \dfrac{1}{2}, \quad x > 0$

19. $\dfrac{4}{x} \cdot \dfrac{x^2}{12} = \dfrac{4x^2}{4 \cdot 3x} = \dfrac{x}{3}, \quad x \neq 0$

21. $\dfrac{7}{8} \cdot \dfrac{2x}{y} \cdot \dfrac{y^2}{14x^2} = \dfrac{7 \cdot 2 \cdot x \cdot y \cdot y}{2 \cdot 2 \cdot 2 \cdot y \cdot 7 \cdot 2 \cdot x \cdot x}$

$= \dfrac{y}{8x}, \quad y \neq 0$

23. $\dfrac{60z}{z + 6} \cdot \dfrac{z^2 - 36}{5} = \dfrac{5 \cdot 12z(z - 6)(z + 6)}{(z + 6)5}$

$= 12z(z - 6), \quad z \neq -6$

25. $\dfrac{u}{u - 3} \cdot \dfrac{3u - u^2}{4u^2} = \dfrac{u}{u - 3} \cdot \dfrac{-u(u - 3)}{4u^2} = -\dfrac{1}{4}, \quad u \neq 0, u \neq 3$

27. $24x^4 \div \dfrac{6x}{5} = 24x^4 \cdot \dfrac{5}{6x} = \dfrac{6 \cdot 4 \cdot 5 \cdot x^4}{6x} = 20x^3, \quad x \neq 0$

29. $25y^2 \div \dfrac{xy}{5} = 25y \cdot y \cdot \dfrac{5}{xy} = \dfrac{125y}{x}, \quad y \neq 0$

31. $\dfrac{x^2 + 3x + 2}{3x^2 + x - 2} \div (x + 2) = \dfrac{(x + 2)(x + 1)}{(3x - 2)(x + 1)} \cdot \dfrac{1}{(x + 2)} = \dfrac{(x + 2)(x + 1)}{(3x - 2)(x + 1)(x + 2)} = \dfrac{1}{3x - 2}, \quad x \neq -2, x \neq -1$

33. $\dfrac{4x}{5} + \dfrac{11x}{5} = \dfrac{15x}{5} = 3x$

35. $\dfrac{15}{3x} - \dfrac{3}{3x} = \dfrac{12}{3x} = \dfrac{4}{x}$

37. $\dfrac{8 - x}{4x} + \dfrac{5}{4x} = \dfrac{8 - x + 5}{4x} = \dfrac{13 - x}{4x} = -\dfrac{x - 13}{4x}$

39. $\dfrac{2(3y + 4)}{2y + 1} + \dfrac{3 - y}{2y + 1} = \dfrac{2(3y + 4) + 3 - y}{2y + 1}$

$= \dfrac{6y + 8 + 3 - y}{2y + 1}$

$= \dfrac{5y + 11}{2y + 1}$

41. $\dfrac{4x}{x+2} + \dfrac{3x-7}{x+2} - \dfrac{9}{x+2} = \dfrac{4x+3x-7-9}{x+2}$

$= \dfrac{7x-16}{x+2}$

43. $\dfrac{3}{5x^2} + \dfrac{4}{10x} = \dfrac{6}{10x^2} + \dfrac{4x}{10x^2} = \dfrac{6+4x}{10x^2}$

$= \dfrac{2(3+2x)}{10x^2} = \dfrac{2x+3}{5x^2}$

45. $\dfrac{1}{x+5} + \dfrac{3}{x-12} = \dfrac{1}{x+5}\left(\dfrac{x-12}{x-12}\right) + \dfrac{3}{x-12}\left(\dfrac{x+5}{x+5}\right)$

$= \dfrac{x-12}{(x+5)(x-12)} + \dfrac{3(x+5)}{(x-12)(x+5)}$

$= \dfrac{x-12+3x+15}{(x+5)(x-12)}$

$= \dfrac{4x+3}{(x+5)(x-12)}$

47. $5x + \dfrac{2}{x-3} - \dfrac{3}{x+2} = \dfrac{5x(x-3)(x+2)}{(x-3)(x+2)} + \dfrac{2}{(x-3)}\left(\dfrac{x+2}{x+2}\right) - \dfrac{3}{(x+2)}\left(\dfrac{x-3}{x-3}\right)$

$= \dfrac{5x^3 - 5x^2 - 30x + 2x + 4 - 3x + 9}{(x-3)(x+2)}$

$= \dfrac{5x^3 - 5x^2 - 31x + 13}{(x-3)(x+2)}$

49. $\dfrac{6}{x-5} - \dfrac{4x+7}{x^2-x-20} = \dfrac{6}{x-5} - \dfrac{4x+7}{(x-5)(x+4)}$

$= \dfrac{6(x+4)}{(x-5)(x+4)} - \dfrac{4x+7}{(x-5)(x+4)}$

$= \dfrac{6x+24-4x-7}{(x-5)(x+4)}$

$= \dfrac{2x+17}{(x-5)(x+4)}$

51. $\dfrac{5}{x+3} - \dfrac{4x}{(x+3)^2} - \dfrac{1}{x-3} = \dfrac{5}{x+3}\left(\dfrac{(x+3)(x-3)}{(x+3)(x-3)}\right) - \dfrac{4x}{(x+3)^2}\left(\dfrac{x-3}{x-3}\right) - \dfrac{1}{x-3}\left(\dfrac{(x+3)^2}{(x+3)^2}\right)$

$= \dfrac{5x^2 - 45 - 4x^2 + 12x - x^2 - 6x - 9}{(x+3)^2(x-3)}$

$= \dfrac{6x-54}{(x+3)^2(x-3)}$

53. $\dfrac{\left(\dfrac{6}{x}\right)}{\left(\dfrac{2}{x^3}\right)} = \dfrac{6}{x} \div \dfrac{2}{x^3} = \dfrac{3 \cdot 2}{x} \cdot \dfrac{x \cdot x^2}{2} = 3x^2, \quad x \neq 0$

55. $\dfrac{\left(\dfrac{x}{x-2}\right)}{\left(\dfrac{2x}{2-x}\right)} = \dfrac{x}{x-2} \cdot \dfrac{2-x}{2x} = \dfrac{x(-1)(x-2)}{(x-2)(2x)}$

$= -\dfrac{1}{2}, \quad x \neq 0,\ x \neq 2$

57. $\dfrac{\left(\dfrac{6x^2}{x^2+2x-35}\right)}{\left(\dfrac{x^3}{x^2-25}\right)} = \dfrac{\dfrac{6x^2}{(x+7)(x-5)}}{\dfrac{x^3}{(x-5)(x+5)}} \cdot \dfrac{(x+7)(x-5)(x+5)}{(x+7)(x-5)(x+5)}$

$= \dfrac{6x^2(x+5)}{x^3(x+7)} = \dfrac{6(x+5)}{x(x+7)}, \quad x \neq 5, x \neq -5$

59. $\dfrac{3t}{\left(5 - \dfrac{2}{t}\right)} \cdot \dfrac{t}{t} = \dfrac{3t^2}{5t-2}, \quad t \neq 0$

61. $$\frac{\left(x - 3 + \frac{2}{x}\right)}{\left(1 - \frac{2}{x}\right)} = \frac{\left(x - 3 + \frac{2}{x}\right)}{\left(1 - \frac{2}{x}\right)} \cdot \frac{x}{x}$$
$$= \frac{x^2 - 3x + 2}{x - 2}, \quad x \neq 0$$
$$= \frac{(x - 2)(x - 1)}{x - 2}, \quad x \neq 0$$
$$= x - 1, \quad x \neq 0, \ x \neq 2$$

63. $$\frac{\left(\frac{1}{a^2 - 16} - \frac{1}{a}\right)}{\left(\frac{1}{a^2 + 4a} + 4\right)} \cdot \frac{a(a - 4)(a + 4)}{a(a - 4)(a + 4)} = \frac{a - (a - 4)(a + 4)}{a - 4 + 4a(a - 4)(a + 4)}$$
$$= \frac{a - (a^2 - 16)}{a - 4 + 4a(a^2 - 16)}$$
$$= \frac{a - a^2 + 16}{a - 4 + 4a^3 - 64a}$$
$$= \frac{-a^2 + a + 16}{4a^3 - 63a - 4}, \quad a \neq 0, a \neq -4$$

65. $$(4x^3 - x) \div 2x = \frac{4x^3 - x}{2x}$$
$$= \frac{4x^3}{2x} - \frac{x}{2x}$$
$$= 2x^2 - \frac{1}{2}, x \neq 0$$

67. $$\frac{3x^3y^2 - x^2y^2 + x^2y}{x^2y} = \frac{3x^3y^2}{x^2y} - \frac{x^2y^2}{x^2y} + \frac{x^2y}{x^2y}$$
$$= 3xy - y + 1, \quad x \neq 0, y \neq 0$$

69.
$$\begin{array}{r l}
 & 5x - 8 + \frac{19}{x + 2} \\
x + 2 \overline{\big)} & 5x^2 + 2x + 3 \\
 & \underline{5x^2 + 10x} \\
 & -8x + 3 \\
 & \underline{-8x - 16} \\
 & 19
\end{array}$$

71.
$$\begin{array}{r l}
 & x^2 - 2, x \neq \pm 1 \\
x^2 - 1 \overline{\big)} & x^4 + 0x^3 - 3x^2 + 2 \\
 & \underline{x^4 \quad\quad - x^2} \\
 & -2x^2 + 2 \\
 & \underline{-2x^2 + 2}
\end{array}$$

73.
$$\begin{array}{r l}
 & x^2 - x - 3 + \frac{-3x^2 + 2x + 3}{x^3 - 2x^2 + x - 1} \\
x^3 - 2x^2 + x - 1 \overline{\big)} & x^5 - 3x^4 + 0x^3 + x^2 + 0x + 6 \\
 & \underline{x^5 - 2x^4 + x^3 - x^2} \\
 & -x^4 - x^3 + 2x^2 + 0x \\
 & \underline{-x^4 + 2x^3 - x^2 + x} \\
 & -3x^3 + 3x^2 - x + 6 \\
 & \underline{-3x^3 + 6x^2 - 3x + 3} \\
 & -3x^2 + 2x + 3
\end{array}$$

75.
$$\begin{array}{r|rrr}
-1 & 1 & 3 & 5 \\
 & & -1 & -2 \\
\hline
 & 1 & 2 & 3
\end{array}$$

$$\frac{x^2 + 3x + 5}{x + 1} = x + 2 + \frac{3}{x + 1}$$

77.
$$\begin{array}{r|rrrr}
-2 & 1 & 7 & 3 & -14 \\
 & & -2 & -10 & 14 \\
\hline
 & 1 & 5 & -7 & 0
\end{array}$$

$$\frac{x^3 - 7x^2 + 3x - 14}{x + 2} = x^2 + 5x - 7, x \neq -2$$

79.

$$\begin{array}{r|rrrrr} 3 & 1 & 0 & -3 & 0 & -25 \\ & & 3 & 9 & 18 & 54 \\ \hline & 1 & 3 & 6 & 18 & 29 \end{array}$$

$$(x^4 - 3x^2 - 25) \div (x - 3) = x^3 + 3x^2 + 6x + 18 + \frac{29}{x - 3}$$

81.

$$\begin{array}{r|rrrr} 2 & 1 & 2 & -5 & -6 \\ & & 2 & 8 & 6 \\ \hline & 1 & 4 & 3 & 0 \end{array}$$

$$x^3 + 2x^2 - 5x - 6 = (x - 2)(x + 1)(x + 3)$$

$$x^2 + 4x + 3$$

$$(x + 3)(x + 1)$$

83.

$$\frac{x}{15} + \frac{3}{5} = 1$$

$$15\left(\frac{x}{15} + \frac{3}{5}\right) = (1)15$$

$$x + 9 = 15$$

$$x = 6$$

85.

$$\frac{3x}{8} = -15 + \frac{x}{4}$$

$$8\left(\frac{3}{8}x\right) = (-15)8 + \left(\frac{x}{4}\right)8$$

$$3x = -120 + 2x$$

$$x = -120$$

Check: $\dfrac{3(-120)}{8} \stackrel{?}{=} -15 + \dfrac{-120}{4}$

$$\frac{-360}{8} \stackrel{?}{=} -15 + -30$$

$$-45 = -45$$

87.

$$\frac{x^2}{6} - \frac{x}{12} = \frac{1}{2}$$

$$12\left(\frac{x^2}{6} - \frac{x}{12}\right) = \left(\frac{1}{2}\right)12$$

$$2x^2 - x = 6$$

$$2x^2 - x - 6 = 0$$

$$(2x + 3)(x - 2) = 0$$

$$2x + 3 = 0 \qquad x - 2 = 0$$

$$x = -\frac{3}{2} \qquad x = 2$$

89. $(3t)\left(8 - \dfrac{12}{t}\right) = \dfrac{1}{3}(3t)$

$$24t - 36 = t$$

$$23t = 36$$

$$t = \frac{36}{23}$$

Check: $8 - \dfrac{12}{\left(\dfrac{36}{23}\right)} \stackrel{?}{=} \dfrac{1}{3}$

$$8 - \frac{23}{3} \stackrel{?}{=} \frac{1}{3}$$

$$\frac{24}{3} - \frac{23}{3} \stackrel{?}{=} \frac{1}{3}$$

$$\frac{1}{3} = \frac{1}{3}$$

91.

$$\frac{2}{y} - \frac{1}{3y} = \frac{1}{3}$$

$$3y\left(\frac{2}{y} - \frac{1}{3y}\right) = \left(\frac{1}{3}\right)3y$$

$$6 - 1 = y$$

$$5 = y$$

Check: $\dfrac{2}{5} - \dfrac{1}{3(5)} \stackrel{?}{=} \dfrac{1}{3}$

$$\frac{2}{5} - \frac{1}{15} \stackrel{?}{=} \frac{1}{3}$$

$$\frac{6}{15} - \frac{1}{15} \stackrel{?}{=} \frac{1}{3}$$

$$\frac{5}{15} \stackrel{?}{=} \frac{1}{3}$$

$$\frac{1}{3} = \frac{1}{3}$$

93.
$$r = 2 + \frac{24}{r}$$
$$r(r) = \left(2 + \frac{24}{r}\right)r$$
$$r^2 = 2r + 24$$
$$r^2 - 2r - 24 = 0$$
$$(r - 6)(r + 4) = 0$$
$$r = 6, r = -4$$

Check: $6 \stackrel{?}{=} 2 + \frac{24}{6}$

$6 \stackrel{?}{=} 2 + 4$

$6 = 6$

Check: $-4 \stackrel{?}{=} 2 + \frac{24}{-4}$

$-4 \stackrel{?}{=} 2 - 6$

$-4 = -4$

95.
$$\frac{t}{4} = \frac{4}{t}$$
$$4t\left(\frac{t}{4}\right) = \left(\frac{4}{t}\right)4t$$
$$t^2 = 16$$
$$t = \pm 4$$

Check: $\frac{4}{4} \stackrel{?}{=} \frac{4}{4}$

$1 = 1$

Check: $\frac{-4}{4} = \frac{4}{-4}$

$-1 = -1$

97.
$$\frac{3}{y+1} - \frac{8}{y} = 1$$
$$y(y+1)\left(\frac{3}{y+1} - \frac{8}{y}\right) = (1)y(y+1)$$
$$3y - 8(y+1) = y(y+1)$$
$$3y - 8y - 8 = y^2 + y$$
$$-5y - 8 = y^2 + y$$
$$0 = y^2 + 6y + 8$$
$$0 = (y+4)(y+2)$$
$$y + 4 = 0 \qquad y + 2 = 0$$
$$y = -4 \qquad y = -2$$

99.
$$\frac{2x}{x-3} - \frac{3}{x} = 0$$
$$x(x-3)\left(\frac{2x}{x-3} - \frac{3}{x}\right) = (0)x(x-3)$$
$$2x(x) - 3(x-3) = 0$$
$$2x^2 - 3x + 9 = 0$$

No real solution

101.
$$\frac{12}{x^2 + x - 12} - \frac{1}{x-3} = -1$$
$$(x-3)(x+4)\left(\frac{12}{x^2 + x - 12} - \frac{1}{x-3}\right) = (-1)(x-3)(x+4)$$
$$12 - (x+4) = -1(x^2 + x - 12)$$
$$12 - x - 4 = -x^2 - x + 12$$
$$(x^2 - 4) = 0$$
$$(x-2)(x+2) = 0$$
$$x - 2 = 0 \qquad x + 2 = 0$$
$$x = 2 \qquad x = -2$$

Check: $\frac{12}{2^2 + 2 - 12} - \frac{1}{2-3} \stackrel{?}{=} -1$

$\frac{12}{-6} - \frac{1}{-1} \stackrel{?}{=} -1$

$-2 + 1 \stackrel{?}{=} -1$

$-1 = -1$

Check: $\frac{12}{(-2)^2 + (-2) - 12} - \frac{1}{(-2) - 3} \stackrel{?}{=} -1$

$\frac{12}{-10} - \frac{1}{-5} \stackrel{?}{=} -1$

$-\frac{6}{5} + \frac{1}{5} \stackrel{?}{=} -1$

$-1 = -1$

103.
$$\frac{5}{x^2-4} - \frac{6}{x-2} = -5$$
$$(x-2)(x+2)\left(\frac{5}{x^2-4} - \frac{6}{x-2}\right) = (-5)(x-2)(x+2)$$
$$5 - 6(x+2) = -5(x^2-4)$$
$$5 - 6x - 12 = -5x^2 + 20$$
$$5x^2 - 6x - 27 = 0$$
$$(5x+9)(x-3) = 0$$
$$5x + 9 = 0 \qquad x - 3 = 0$$
$$x = -\frac{9}{5} \qquad x = 3$$

Check: $\dfrac{5}{\left(-\frac{9}{5}\right)^2 - 4} - \dfrac{6}{-\frac{9}{5} - 2} \stackrel{?}{=} -5$

$$\frac{5}{-\frac{19}{25}} - \frac{6}{-\frac{19}{5}} \stackrel{?}{=} -5$$
$$-\frac{125}{19} + \frac{30}{19} \stackrel{?}{=} -5$$
$$-\frac{95}{19} \stackrel{?}{=} -5$$
$$-5 = -5$$

Check: $\dfrac{5}{3^2 - 4} - \dfrac{6}{3-2} \stackrel{?}{=} -5$

$$\frac{5}{5} - \frac{6}{1} \stackrel{?}{=} -5$$
$$-5 = -5$$

105. *Verbal Model*: [Your time] = [Your friend's time]

[Your distance / Your rate] = [Friend's distance / Friend's rate]

Labels: Your distance = 24

Your rate = r

Friend's distance = 15

Friend's rate = $r - 6$

Equation:
$$\frac{24}{r} = \frac{15}{r-6}$$
$$r(r-6)\left(\frac{24}{r}\right) = \left(\frac{15}{r-6}\right)r(r-6)$$
$$24(r-6) = 15r$$
$$24r - 144 = 15r$$
$$9r = 144$$
$$r = 16 \text{ miles per hour} \qquad r - 6 = 10 \text{ miles per hour}$$

107.
$$d = kF$$
$$4 = k(100)$$
$$\frac{4}{100} = k$$
$$\frac{1}{25} = k$$
$$6 = \frac{1}{25}F$$
$$150 \text{ pounds} = F$$

109.
$$t = \frac{k}{r}$$
$$3 = \frac{k}{65}$$
$$3(65) = k$$
$$195 = k$$
$$t = \frac{195}{80}$$
$$t \approx 2.44 \text{ hours}$$

111. $D = \frac{ka}{p^2}$

$900 = \frac{k(20{,}000)}{55^2}$

$\frac{900(55^2)}{20{,}000} = k$

$136.125 = k$

$900 = \frac{136.125(25{,}000)}{p^2}$

$p^2 = \frac{136.125(25{,}000)}{900}$

$p^2 = 3781.25$

$p = \$61.49$

113. $C = khw^2$

$28.80 = k(16)(6)^2$ $\quad C = 0.05(14)(8)^2$

$28.80 = k(576)$ $\quad C = \$44.80$

$\frac{28.80}{576} = k$

$0.05 = k$

Chapter Test for Chapter 6

1. $f(x) = \frac{x+1}{x^2 - 6x + 5}$

$x^2 - 6x + 5 \neq 0$

$(x-1)(x-5) \neq 0$

$x - 1 \neq 0 \quad x - 5 \neq 0$

$x \neq 1 \quad x \neq 5$

Domain: $(-\infty, 1) \cup (1, 5) \cup (5, \infty)$

2. $\frac{4 - 2x}{x - 2} = \frac{-2(x-2)}{x-2} = -2, \quad x \neq 2$

3. $\frac{2a^2 - 5a - 12}{5a - 20} = \frac{(2a+3)(a-4)}{5(a-4)}$

$= \frac{2a+3}{5}, \quad a \neq 4$

4. Least common multiple of x^2, $3x^3$, and $(x+4)^2$:

$3x^3(x+4)^2$

5. $\frac{4z^3}{5} \cdot \frac{25}{12z^2} = \frac{4 \cdot z^2 \cdot z \cdot 5 \cdot 5}{5 \cdot 4 \cdot 3 \cdot z^2}$

$= \frac{5z}{3}, \quad z \neq 0$

6. $\frac{y^2 + 8y + 16}{2(y-2)} \cdot \frac{8y - 16}{(y+4)^3} = \frac{(y+4)^2 \cdot 8(y-2)}{2(y-2)(y+4)^2(y+4)}$

$= \frac{4}{y+4}, \quad y \neq 2$

7. $\frac{(2xy^2)^3}{15} \div \frac{12x^3}{21} = \frac{(2xy^2)^3}{15} \cdot \frac{21}{12x^3} = \frac{8x^3y^6 \cdot 7 \cdot 3}{5 \cdot 3 \cdot 4 \cdot 3x^3} = \frac{14y^6}{15}, \quad x \neq 0$

8. $(4x^2 - 9) \div \frac{2x+3}{2x^2 - x - 3} = (2x-3)(2x+3) \cdot \frac{(2x-3)(x+1)}{2x+3} = (2x-3)^2(x+1), \quad x \neq -\frac{3}{2}, x \neq -1$

9. $\frac{3}{x-3} + \frac{x-2}{x-3} = \frac{3 + x - 2}{x-3} = \frac{x+1}{x-3}$

10. $2x + \frac{1 - 4x^2}{x+1} = 2x\left(\frac{x+1}{x+1}\right) + \frac{1-4x^2}{x+1} = \frac{2x^2 + 2x}{x+1} + \frac{1 - 4x^2}{x+1} = \frac{-2x^2 + 2x + 1}{x+1}$

11. $\frac{5x}{x+2} - \frac{2}{x^2 - x - 6} = \frac{5x}{x+2} - \frac{2}{(x-3)(x+2)} = \frac{5x}{x+2}\left(\frac{x-3}{x-3}\right) - \frac{2}{(x-3)(x+2)} = \frac{5x^2 - 15x - 2}{(x+2)(x-3)}$

12. $$\frac{3}{x}-\frac{5}{x^2}+\frac{2x}{x^2+2x+1}=\frac{3}{x}-\frac{5}{x^2}+\frac{2x}{(x+1)^2}$$
$$=\frac{3}{x}\left[\frac{x(x+1)^2}{x(x+1)^2}\right]-\frac{5}{x^2}\left[\frac{(x+1)^2}{(x+1)^2}\right]+\frac{2x}{(x+1)^2}\left(\frac{x^2}{x^2}\right)$$
$$=\frac{3x(x^2+2x+1)-5(x^2+2x+1)+2x^3}{x^2(x+1)^2}$$
$$=\frac{3x^3+6x^2+3x-5x^2-10x-5+2x^3}{x^2(x+1)^2}$$
$$=\frac{5x^3+x^2-7x-5}{x^2(x+1)^2}$$

13. $$\frac{\left(\frac{3x}{x+2}\right)}{\left(\frac{12}{x^3+2x^2}\right)}=\frac{3x}{x+2}\div\frac{12}{x^3+2x^2}=\frac{3x}{x+2}\cdot\frac{x^2(x+2)}{12}=\frac{x^3}{4},\quad x\neq 0,-2$$

14. $$\frac{\left(9x-\frac{1}{x}\right)}{\left(\frac{1}{x}-3\right)}=\frac{\left(9x-\frac{1}{x}\right)}{\left(\frac{1}{x}-3\right)}\cdot\frac{x}{x}$$
$$=\frac{9x(x)-\frac{1}{x}(x)}{\frac{1}{x}(x)-3(x)}$$
$$=\frac{9x^2-1}{1-3x}$$
$$=\frac{(3x-1)(3x+1)}{-1(-1+3x)}$$
$$=-(3x+1),\quad x\neq 0,\frac{1}{3}$$

15. $$\frac{\left(\frac{3}{x^2}+\frac{1}{y}\right)}{\left(\frac{1}{x+y}\right)}=\frac{\left(\frac{3}{x^2}+\frac{1}{y}\right)}{\left(\frac{1}{x+y}\right)}\cdot\frac{x^2y(x+y)}{x^2y(x+y)}$$
$$=\frac{3y(x+y)+x^2(x+y)}{x^2y}$$
$$=\frac{(3y+x^2)(x+y)}{x^2y},\quad x\neq -y$$

16. $$\frac{6x^2-4x+8}{2x}=\frac{6x^2}{2x}-\frac{4x}{2x}+\frac{8}{2x}=3x-2+\frac{4}{x}$$

17. $$\frac{2x^4-15x^2-7}{x-3}$$

3	2	0	−15	0	−7
		6	18	9	27
	2	6	3	9	20

$$\frac{2x^4-15x^2-7}{x-3}=2x^3+6x^2+3x+9+\frac{20}{x-3}$$

18. $$\frac{t^4+t^2-6t}{t^2-2}=$$
$$\begin{array}{r}t^2 \qquad + 3 - \dfrac{6t-6}{t^2-2}\\ t^2-2\overline{\big)\,t^4+0t^3+t^2-6t+0}\\ t^4 \qquad -2t^2 \qquad\qquad\\ \hline 3t^2-6t \qquad\\ 3t^2 \qquad -6\\ \hline -6t+6\end{array}$$

19.
$$\frac{3}{h+2} = \frac{1}{6}$$
$$6(h+2)\left(\frac{3}{h+2}\right) = \left(\frac{1}{6}\right)6(h+2)$$
$$18 = h + 2$$
$$16 = h$$

20.
$$\frac{2}{x+5} - \frac{3}{x+3} = \frac{1}{x}$$
$$2x(x+3) - 3x(x+5) = (x+5)(x+3)$$
$$2x^2 + 6x - 3x^2 - 15x = x^2 + 3x + 5x + 15$$
$$-2x^2 - 17x - 15 = 0$$
$$2x^2 + 17x + 15 = 0$$
$$(2x+15)(x+1) = 0$$
$$2x + 15 = 0 \qquad x + 1 = 0$$
$$2x = 15 \qquad x = -1$$
$$x = -\frac{15}{2}$$

Check: $\dfrac{2}{-\frac{15}{2}+5} - \dfrac{3}{-\frac{15}{2}+3} \stackrel{?}{=} \dfrac{1}{-\frac{15}{2}}$
$$-\frac{12}{15} + \frac{10}{15} \stackrel{?}{=} -\frac{2}{15}$$
$$-\frac{2}{15} = -\frac{2}{15}$$

Check: $\dfrac{2}{-1+5} - \dfrac{3}{-1+3} \stackrel{?}{=} -\dfrac{1}{1}$
$$\frac{2}{4} - \frac{3}{2} \stackrel{?}{=} 1$$
$$\frac{1}{2} - \frac{3}{2} \stackrel{?}{=} -1$$
$$-1 = -1$$

21. $\dfrac{1}{x+1} + \dfrac{1}{x-1} = \dfrac{2}{x^2-1}$
$$x - 1 + x + 1 = 2$$
$$2x = 2$$
$$x = 1$$

Check: $\dfrac{1}{1+1} + \dfrac{1}{1-1} \neq \dfrac{2}{1-1}$

Division by zero is undefined. Solution is extraneous, so equation has no solution.

22.
$$v = k\sqrt{u}$$
$$\tfrac{3}{2} = k\sqrt{36}$$
$$\tfrac{3}{2} = k \cdot 6$$
$$\tfrac{1}{6} \cdot \tfrac{3}{2} = k$$
$$\tfrac{1}{4} = k$$
$$v = \tfrac{1}{4}\sqrt{u}$$

23.
$$P = \frac{k}{V}$$
$$1 = \frac{k}{180}$$
$$180 = k$$
$$0.75 = \frac{180}{V}$$
$$V = \frac{180}{0.75}$$
$$V = 240 \text{ cubic meters}$$

CHAPTER 7
Radicals and Complex Numbers

Section 7.1 Radicals and Rational Exponents 209

Section 7.2 Simplifying Radical Expressions...211

Section 7.3 Adding and Subtracting Radical Expressions 214

Mid-Chapter Quiz..217

Section 7.4 Multiplying and Dividing Radical Expressions 218

Section 7.5 Radical Equations and Applications 223

Section 7.6 Complex Numbers...229

Review Exercises 232

Chapter Test 237

Cumulative Test for Chapters 5–7..238

CHAPTER 7
Radicals and Complex Numbers

Section 7.1 Radicals and Rational Exponents

1. $\sqrt{64} = 8$ because $8 \cdot 8 = 64$.

3. $-\sqrt{100} = -10$ because $10 \cdot 10 = 100$.

5. $\sqrt{-25}$ is not a real number because no real number multiplied by itself yields -25.

7. $-\sqrt{-1}$ is not a real number because no real number multiplied by itself yields -4.

9. $\sqrt[3]{27} = 3$ because $3^3 = 27$.

11. $\sqrt[3]{-27} = -3$ because $-3 \cdot -3 \cdot -3 = -27$.

13. $-\sqrt[3]{1} = -1$ because $1^3 = 1$.

15. $-\sqrt[3]{-27} = 3$ because $(-3)^3 = 27$.

17. 25 is a perfect square because $5^2 = 25$.

19. $\frac{1}{16}$ is a perfect square because $\left(\frac{1}{4}\right)^2 = \frac{1}{16}$.

21. 49: Perfect square because $7^2 = 49$.

23. 1728: Perfect cube because $12^3 = 1728$.

25. 96: Neither because there is no integer that can be squared or cubed and yield 96.

27. $l = 11$ in. because $\sqrt{121} = 11$.

29. $l = 6$ ft because $\sqrt[3]{216} = 6$.

31. $\sqrt{8^2} = |8| = 8$

(index is even)

33. $\sqrt[3]{(5)^3} = 5$

(index is odd)

35. $\sqrt[3]{(-7)^3} = -7$

(index is odd)

37. $\sqrt{(-10)^2} = |-10| = 10$

(index is even)

39. $\sqrt{-\left(\frac{3}{10}\right)^2}$ is not a real number

(even root of a negative number)

41. $-\sqrt[3]{\left(\frac{1}{5}\right)^3} = -\frac{1}{5}$

(index is odd)

43. $-\sqrt[4]{2^4} = -2$

(inverse property of powers and roots)

45. $-\sqrt[5]{7^5} = -7$

(inverse property of powers and roots)

47.

Radical Form	*Rational Exponent Form*
$\sqrt{36} = 6$	$36^{1/2} = 6$

49.

Radical Form	*Rational Exponent Form*
$\sqrt[4]{256^3} = 64$	$256^{3/4} = 64$

51. $x^{1/3} = \sqrt[3]{x}$

53. $y^{2/5} = \sqrt[5]{y^2}$ or $\left(\sqrt[5]{y}\right)^2$

55. $27^{2/3} = \left(\sqrt[3]{27}\right)^2 = 3^2 = 9$

57. $\left(3^3\right)^{2/3} = (27)^{2/3} = \left(\sqrt[3]{27}\right)^2 = (3)^2 = 9$

59. $32^{-2/5} = \dfrac{1}{\left(\sqrt[5]{32}\right)^2} = \dfrac{1}{2^2} = \dfrac{1}{4}$

Root is 5. Power is 2.

61. $\left(\frac{8}{27}\right)^{2/3} = \left(\sqrt[3]{\frac{8}{27}}\right)^2 = \left(\frac{2}{3}\right)^2 = \frac{4}{9}$

Root is 3. Power is 2.

63. $-36^{1/2} = -\sqrt{36} = -6$

Root is 2. Power is 1.

65. $(-27)^{-2/3} = \dfrac{1}{(-27)^{2/3}} = \dfrac{1}{\left(\sqrt[3]{-27}\right)^2} = \dfrac{1}{9}$

Root is 3. Power is 2.

67. $x\sqrt[3]{x^6} = x \cdot x^{6/3} = x \cdot x^2 = x^3$

Root is 3. Power is 6.

69. $\dfrac{\sqrt[4]{t}}{\sqrt{t^5}} = \dfrac{t^{1/4}}{t^{5/2}} = t^{1/4-5/2} = t^{1/4-10/4} = t^{-9/4} = \dfrac{1}{t^{9/4}}$

71. $\sqrt[4]{x^3y} = \left(x^3y\right)^{1/4} = x^{3/4}y^{1/4}$

73. $\sqrt{\sqrt[4]{y}} = \left(y^{1/4}\right)^{1/2} = y^{1/4 \cdot 1/2} = y^{1/8} = \sqrt[8]{y}$

75.
$$\begin{aligned}\frac{(x+y)^{3/4}}{\sqrt[4]{x+y}} &= \frac{(x+y)^{3/4}}{(x+y)^{1/4}}\\ &= (x+y)^{3/4-1/4}\\ &= (x+y)^{2/4}\\ &= (x+y)^{1/2}\\ &= \sqrt{x+y}\end{aligned}$$

77. $f(x) = \sqrt{2x+9}$

(a) $f(0) = \sqrt{2(0)+9} = \sqrt{9} = 3$

(b) $f(8) = \sqrt{2(8)+9} = \sqrt{25} = 5$

(c) $f(-6) = \sqrt{2(-6)+9} = \sqrt{-3} =$ not a real number

(d) $f(36) = \sqrt{2(36)+9} = \sqrt{81} = 9$

79. $f(x) = \sqrt[5]{x}$

Domain: $(-\infty, \infty)$

81. $f(x) = 3\sqrt{x},\ x \ge 0$

Domain: $[0, \infty)$

83. $h(x) = \sqrt{2x+9}$

$$\begin{aligned}2x + 9 &\ge 0\\ 2x &\ge -9\\ x &\ge -\tfrac{9}{2}\end{aligned}$$

Domain: $\left[-\frac{9}{2}, \infty\right)$

85. The principal cube root of x represented in radical form is $\sqrt[3]{x}$ and rational exponent form is $x^{1/3}$.

87. The rule of exponents used to write $x^{1/3} \cdot x^{3/4}$ as $x^{1/3+3/4}$ is the product rule of exponents.

89. The square roots of $\frac{9}{16}$ are $-\frac{3}{4}$ and $\frac{3}{4}$ because $\left(-\frac{3}{4}\right)\left(-\frac{3}{4}\right) = \frac{9}{16}$ and $\left(\frac{3}{4}\right)\left(\frac{3}{4}\right) = \frac{9}{16}$.

91. The square roots of 0.16 are -0.4 and 0.4 because $(-0.4)(-0.4) = 0.16$ and $(0.4)(0.4) = 0.16$.

93. The cube root of $\frac{1}{1000}$ is $\frac{1}{10}$ because $\left(\frac{1}{10}\right)\left(\frac{1}{10}\right)\left(\frac{1}{10}\right) = \frac{1}{1000}$.

95. The cube root of 0.001 is 0.1 because $(0.1)(0.1)(0.1) = 0.001$.

97. $u^2\sqrt[3]{u} = u^2 \cdot u^{1/3} = u^{2+1/3} = u^{7/3}$

Root is 3. Power is 7.

99. $\dfrac{\sqrt{x}}{\sqrt{x^3}} = \dfrac{x^{1/2}}{x^{3/2}} = x^{1/2-3/2} = x^{-1} = \dfrac{1}{x}$

101.
$$\begin{aligned}\sqrt[4]{y^3} \cdot \sqrt[3]{y} &= y^{3/4} \cdot y^{1/3} = y^{3/4+1/3}\\ &= y^{9/12+4/12} = y^{13/12}\end{aligned}$$

103.
$$\begin{aligned}z^2\sqrt{y^5z^4} &= z^2 \cdot \left(y^5z^4\right)^{1/2}\\ &= z^2y^{5/2}z^2\\ &= z^{2+2}y^{5/2}\\ &= z^4y^{5/2}\end{aligned}$$

105. $\sqrt[4]{\sqrt{x^3}} = \sqrt[4]{x^{3/2}} = \left(x^{3/2}\right)^{1/4} = x^{3/2 \cdot 1/4} = x^{3/8}$

107.
$$\begin{aligned}\frac{(3u-2v)^{2/3}}{\sqrt{(3u-2v)^3}} &= \frac{(3u-2v)^{2/3}}{(3u-2v)^{3/2}}\\ &= (3u-2v)^{2/3-3/2}\\ &= (3u-2v)^{4/6-9/6}\\ &= (3u-2v)^{-5/6}\\ &= \frac{1}{(3u-2v)^{5/6}}\end{aligned}$$

109. $r = 1 - \left(\dfrac{25{,}000}{75{,}000}\right)^{1/8} = 1 - \left(\dfrac{1}{3}\right)^{1/8} \approx 0.128 \approx 12.8\%$

111. *Verbal Model:* Area = Side · Side

Labels: Area = 529

Side = x

Equation:
$$\begin{aligned}529 &= x \cdot x\\ 529 &= x^2\\ \sqrt{529} &= x\\ 23 &= x\end{aligned}$$

23 feet × 23 feet

113. If a and b are real numbers, n is an integer greater than or equal to 2, and $a = b^n$, then b is the nth root of a.

115. No. $\sqrt{2}$ is an irrational number. Its decimal representation is a nonterminating, nonrepeating decimal.

117. (a) "Last digits:"

1 (Perfect square 81)

4 (Perfect square 64)

5 (Perfect square 25)

6 (Perfect square 36)

9 (Perfect square 49)

0 (Perfect square 100)

(b) Yes, 4,322,788,986 ends in a 6, but it is not a perfect square.

119.
$$\begin{aligned} \frac{a}{5} &= \frac{a-3}{2} \\ 10\left(\frac{a}{5}\right) &= \left(\frac{a-3}{2}\right)^{10} \\ 2a &= 5(a-3) \\ 2a &= 5a - 15 \\ -3a &= -15 \\ a &= 5 \end{aligned}$$

121.
$$\begin{aligned} \frac{2}{u+4} &= \frac{5}{8} \\ 8(u+4)\left(\frac{2}{u+4}\right) &= \left(\frac{5}{8}\right)8(u+4) \\ 16 &= 5(u+4) \\ 16 &= 5u + 20 \\ -4 &= 5u \\ -\frac{4}{5} &= u \end{aligned}$$

123. $s = kt^2$

125. $a = kbc$

Section 7.2 Simplifying Radical Expressions

1. $\sqrt{28} = \sqrt{4 \cdot 7} = \sqrt{4} \cdot \sqrt{7} = 2\sqrt{7}$

The figure supports the answer.

For a square, $A = s^2$, so $\sqrt{A} = s$.

So, the figure shows $\sqrt{A} = \sqrt{28} = s = 2\sqrt{7}$.

3. $\sqrt{8} = \sqrt{4 \cdot 2} = \sqrt{2^2 \cdot 2} = 2\sqrt{2}$

5. $\sqrt{18} = \sqrt{9 \cdot 2} = \sqrt{3^2 \cdot 2} = 3\sqrt{2}$

7. $\sqrt{45} = \sqrt{9 \cdot 5} = \sqrt{3^2 \cdot 5} = 3\sqrt{5}$

9. $\sqrt{96} = \sqrt{16 \cdot 6} = \sqrt{4^2 \cdot 6} = 4\sqrt{6}$

11. $\sqrt{153} = \sqrt{9 \cdot 17} = \sqrt{3^2 \cdot 17} = 3\sqrt{17}$

13. $\sqrt{1183} = \sqrt{169 \cdot 7} = \sqrt{13^2 \cdot 7} = 13\sqrt{7}$

15. $\sqrt{4y^2} = \sqrt{2^2 y^2} = \sqrt{2^2} \cdot \sqrt{y^2} = 2|y|$

17. $\sqrt{9x^5} = \sqrt{3^2 x^4 \cdot x} = 3 \cdot x^2 \cdot \sqrt{x} = 3x^2\sqrt{x}$

19. $\sqrt{48y^4} = \sqrt{16 \cdot 3 \cdot y^4} = 4y^2\sqrt{3}$

21. $\sqrt{117y^6} = \sqrt{9 \cdot 13 \cdot y^6} = 3|y^3|\sqrt{13}$

23. $\sqrt{120x^2y^3} = \sqrt{4 \cdot 30 \cdot x^2 \cdot y^2 \cdot y} = 2|x|y\sqrt{30y}$

25. $\sqrt{192a^5b^7} = \sqrt{64 \cdot 3 \cdot a^4 \cdot a \cdot b^6 \cdot b} = 8a^2b^3\sqrt{3ab}$

27. $\sqrt[3]{48} = \sqrt[3]{16 \cdot 3} = \sqrt[3]{2^4 \cdot 3} = 2\sqrt[3]{3 \cdot 2} = 2\sqrt[3]{6}$

29. $\sqrt[3]{112} = \sqrt[3]{8 \cdot 14} = \sqrt[3]{8} \cdot \sqrt[3]{14} = 2\sqrt[3]{14}$

31. $\sqrt[4]{x^7} = \sqrt[4]{x^4x^3} = \sqrt[4]{x^4} \cdot \sqrt[4]{x^3} = x\sqrt[4]{x^3}$

33. $\sqrt[4]{x^6} = \sqrt[4]{x^4x^2} = \sqrt[4]{x^4} \cdot \sqrt[4]{x^2} = |x|\sqrt[4]{x^2}$

35. $\sqrt[3]{40x^5} = \sqrt[3]{8 \cdot 5 \cdot x^3 \cdot x^2} = 2x\sqrt[3]{5x^2}$

37.
$$\begin{aligned} \sqrt[4]{324y^6} &= \sqrt[4]{81 \cdot 4 \cdot y^4 \cdot y^2} \\ &= 3|y|\sqrt[4]{4y^2} \\ &= 3|y|\sqrt[4]{2^2y^2} \\ &= 3|y|\sqrt{2y} \end{aligned}$$

39. $\sqrt[3]{x^4y^3} = \sqrt[3]{x^3 \cdot x \cdot y^3} = xy\sqrt[3]{x}$

41. $\sqrt[4]{4x^4y^6} = \sqrt[4]{4x^4 \cdot y^4 \cdot y^2} = |xy|\sqrt[4]{4y^2}$

43. $\sqrt[5]{32x^5y^6} = \sqrt[5]{2^5 \cdot x^5 \cdot y^5 \cdot y} = 2xy\sqrt[5]{y}$

45. $\sqrt{\dfrac{16}{9}} = \dfrac{\sqrt{16}}{\sqrt{9}} = \dfrac{4}{3}$

47. $\sqrt[3]{\dfrac{35}{64}} = \dfrac{\sqrt[3]{35}}{4}$

49. $\dfrac{\sqrt{39y^2}}{\sqrt{3}} = \sqrt{\dfrac{39}{3}y^2} = \sqrt{13y^2} = |y|\sqrt{13}$

51. $\sqrt{\dfrac{32a^4}{b^2}} = \dfrac{\sqrt{16 \cdot 2 \cdot a^4}}{\sqrt{b^2}} = \dfrac{4a^2\sqrt{2}}{|b|}$

53. $\sqrt[5]{\dfrac{32x^2}{y^5}} = \sqrt[5]{\dfrac{2^5x^2}{y^5}} = \dfrac{2}{y}\sqrt[5]{x^2}$

55. $\sqrt{\dfrac{1}{3}} = \dfrac{1}{\sqrt{3}} \cdot \dfrac{\sqrt{3}}{\sqrt{3}} = \dfrac{\sqrt{3}}{3}$

57. $\dfrac{1}{\sqrt{7}} = \dfrac{1}{\sqrt{7}} \cdot \dfrac{\sqrt{7}}{\sqrt{7}} = \dfrac{\sqrt{7}}{7}$

59. $\sqrt[4]{\dfrac{5}{4}} = \dfrac{\sqrt[4]{5}}{\sqrt[4]{2^2}} \cdot \dfrac{\sqrt[4]{2^2}}{\sqrt[4]{2^2}} = \dfrac{\sqrt[4]{5 \cdot 2^2}}{\sqrt[4]{2^4}} = \dfrac{\sqrt[4]{20}}{2}$

61. $\dfrac{6}{\sqrt[3]{32}} = \dfrac{6}{\sqrt[3]{2^3 \cdot 2^2}}$

$= \dfrac{6}{2\sqrt[3]{2^2}} \cdot \dfrac{\sqrt[3]{2}}{\sqrt[3]{2}} = \dfrac{6\sqrt[3]{2}}{2\sqrt[3]{2^3}} = \dfrac{6\sqrt[3]{2}}{4} = \dfrac{3\sqrt[3]{2}}{2}$

63. $\dfrac{1}{\sqrt{y}} = \dfrac{1}{\sqrt{y}} \cdot \dfrac{\sqrt{y}}{\sqrt{y}} = \dfrac{\sqrt{y}}{\sqrt{y^2}} = \dfrac{\sqrt{y}}{y}$

65. $\sqrt{\dfrac{4}{x}} = \dfrac{\sqrt{4}}{\sqrt{x}} = \dfrac{2}{\sqrt{x}} \cdot \dfrac{\sqrt{x}}{\sqrt{x}} = \dfrac{2\sqrt{x}}{x}$

67. $\dfrac{1}{x\sqrt{2}} = \dfrac{1}{x\sqrt{2}} \cdot \dfrac{\sqrt{2}}{\sqrt{2}} = \dfrac{\sqrt{2}}{2x}$

69. $\dfrac{6}{\sqrt{3b^3}} = \dfrac{6}{b\sqrt{3b}} \cdot \dfrac{\sqrt{3b}}{\sqrt{3b}} = \dfrac{6\sqrt{3b}}{3b^2} = \dfrac{2\sqrt{3b}}{b^2}$

71. $\sqrt[3]{\dfrac{2x}{3y}} = \dfrac{\sqrt[3]{2x}}{\sqrt[3]{3y}} \cdot \dfrac{\sqrt[3]{3^2y^2}}{\sqrt[3]{3^2y^2}} = \dfrac{\sqrt[3]{2x \cdot 3^2y^2}}{\sqrt[3]{3^3y^3}} = \dfrac{\sqrt[3]{18xy^2}}{3y}$

73. $\sqrt[3]{\dfrac{24x^3y^4}{25z}} = \dfrac{\sqrt[3]{(2^3)(3)(x^3)(y^3)(y)}}{\sqrt[3]{25z}} \cdot \dfrac{\sqrt[3]{5z^2}}{\sqrt[3]{5z^2}}$

$= \dfrac{2xy\sqrt[3]{3y} \cdot \sqrt[3]{5z^2}}{\sqrt[3]{5^3z^3}}$

$= \dfrac{2xy\sqrt[3]{15yz^2}}{5z}$

75. $c = \sqrt{a^2 + b^2}$

$= \sqrt{6^2 + 3^2}$

$= \sqrt{36 + 9}$

$= \sqrt{45}$

$= \sqrt{9 \cdot 5}$

$= 3\sqrt{5}$

3

6

77. $c = \sqrt{a^2 + b^2}$

$= \sqrt{\left(4\sqrt{2}\right)^2 + 7^2}$

$= \sqrt{32 + 49}$

$= \sqrt{81}$

$= 9$

79. *Verbal Model:* $\boxed{\text{Hypotenuse}}^2 = \boxed{\text{Leg 1}}^2 + \boxed{\text{Leg 2}}^2$

Labels: Hypotenuse $= c$

Leg 1 $= 26$

Leg 2 $= 10$

Equation: $c^2 = 26^2 + 10^2$

$c^2 = 676 + 100$

$c^2 = 776$

$c = \sqrt{776} \approx 27.86$ feet

81. Product Rule for Radicals:

If a and b are real numbers, variables, or algebraic expressions, and if the nth roots of a and b are real, then $\sqrt[n]{ab} = \sqrt[n]{a} \cdot \sqrt[n]{b}$.

Example: $\sqrt[3]{108} = \sqrt[3]{27} \cdot \sqrt[3]{4} = 3\sqrt[3]{4}$

83. Three conditions that must be true for a radical expression to be in simplest form:

- All possible nth-powered factors have been removed from each radical.
- No radical contains a fraction.
- No denominator of a fraction contains a radical.

85. $\sqrt{0.04} = \sqrt{4 \cdot 0.01} = \sqrt{4}\sqrt{0.01} = 2 \cdot 0.1 = 0.2$

87. $\sqrt{0.0072} = \sqrt{36 \cdot 2 \cdot 0.0001}$

$= \sqrt{36} \cdot \sqrt{2} \cdot \sqrt{0.0001}$

$= 6 \cdot 0.01\sqrt{2}$

$= 0.06\sqrt{2}$

89. $\sqrt{\dfrac{60}{3}} = \sqrt{20} = \sqrt{4 \cdot 5} = \sqrt{2^2 \cdot 5} = 2\sqrt{5}$

91. $\sqrt{\dfrac{13}{25}} = \dfrac{\sqrt{13}}{5}$

93. $\sqrt[3]{\dfrac{54a^4}{b^9}} = \sqrt[3]{\dfrac{3^3 \cdot 2 \cdot a^3 \cdot a}{b^9}} = \dfrac{3a}{b^3}\sqrt[3]{2a}$

95. $\sqrt{4 \times 10^{-4}} = \sqrt{4 \cdot 0.0001}$

$= \sqrt{4} \cdot \sqrt{0.0001}$

$= 2 \cdot 0.01$

$= 2 \cdot \frac{1}{100}$

$= \frac{1}{50}$

97. $\sqrt[3]{2.4 \times 10^6} = \sqrt[3]{24 \times 10^5}$

$= \sqrt[3]{(2^3)(3)(10^3)(10^2)}$

$= 2 \cdot 10 \cdot \sqrt[3]{3 \cdot 10^2}$

$= 20\sqrt[3]{300}$

99. $f = \dfrac{1}{100}\sqrt{\dfrac{400 \times 10^6}{5}}$

$\approx 8.9443 \times 10^1$

≈ 89.443

≈ 89.44 cycles per second

101. (a) *Verbal Model:* $\boxed{\text{Hypotenuse}}^2 = \boxed{\text{Leg 1}}^2 + \boxed{\text{Leg 2}}^2$

Labels: Hypotenuse $= c$

Leg 1 $= \frac{30}{2} = 15$

Leg 2 $= 8$

Equation: $c^2 = 15^2 + 8^2$

$c^2 = 225 + 64$

$c^2 = 289$

$c = \sqrt{289}$

$c = 17$ ft

(b) Area $= 2 \cdot 17 \cdot 40$

$= 1360$ sq ft

103. $\sqrt{8}$ is not in simplest form because the perfect square factor 4 needs to be removed from the radical.

105. The display appears to be approaching 1.

107. Because $\sqrt[3]{u} = u^{1/3}$ and $\sqrt[4]{u} = u^{1/4}$, when $\sqrt[3]{u}$ and $\sqrt[4]{u}$ are multiplied, the rational exponents need to be added together. Therefore,

$\sqrt[3]{u} \cdot \sqrt[4]{u} = u^{1/3} \cdot u^{1/4} = u^{4/12+3/12} = u^{7/12} = \sqrt[12]{u^7}.$

109. $\begin{cases} 2x + 3y = 12 \\ 4x - y = 10 \end{cases} \Rightarrow \begin{aligned} 3y &= -2x + 12 \\ -y &= -4x + 10 \end{aligned} \Rightarrow \begin{aligned} y &= -\tfrac{2}{3}x + 4 \\ y &= 4x - 10 \end{aligned}$

Solution: (3, 2)

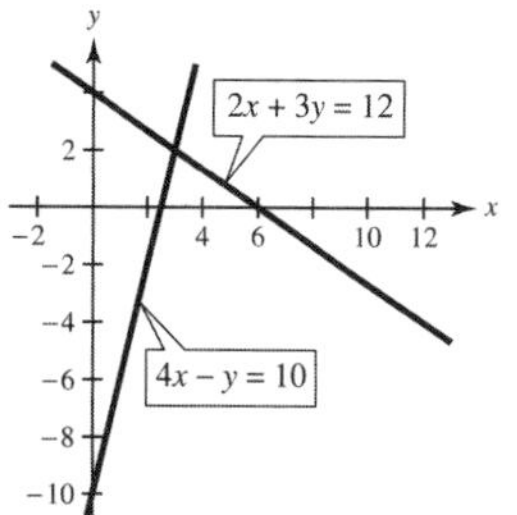

111. $\begin{cases} y = x + 2 \\ y - x = 8 \end{cases}$

$(x + 2) - x = 8$

$2 \neq 8$

No solution

113. $\begin{cases} x + 4y + 3z = 2 \\ 2x + y + z = 10 \\ -x + y + 2z = 8 \end{cases}$

$$\begin{aligned} x + 4y + 3z &= 2 \\ -7y - 5z &= 6 \\ 5y + 5z &= 10 \end{aligned}$$

$$\begin{aligned} x + 4y + 3z &= 2 \\ -7y - 5z &= 6 \\ -2y &= 16 \\ y &= -8 \end{aligned} \qquad \begin{aligned} -7(-8) - 5z &= 6 \\ 56 - 5z &= 6 \\ -5z &= -50 \\ z &= 10 \end{aligned} \qquad \begin{aligned} x + 4(-8) + 3(10) &= 2 \\ x - 32 + 30 &= 2 \\ x - 2 &= 2 \\ x &= 4 \end{aligned}$$

$(4, -8, 10)$

Section 7.3 Adding and Subtracting Radical Expressions

1. $3\sqrt{2} - \sqrt{2} = 2\sqrt{2}$

3. $4\sqrt[3]{y} + 9\sqrt[3]{y} = 13\sqrt[3]{y}$

5. $\sqrt{7} + 3\sqrt{7} - 2\sqrt{7} = (1 + 3 - 2)\sqrt{7} = 2\sqrt{7}$

7. $8\sqrt{2} + 6\sqrt{2} - 5\sqrt{2} = (8 + 6 - 5)\sqrt{2} = 9\sqrt{2}$

9. $2\sqrt{2} + 5\sqrt{2} - \sqrt{2} + 3\sqrt{2} = (2 + 5 - 1 + 3)\sqrt{2} = 9\sqrt{2}$

11. $\sqrt[4]{5} - 6\sqrt[4]{13} + 3\sqrt[4]{5} - \sqrt[4]{13} = (1 + 3)\sqrt[4]{5} + (-6 - 1)\sqrt[4]{13} = 4\sqrt[4]{5} - 7\sqrt[4]{13}$

13. $9\sqrt[3]{7} - \sqrt{3} + 4\sqrt[3]{7} + 2\sqrt{3} = \left(9\sqrt[3]{7} + 4\sqrt[3]{7}\right) + \left(-\sqrt{3} + 2\sqrt{3}\right) = 13\sqrt[3]{7} + \sqrt{3}$

15. $7\sqrt{x} + 5\sqrt[3]{9} - 3\sqrt{x} + 3\sqrt[3]{9} = (5 + 3)\sqrt[3]{9} + (7 - 3)\sqrt{x} = 8\sqrt[3]{9} + 4\sqrt{x}$

17. $3\sqrt[4]{x} + 5\sqrt[4]{x} - 3\sqrt{x} - 11\sqrt[4]{x} = (3 + 5 - 11)\sqrt[4]{x} - 3\sqrt{x} = -3\sqrt[4]{x} - 3\sqrt{x}$

19. $$\begin{aligned} 8\sqrt{27} - 3\sqrt{3} &= 8\sqrt{9 \cdot 3} - 3\sqrt{3} \\ &= 8(3)\sqrt{3} - 3\sqrt{3} \\ &= 24\sqrt{3} - 3\sqrt{3} \\ &= 21\sqrt{3} \end{aligned}$$

21. $$\begin{aligned} 3\sqrt{45} + 7\sqrt{20} &= 3\sqrt{9 \cdot 5} + 7\sqrt{4 \cdot 5} \\ &= 3(3)\sqrt{5} + 7(2)\sqrt{5} \\ &= 9\sqrt{5} + 14\sqrt{5} \\ &= 23\sqrt{5} \end{aligned}$$

23. $\sqrt{16x} - \sqrt{9x} = 4\sqrt{x} - 3\sqrt{x} = \sqrt{x}$

25. $5\sqrt{9x} - 3\sqrt{x} = 15\sqrt{x} - 3\sqrt{x} = 12\sqrt{x}$

27. $\sqrt{18y} + 4\sqrt{72y} = 3\sqrt{2y} + 24\sqrt{2y} = 27\sqrt{2y}$

29. $6\sqrt{x^3} - x\sqrt{9x} = 6x\sqrt{x} - 3x\sqrt{x} = 3x\sqrt{x}$

31. $9\sqrt{\dfrac{40}{x^4}} - 2\sqrt{\dfrac{90}{x^4}} = \dfrac{18\sqrt{10}}{x^2} - \dfrac{6\sqrt{10}}{x^2} = \dfrac{12\sqrt{10}}{x^2}$

33. $\sqrt{48x^5} + 2x\sqrt{27x^3} - 3x^2\sqrt{3x} = 4x^2\sqrt{3x} + 6x^2\sqrt{3x} - 3x^2\sqrt{3x} = 7x^2\sqrt{3x}$

35. $$\begin{aligned} 2\sqrt[3]{54} + 12\sqrt[3]{16} &= 2\sqrt[3]{27 \cdot 2} + 12\sqrt[3]{8 \cdot 2} \\ &= 6\sqrt[3]{2} + 24\sqrt[3]{2} \\ &= 30\sqrt[3]{2} \end{aligned}$$

37. $$\begin{aligned} \sqrt[3]{6x^4} + \sqrt[3]{48x} &= \sqrt[3]{6 \cdot x^3 \cdot x} + \sqrt[3]{8 \cdot 6 \cdot x} \\ &= x\sqrt[3]{6x} + 2\sqrt[3]{6x} \\ &= (x + 2)\sqrt[3]{6x} \end{aligned}$$

39. $5\sqrt[3]{24u^2} + 2\sqrt[3]{81u^5} = 5\sqrt[3]{2^3 \cdot 3u^2} + 2\sqrt[3]{3^3 \cdot 3u^3 \cdot u^2}$
$= 5(2)\sqrt[3]{3u^2} + 2(3u)\sqrt[3]{3u^2}$
$= 10\sqrt[3]{3u^2} + 6u\sqrt[3]{3u^2}$
$= (10 + 6u)\sqrt[3]{3u^2}$

41. $\sqrt[3]{3x^4} + \sqrt[3]{81x} + \sqrt[3]{24x^4} = x\sqrt[3]{3x} + 3\sqrt[3]{3x} + 2x\sqrt[3]{3x} = (3x + 3)\sqrt[3]{3x}$

43. $\sqrt[3]{\frac{6x}{24}} + 4\sqrt[3]{2x} = \frac{1}{2}\sqrt[3]{\frac{6x}{3}} + 4\sqrt[3]{2x} = \frac{1}{2}\sqrt[3]{2x} + 4\sqrt[3]{2x} = \frac{9}{2}\sqrt[3]{2x}$

45. $h = \sqrt[3]{x} + \sqrt[3]{8x} = \sqrt[3]{x} + 2\sqrt[3]{x} = 3\sqrt[3]{x}$ ft

47. $\sqrt{5} - \frac{3}{\sqrt{5}} = \sqrt{5} - \left(\frac{3}{\sqrt{5}} \cdot \frac{\sqrt{5}}{\sqrt{5}}\right) = \sqrt{5} - \frac{3\sqrt{5}}{5}$
$= \left(1 - \frac{3}{5}\right)\sqrt{5}$
$= \frac{2}{5}\sqrt{5}$

49. $\sqrt{32} + \sqrt{\frac{1}{2}} = \sqrt{16 \cdot 2} + \frac{\sqrt{1}}{\sqrt{2}} \cdot \frac{\sqrt{2}}{\sqrt{2}}$
$= 4\sqrt{2} + \frac{\sqrt{2}}{2}$
$= \frac{8\sqrt{2}}{2} + \frac{1\sqrt{2}}{2}$
$= \frac{9\sqrt{2}}{2}$

51. $\sqrt{18y} - \frac{y}{\sqrt{2y}} = 3\sqrt{2y} - \frac{y}{\sqrt{2y}} \cdot \frac{\sqrt{2y}}{\sqrt{2y}}$
$= 3\sqrt{2y} - \frac{y\sqrt{2y}}{2y}$
$= \frac{5y\sqrt{2y}}{2y}$
$= \frac{5\sqrt{2y}}{2}$

53. $\frac{2}{\sqrt{3x}} + \sqrt{3x} = \frac{2}{\sqrt{3x}} \cdot \frac{\sqrt{3x}}{\sqrt{3x}} + \sqrt{3x}$
$= \frac{2\sqrt{3x}}{3x} + \frac{3x\sqrt{3x}}{3x}$
$= \frac{(3x + 2)\sqrt{3x}}{3x}$

55. $\sqrt{7y^3} - \sqrt{\frac{9}{7y^3}} = \sqrt{7y^2 \cdot y} - \frac{3}{\sqrt{7y^2 \cdot y}}$
$= y\sqrt{7y} - \frac{3}{y\sqrt{7y}} \cdot \frac{\sqrt{7y}}{\sqrt{7y}}$
$= y\sqrt{7y} - \frac{3\sqrt{7y}}{y \cdot 7y}$
$= y\sqrt{7y} - \frac{3\sqrt{7y}}{7y^2}$
$= \frac{7y^2}{7y^2}\left(\frac{y\sqrt{7y}}{1}\right) - \frac{3\sqrt{7y}}{7y^2}$
$= \frac{7y^3\sqrt{7y}}{7y^2} - \frac{3\sqrt{7y}}{7y^2}$
$= \frac{\sqrt{7y}(7y^3 - 3)}{7y^2}$

57. $P = \sqrt{54x} + \sqrt{96x} + \sqrt{150x}$
$= \sqrt{9 \cdot 6x} + \sqrt{16 \cdot 6x} + \sqrt{25 \cdot 6x}$
$= 3\sqrt{6x} + 4\sqrt{6x} + 5\sqrt{6x}$
$= 12\sqrt{6x}$

59. $P = \sqrt{175x} + \sqrt{63x} + \sqrt{252x} + \sqrt{112x}$
$= 5\sqrt{7x} + 3\sqrt{7x} + 6\sqrt{7x} + 4\sqrt{7x}$
$= 18\sqrt{7x}$

61. $P = x\sqrt{128} + \sqrt{162x} + \sqrt{200x}$
$= 8x\sqrt{2} + 9\sqrt{2x} + 10\sqrt{2x}$
$= 19\sqrt{2x} + 8x\sqrt{2}$

63. $P = \sqrt{12x} + \sqrt{48}x + \sqrt{27x} + \sqrt{75}x$
$= \sqrt{4 \cdot 3x} + \sqrt{16 \cdot 3}x + \sqrt{9 \cdot 3x} + \sqrt{25 \cdot 3}x$
$= 2\sqrt{3x} + 4\sqrt{3}x + 3\sqrt{3x} + 5\sqrt{3}x$
$= \left(2\sqrt{3x} + 3\sqrt{3x}\right) + \left(4\sqrt{3} + 5\sqrt{3}\right)x$
$= 5\sqrt{3x} + \left(9\sqrt{3}\right)x$
$= 9x\sqrt{3} + 5\sqrt{3x}$

65. (a) $T = G + S$

$$= \left(-3856 + 102{,}201\sqrt{t} - 58{,}994t + 8854\sqrt{t^3}\right) + \left(2{,}249{,}527 - 2{,}230{,}479\sqrt{t} + 742{,}197t - 82{,}167\sqrt{t^3}\right)$$

$$= 2{,}245{,}671 - 2{,}127{,}778\sqrt{t} + 683{,}203t - 73{,}313\sqrt{t^3}$$

(b) When $t = 9$:

$$T = 2{,}245{,}671 - 2{,}127{,}778\sqrt{9} + 683{,}203(9) - 73{,}313\sqrt{9^3}$$

$$\approx 31{,}713 \text{ people}$$

67. The expression that was written in simplest form is $6\sqrt{6x} + \sqrt{6x} + 2x\sqrt{6} - 3\sqrt{6}$.

69. The like radicals in $6\sqrt{6x} + \sqrt{6x} + 2x\sqrt{6} - 3\sqrt{6}$ are $6\sqrt{6x}$ and $\sqrt{6x}$, and $2x\sqrt{6}$ and $-3\sqrt{6}$.

71. $3\sqrt{x+1} + 10\sqrt{x+1} = 13\sqrt{x+1}$

73. $\sqrt[3]{16t^4} - \sqrt[3]{54t^4} = \sqrt[3]{2^3 \cdot 2t^3 \cdot t} - \sqrt[3]{3^3 \cdot 2t^3 \cdot t}$

$$= 2t\sqrt[3]{2t} - 3t\sqrt[3]{2t}$$

$$= -t\sqrt[3]{2t}$$

75. $\sqrt{5a} + 2\sqrt{45a^3} = \sqrt{5a} + 2\sqrt{9 \cdot 5 \cdot a^2 \cdot a}$

$$= \sqrt{5a} + 2(3)(a)\sqrt{5a}$$

$$= \sqrt{5a} + 6a\sqrt{5a}$$

$$= (1 + 6a)\sqrt{5a}$$

$$= (6a + 1)\sqrt{5a}$$

77. $\sqrt{9x-9} + \sqrt{x-1} = \sqrt{9(x-1)} + \sqrt{x-1}$

$$= 3\sqrt{x-1} + \sqrt{x-1}$$

$$= 4\sqrt{x-1}$$

79. $\sqrt{x^3 - x^2} + \sqrt{4x-4} = \sqrt{x^2(x-1)} + \sqrt{4(x-1)}$

$$= x\sqrt{x-1} + 2\sqrt{x-1}$$

$$= (x+2)\sqrt{x-1}$$

81. $2\sqrt[3]{a^4b^2} + 3a\sqrt[3]{ab^2} = 2\sqrt[3]{a^3 \cdot a \cdot b^2} + 3a\sqrt[3]{ab^2}$

$$= 2a\sqrt[3]{ab^2} + 3a\sqrt[3]{ab^2}$$

$$= 5a\sqrt[3]{ab^2}$$

83. $\sqrt{4r^7s^5} + 3r^2\sqrt{r^3s^5} - 2rs\sqrt{r^5s^3} = \sqrt{4r^6 \cdot r \cdot s^4 \cdot s} + 3r^2\sqrt{r^2 \cdot r \cdot s^4 \cdot s} - 2rs\sqrt{r^4 \cdot r \cdot s^2 \cdot s}$

$$= 2r^3s^2\sqrt{rs} + 3r^3s^2\sqrt{rs} - 2r^3s^2\sqrt{rs}$$

$$= 3r^3s^2\sqrt{rs}$$

85. $\sqrt[3]{128x^9y^{10}} - 2x^2y\sqrt[3]{16x^3y^7} = \sqrt[3]{64 \cdot 2 \cdot x^9 \cdot y^9 \cdot y} - 2x^2y\sqrt[3]{8 \cdot 2 \cdot x^3 \cdot y^6 \cdot y}$

$$= 4x^3y^3\sqrt[3]{2y} - 4x^3y^3\sqrt[3]{2y} = 0$$

87. $\sqrt{7} + \sqrt{18} > \sqrt{7 + 18}$

89. $5 < \sqrt{9^2 - 4^2}$

$$5 < \sqrt{81 - 16}$$

$$5 < \sqrt{65}$$

$$5 < 8.06$$

91. (a) $c = \sqrt{a^2 + b^2}$

$$c = \sqrt{(15)^2 + (5)^2}$$

$$c = \sqrt{225 + 25}$$

$$c = \sqrt{250}$$

$$c = \sqrt{25 \cdot 10}$$

$$c = 5\sqrt{10}$$

5

15

(b) Length $= 2(40) + 4\left(5\sqrt{10}\right) = 80 + 20\sqrt{10} \approx 143$ ft

93. No; $\sqrt{5} + \left(-\sqrt{5}\right) = 0.$

95. Yes. $\sqrt{2x} + \sqrt{2x} = (1 + 1)\sqrt{2x}$

$$= 2\sqrt{2x} = \sqrt{4} \cdot \sqrt{2x} = \sqrt{8x}$$

97. (a) The student combined terms with unlike radicands and added the radicands; can be simplified no further.

(b) The student combined terms with unlike indices; can be simplified no further.

99. $\dfrac{7z - 2}{2z} - \dfrac{4z + 1}{2z} = \dfrac{7z - 2 - 4z - 1}{2z} = \dfrac{3z - 3}{2z}$

$$= \frac{3(z - 1)}{2z}$$

101. $\dfrac{2x + 3}{x - 3} + \dfrac{6 - 5x}{x - 3} = \dfrac{2x + 3 + 6 - 5x}{x - 3} = \dfrac{-3x + 9}{x - 3}$

$$= \frac{-3(x - 3)}{x - 3} = -3, \quad x \neq 3$$

103. $\dfrac{2v}{v - 5} - \dfrac{3}{5 - v} = \dfrac{2v}{v - 5} - \dfrac{3}{-1(v - 5)}$

$$= \frac{2v}{v - 5} + \frac{3}{v - 5} = \frac{2v + 3}{v - 5}$$

105. $\dfrac{\left(\dfrac{2}{3}\right)}{\left(\dfrac{4}{15}\right)} = \dfrac{2}{3} \cdot \dfrac{15}{4} = \dfrac{2(15)}{3(4)} = \dfrac{5}{2}$

107. $\dfrac{3w - 9}{\left(\dfrac{w^2 - 10w + 21}{w + 1}\right)} = 3w - 9 \cdot \dfrac{w + 1}{w^2 - 10w + 21}$

$$= \frac{3(w - 3)(w + 1)}{(w - 7)(w - 3)}$$

$$= \frac{3(w + 1)}{w - 7}, \quad w \neq -1,\ w \neq 3$$

Mid-Chapter Quiz for Chapter 7

1. $\sqrt{225} = 15$ because $15 \cdot 15 = 225.$

2. $\sqrt[4]{\frac{81}{16}} = \frac{3}{2}$ because $\frac{3}{2} \cdot \frac{3}{2} \cdot \frac{3}{2} \cdot \frac{3}{2} = \frac{81}{16}.$

3. $49^{1/2} = \sqrt{49} = 7$ because $7 \cdot 7 = 49.$

4. $(-27)^{2/3} = \sqrt[3]{(-27)^2} = \left(\sqrt[3]{-27}\right)^2 = (-3)^2 = 9$

5. $f(x) = \sqrt{3x - 5}$

(a) $f(0) = \sqrt{3(0) - 5} = \sqrt{-5} =$ not a real number

(b) $f(2) = \sqrt{3(2) - 5} = \sqrt{1} = 1$

(c) $f(10) = \sqrt{3(10) - 5} = \sqrt{25} = 5$

6. $g(x) = \sqrt{9 - x}$

(a) $g(-7) = \sqrt{9 - (-7)} = \sqrt{9 + 7} = \sqrt{16} = 4$

(b) $g(5) = \sqrt{9 - 5} = \sqrt{4} = 2$

(c) $g(9) = \sqrt{9 - 9} = \sqrt{0} = 0$

7. $g(x) = \dfrac{12}{\sqrt[3]{x}}$

Domain: $(-\infty, 0) \cup (0, \infty)$

$\sqrt[3]{x} \neq 0$

$x \neq 0$

8. $h(x) = \sqrt{3x + 10}$; Domain: $\left[-\frac{10}{3}, \infty\right)$

$3x + 10 \geq 0$

$3x \geq -10$

$x \geq -\frac{10}{3}$

9. $\sqrt{27x^2} = \sqrt{9 \cdot 3 \cdot x^2} = 3|x|\sqrt{3}$

10. $\sqrt[4]{32x^8} = \sqrt[4]{16 \cdot 2 \cdot (x^2)^4} = 2x^2\sqrt[4]{2}$

11. $\sqrt{\dfrac{4u^3}{9}} = \dfrac{\sqrt{4 \cdot u^2 \cdot u}}{\sqrt{9}} = \dfrac{2u\sqrt{u}}{3}$

12. $\sqrt[3]{\dfrac{16}{u^6}} = \dfrac{\sqrt[3]{16}}{\sqrt[3]{u^6}} = \dfrac{\sqrt[3]{16}}{u^2} = \dfrac{\sqrt[3]{8 \cdot 2}}{u^2} = \dfrac{2\sqrt[3]{2}}{u^2}$

13. $\sqrt{125x^3y^2z^4} = \sqrt{25 \cdot 5 \cdot x^2 \cdot x \cdot y^2 \cdot z^4}$

$$= 5x|y|z^2\sqrt{5x}$$

14. $2a\sqrt[3]{16a^3b^5} = 2a\sqrt[3]{8 \cdot 2 \cdot a^3 \cdot b^3 \cdot b^2}$

$$= 2a \cdot 2ab\sqrt[3]{2b^2}$$

$$= 4a^2b\sqrt[3]{2b^2}$$

15. $\dfrac{24}{\sqrt{12}} = \dfrac{24}{\sqrt{4 \cdot 3}} = \dfrac{24}{2\sqrt{3}} = \dfrac{12}{\sqrt{3}} \cdot \dfrac{\sqrt{3}}{\sqrt{3}} = \dfrac{12\sqrt{3}}{3} = 4\sqrt{3}$

16. $\dfrac{21x^2}{\sqrt{7x}} = \dfrac{21x^2}{\sqrt{7x}} \cdot \dfrac{\sqrt{7x}}{\sqrt{7x}} = \dfrac{21x^2\sqrt{7x}}{7x} = 3x\sqrt{7x}, \quad x \neq 0$

17. $2\sqrt{3} - 4\sqrt{7} + \sqrt{3} = \left(2\sqrt{3} + \sqrt{3}\right) - 4\sqrt{7} = 3\sqrt{3} - 4\sqrt{7}$

18. $\sqrt{200y} - 3\sqrt{8y} = \sqrt{100 \cdot 2y} - 3\sqrt{4 \cdot 2y} = 10\sqrt{2y} - 6\sqrt{2y} = 4\sqrt{2y}$

19. $5\sqrt{12} + 2\sqrt{3} - \sqrt{75} = 5\sqrt{4 \cdot 3} + 2\sqrt{3} - \sqrt{25 \cdot 3} = 10\sqrt{3} + 2\sqrt{3} - 5\sqrt{3} = 7\sqrt{3}$

20. $\sqrt{25x + 50} - \sqrt{x + 2} = \sqrt{25(x + 2)} - \sqrt{x + 2} = 5\sqrt{x + 2} - \sqrt{x + 2} = 4\sqrt{x + 2}$

21. $6x\sqrt[3]{5x^2} + 2\sqrt[3]{40x^4} = 6x\sqrt[3]{5x^2} + 2\sqrt[3]{8 \cdot 5 \cdot x^3 \cdot x} = 6x\sqrt[3]{5x^2} + 4x\sqrt[3]{5x}$

22. $3\sqrt{x^3y^4z^5} + 2xy^2\sqrt{xz^5} - xz^2\sqrt{xy^4z} = 3\sqrt{x^2 \cdot x \cdot y^4 \cdot z^4 \cdot z} + 2xy^2\sqrt{x \cdot z^4 \cdot z} - xz^2\sqrt{xy^4z}$

$= 3xy^2z^2\sqrt{xz} + 2xy^2z^2\sqrt{xz} - xy^2z^2\sqrt{xz}$

$= 4xy^2z^2\sqrt{xz}$

23. $C = \sqrt{2^2 + 2^2} = \sqrt{4 + 4} = \sqrt{8}$

Equation:

$P = 2(7) + 2\left(4\frac{1}{2}\right) + 4\left(\sqrt{8}\right) = 14 + 9 + 8\sqrt{2} = 23 + 8\sqrt{2}$ inches

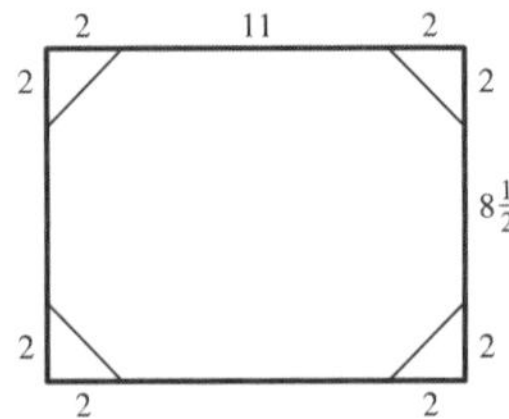

Section 7.4 Multiplying and Dividing Radical Expressions

1. $\sqrt{2} \cdot \sqrt{8} = \sqrt{2 \cdot 8} = \sqrt{16} = 4$

3. $\sqrt{3} \cdot \sqrt{15} = \sqrt{3 \cdot 15} = \sqrt{45} = \sqrt{9 \cdot 5} = 3\sqrt{5}$

5. $\sqrt[3]{12} \cdot \sqrt[3]{6} = \sqrt[3]{12 \cdot 6} = \sqrt[3]{8 \cdot 9} = 2\sqrt[3]{9}$

7. $\sqrt[4]{8} \cdot \sqrt[4]{2} = \sqrt[4]{8 \cdot 2} = \sqrt[4]{16} = 2$

9. $\sqrt{7}\left(3 - \sqrt{7}\right) = 3\sqrt{7} - 7$

11. $\sqrt{2}\left(\sqrt{20} + 8\right) = \sqrt{2}\sqrt{20} + 8\sqrt{2}$

$= \sqrt{40} + 8\sqrt{2}$

$= 2\sqrt{10} + 8\sqrt{2}$

13. $\sqrt{6}\left(\sqrt{12} - \sqrt{3}\right) = \sqrt{6}\sqrt{12} - \sqrt{6}\sqrt{3}$

$= \sqrt{72} - \sqrt{18}$

$= \sqrt{36 \cdot 2} - \sqrt{9 \cdot 2}$

$= 6\sqrt{2} - 3\sqrt{2}$

$= 3\sqrt{2}$

15. $4\sqrt{3}\left(\sqrt{3} - \sqrt{5}\right) = 4\sqrt{3} \cdot \sqrt{3} - 4\sqrt{3} \cdot \sqrt{5}$

$= 4 \cdot 3 - 4\sqrt{3 \cdot 5}$

$= 12 - 4\sqrt{15}$

17. $\sqrt{y}\left(\sqrt{y} + 4\right) = \left(\sqrt{y}\right)^2 + 4\sqrt{y} = y + 4\sqrt{y}$

19. $\sqrt{a}\left(4 - \sqrt{a}\right) = \sqrt{a} \cdot 4 - \sqrt{a}\sqrt{a} = 4\sqrt{a} - a$

21. $\left(\sqrt{5} + 3\right)\left(\sqrt{3} - 5\right) = \sqrt{15} - 5\sqrt{5} + 3\sqrt{3} - 15$

23. $\left(\sqrt{20} + 2\right)^2 = \left(\sqrt{20}\right)^2 + 2 \cdot \sqrt{20} \cdot 2 + 2^2$

$= 20 + 4\sqrt{20} + 4$

$= 24 + 4\sqrt{4 \cdot 5}$

$= 24 + 8\sqrt{5}$

25. $\left(\sqrt{5}-\sqrt{3}\right)\left(\sqrt{5}-\sqrt{3}\right)=\left(\sqrt{5}\right)^2-2\cdot\sqrt{5}\cdot\sqrt{3}+\left(\sqrt{3}\right)^2=5-2\sqrt{15}+3=8-2\sqrt{15}$

27. $\left(10+\sqrt{2x}\right)^2=10^2+2\cdot10\cdot\sqrt{2x}+\left(\sqrt{2x}\right)^2=100+20\sqrt{2x}+2x$

29. $\left(9\sqrt{x}+2\right)\left(5\sqrt{x}-3\right)=\left(9\sqrt{x}\right)\left(5\sqrt{x}\right)-27\sqrt{x}+10\sqrt{x}-6=45x-17\sqrt{x}-6$

31. $$\begin{aligned}\text{Area}&=(\text{length})(\text{width})\\&=\left(5+\sqrt{3}\right)\left(6-\sqrt{3}\right)\\&=30-5\sqrt{3}+6\sqrt{3}-3\\&=27+\sqrt{3}\end{aligned}$$

33. $$\begin{aligned}\text{Area}&=(\text{length})(\text{width})\\&=\left(2+\sqrt{x}\right)\left(4-\sqrt{x}\right)\\&=8-2\sqrt{x}+4\sqrt{x}-x\\&=8+2\sqrt{x}-x\end{aligned}$$

35. $2+\sqrt{5}$, conjugate $=2-\sqrt{5}$

$$\begin{aligned}\text{Product}&=\left(2+\sqrt{5}\right)\left(2-\sqrt{5}\right)\\&=2^2-\left(\sqrt{5}\right)^2\\&=4-5=-1\end{aligned}$$

37. $\sqrt{11}-\sqrt{3}$, conjugate $=\sqrt{11}+\sqrt{3}$

$$\begin{aligned}\text{Product}&=\left(\sqrt{11}-\sqrt{3}\right)\left(\sqrt{11}+\sqrt{3}\right)\\&=\left(\sqrt{11}\right)^2-\left(\sqrt{3}\right)^2\\&=11-3=8\end{aligned}$$

39. $\sqrt{15}+3$, conjugate $=\sqrt{15}-3$

$$\begin{aligned}\text{Product}&=\left(\sqrt{15}+3\right)\left(\sqrt{15}-3\right)\\&=\sqrt{15}\cdot\sqrt{15}-3\sqrt{15}+3\sqrt{15}-9\\&=15-9=6\end{aligned}$$

41. $\sqrt{x}-3$, conjugate $=\sqrt{x}+3$

$$\begin{aligned}\text{Product}&=\left(\sqrt{x}-3\right)\left(\sqrt{x}+3\right)\\&=\left(\sqrt{x}\right)^2-3^2\\&=x-9\end{aligned}$$

43. $\sqrt{2u}-\sqrt{3}$, conjugate $=\sqrt{2u}+\sqrt{3}$

$$\begin{aligned}\text{Product}&=\left(\sqrt{2u}-\sqrt{3}\right)\left(\sqrt{2u}+\sqrt{3}\right)\\&=\left(\sqrt{2u}\right)^2-\left(\sqrt{3}\right)^2\\&=2u-3\end{aligned}$$

45. $2\sqrt{2}+\sqrt{4}$, conjugate $=2\sqrt{2}-\sqrt{4}$

$$\begin{aligned}\text{Product}&=\left(2\sqrt{2}+\sqrt{4}\right)\left(2\sqrt{2}-\sqrt{4}\right)\\&=\left(2\sqrt{2}\right)^2-\left(\sqrt{4}\right)^2\\&=4\cdot2-4\\&=8-4=4\end{aligned}$$

47. $\sqrt{x}+\sqrt{y}$, conjugate $=\sqrt{x}-\sqrt{y}$

$$\begin{aligned}\text{Product}&=\left(\sqrt{x}+\sqrt{y}\right)\left(\sqrt{x}-\sqrt{y}\right)\\&=\left(\sqrt{x}\right)^2-\left(\sqrt{y}\right)^2\\&=x-y\end{aligned}$$

49. $$\begin{aligned}\text{Area}&=(\text{length})(\text{width})\\&=\left(5+\sqrt{2}\right)\left(5-\sqrt{2}\right)\\&=5^2-\left(\sqrt{2}\right)^2\\&=25-2\\&=23\end{aligned}$$

51. $$\begin{aligned}\frac{1}{2+\sqrt{5}}&=\frac{1}{2+\sqrt{5}}\cdot\frac{2-\sqrt{5}}{2-\sqrt{5}}\\&=\frac{2-\sqrt{5}}{2^2-\left(\sqrt{5}\right)^2}\\&=\frac{2-\sqrt{5}}{4-5}\\&=\frac{2-\sqrt{5}}{-1}\\&=\sqrt{5}-2\end{aligned}$$

53. $$\begin{aligned}\frac{\sqrt{2}}{2-\sqrt{6}}&=\frac{\sqrt{2}}{2-\sqrt{6}}\cdot\frac{2+\sqrt{6}}{2+\sqrt{6}}\\&=\frac{\sqrt{2}+\left(2+\sqrt{6}\right)}{2^2-\left(\sqrt{6}\right)^2}\\&=\frac{2\sqrt{2}+\sqrt{12}}{4-6}\\&=\frac{2\sqrt{2}+2\sqrt{3}}{-2}\\&=-\sqrt{2}-\sqrt{3}\end{aligned}$$

55. $$\frac{5}{9-\sqrt{6}} = \frac{5}{9-\sqrt{6}} \cdot \frac{9+\sqrt{6}}{9+\sqrt{6}} = \frac{5(9+\sqrt{6})}{9^2-(\sqrt{6})^2} = \frac{5(9+\sqrt{6})}{81-6} = \frac{5(9+\sqrt{6})}{75} = \frac{9+\sqrt{6}}{15}$$

57. $$\frac{6}{\sqrt{11}-2} = \frac{6}{\sqrt{11}-2} \cdot \frac{\sqrt{11}+2}{\sqrt{11}+2} = \frac{6(\sqrt{11}+2)}{(\sqrt{11})^2-2^2} = \frac{6(\sqrt{11}+2)}{11-4} = \frac{6(\sqrt{11}+2)}{7}$$

59. $$\frac{7}{\sqrt{3}+5} = \frac{7}{\sqrt{3}+5} \cdot \frac{\sqrt{3}-5}{\sqrt{3}-5} = \frac{7(\sqrt{3}-5)}{(\sqrt{3})^2-5^2} = \frac{7(\sqrt{3}-5)}{3-25} = \frac{7(\sqrt{3}-5)}{-22} = \frac{7(5-\sqrt{3})}{22}$$

61. $$\frac{\sqrt{5}}{\sqrt{6}-\sqrt{5}} = \frac{\sqrt{5}}{\sqrt{6}-\sqrt{5}} \cdot \frac{\sqrt{6}+\sqrt{5}}{\sqrt{6}+\sqrt{5}} = \frac{\sqrt{5}(\sqrt{6}+\sqrt{5})}{(\sqrt{6})^2-(\sqrt{5})^2} = \frac{\sqrt{30}+\sqrt{25}}{6-5} = \frac{\sqrt{30}+5}{1} = \sqrt{30}+5$$

63. $$\frac{4\sqrt{3}}{\sqrt{5}+\sqrt{3}} = \frac{4\sqrt{3}}{\sqrt{5}+\sqrt{3}} \cdot \frac{\sqrt{5}-\sqrt{3}}{\sqrt{5}-\sqrt{3}} = \frac{4\sqrt{3}(\sqrt{5}-\sqrt{3})}{(\sqrt{5})^2-(\sqrt{3})^2} = \frac{4\sqrt{15}+4\sqrt{9}}{5-3} = \frac{4\sqrt{15}-12}{2} = 2\sqrt{15}-6$$

65. $$1 \div \left(\sqrt{x}-\sqrt{x+3}\right) = \frac{1}{\sqrt{x}-\sqrt{x+3}} = \frac{1}{\sqrt{x}-\sqrt{x+3}} \cdot \frac{\sqrt{x}+\sqrt{x+3}}{\sqrt{x}+\sqrt{x+3}} = \frac{\sqrt{x}+\sqrt{x+3}}{(\sqrt{x})^2-(\sqrt{x+3})^2} = \frac{\sqrt{x}+\sqrt{x+3}}{x-(x+3)} = -\frac{\sqrt{x}+\sqrt{x+3}}{3}$$

67. $\dfrac{3x}{\sqrt{15}-\sqrt{3}} = \dfrac{3x}{\sqrt{15}-\sqrt{3}}\cdot\dfrac{\sqrt{15}+\sqrt{3}}{\sqrt{15}+\sqrt{3}}$

$= \dfrac{3x\left(\sqrt{15}+\sqrt{3}\right)}{\left(\sqrt{15}\right)^2-\left(\sqrt{3}\right)^2}$

$= \dfrac{3x\left(\sqrt{15}+\sqrt{3}\right)}{15-3}$

$= \dfrac{3x\left(\sqrt{15}+\sqrt{3}\right)}{12}$

$= \dfrac{x\sqrt{15}+x\sqrt{3}}{4}$ or $\dfrac{\left(\sqrt{15}+\sqrt{3}\right)x}{4}$

69. $8 \div \left(\sqrt{x}-3\right) = \dfrac{8}{\sqrt{x}-3}$

$= \dfrac{8}{\sqrt{x}-3}\cdot\dfrac{\sqrt{x}+3}{\sqrt{x}+3}$

$= \dfrac{8\left(\sqrt{x}+3\right)}{\left(\sqrt{x}\right)^2-3^2}$

$= \dfrac{8\sqrt{x}+24}{x-9}$

71. $\dfrac{\sqrt{5t}}{\sqrt{5}-\sqrt{t}} = \dfrac{\sqrt{5t}}{\sqrt{5}-\sqrt{t}}\cdot\dfrac{\sqrt{5}+\sqrt{t}}{\sqrt{5}+\sqrt{t}}$

$= \dfrac{\sqrt{5t}\cdot\sqrt{5}+\sqrt{5t}\cdot\sqrt{t}}{5-t}$

$= \dfrac{5\sqrt{t}+t\sqrt{5}}{5-t}$

73. $\dfrac{8a}{\sqrt{3a}+\sqrt{a}} = \dfrac{8a}{\sqrt{3a}+\sqrt{a}}\cdot\dfrac{\sqrt{3a}-\sqrt{a}}{\sqrt{3a}-\sqrt{a}}$

$= \dfrac{8a\left(\sqrt{3a}-\sqrt{a}\right)}{\left(\sqrt{3a}\right)^2-\left(\sqrt{a}\right)^2}$

$= \dfrac{8a\left(\sqrt{3a}-\sqrt{a}\right)}{3a-a}$

$= \dfrac{8a\left(\sqrt{3a}-\sqrt{a}\right)}{2a}$

$= 4\left(\sqrt{3a}-\sqrt{a}\right),\quad a \neq 0$

75. $\left(\sqrt{7}+2\right) \div \left(\sqrt{7}-2\right) = \dfrac{\sqrt{7}+2}{\sqrt{7}-2}\cdot\dfrac{\sqrt{7}+2}{\sqrt{7}+2}$

$= \dfrac{\left(\sqrt{7}\right)^2+2\sqrt{7}+2\sqrt{7}+4}{\left(\sqrt{7}\right)^2-2^2}$

$= \dfrac{7+4\sqrt{7}+4}{7-4}$

$= \dfrac{11+4\sqrt{7}}{3}$

77. $\left(\sqrt{x}-5\right) \div \left(2\sqrt{x}-1\right) = \dfrac{\sqrt{x}-5}{2\sqrt{x}-1}\cdot\dfrac{2\sqrt{x}+1}{2\sqrt{x}+1}$

$= \dfrac{2x+\sqrt{x}-10\sqrt{x}-5}{\left(2\sqrt{x}\right)^2-1^2}$

$= \dfrac{2x-9\sqrt{x}-5}{4x-1}$

79. $w = \dfrac{\text{Area}}{\text{Length}}$

$= \dfrac{10}{\sqrt{12}+\sqrt{5}}$

$= \dfrac{10}{\sqrt{12}+\sqrt{5}}\cdot\dfrac{\sqrt{12}-\sqrt{5}}{\sqrt{12}-\sqrt{5}}$

$= \dfrac{10\left(\sqrt{12}-\sqrt{5}\right)}{\left(\sqrt{12}\right)^2-\left(\sqrt{5}\right)^2}$

$= \dfrac{10\left(\sqrt{12}-\sqrt{5}\right)}{12-5}$

$= \dfrac{10\left(2\sqrt{3}-\sqrt{5}\right)}{7}$

$= \dfrac{20\sqrt{3}-10\sqrt{5}}{7}$

81. The expression $\left(1+\sqrt{2}\right)\left(3+\sqrt{2}\right)$ is simplified by the FOIL method.

83. The relationship between $\sqrt{2}+1$ and $\sqrt{2}-1$ is that they are conjugates.

85. $\left(2\sqrt{2x}-\sqrt{5}\right)\left(2\sqrt{2x}+\sqrt{5}\right) = \left(2\sqrt{2x}\right)^2-\left(\sqrt{5}\right)^2 = 4\cdot 2x - 5 = 8x-5$

87. $\left(\sqrt[3]{t}+1\right)\left(\sqrt[3]{t^2}+4\sqrt[3]{t}-3\right) = \sqrt[3]{t}\sqrt[3]{t^2}+\sqrt[3]{t}\cdot 4\sqrt[3]{t}-3\sqrt[3]{t}+\sqrt[3]{t^2}+4\sqrt[3]{t}-3$

$= \sqrt[3]{t^3}+4\sqrt[3]{t^2}-3\sqrt[3]{t}+\sqrt[3]{t^2}+4\sqrt[3]{t}-3$

$= t+5\sqrt[3]{t^2}+\sqrt[3]{t}-3$

89. $2\sqrt[3]{x^4y^5}\left(\sqrt[3]{8x^{12}y^4} + \sqrt[3]{16xy^9}\right) = 2\sqrt[3]{x^4y^5} \cdot \sqrt[3]{8x^{12}y^4} + 2\sqrt[3]{x^4y^5} \cdot \sqrt[3]{16xy^9}$

$$= 2\sqrt[3]{8x^{16}y^9} + 2\sqrt[3]{16x^5y^{14}}$$
$$= 2\sqrt[3]{8x^{15} \cdot x \cdot y^9} + 2\sqrt[3]{8 \cdot 2x^3 \cdot x^2 \cdot y^{12} \cdot y^2}$$
$$= 4x^5y^3\sqrt[3]{x} + 4xy^4\sqrt[3]{2x^2y^2}$$
$$= 4xy^3\left(x^4\sqrt[3]{x} + y\sqrt[3]{2x^2y^2}\right)$$

91. $f(x) = x^2 - 6x + 1$

(a) $f(2 - \sqrt{3}) = (2 - \sqrt{3})^2 - 6(2 - \sqrt{3}) + 1 = 4 - 4\sqrt{3} + 3 - 12 + 6\sqrt{3} + 1 = 2\sqrt{3} - 4$

(b) $f(3 - 2\sqrt{2}) = (3 - 2\sqrt{2})^2 - 6(3 - 2\sqrt{2}) + 1 = 9 - 12\sqrt{2} + 8 - 18 + 12\sqrt{2} + 1 = 0$

93. $\dfrac{\sqrt{10}}{\sqrt{3x}} = \dfrac{\sqrt{10}}{\sqrt{3x}} \cdot \dfrac{\sqrt{10}}{\sqrt{10}} = \dfrac{10}{\sqrt{30x}}$

95. $\dfrac{\sqrt{7} + \sqrt{3}}{5} = \dfrac{\sqrt{7} + \sqrt{3}}{5} \cdot \dfrac{\sqrt{7} - \sqrt{3}}{\sqrt{7} - \sqrt{3}} = \dfrac{(\sqrt{7})^2 - (\sqrt{3})^2}{5(\sqrt{7} - \sqrt{3})} = \dfrac{7 - 3}{5(\sqrt{7} - \sqrt{3})} = \dfrac{4}{5(\sqrt{7} - \sqrt{3})}$

97. Area $= h \cdot w$

$$= \sqrt{24^2 - (8\sqrt{3})^2} \cdot 8\sqrt{3}$$
$$= \sqrt{576 - 192} \cdot 8\sqrt{3}$$
$$= \sqrt{384} \cdot 8\sqrt{3}$$
$$= 8\sqrt{1152}$$
$$= 8\sqrt{2^7 \cdot 3^2}$$
$$= 8 \cdot 2^3 \cdot 3\sqrt{2}$$
$$= 192\sqrt{2} \text{ square inches}$$

99. $\dfrac{\text{Diameter of ball}}{\text{Diameter of hoop}} = \dfrac{2\sqrt{\dfrac{70}{\pi}}}{2\sqrt{\dfrac{254}{\pi}}} = \sqrt{\dfrac{\dfrac{70}{\pi}}{\dfrac{254}{\pi}}}$

$$= \sqrt{\frac{70}{254}} = \sqrt{\frac{35}{127}}$$
$$= \frac{\sqrt{35}}{\sqrt{127}} \cdot \frac{\sqrt{127}}{\sqrt{127}} = \frac{\sqrt{4445}}{127}$$

Area of a circle $= \pi \cdot r^2$ (r = radius)

$$A = \pi\left(\frac{d}{2}\right)^2 \quad (d = \text{diameter})$$
$$A = \frac{\pi d^2}{4}$$
$$\frac{4A}{\pi} = d^2$$
$$\sqrt{\frac{4A}{\pi}} = d$$
$$2\sqrt{\frac{A}{\pi}} = d$$

101. (a) If either a or b (or both) equal zero, the expression $a\sqrt{b}$ is zero and therefore rational.

(b) If the product of a and b is a perfect square, then the expression is rational.

103. Conjugate of $\sqrt{a} + \sqrt{b}$ is $\sqrt{a} - \sqrt{b}$.

$$\text{Product} = (\sqrt{a} + \sqrt{b})(\sqrt{a} - \sqrt{b})$$
$$= (\sqrt{a})^2 - (\sqrt{b})^2 = a - b$$

Conjugate of $\sqrt{b} + \sqrt{a}$ is $\sqrt{b} - \sqrt{a}$.

$$\text{Product} = (\sqrt{b} + \sqrt{a})(\sqrt{b} - \sqrt{a})$$
$$= (\sqrt{b})^2 - (\sqrt{a})^2 = b - a$$

By changing the order of the terms, the conjugate and the product both change by a factor of -1.

105. $3x - 18 = 0$

$$3x = 18$$
$$x = 6$$

107. $3x - 4 = 3x$

$$-4 = 0$$

no solution

109. $x^2 - 144 = 0$

$$(x + 12)(x - 12) = 0$$
$$x + 12 = 0 \qquad x - 12 = 0$$
$$x = -12 \qquad x = 12$$

111. $x^2 + 2x - 15 = 0$

$(x + 5)(x - 3) = 0$

$x + 5 = 0 \qquad x - 3 = 0$

$x = -5 \qquad x = 3$

113. $\sqrt{32x^2y^5} = \sqrt{16 \cdot 2 \cdot x^2 \cdot y^4 \cdot y} = 4|x|y^2\sqrt{2y}$

115. $\sqrt[4]{32x^2y^5} = \sqrt[4]{16 \cdot 2 \cdot x^2 \cdot y^4 \cdot y} = 2y\sqrt[4]{2x^2y}$

Section 7.5 Radical Equations and Applications

1. (a) $x = -4$ $\quad \sqrt{-4} - 10 \neq 0$ $\quad$ Not a solution

(b) $x = -100$ $\quad \sqrt{-100} - 10 \neq 0$ $\quad$ Not a solution

(c) $x = \sqrt{10}$ $\quad \sqrt{\sqrt{10}} - 10 \neq 0$ $\quad$ Not a solution

(d) $x = 100$ $\quad \sqrt{100} - 10 = 0$ $\quad$ A solution

3. (a) $x = -60$ $\quad \sqrt[3]{-60 - 4} \neq 4$ $\quad$ Not a solution

(b) $x = 68$ $\quad \sqrt[3]{68 - 4} = 4$ $\quad$ A solution

(c) $x = 20$ $\quad \sqrt[3]{20 - 4} \neq 4$ $\quad$ Not a solution

(d) $x = 0$ $\quad \sqrt[3]{0 - 4} \neq 4$ $\quad$ Not a solution

5. $\sqrt{x} = 12$

$\left(\sqrt{x}\right)^2 = 12^2$

$x = 144$

Check: $\sqrt{144} \stackrel{?}{=} 12$

$12 = 12$

7. $\sqrt{y} = 7$

$\left(\sqrt{y}\right)^2 = 7^2$

$y = 49$

Check: $\sqrt{49} \stackrel{?}{=} 7$

$7 = 7$

9. $\sqrt[3]{z} = 3$

$\left(\sqrt[3]{z}\right)^3 = 3^3$

$z = 27$

Check: $\sqrt[3]{27} \stackrel{?}{=} 3$

$3 = 3$

11. $\sqrt{y} - 7 = 0$

$\sqrt{y} = 7$

$\left(\sqrt{y}\right)^2 = 7^2$

$y = 49$

Check: $\sqrt{49} - 7 \stackrel{?}{=} 0$

$7 - 7 \stackrel{?}{=} 0$

$0 = 0$

13. $\sqrt{x} - 8 = 0$

$\sqrt{x} = 8$

$\left(\sqrt{x}\right)^2 = 8^2$

$x = 64$

Check: $\sqrt{64} - 8 \stackrel{?}{=} 0$

$8 - 8 \stackrel{?}{=} 0$

$0 = 0$

15. $\sqrt{u} + 13 = 0$

$\sqrt{u} = -13$

$\left(\sqrt{u}\right)^2 = (-13)^2$

$u = 169$

Check: $\sqrt{169} + 13 \stackrel{?}{=} 0$

$13 + 13 \neq 0$

No solution

17. $\sqrt{10x} = 30$

$\left(\sqrt{10x}\right)^2 = 30^2$

$10x = 900$

$x = 90$

Check: $\sqrt{10 \cdot 90} \stackrel{?}{=} 30$

$\sqrt{900} \stackrel{?}{=} 30$

$30 = 30$

19. $\sqrt{-3x} = 9$

$\left(\sqrt{-3x}\right)^2 = 9^2$

$-3x = 81$

$x = -27$

Check: $\sqrt{-3(-27)} \stackrel{?}{=} 9$

$\sqrt{81} \stackrel{?}{=} 9$

$9 = 9$

21. $\sqrt{3y + 1} = 4$

$\left(\sqrt{3y + 1}\right)^2 = 4^2$

$3y + 1 = 16$

$3y = 15$

$y = 5$

Check: $\sqrt{3(5) + 1} \stackrel{?}{=} 4$

$\sqrt{16} \stackrel{?}{=} 4$

$4 = 4$

23. $\sqrt{9 - 2x} = -9$

$\left(\sqrt{9 - 2x}\right)^2 = (-9)^2$

$9 - 2x = 81$

$-2x = 72$

$x = -36$

Check: $\sqrt{9 - 2(-36)} \stackrel{?}{=} -9$

$\sqrt{9 + 72} \stackrel{?}{=} -9$

$\sqrt{81} \stackrel{?}{=} -9$

$9 \neq -9$

No solution

25. $\sqrt[3]{y + 1} - 2 = 4$

$\sqrt[3]{y + 1} = 6$

$y + 1 = 6^3$

$y + 1 = 216$

$y = 215$

Check: $\sqrt[3]{215 + 1} - 2 \stackrel{?}{=} 4$

$\sqrt[3]{215 + 1} \stackrel{?}{=} 6$

$\sqrt[3]{216} \stackrel{?}{=} 6$

$6 = 6$

27. $\sqrt{x + 3} = \sqrt{2x - 1}$

$\left(\sqrt{x + 3}\right)^2 = \left(\sqrt{2x - 1}\right)^2$

$x + 3 = 2x - 1$

$4 = x$

Check: $\sqrt{4 + 3} \stackrel{?}{=} \sqrt{2(4) - 1}$

$\sqrt{7} = \sqrt{7}$

29. $\sqrt{3x + 4} = \sqrt{4x + 3}$

$\left(\sqrt{3x + 4}\right)^2 = \left(\sqrt{4x + 3}\right)^2$

$3x + 4 = 4x + 3$

$1 = x$

Check: $\sqrt{3(1) + 4} \stackrel{?}{=} \sqrt{4(1) + 3}$

$\sqrt{7} = \sqrt{7}$

31. $\sqrt{3y - 5} - 3\sqrt{y} = 0$

$\sqrt{3y - 5} = 3\sqrt{y}$

$\left(\sqrt{3y - 5}\right)^2 = \left(3\sqrt{y}\right)^2$

$3y - 5 = 9y$

$-5 = 6y$

$-\frac{5}{6} = y$

Check: $\sqrt{3\left(-\frac{5}{6}\right) - 5} - 3\sqrt{-\frac{5}{6}} \stackrel{?}{=} 0$

No solution

33. $\sqrt[3]{3x - 4} = \sqrt[3]{x + 10}$

$\left(\sqrt[3]{3x - 4}\right)^3 = \left(\sqrt[3]{x + 10}\right)^3$

$3x - 4 = x + 10$

$2x = 14$

$x = 7$

Check: $\sqrt[3]{3(7) - 4} \stackrel{?}{=} \sqrt[3]{7 + 10}$

$\sqrt[3]{17} = \sqrt[3]{17}$

35. $\sqrt[3]{2x + 15} - \sqrt[3]{x} = 0$

$\sqrt[3]{2x + 15} = \sqrt[3]{x}$

$\left(\sqrt[3]{2x + 15}\right)^3 = \left(\sqrt[3]{x}\right)^3$

$2x + 15 = x$

$x = -15$

Check: $\sqrt[3]{2(-15) + 15} - \sqrt[3]{-15} \stackrel{?}{=} 0$

$\sqrt[3]{-15} - \sqrt[3]{-15} \stackrel{?}{=} 0$

$0 = 0$

37. $\sqrt{x^2 - 2} = x + 4$

$$\left(\sqrt{x^2 - 2}\right)^2 = (x + 4)^2$$
$$x^2 - 2 = x^2 + 8x + 16$$
$$-2 = 8x + 16$$
$$-18 = 8x$$
$$-\tfrac{18}{8} = x$$
$$-\tfrac{9}{4} = x$$

Check: $\sqrt{\left(-\frac{9}{4}\right)^2 - 2} \stackrel{?}{=} -\frac{9}{4} + 4$

$$\sqrt{\tfrac{81}{16} - \tfrac{32}{16}} \stackrel{?}{=} -\tfrac{9}{4} + \tfrac{16}{4}$$
$$\sqrt{\tfrac{49}{16}} \stackrel{?}{=} \tfrac{7}{4}$$
$$\tfrac{7}{4} = \tfrac{7}{4}$$

39. $\sqrt{2x} = x - 4$

$$\left(\sqrt{2x}\right)^2 = (x - 4)^2$$
$$2x = x^2 - 8x + 16$$
$$0 = x^2 - 10x + 16$$
$$0 = (x - 8)(x - 2)$$
$$8 = x, \qquad x = 2, \text{ Not a solution}$$

Check: $\sqrt{2(8)} \stackrel{?}{=} 8 - 4$

$$\sqrt{16} \stackrel{?}{=} 4$$
$$4 = 4$$
$$\sqrt{2(2)} \stackrel{?}{=} 2 - 4$$
$$\sqrt{4} \stackrel{?}{=} -2$$
$$2 \neq -2$$

41. $\sqrt{8x + 1} = x + 2$

$$\left(\sqrt{8x + 1}\right)^2 = (x + 2)^2$$
$$8x + 1 = x^2 + 4x + 4$$
$$0 = x^2 - 4x + 3$$
$$0 = (x - 3)(x - 1)$$
$$3 = x, \qquad x = 1$$

Check: $\sqrt{8(3) + 1} \stackrel{?}{=} 3 + 2$

$$\sqrt{25} \stackrel{?}{=} 5$$
$$5 = 5$$
$$\sqrt{8(1) + 1} \stackrel{?}{=} 1 + 2$$
$$\sqrt{9} \stackrel{?}{=} 3$$
$$3 = 3$$

43. $\sqrt{z + 2} = 1 + \sqrt{z}$

$$\left(\sqrt{z + 2}\right)^2 = \left(1 + \sqrt{z}\right)^2$$
$$z + 2 = 1 + 2\sqrt{z} + z$$
$$1 = 2\sqrt{z}$$
$$1^2 = \left(2\sqrt{z}\right)^2$$
$$1 = 4z$$
$$\tfrac{1}{4} = z$$

Check: $\sqrt{\frac{1}{4} + 2} \stackrel{?}{=} 1 + \sqrt{\frac{1}{4}}$

$$\sqrt{\tfrac{9}{4}} \stackrel{?}{=} 1 + \tfrac{1}{2}$$
$$\tfrac{3}{2} = \tfrac{3}{2}$$

45. $\sqrt{2t + 3} = 3 - \sqrt{2t}$

$$\left(\sqrt{2t + 3}\right)^2 = \left(3 - \sqrt{2t}\right)^2$$
$$2t + 3 = 9 - 6\sqrt{2t} + 2t$$
$$-6 = -6\sqrt{2t}$$
$$1 = \sqrt{2t}$$
$$1^2 = \left(\sqrt{2t}\right)^2$$
$$1 = 2t$$
$$\tfrac{1}{2} = t$$

Check: $\sqrt{2\left(\frac{1}{2}\right) + 3} \stackrel{?}{=} 3 - \sqrt{2\left(\frac{1}{2}\right)}$

$$\sqrt{1 + 3} \stackrel{?}{=} 3 - \sqrt{1}$$
$$\sqrt{4} \stackrel{?}{=} 3 - 1$$
$$2 = 2$$

47. $\sqrt{x + 5} - \sqrt{x} = 1$

$$\sqrt{x + 5} = 1 + \sqrt{x}$$
$$\left(\sqrt{x + 5}\right)^2 = \left(1 + \sqrt{x}\right)^2$$
$$x + 5 = 1 + 2\sqrt{x} + x$$
$$4 = 2\sqrt{x}$$
$$2 = \sqrt{x}$$
$$2^2 = \left(\sqrt{x}\right)^2$$
$$4 = x$$

Check: $\sqrt{4 + 5} - \sqrt{4} \stackrel{?}{=} 1$

$$\sqrt{9} - \sqrt{4} \stackrel{?}{=} 1$$
$$3 - 2 \stackrel{?}{=} 1$$
$$1 = 1$$

49.
$$\sqrt{x-6}+3=\sqrt{x+9}$$
$$\left(\sqrt{x-6}+3\right)^2=\left(\sqrt{x+9}\right)^2$$
$$x-6+6\sqrt{x-6}+9=x+9$$
$$6\sqrt{x-6}+3=9$$
$$6\sqrt{x-6}=6$$
$$\sqrt{x-6}=1$$
$$\left(\sqrt{x-6}\right)^2=1^2$$
$$x-6=1$$
$$x=7$$

Check: $\sqrt{7-6}+3\stackrel{?}{=}\sqrt{7+9}$
$$1+3\stackrel{?}{=}\sqrt{16}$$
$$4=4$$

51.
$$t=\sqrt{\frac{d}{16}}$$
$$2=\sqrt{\frac{d}{16}}$$
$$2^2=\left(\sqrt{\frac{d}{16}}\right)^2$$
$$4=\frac{d}{16}$$
$$64\text{ feet}=d$$

53.
$$v=\sqrt{2gh}$$
$$32\sqrt{5}=\sqrt{2(32)h}$$
$$\left(32\sqrt{5}\right)^2=\left(\sqrt{64h}\right)^2$$
$$5120=64h$$
$$h=80\text{ ft}$$

55.
$$c^2=a^2+b^2$$
$$15^2=x^2+12^2$$
$$225=x^2+144$$
$$81=x^2$$
$$\sqrt{81}=x^2$$
$$9=x$$

15
x
12

57.
$$c^2=a^2+b^2$$
$$13^2=x^2+5^2$$
$$169=x^2+25$$
$$144=x^2$$
$$\sqrt{144}=x$$
$$12=x$$

13
5
x

59.
$$4^2+x^2=20^2$$
$$16+x^2=400$$
$$x^2=384$$
$$x=\sqrt{384}$$
$$x=8\sqrt{6}$$
$$x\approx 19.6\text{ ft}$$

61. Each side is squared to eliminate the radical.

63. A reason to check the solution in the original equation is to determine whether the solution is extraneous.

65.
$$3y^{1/3}=18$$
$$y^{1/3}=6$$
$$\sqrt[3]{y}=6$$
$$\left(\sqrt[3]{y}\right)^3=6^3$$
$$y=216$$

Check: $3(216)^{1/3}\stackrel{?}{=}18$
$$3\sqrt[3]{216}\stackrel{?}{=}18$$
$$3\cdot 6\stackrel{?}{=}18$$
$$18=18$$

67.
$$(x+4)^{2/3}=4$$
$$\sqrt[3]{(x+4)^2}=4$$
$$\left(\sqrt[3]{(x+4)^2}\right)^3=(4)^3$$
$$(x+4)^2=64$$
$$x+4=\pm\sqrt{64}$$
$$x=-4\pm 8$$
$$=4,-12$$

Check: $(4+4)^{2/3}\stackrel{?}{=}4$
$$8^{2/3}\stackrel{?}{=}4$$
$$2^2=4$$
$$(-12+4)^{2/3}\stackrel{?}{=}4$$
$$(-8)^{2/3}\stackrel{?}{=}4$$
$$(-2)^2=4$$

69.
$$1 = \sqrt{x} - \sqrt{x - 9}$$
$$1 + \sqrt{x - 9} = \sqrt{x}$$
$$\left(1 + \sqrt{x - 9}\right)^2 = \left(\sqrt{x}\right)^2$$
$$1 + 2\sqrt{x - 9} + x - 9 = x$$
$$2\sqrt{x - 9} - 8 = 0$$
$$2\sqrt{x - 9} = 8$$
$$\sqrt{x - 9} = 4$$
$$\left(\sqrt{x - 9}\right)^2 = 4^2$$
$$x - 9 = 16$$
$$x = 25$$

Check: $1 \stackrel{?}{=} \sqrt{25} - \sqrt{25 - 9}$

$1 \stackrel{?}{=} 5 - \sqrt{16}$

$1 \stackrel{?}{=} 5 - 4$

$1 = 1$

71.
$$-3 = \sqrt{x - 2} - \sqrt{4x + 1}$$
$$\sqrt{4x + 1} - 3 = \sqrt{x - 2}$$
$$\left(\sqrt{4x + 1} - 3\right)^2 = \left(\sqrt{x - 2}\right)^2$$
$$4x + 1 - 6\sqrt{4x + 1} + 9 = x - 2$$
$$-6\sqrt{4x + 1} + 10 = -3x - 2$$
$$-6\sqrt{4x + 1} = -3x - 12$$
$$2\sqrt{4x + 1} = x + 4$$
$$\left(2\sqrt{4x + 1}\right)^2 = (x + 4)^2$$
$$4(4x + 1) = x^2 + 8x + 16$$
$$16x + 4 = x^2 + 8x + 16$$
$$0 = x^2 - 8x + 12$$
$$0 = (x - 6)(x - 2)$$
$$x = 6, \quad x = 2$$

Check: $-3 \stackrel{?}{=} \sqrt{6 - 2} - \sqrt{4(6) + 1}$

$-3 \stackrel{?}{=} \sqrt{4} - \sqrt{25}$

$-3 \stackrel{?}{=} 2 - 5$

$-3 = -3$

$-3 \stackrel{?}{=} \sqrt{2 - 2} - \sqrt{4(2) + 1}$

$-3 \stackrel{?}{=} \sqrt{0} - \sqrt{9}$

$-3 \stackrel{?}{=} 0 - 3$

$-3 = -3$

73.
$$0 = \sqrt{x + 5} - 3 + \sqrt{x}$$
$$3 - \sqrt{x} = \sqrt{x + 5}$$
$$\left(3 - \sqrt{x}\right)^2 = \left(\sqrt{x + 5}\right)^2$$
$$9 - 6\sqrt{x} + x = x + 5$$
$$-6\sqrt{x} = -4$$
$$3\sqrt{x} = 2$$
$$\left(3\sqrt{x}\right)^2 = (2)^2$$
$$9x = 4$$
$$x = \tfrac{4}{9}$$

Check: $0 \stackrel{?}{=} \sqrt{\tfrac{4}{9} + 5} - 3 + \sqrt{\tfrac{4}{9}}$

$0 \stackrel{?}{=} \sqrt{\tfrac{4}{9} + \tfrac{45}{9}} - 3 + \tfrac{2}{3}$

$0 \stackrel{?}{=} \sqrt{\tfrac{49}{9}} - 3 + \tfrac{2}{3}$

$0 \stackrel{?}{=} \tfrac{7}{3} - \tfrac{9}{3} + \tfrac{2}{3}$

$0 = 0$

75.
$$0 = \sqrt{3x - 2} - 1 - \sqrt{2x - 3}$$
$$\sqrt{2x - 3} = \sqrt{3x - 2} - 1$$
$$\left(\sqrt{2x - 3}\right)^2 = \left(\sqrt{3x - 2} - 1\right)^2$$
$$2x - 3 = 3x - 2 - 2\sqrt{3x - 2} + 1$$
$$-x - 2 = -2\sqrt{3x - 2}$$
$$x + 2 = 2\sqrt{3x - 2}$$
$$(x + 2)^2 = \left(2\sqrt{3x - 2}\right)^2$$
$$x^2 + 4x + 4 = 4(3x - 2)$$
$$x^2 + 4x + 4 = 12x - 8$$
$$x^2 - 8x + 12 = 0$$
$$(x - 2)(x - 6) = 0$$
$$x - 2 = 0 \qquad x - 6 = 0$$
$$x = 2 \qquad x = 6$$

Check: $0 \stackrel{?}{=} \sqrt{3(2) - 2} - 1 - \sqrt{2(2) - 3}$

$0 \stackrel{?}{=} \sqrt{4} - 1 - \sqrt{1}$

$0 \stackrel{?}{=} 2 - 1 - 1$

$0 = 0$

Check: $0 \stackrel{?}{=} \sqrt{3(6) - 2} - 1 - \sqrt{2(6) - 3}$

$0 \stackrel{?}{=} \sqrt{16} - 1 - \sqrt{9}$

$0 \stackrel{?}{=} 4 - 1 - 3$

$0 = 0$

77. Looking at the graph, we see that there are approximately 30,000 passengers.

Check:

$$\begin{aligned}
2.5 &= \sqrt{0.2x+1} \\
2.5^2 &= \left(\sqrt{0.2x+1}\right)^2 \\
6.25 &= 0.2x+1 \\
5.25 &= 0.2x \\
\frac{5.25}{0.2} &= x
\end{aligned}$$

26.25 thousand = number of passengers

26,250 = number of passengers

79.

$$\begin{aligned}
S &= \pi r\sqrt{r^2+h^2} \\
\frac{S}{\pi r} &= \sqrt{r^2+h^2} \\
\left(\frac{S}{\pi r}\right)^2 &= \left(\sqrt{r^2+h^2}\right)^2 \\
\frac{S^2}{\pi^2 r^2} &= r^2+h^2 \\
\frac{S^2}{\pi^2 r^2} - r^2 &= h^2 \\
\frac{S^2-\pi^2 r^4}{\pi^2 r^2} &= h^2 \\
\sqrt{\frac{S^2-\pi^2 r^4}{\pi^2 r^2}} &= h \\
\frac{\sqrt{S^2-\pi^2 r^4}}{\pi r} &= h
\end{aligned}$$

$$\begin{aligned}
h &= \frac{\sqrt{\left(364\pi\sqrt{2}\right)^2 - \pi^2(14)^4}}{\pi(14)} \\
&= \frac{\sqrt{264{,}992\pi^2 - 38{,}416\pi^2}}{14\pi} \\
&= \frac{\sqrt{226{,}576\pi^2}}{14\pi} \\
&= \frac{476\pi}{14\pi} \\
&= 34 \text{ cm}
\end{aligned}$$

81. $\left(\sqrt{x}+\sqrt{6}\right)^2 \neq \left(\sqrt{x}\right)^2 + \left(\sqrt{6}\right)^2$

$\left(\sqrt{x}+\sqrt{6}\right)^2$ must be multiplied by FOIL.

83. Substitute $x = 20$ into the equation, then choose any value of a such that $a \le 20$ and solve the resulting equation for b.

Example:

$$\begin{aligned}
20 + \sqrt{20-4} &= b \qquad \text{let } a = 4 \\
20 + \sqrt{16} &= b \\
20 + 4 &= b \\
24 &= b
\end{aligned}$$

85. L_1 and L_2 are parallel because $m_1 = 4$ and $m_2 = 4$. $m_1 = m_2$

87. L_1 and L_2 are perpendicular because $m_1 = -1$ and $m_2 = 1$. $m_1 \cdot m_2 = -1$

89.

$$\begin{aligned}
4x - y &= 10 \\
-7x - 2y &= -25
\end{aligned}
\Rightarrow
\left[\begin{array}{cc:c} 4 & -1 & 10 \\ -7 & -2 & -25 \end{array}\right]
\overset{\frac{1}{4}R_1}{}
\left[\begin{array}{cc:c} 1 & -\frac{1}{4} & \frac{5}{2} \\ -7 & -2 & -25 \end{array}\right]$$

$$\overset{7R_1+R_2}{}
\left[\begin{array}{cc:c} 1 & -\frac{1}{4} & \frac{5}{2} \\ 0 & -\frac{15}{4} & -\frac{15}{2} \end{array}\right]
\overset{-\frac{4}{15}R_2}{}
\left[\begin{array}{cc:c} 1 & -\frac{1}{4} & \frac{5}{2} \\ 0 & 1 & 2 \end{array}\right]$$

$$\overset{\frac{1}{4}R_2+R_1}{}
\left[\begin{array}{cc:c} 1 & 0 & 3 \\ 0 & 1 & 2 \end{array}\right]$$

(3, 2)

91. $a^{3/5} \cdot a^{1/5} = a^{3/5+1/5} = a^{4/5}$

93. $\left(\dfrac{x^{1/2}}{x^{1/8}}\right)^4 = \left(x^{1/2-1/8}\right)^4 = \left(x^{4/8-1/8}\right)^4 = \left(x^{3/8}\right)^4$

$= x^{3/8 \cdot 4/1} = x^{3/2}$

Section 7.6 Complex Numbers

1. $\sqrt{-4} = \sqrt{-1 \cdot 4} = \sqrt{-1} \cdot \sqrt{4} = 2i$

3. $-\sqrt{-144} = -\sqrt{144 \cdot -1} = -\sqrt{144} \cdot \sqrt{-1} = -12i$

5. $\sqrt{-\frac{4}{25}} = \sqrt{\frac{4}{25} \cdot -1} = \sqrt{\frac{4}{25}} \cdot \sqrt{-1} = \frac{2}{5}i$

7. $-\sqrt{-\frac{36}{121}} = -\sqrt{\frac{36}{121} \cdot -1} = -\sqrt{\frac{36}{121}} \cdot \sqrt{-1} = -\frac{6}{11}i$

9. $\sqrt{-8} = \sqrt{4 \cdot 2 \cdot -1} = \sqrt{4} \cdot \sqrt{2} \cdot \sqrt{-1} = 2\sqrt{2}i$

11. $\sqrt{-7} = \sqrt{7 \cdot -1} = \sqrt{7} \cdot \sqrt{-1} = \sqrt{7}i$

13. $\dfrac{\sqrt{-12}}{\sqrt{-3}} = \dfrac{\sqrt{4 \cdot 3 \cdot -1}}{\sqrt{3 \cdot -1}}$

$= \dfrac{\sqrt{4} \cdot \sqrt{3} \cdot \sqrt{-1}}{\sqrt{3} \cdot \sqrt{-1}}$

$= \sqrt{4} = 2$ or

$\dfrac{\sqrt{-12}}{\sqrt{-3}} = \sqrt{\dfrac{-12}{-3}} = \sqrt{4} = 2$

15. $\sqrt{-\dfrac{18}{25}} = \sqrt{-1} \cdot \dfrac{\sqrt{18}}{\sqrt{25}} = i \cdot \dfrac{\sqrt{9 \cdot 2}}{5} = \dfrac{3\sqrt{2}}{5}i$

17. $\sqrt{-0.09} = \sqrt{0.09 \cdot -1} = \sqrt{0.09} \cdot \sqrt{-1} = 0.3i$

19. $\sqrt{-16} + \sqrt{-36} = 4i + 6i = (4 + 6)i = 10i$

21. $\sqrt{-50} - \sqrt{-8} = 5i\sqrt{2} - 2i\sqrt{2}$

$= \left(5\sqrt{2} - 2\sqrt{2}\right)i$

$= 3\sqrt{2}i$

23. $\sqrt{-12}\sqrt{-2} = 2\sqrt{3}i \cdot \sqrt{2}i = 2\sqrt{6}i^2 = -2\sqrt{6}$

25. $\sqrt{-18}\sqrt{-3} = \left(3i\sqrt{2}\right)\left(i\sqrt{3}\right)$

$= 3\sqrt{6} \cdot i^2$

$= -3\sqrt{6}$

27. $\sqrt{-0.16}\sqrt{-1.21} = (0.4i)(1.1i) = 0.44i^2 = -0.44$

29. $\sqrt{-3}\left(\sqrt{-3} + \sqrt{-4}\right) = i\sqrt{3}\left(i\sqrt{3} + 2i\right)$

$= \left(i\sqrt{3}\right)^2 + 2\sqrt{3}i^2$

$= -3 - 2\sqrt{3}$

31. $\sqrt{-2}\left(3 - \sqrt{-8}\right) = i\sqrt{2}\left(3 - 2i\sqrt{2}\right)$

$= 3\sqrt{2}i - 2i^2(2)$

$= 4 + 3\sqrt{2}i$

33. $\left(\sqrt{-16}\right)^2 = (4i)^2 = 16i^2 = -16$

35. $\sqrt{1} + \sqrt{-25} = 1 + 5i$

37. $\sqrt{27} - \sqrt{-8} = 3\sqrt{3} - 2\sqrt{2}i \neq 3\sqrt{3} + 2\sqrt{2}i$

39. $3 - 4i = a + bi$

$a = 3, \quad b = -4$

41. $5 - 4i = (a + 3) + (b - 1)i$

$a + 3 = 5 \qquad b - 1 = -4$

$a = 2 \qquad b = -3$

43. $-4 - \sqrt{-8} = a + bi$

$-4 - 2i\sqrt{2} = a + bi$

$-4 = a \qquad -2i\sqrt{2} = bi$

$-2\sqrt{2} = b$

45. $\sqrt{a} + \sqrt{-49} = 8 + bi$

$\sqrt{a} = 8 \qquad \sqrt{-49} = bi$

$a = 64 \qquad 7i = bi$

$7 = b$

47. $(-4 - 7i) + (-10 - 33i) = (-4 - 10) + (-7 - 33)i$

$= -14 - 40i$

49. $13i - (14 - 7i) = (-14) + (13 + 7)i$

$= -14 + 20i$

51. $6 - (3 - 4i) + 2i = 6 - 3 + 4i + 2i = 3 + 6i$

53. $15i - (3 - 25i) + \sqrt{-81} = 15i - 3 + 25i + 9i$

$= -3 + (15 + 25 + 9)i$

$= -3 + 49i$

55. $(3i)(-8i) = -24i^2 = -24(-1) = 24$

57. $(9 - 2i)(\sqrt{-4}) = (9 - 2i)(2i)$
$= 18i - 4i^2$
$= 18i + 4$
$= 4 + 18i$

59. $(4 + 3i)(-7 + 4i) = -28 + 16i - 21i + 12i^2$
$= -28 - 12 - 5i$
$= -40 - 5i$

61. $(6 + 3i)(6 - 3i) = 6^2 - (3i)^2$
$= 36 - (9)(-1)$
$= 36 + 9$
$= 45$

63. $-2 - 8i$, conjugate $= -2 + 8i$
Product $= (-2 - 8i)(-2 + 8i)$
$= (-2)^2 - (8i)^2$
$= 4 - 64i^2 = 4 + 64 = 68$

65. $2 + i$, conjugate $= 2 - i$
Product $= (2 + i)(2 - i)$
$= 2^2 - i^2 = 4 + 1 = 5$

67. $10i$, conjugate $= -10i$
Product $= (10i)(-10i) = -(10i)^2 = -100i^2 = 100$

69. -12, conjugate: $-12 - 0i$
Product: $(-12 + 0i)(-12 - 0i) = 144$

71. $\dfrac{2 + i}{-5i} = \dfrac{2 + i}{-5i} \cdot \dfrac{i}{i}$
$= \dfrac{(2 + i)i}{-5 \cdot i^2} = \dfrac{2i + i^2}{-5(-1)} = \dfrac{-1 + 2i}{5} = -\dfrac{1}{5} + \dfrac{2}{5}i$

73. $\dfrac{-12}{2 + 7i} = \dfrac{-12}{2 + 7i} \cdot \dfrac{2 - 7i}{2 - 7i}$
$= \dfrac{-12(2 - 7i)}{4 + 49}$
$= \dfrac{-12(2 - 7i)}{53}$
$= \dfrac{-24 + 84i}{53}$
$= -\dfrac{24}{53} + \dfrac{84}{53}i$

75. $\dfrac{5 - i}{5 + i} = \dfrac{5 - i}{5 + i} \cdot \dfrac{5 - i}{5 - i}$
$= \dfrac{25 - 2(5i) + i^2}{25 - i^2}$
$= \dfrac{25 - 10i - 1}{25 - (-1)}$
$= \dfrac{24 - 10i}{26}$
$= \dfrac{24}{26} - \dfrac{10}{26}i$
$= \dfrac{12}{13} - \dfrac{5}{13}i$

77. $\dfrac{4 - i}{3 + i} = \dfrac{4 - i}{3 + i} \cdot \dfrac{3 - i}{3 - i}$
$= \dfrac{(4 - i)(3 - i)}{(3 + i)(3 - i)}$
$= \dfrac{12 - 7i + i^2}{3^2 + 1^2}$
$= \dfrac{12 - 7i - 1}{10}$
$= \dfrac{11 - 7i}{10}$
$= \dfrac{11}{10} - \dfrac{7}{10}i$

79. $\dfrac{4 + 5i}{3 - 7i} = \dfrac{4 + 5i}{3 - 7i} \cdot \dfrac{3 + 7i}{3 + 7i}$
$= \dfrac{(4 + 5i)(3 + 7i)}{9 + 49}$
$= \dfrac{12 + 28i + 15i - 35}{58}$
$= \dfrac{-23 + 43i}{58}$
$= -\dfrac{23}{58} + \dfrac{43}{58}i$

81. $\sqrt{-2}$ in i-form is $\sqrt{2}i$.

83. Use the FOIL Method to multiply $(1 + i)$ and $(3 - 2i)$.

85. $\sqrt{-48} + \sqrt{-12} - \sqrt{-27} = \sqrt{16 \cdot 3 \cdot -1} + \sqrt{4 \cdot 3 \cdot -1} - \sqrt{9 \cdot 3 \cdot -1}$
$= 4i\sqrt{3} + 2i\sqrt{3} - 3i\sqrt{3}$
$= (4i + 2i - 3i)\sqrt{3}$
$= 3\sqrt{3}i$

87. $(-5i)(-i)(\sqrt{-49}) = -5i \cdot -i \cdot 7i = 35i^3$
$= 35 \cdot i^2 \cdot i = 35 \cdot (-1) \cdot i = -35i$

89. $(-3i)^3 = -27i^3 = 27i$

91. $(-2 + \sqrt{-5})(-2 - \sqrt{-5}) = (-2 + i\sqrt{5})(-2 - i\sqrt{5})$
$= 4 + 2i\sqrt{5} - 2i\sqrt{5} - 5i^2$
$= 4 + 5$
$= 9$

93. $(2 + 5i)^2 = 2^2 + 2(2)(5i) + (5i)^2$
$= 4 + 20i + 25i^2$
$= 4 - 25 + 20i$
$= -21 + 20i$

95. $(3 + i)^3 = (3 + i)^2(3 + i)$
$= (9 + 6i + i^2)(3 + i)$
$= (9 + 6i - 1)(3 + i)$
$= (8 + 6i)(3 + i)$
$= 24 + 8i + 18i + 6i^2$
$= 24 + (8 + 18)i + 6(-1)$
$= (24 - 6) + 26i$
$= 18 + 26i$

97. $i^9 = (i^4)^2 i = (1)^2 i = i$

99. $i^{42} = (i^4)^{10}(i^2) = (1)^{10}(-1) = -1$

101. $i^{35} = (i^4)^8 \cdot i^2 \cdot i = (1)^8 \cdot (-1) \cdot i = -i$

103. $(-i)^6 = i^6 = i^4 \cdot i^2 = 1 \cdot -1 = -1$

105. $\dfrac{5}{3+i} + \dfrac{1}{3-i} = \dfrac{5}{3+i} \cdot \dfrac{3-i}{3-i} + \dfrac{1}{3-i} \cdot \dfrac{3+i}{3+i}$
$= \dfrac{5(3-i)}{9+1} + \dfrac{3+i}{9+1}$
$= \dfrac{15 - 5i + 3 + i}{10}$
$= \dfrac{18 - 4i}{10}$
$= \dfrac{18}{10} - \dfrac{4}{10}i$
$= \dfrac{9}{5} - \dfrac{2}{5}i$

107. $\dfrac{3i}{1+i} + \dfrac{2}{2+3i} = \dfrac{3i}{1+i} \cdot \dfrac{1-i}{1-i} + \dfrac{2}{2+3i} \cdot \dfrac{2-3i}{2-3i}$
$= \dfrac{3i(1-i)}{1+1} + \dfrac{2(2-3i)}{4+9}$
$= \dfrac{3i(1-i)}{2} + \dfrac{2(2-3i)}{13}$
$= \dfrac{3i \cdot 13(1-i) + 2 \cdot 2(2-3i)}{26}$
$= \dfrac{39i(1-i) + 4(2-3i)}{26}$
$= \dfrac{39i + 39 + 8 - 12i}{26}$
$= \dfrac{47 + 27i}{26}$
$= \dfrac{47}{26} + \dfrac{27}{26}i$

109. $(a + bi) + (a - bi) = (a + a) + (b - b)i = 2a + 0i$

111. $(a + bi) - (a - bi) = (a - a) + (b + b)i = 2bi$

113. (a) $\left(\dfrac{-5 + 5\sqrt{3}i}{2}\right)^3 = \left(\dfrac{-5}{2} + \dfrac{5}{2}\sqrt{3}i\right)^2\left(\dfrac{-5}{2} + \dfrac{5}{2}\sqrt{3}i\right)$
$= \left(\dfrac{25}{4} - \dfrac{25}{2}\sqrt{3}i + \dfrac{25}{4}(3)i^2\right)\left(\dfrac{-5}{2} + \dfrac{5}{2}\sqrt{3}i\right)$
$= \left(\dfrac{25}{4} - \dfrac{25}{2}\sqrt{3}i - \dfrac{75}{4}\right)\left(\dfrac{-5}{2} + \dfrac{5}{2}\sqrt{3}i\right)$
$= \left(\dfrac{-50}{4} - \dfrac{25}{2}\sqrt{3}i\right)\left(\dfrac{-5}{2} + \dfrac{5}{2}\sqrt{3}i\right)$
$= \left(\dfrac{-25}{2} - \dfrac{25}{2}\sqrt{3}i\right)\left(\dfrac{-5}{2} + \dfrac{5}{2}\sqrt{3}i\right)$
$= \dfrac{125}{4} - \dfrac{125}{4}\sqrt{3}i + \dfrac{125}{4}\sqrt{3}i - \dfrac{125}{4}(3)i^2$
$= \dfrac{125}{4} + \dfrac{375}{4} = \dfrac{500}{4} = 125$

(b) Use the same method as part (a).

115. (a) $1, \dfrac{-1+\sqrt{3}i}{2}, \dfrac{-1-\sqrt{3}i}{2}$

(b) $2, \dfrac{-2+2\sqrt{3}i}{2} = -1+\sqrt{3}i,$

$\dfrac{-2-2\sqrt{3}i}{2} = -1-\sqrt{3}i$

(c) $4, \dfrac{-4+4\sqrt{3}i}{2} = -2+2\sqrt{3}i,$

$\dfrac{-4-4\sqrt{3}i}{2} = -2-2\sqrt{3}i$

117. Exercise 109: The sum of complex conjugates of the form $a + bi$ and $a - bi$ is twice the real number a, or $2a$.

Exercise 110: The product of complex conjugates of the form $a + bi$ and $a - bi$ is the sum of squares of a and b, or $a^2 + b^2$.

Exercise 111: The difference of complex conjugates of the form $a + bi$ and $a - bi$ is twice the imaginary number bi, or $2bi$.

Exercise 112: The sum of the squares of complex conjugates of the form $a + bi$ and $a - bi$ is the difference of twice the squares of a and b, or $2a^2 - 2b^2$.

119. The numbers must be written in i-form first.

$\sqrt{-3}\sqrt{-3} = (\sqrt{3}i)(\sqrt{3}i) = 3i^2 = 3(-1) = -3$

121. To simplify the quotient, multiply the numerator and the denominator by $-bi$. This will yield a positive real number in the denominator. The number i can also be used to simplify the quotient. The denominator will be the opposite of b, but the resulting number will be the same.

123. $(x-5)(x+7) = 0$

$x - 5 = 0 \qquad x + 7 = 0$

$x = 5 \qquad x = -7$

125. $3y(y-3)(y+4) = 0$

$3y = 0 \qquad y - 3 = 0 \qquad y + 4 = 0$

$y = 0 \qquad y = 3 \qquad y = -4$

127. $\sqrt{x} = 9$

$(\sqrt{x})^2 = (9)^2$

$x = 81$

Check: $\sqrt{81} \stackrel{?}{=} 9$

$9 = 9$

129. $\sqrt{x} - 5 = 0$

$\sqrt{x} = 5$

$(\sqrt{x})^2 = (5)^2$

$x = 25$

Check: $\sqrt{25} - 5 \stackrel{?}{=} 0$

$5 - 5 \stackrel{?}{=} 0$

$0 = 0$

Review Exercises for Chapter 7

1. $-\sqrt{81} = -9$ because $9 \cdot 9 = 81$.

3. $-\sqrt[3]{64} = -4$ because $4 \cdot 4 \cdot 4 = 64$.

5. $-\sqrt{\left(\frac{3}{4}\right)^2} = -\frac{3}{4}$

7. $\sqrt[3]{-\left(\frac{1}{5}\right)^3} = -\frac{1}{5}$ (inverse property of powers and roots)

9. $\sqrt{-2^2} = 2i$ (not a real number)

11. $l = \sqrt{A}$

$= \sqrt{\dfrac{16}{81}}$

$= \dfrac{4}{9}$ cm

13.

Radical Form	Rational Exponent Form
$\sqrt[3]{27} = 3$	$27^{1/3} = 3$

15. $\sqrt[3]{216} = 6$

17. $27^{4/3} = (\sqrt[3]{27})^4 = 3^4 = 81$

19. $(-25)^{3/2} = (\sqrt{-25})^3$ (not a real number)

21. $8^{-4/3} = \frac{1}{8^{4/3}} = \frac{1}{\left(\sqrt[3]{8}\right)^4} = \frac{1}{2^4} = \frac{1}{16}$

23. $x^{3/4} \cdot x^{-1/6} = x^{3/4+(-1/6)}$
$= x^{9/12+(-2/12)}$
$= x^{7/12}$

25. $z\sqrt[3]{z^2} = z \cdot z^{2/3}$
$= z^{1+2/3}$
$= z^{5/3}$

27. $\frac{\sqrt[4]{x^3}}{\sqrt{x^4}} = \frac{x^{3/4}}{x^{4/2}}$
$= x^{3/4-2} = x^{3/4-8/4}$
$= x^{-5/4} = \frac{1}{x^{5/4}}$

29. $\sqrt[3]{a^3b^2} = a\sqrt[3]{b^2} = ab^{2/3}$

31. $\sqrt[4]{\sqrt{x}} = \sqrt[4]{x^{1/2}} = \left(x^{1/2}\right)^{1/4} = x^{1/8}$

33. $\frac{(3x+2)^{2/3}}{\sqrt[3]{3x+2}} = \frac{(3x+2)^{2/3}}{(3x+2)^{1/3}}$
$= (3x+2)^{2/3-1/3}$
$= (3x+2)^{1/3}$
$= \sqrt[3]{3x+2}, x \neq -\frac{2}{3}$

35. $f(x) = \sqrt{x-2}$

(a) $f(-7) = \sqrt{-7-2} = \sqrt{-9}$ (not a real number)

(b) $f(51) = \sqrt{51-2} = \sqrt{49} = 7$

37. $g(x) = \sqrt[3]{2x-1}$

(a) $g(0) = \sqrt[3]{2(0)-1} = \sqrt[3]{-1} = -1$

(b) $g(14) = \sqrt[3]{2(14)-1} = \sqrt[3]{27} = 3$

39. $f(x) = \sqrt[7]{x}$

Domain: $(-\infty, \infty)$

41. $g(x) = \sqrt{6x}$

Domain: $[0, \infty)$

43. $f(x) = \sqrt{9-2x}$

Domain: $\left(-\infty, \frac{9}{2}\right]$

$9 - 2x \geq 0$
$-2x \geq -9$
$x \leq \frac{9}{2}$

45. $\sqrt{63} = \sqrt{9 \cdot 7} = \sqrt{9}\sqrt{7} = 3\sqrt{7}$

47. $\sqrt{242} = \sqrt{121 \cdot 2} = \sqrt{121}\sqrt{2} = 11\sqrt{2}$

49. $\sqrt{36u^5v^2} = \sqrt{6^2 \cdot u^4 \cdot u \cdot v^2} = 6u^2|v|\sqrt{u}$

51. $\sqrt{0.25x^4y} = \sqrt{25 \times 10^{-2}x^4y}$
$= 5 \times 10^{-1}x^2\sqrt{y}$
$= 0.5x^2\sqrt{y}$

53. $\sqrt[3]{48a^3b^4} = \sqrt[3]{8 \cdot 6a^3b^3b}$
$= 2ab\sqrt[3]{6b}$

55. $\sqrt{\frac{5}{6}} = \sqrt{\frac{5}{6}} \cdot \frac{\sqrt{6}}{\sqrt{6}} = \frac{\sqrt{30}}{6}$

57. $\frac{2}{\sqrt[3]{2x}} = \frac{2}{\sqrt[3]{2x}} \cdot \frac{\sqrt[3]{2^2x^2}}{\sqrt[3]{2^2x^2}} = \frac{2\sqrt[3]{4x^2}}{\sqrt[3]{8x^3}} = \frac{2\sqrt[3]{4x^2}}{2x} = \frac{\sqrt[3]{4x^2}}{x}$

59. $c = \sqrt{a^2 + b^2}$
$c = \sqrt{9^2 + 8^2}$
$c = \sqrt{81 + 64}$
$c = \sqrt{145}$

61. $c = \sqrt{a^2 + b^2}$
$= \sqrt{15^2 + 8^2}$
$= \sqrt{225 + 64}$
$= \sqrt{289}$
$= 17$

63. $2\sqrt{24} + 7\sqrt{6} - \sqrt{54} = 2\sqrt{4 \cdot 6} + 7\sqrt{6} - \sqrt{9 \cdot 6} = 4\sqrt{6} + 7\sqrt{6} - 3\sqrt{6} = (4 + 7 - 3)\sqrt{6} = 8\sqrt{6}$

65. $5\sqrt{x} - \sqrt[3]{x} + 9\sqrt{x} - 8\sqrt[3]{x} = 5\sqrt{x} + 9\sqrt{x} - \sqrt[3]{x} - 8\sqrt[3]{x} = (5 + 9)\sqrt{x} + (-1 - 8)\sqrt[3]{x} = 14\sqrt{x} - 9\sqrt[3]{x}$

67. $10\sqrt[4]{y+3} - 3\sqrt[4]{y+3} = (10 - 3)\sqrt[4]{y+3} = 7\sqrt[4]{y+3}$

69. $2x\sqrt[3]{24x^2y} - \sqrt[3]{3x^5y} = 2x\sqrt[3]{8 \cdot 3 \cdot x^2y} - \sqrt[3]{3 \cdot x^3 \cdot x^2 \cdot y} = 4x\sqrt[3]{3x^2y} - x\sqrt[3]{3x^2y} = 3x\sqrt[3]{3x^2y}$

71. $c = \sqrt{a^2 + b^2}$

$c = \sqrt{\left(2\sqrt{x}\right)^2 + \left(2\sqrt{3x}\right)^2}$

$c = \sqrt{4x + 12x}$

$c = \sqrt{16x}$

$c = 4\sqrt{x}$

Perimeter $= 2\sqrt{x} + 2\sqrt{3x} + 4\sqrt{x} = 6\sqrt{x} + 2\sqrt{3x}$ ft

73. $\sqrt{15} \cdot \sqrt{20} = \sqrt{15 \cdot 20}$

$= \sqrt{300}$

$= \sqrt{100 \cdot 3}$

$= 10\sqrt{3}$

75. $\sqrt{10}\left(\sqrt{2} + \sqrt{5}\right) = \sqrt{10}\sqrt{2} + \sqrt{10}\sqrt{5}$

$= \sqrt{20} + \sqrt{50}$

$= \sqrt{4 \cdot 5} + \sqrt{25 \cdot 2}$

$= 2\sqrt{5} + 5\sqrt{2}$

77. $\left(\sqrt{3} - \sqrt{x}\right)\left(\sqrt{3} + \sqrt{x}\right) = 3 - \sqrt{3x} + \sqrt{3x} - x$

$= 3 - x$

79. $3 - \sqrt{7}$

Conjugate: $3 + \sqrt{7}$

$\left(3 - \sqrt{7}\right)\left(3 + \sqrt{7}\right) = 3^2 - \left(\sqrt{7}\right)^2$

$= 9 - 7$

$= 2$

81. $\sqrt{x} + 20$

Conjugate: $\sqrt{x} - 20$

$\left(\sqrt{x} + 20\right)\left(\sqrt{x} - 20\right) = \left(\sqrt{x}\right)^2 - 20^2$

$= x - 400$

83. $\dfrac{\sqrt{2} - 1}{\sqrt{3} - 4} = \dfrac{\sqrt{2} - 1}{\sqrt{3} - 4} \cdot \dfrac{\sqrt{3} + 4}{\sqrt{3} + 4}$

$= \dfrac{\sqrt{6} + 4\sqrt{2} - \sqrt{3} - 4}{\left(\sqrt{3}\right)^2 - 4^2}$

$= \dfrac{\sqrt{6} + 4\sqrt{2} - \sqrt{3} - 4}{3 - 16}$

$= \dfrac{\sqrt{6} + 4\sqrt{2} - \sqrt{3} - 4}{-13}$

$= -\dfrac{\sqrt{6} + 4\sqrt{2} - \sqrt{3} - 4}{13}$

or $-\dfrac{\left(\sqrt{2} - 1\right)\left(\sqrt{3} + 4\right)}{13}$

85. $\dfrac{\sqrt{x} + 10}{\sqrt{x} - 10} = \dfrac{\sqrt{x} + 10}{\sqrt{x} - 10} \cdot \dfrac{\sqrt{x} + 10}{\sqrt{x} + 10}$

$= \dfrac{x + 10\sqrt{x} + 10\sqrt{x} + 100}{\left(\sqrt{x}\right)^2 - 10^2}$

$= \dfrac{x + 20\sqrt{x} + 100}{x - 100}$

87. $\sqrt{2x} - 8 = 0$

$\sqrt{2x} = 8$

$\left(\sqrt{2x}\right)^2 = 8^2$

$2x = 64$

$x = 32$

Check: $\sqrt{2(32)} - 8 \stackrel{?}{=} 0$

$\sqrt{64} - 8 \stackrel{?}{=} 0$

$8 - 8 \stackrel{?}{=} 0$

$0 = 0$

89. $\sqrt[4]{3x-1}+6=3$

$$\sqrt[4]{3x-1}=-3$$
$$\left(\sqrt[4]{3x-1}\right)^4=(-3)^4$$
$$3x-1=81$$
$$3x=82$$
$$x=\tfrac{82}{3}$$

Not a real solution

Check: $\sqrt[4]{3\left(\frac{82}{3}\right)-1}+6\stackrel{?}{=}3$

$$\sqrt[4]{82-1}+6\stackrel{?}{=}3$$
$$\sqrt[4]{81}+6\stackrel{?}{=}3$$
$$3+6\stackrel{?}{=}3$$
$$9\neq 3$$

91. $\sqrt[3]{5x+2}-\sqrt[3]{7x-8}=0$

$$\sqrt[3]{5x+2}=\sqrt[3]{7x-8}$$
$$\left(\sqrt[3]{5x+2}\right)^3=\left(\sqrt[3]{7x-8}\right)^3$$
$$5x+2=7x-8$$
$$10=2x$$
$$5=x$$

Check: $\sqrt[3]{5(5)+2}-\sqrt[3]{7(5)-8}\stackrel{?}{=}0$

$$\sqrt[3]{27}-\sqrt[3]{27}\stackrel{?}{=}0$$
$$0=0$$

93. $\sqrt{2(x+5)}=x+5$

$$\left(\sqrt{2(x+5)}\right)^2=(x+5)^2$$
$$2(x+5)=x^2+10x+25$$
$$2x+10=x^2+10x+25$$
$$0=x^2+8x+15$$
$$0=(x+5)(x+3)$$
$$-5=x,\quad x=-3$$

Check: $\sqrt{2(-5+5)}\stackrel{?}{=}-5+5$

$$\sqrt{0}\stackrel{?}{=}0$$
$$0=0$$
$$\sqrt{2(-3+5)}\stackrel{?}{=}-3+5$$
$$\sqrt{4}\stackrel{?}{=}2$$
$$2=2$$

95. $\sqrt{1+6x}=2-\sqrt{6x}$

$$\left(\sqrt{1+6x}\right)^2=\left(2-\sqrt{6x}\right)^2$$
$$1+6x=4-4\sqrt{6x}+6x$$
$$1=4-4\sqrt{6x}$$
$$-3=-4\sqrt{6x}$$
$$(3)^2=\left(4\sqrt{6x}\right)^2$$
$$9=16(6x)$$
$$\frac{9}{96}=x$$
$$\frac{3}{32}=x$$

Check: $\sqrt{1+6\left(\frac{3}{32}\right)}\stackrel{?}{=}2-\sqrt{6\left(\frac{3}{32}\right)}$

$$\sqrt{\frac{32}{32}+\frac{18}{32}}\stackrel{?}{=}2-\sqrt{\frac{18}{32}}$$
$$\sqrt{\frac{50}{32}}\stackrel{?}{=}2-\sqrt{\frac{9\cdot 2}{16\cdot 2}}$$
$$\sqrt{\frac{25\cdot 2}{16\cdot 2}}\stackrel{?}{=}2-\sqrt{\frac{9\cdot 2}{16\cdot 2}}$$
$$\sqrt{\frac{25}{16}}\stackrel{?}{=}2-\sqrt{\frac{9}{16}}$$
$$\frac{5}{4}\stackrel{?}{=}2-\frac{3}{4}$$
$$\frac{5}{4}\stackrel{?}{=}\frac{8}{4}-\frac{3}{4}$$
$$\frac{5}{4}=\frac{5}{4}$$

97. $a^2+b^2=c^2$

$$(36.8)^2+b^2=(42)^2$$
$$1354.24+b^2=1764$$
$$b^2=409.76$$
$$b\approx 20.24\text{ in.}$$

99.
$$\begin{aligned} t &= 2\pi\sqrt{\frac{L}{32}} \\ 1.9 &= 2\pi\sqrt{\frac{L}{32}} \\ \frac{1.9}{2\pi} &= \sqrt{\frac{L}{32}} \\ \left(\frac{1.9}{2\pi}\right)^2 &= \left(\sqrt{\frac{L}{32}}\right)^2 \\ \frac{3.61}{4\pi^2} &= \frac{L}{32} \\ \frac{3.61(32)}{4\pi^2} &= L \\ 2.93 \text{ feet} &\approx L \end{aligned}$$

101.
$$\begin{aligned} v &= \sqrt{2gh} \\ 64 &= \sqrt{2gh} \\ 64^2 &= \left(\sqrt{2gh}\right)^2 \\ 4096 &= 2(32)h \\ \tfrac{4096}{64} &= h \\ 64 \text{ feet} &= h \end{aligned}$$

103. $\sqrt{-48} = \sqrt{16 \cdot 3 \cdot -1} = 4\sqrt{3}i$

105.
$$\begin{aligned} 10 - 3\sqrt{-27} &= 10 - 3\sqrt{-1 \cdot 9 \cdot 3} \\ &= 10 - 3\sqrt{-1} \cdot \sqrt{9} \cdot \sqrt{3} \\ &= 10 - 9\sqrt{3}i \end{aligned}$$

107.
$$\begin{aligned} \tfrac{3}{4} - 5\sqrt{-\tfrac{3}{25}} &= \tfrac{3}{4} - 5\sqrt{\tfrac{3}{25} \cdot -1} \\ &= \tfrac{3}{4} - \tfrac{5}{5}i\sqrt{3} \\ &= \tfrac{3}{4} - \sqrt{3}i \end{aligned}$$

109.
$$\begin{aligned} 8.4 + 20\sqrt{-0.81} &= 8.4 + 20\sqrt{-1 \cdot 0.81} \\ &= 8.4 + 20\sqrt{-1}\sqrt{0.81} \\ &= 8.4 + 20(0.9)i \\ &= 8.4 + 18i \end{aligned}$$

111.
$$\begin{aligned} \sqrt{-9} - \sqrt{-1} &= \sqrt{-1 \cdot 9} - \sqrt{-1} \\ &= \sqrt{-1}\sqrt{9} - \sqrt{-1} \\ &= -3i - i \\ &= 2i \end{aligned}$$

113. $\sqrt{-81} + \sqrt{-36} = 9i + 6i = 15i$

115.
$$\begin{aligned} \sqrt{-10}\left(\sqrt{-4} - \sqrt{-7}\right) &= i\sqrt{10}\left(2i - i\sqrt{7}\right) \\ &= 2i^2\sqrt{10} - i^2\sqrt{70} \\ &= -2\sqrt{10} + \sqrt{70} \end{aligned}$$

117. $12 - 5i = (a + 2) + (b - 1)i$

$$\begin{aligned} a + 2 &= 12 \qquad & b - 1 &= -5 \\ a &= 10 & b &= -4 \end{aligned}$$

119.
$$\begin{aligned} \sqrt{-49} + 4 &= a + bi \\ 7i + 4 &= a + bi \end{aligned}$$
$a = 4, \quad b = 7$

121.
$$\begin{aligned} (-4 + 5i) - (-12 + 8i) &= (-4 + 12) + (5 - 8)i \\ &= 8 - 3i \end{aligned}$$

123. $(4 - 3i)(4 + 3i) = 4^2 - (3i)^2 = 16 + 9 = 25$

125.
$$\begin{aligned} (6 - 5i)^2 &= 6^2 - 2(6)(5i) + (5i)^2 \\ &= 36 - 60i - 25 \\ &= 11 - 60i \end{aligned}$$

127. $\dfrac{7}{3i} = \dfrac{7}{3i} \cdot \dfrac{-i}{-i} = \dfrac{-7i}{-3i^2} = -\dfrac{7i}{3}$

129.
$$\begin{aligned} \frac{-3i}{4 - 6i} &= \frac{-3i}{4 - 6i} \cdot \frac{4 + 6i}{4 + 6i} = \frac{-12i - 18i^2}{16 - 36i^2} \\ &= \frac{-12i + 18}{16 + 36} \\ &= \frac{18 - 12i}{52} \\ &= \frac{18}{52} - \frac{12}{52}i \\ &= \frac{9}{26} - \frac{3}{13}i \end{aligned}$$

131.
$$\begin{aligned} \frac{3 - 5i}{6 + i} &= \frac{3 - 5i}{6 + i} \cdot \frac{6 - i}{6 - i} \\ &= \frac{18 - 3i - 30i + 5i^2}{6^2 - i^2} \\ &= \frac{18 - 33i - 5}{36 + 1} \\ &= \frac{13 - 33i}{37} \\ &= \frac{13}{37} - \frac{33}{37}i \end{aligned}$$

Chapter Test for Chapter 7

1. (a) $16^{3/2} = \left(\sqrt{16}\right)^3 = 4^3 = 64$

(b) $\sqrt{5}\sqrt{20} = \sqrt{5 \cdot 20} = \sqrt{100} = 10$

2. (a) $125^{-2/3} = \dfrac{1}{125^{2/3}} = \dfrac{1}{\left(\sqrt[3]{125}\right)^2} = \dfrac{1}{5^2} = \dfrac{1}{25}$

(b) $\sqrt{3}\sqrt{12} = \sqrt{3 \cdot 12} = \sqrt{36} = 6$

3. $f(x) = \sqrt{9 - 5x}$

(a) $f(-8) = \sqrt{9 - 5(-8)} = \sqrt{9 + 40} = \sqrt{49} = 7$

(b) $f(0) = \sqrt{9 - 5(0)} = \sqrt{9} = 3$

4. $g(x) = \sqrt{7x - 3}$

$$7x - 3 \ge 0$$
$$7x \ge 3$$
$$x \ge \tfrac{3}{7}$$

$\left[\tfrac{3}{7}, \infty\right)$

5. (a) $\left(\dfrac{x^{1/2}}{x^{1/3}}\right)^2 = \dfrac{x}{x^{2/3}} = x^{1-2/3} = x^{1/3},\ x \ne 0$

(b) $5^{1/4} \cdot 5^{7/4} = 5^{1/4+7/4} = 5^{8/4} = 5^2 = 25$

6. (a) $\sqrt{\dfrac{32}{9}} = \sqrt{\dfrac{16 \cdot 2}{9}} = \dfrac{4}{3}\sqrt{2}$

(b) $\sqrt[3]{24} = \sqrt[3]{8 \cdot 3} = 2\sqrt[3]{3}$

7. (a) $\sqrt{24x^3} = \sqrt{4 \cdot 6 \cdot x^2 \cdot x} = 2x\sqrt{6x}$

(b) $\sqrt[4]{16x^5y^8} = \sqrt[4]{16x^4xy^8} = 2xy^2\sqrt[4]{x}$

8. $\dfrac{2}{\sqrt[3]{9y}} = \dfrac{2}{\sqrt[3]{9y}} \cdot \dfrac{\sqrt[3]{3y^2}}{\sqrt[3]{3y^2}} = \dfrac{2\sqrt[3]{3y^2}}{\sqrt[3]{27y^3}} = \dfrac{2\sqrt[3]{3y^2}}{3y}$

9.
$$\frac{10}{\sqrt{6} - \sqrt{2}} = \frac{10}{\sqrt{6} - \sqrt{2}} \cdot \frac{\sqrt{6} + \sqrt{2}}{\sqrt{6} + \sqrt{2}}$$
$$= \frac{10\left(\sqrt{6} + \sqrt{2}\right)}{\left(\sqrt{6}\right)^2 - \left(\sqrt{2}\right)^2}$$
$$= \frac{10\left(\sqrt{6} + \sqrt{2}\right)}{6 - 2}$$
$$= \frac{10\left(\sqrt{6} + \sqrt{2}\right)}{4}$$
$$= \frac{5\left(\sqrt{6} + \sqrt{2}\right)}{2}$$

10.
$$6\sqrt{18x} - 3\sqrt{32x} = 6\sqrt{9 \cdot 2x} - 3\sqrt{16 \cdot 2x}$$
$$= 18\sqrt{2x} - 12\sqrt{2x}$$
$$= 6\sqrt{2x}$$

11.
$$\sqrt{5}\left(\sqrt{15x} + 3\right) = \sqrt{75x} + 3\sqrt{5}$$
$$= \sqrt{25 \cdot 3x} + 3\sqrt{5}$$
$$= 5\sqrt{3x} + 3\sqrt{5}$$

12. $\left(4 - \sqrt{2x}\right)^2 = 16 - 8\sqrt{2x} + 2x$

13.
$$7\sqrt{27} + 14y\sqrt{12} = 7\sqrt{9 \cdot 3} + 14y\sqrt{4 \cdot 3}$$
$$= 21\sqrt{3} + 28y\sqrt{3}$$
$$= 7\sqrt{3}(3 + 4y)$$

14.
$$\sqrt{6z} + 5 = 17$$
$$\sqrt{6z} = 12$$
$$\left(\sqrt{6z}\right)^2 = 12^2$$
$$6z = 144$$
$$z = \frac{144}{6} = 24$$

Check:
$$\sqrt{6(24)} + 5 \stackrel{?}{=} 17$$
$$\sqrt{144} + 5 \stackrel{?}{=} 17$$
$$12 + 5 \stackrel{?}{=} 17$$
$$17 = 17$$

15.
$$\sqrt{x^2 - 1} = x - 2$$
$$\left(\sqrt{x^2 - 1}\right)^2 = (x - 2)^2$$
$$x^2 - 1 = x^2 - 4x + 4$$
$$4x = 5$$
$$x = \tfrac{5}{4}$$

No solution

Check:
$$\sqrt{\left(\tfrac{5}{4}\right)^2 - 1} \stackrel{?}{=} \tfrac{5}{4} - 2$$
$$\sqrt{\tfrac{25}{16} - \tfrac{16}{16}} \stackrel{?}{=} \tfrac{5}{4} - \tfrac{8}{4}$$
$$\sqrt{\tfrac{9}{16}} \stackrel{?}{=} -\tfrac{3}{4}$$
$$\tfrac{3}{4} \ne -\tfrac{3}{4}$$

16. $\sqrt{x} - x + 6 = 0$

$$\left(\sqrt{x}\right)^2 = (x - 6)^2$$
$$x = x^2 - 12x + 36$$
$$0 = x^2 - 13x + 36$$
$$0 = (x - 9)(x - 4)$$

$0 = x - 9$ $\quad$ $0 = x - 4$

$9 = x$ $\quad$ $4 = x$

Not a solution

Check: $\sqrt{9} - 9 + 6 \stackrel{?}{=} 0$

$$3 - 9 + 6 \stackrel{?}{=} 0$$
$$0 = 0$$
$$\sqrt{4} - 4 + 6 \stackrel{?}{=} 0$$
$$2 - 4 + 6 \stackrel{?}{=} 0$$
$$4 \neq 0$$

17. $(2 + 3i) - \sqrt{-25} = 2 + 3i - 5i = 2 - 2i$

18. $(3 - 5i)^2 = 3^2 - 2(3)(5i) + (5i)^2$

$$= 9 - 30i + 25i^2$$
$$= 9 - 30i - 25$$
$$= -16 - 30i$$

19. $\sqrt{-16}\left(1 + \sqrt{-4}\right) = 4i(1 + 2i) = 4i + 8i^2 = -8 + 4i$

20. $(3 - 2i)(1 + 5i) = 3 + 13i - 10i^2$

$$= 3 + 13i + 10$$
$$= 13 + 13i$$

21. $\dfrac{5 - 2i}{3 + i} = \dfrac{5 - 2i}{3 + i} \cdot \dfrac{3 - i}{3 - i}$

$$= \frac{(5 - 2i)(3 - i)}{9 + 1}$$
$$= \frac{15 - 5i - 6i + 2i^2}{10}$$
$$= \frac{15 - 11i - 2}{10}$$
$$= \frac{13}{10} - \frac{11}{10}i$$

22.
$$v = \sqrt{2gh}$$
$$96 = \sqrt{2(32)h}$$
$$96^2 = \left(\sqrt{64h}\right)^2$$
$$9216 = 64h$$
$$\frac{9216}{64} = h$$
$$144 \text{ feet} = h$$

Cumulative Test for Chapters 5–7

1. $\left(-2x^5y^{-2}z^0\right)^{-1} = -2^{-1}x^{-5}y^2 = -\dfrac{y^2}{2x^5},\quad z \neq 0, y \neq 0$

2. $\dfrac{12s^5t^{-2}}{20s^{-2}t^{-1}} = \dfrac{3s^{5+2}t^{-2+1}}{5} = \dfrac{3s^7}{5t},\quad s \neq 0$

3. $\left(\dfrac{2x^{-4}y^3}{3x^5y^{-3}z^0}\right)^{-2} = \left(\dfrac{2x^{(-4)+(-5)}y^{3+3}}{3}\right)^{-2}$

$$= \left(\frac{2x^{-9}y^6}{3}\right)^{-2} = \left(\frac{2y^6}{3x^9}\right)^{-2} = \left(\frac{3x^9}{2y^6}\right)^2$$
$$= \frac{9x^{18}}{4y^{12}},\ x \neq 0, z \neq 0$$

4. $\left(5 \times 10^3\right)^2 = 5^2 \times 10^{3\cdot 2} = 25 \times 10^6 = 2.5 \times 10^7$

5. $\left(x^5 + 2x^3 + x^2 - 10x\right) - \left(2x^3 - x^2 + x - 4\right) = x^5 + 2x^3 + x^2 - 10x - 2x^3 + x^2 - x + 4$

$$= x^5 + \left(2x^3 - 2x^3\right) + \left(x^2 + x^2\right) + (-10x - x) + 4$$
$$= x^5 + 2x^2 - 11x + 4$$

6. $-3\left(3x^3 - 4x^2 + x\right) + 3x\left(2x^2 + x - 1\right) = -9x^3 + 12x^2 - 3x + 6x^3 + 3x^2 - 3x$

$$= \left(-9x^3 + 6x^3\right) + \left(12x^2 + 3x^2\right) + (-3x - 3x)$$
$$= -3x^3 + 15x^2 - 6x$$

7. $(x + 8)(3x - 2) = 3x^2 - 2x + 24x - 16 = 3x^2 + 22x - 16$

8. $(3x + 2)(3x^2 - x + 1) = 9x^3 - 3x^2 + 3x + 6x^2 - 2x + 2$
$= 9x^3 + (-3x^2 + 6x^2) + (3x - 2x) + 2$
$= 9x^3 + 3x^2 + x + 2$

9. $2x^2 - 11x + 15 = (2x - 5)(x - 3)$

10. $9x^2 - 144 = (3x - 12)(3x + 12)$
$= 3(x - 4)3(x + 4)$
$= 9(x - 4)(x + 4)$

11. $y^3 - 3y^2 - 9y + 27 = (y^3 - 3y^2) + (-9y + 27)$
$= y^2(y - 3) - 9(y - 3)$
$= (y - 3)(y^2 - 9)$
$= (y - 3)(y - 3)(y + 3)$
$= (y - 3)^2(y + 3)$

12. $8t^3 - 40t^2 + 50t = 2t(4t^2 - 20t + 25)$
$= 2t(2t - 5)(2t - 5)$
$= 2t(2t - 5)^2$

13. $3x^2 + x - 24 = 0$
$(3x - 8)(x + 3) = 0$

$3x - 8 = 0 \qquad x + 3 = 0$
$3x = 8 \qquad x = -3$
$x = \frac{8}{3}$

14. $6x^3 - 486x = 0$
$6x(x^2 - 81) = 0$
$6x(x + 9)(x - 9) = 0$

$6x = 0 \qquad x + 9 = 0 \qquad x - 9 = 0$
$x = 0 \qquad x = -9 \qquad x = 9$

15. $\dfrac{x^2 + 8x + 16}{18x^2} \cdot \dfrac{2x^4 + 4x^3}{x^2 - 16} = \dfrac{(x + 4)^2}{18x^2} \cdot \dfrac{2x^3(x + 2)}{(x - 4)(x + 4)} = \dfrac{x(x + 4)(x + 2)}{9(x - 4)}, \; x \neq -4, \; x \neq 0$

16. $\dfrac{x^2 + 4x}{2x^2 - 7x + 3} \div \dfrac{x^2 - 16}{x - 3} = \dfrac{x(x + 4)}{(2x - 1)(x - 3)} \cdot \dfrac{x - 3}{(x - 4)(x + 4)}$
$= \dfrac{x(x + 4)(x - 3)}{(2x - 1)(x - 3)(x - 4)(x + 4)}$
$= \dfrac{x}{(2x - 1)(x - 4)}, \quad x \neq -4, x \neq 3$

17. $\dfrac{5x}{x + 2} - \dfrac{2}{x^2 - x - 6} = \dfrac{5x}{x + 2} - \dfrac{2}{(x - 3)(x + 2)} = \dfrac{5x(x - 3)}{(x + 2)(x - 3)} - \dfrac{2}{(x - 3)(x + 2)} = \dfrac{5x^2 - 15x - 2}{(x + 2)(x - 3)}$

18. $\frac{2}{x} - \frac{x}{x^3 + 3x^2} + \frac{1}{x+3} = \frac{2}{x} - \frac{x}{x^2(x+3)} + \frac{1}{x+3}$

$= \frac{2}{x} - \frac{1}{x(x+3)} + \frac{1}{x+3}$

$= \frac{2}{x}\left(\frac{x+3}{x+3}\right) - \frac{1}{x(x+3)}\left(\frac{1}{1}\right) + \frac{1}{x+3}\left(\frac{x}{x}\right)$

$= \frac{2x+6}{x(x+3)} - \frac{1}{x(x+3)} + \frac{x}{x(x+3)}$

$= \frac{2x + 6 - 1 + x}{x(x+3)}$

$= \frac{3x+5}{x(x+3)}$

19. $\frac{\left(\frac{3x}{x+2}\right)}{\left(\frac{12}{x^3+2x^2}\right)} = \frac{3x}{x+2} \div \frac{12}{x^3+2x^2} = \frac{3x}{x+2} \cdot \frac{x^2(x+2)}{12} = \frac{(3x)(x^2)(x+2)}{(x+2)12} = \frac{x^3}{4}, \quad x \neq -2, x \neq 0$

20. $\frac{\left(\frac{x}{y} - \frac{y}{x}\right)}{\left(\frac{x-y}{xy}\right)} = \frac{\left(\frac{x}{y} - \frac{y}{x}\right)}{\left(\frac{x-y}{xy}\right)} \cdot \frac{xy}{xy} = \frac{x^2 - y^2}{x - y} = \frac{(x-y)(x+y)}{x-y} = x + y, \quad x \neq 0, y \neq 0, x \neq y$

21.

–4	2	7	0	–5
		–8	4	–16
	2	–1	4	–21

$(2x^3 + 7x^2 - 5) \div (x + 4) = 2x^2 - x + 4 - \frac{21}{x+4}$

22.

$$\begin{array}{r} 2x^3 - 2x^2 - x - \frac{4}{2x-1} \\ 2x - 1 \overline{)4x^4 - 6x^3 + 0x^2 + x - 4} \\ \underline{4x^4 - 2x^3} \qquad\qquad\qquad \\ -4x^3 + 0x^2 \qquad\quad \\ \underline{-4x^3 + 2x^2} \qquad\quad \\ -2x^2 + x \quad \\ \underline{-2x^2 + x} \quad \end{array}$$

23. $\frac{1}{x} + \frac{4}{10 - x} = 1$

$x(10 - x)\left(\frac{1}{x} + \frac{4}{10 - x}\right) = (1)x(10 - x)$

$10 - x + 4x = 10x - x^2$

$x^2 - 7x + 10 = 0$

$(x - 5)(x - 2) = 0$

$x = 5, \quad x = 2$

Check: $\frac{1}{5} + \frac{4}{10 - 5} \stackrel{?}{=} 1$ $\qquad$ $\frac{1}{2} + \frac{4}{10 - 2} \stackrel{?}{=} 1$

$\frac{1}{5} + \frac{4}{5} \stackrel{?}{=} 1$ $\qquad$ $\frac{1}{2} + \frac{4}{8} \stackrel{?}{=} 1$

$\frac{5}{5} \stackrel{?}{=} 1$ $\qquad$ $\frac{1}{2} + \frac{1}{2} \stackrel{?}{=} 1$

$1 = 1$ $\qquad$ $1 = 1$

24. $$\frac{x-3}{x}+1=\frac{x-4}{x-6}$$

$$x(x-6)\left(\frac{x-3}{x}+1\right)=\left(\frac{x-4}{x-6}\right)x(x-6)$$

$$(x-6)(x-3)+x(x-6)=x(x-4)$$

$$x^2-9x+18+x^2-6x=x^2-4x$$

$$x^2-11x+18=0$$

$$(x-9)(x-2)=0$$

$$x=9, \quad x=2$$

Check:

$$\frac{9-3}{9}+1\stackrel{?}{=}\frac{9-4}{9-6} \qquad \frac{2-3}{2}+1\stackrel{?}{=}\frac{2-4}{2-6}$$

$$\frac{6}{9}+1\stackrel{?}{=}\frac{5}{3} \qquad \frac{-1}{2}+\frac{2}{2}\stackrel{?}{=}\frac{-2}{-4}$$

$$\frac{2}{3}+\frac{3}{3}\stackrel{?}{=}\frac{5}{3} \qquad \frac{1}{2}=\frac{1}{2}$$

$$\frac{5}{3}=\frac{5}{3}$$

25. $\sqrt{24x^2y^3}=\sqrt{4\cdot 6\cdot x^2\cdot y^2\cdot y}=2|x|y\sqrt{6y}$

26. $\sqrt[3]{80a^{15}b^8}=\sqrt[3]{8\cdot 10\cdot a^{15}\cdot b^6\cdot b^2}=2a^5b^2\sqrt[3]{10b^2}$

27. $\left(12a^{-4}b^6\right)^{1/2}=\sqrt{12}a^{-2}|b^3|=\dfrac{\sqrt{4\cdot 3}|b^3|}{a^2}=\dfrac{2\sqrt{3}|b^3|}{a^2}$

28. $\left(\dfrac{t^{1/2}}{t^{1/4}}\right)^2=\dfrac{t}{t^{1/2}}=t^{1-1/2}=t^{1/2}=\sqrt{t},\ t\neq 0$

29.
$$\begin{aligned}10\sqrt{20x}+3\sqrt{125x}&=10\sqrt{4\cdot 5x}+3\sqrt{25\cdot 5x}\\&=20\sqrt{5x}+15\sqrt{5x}\\&=35\sqrt{5x}\end{aligned}$$

30. $\left(\sqrt{2x}-3\right)^2=2x-6\sqrt{2x}+9$

31.
$$\begin{aligned}\frac{3}{\sqrt{10}-\sqrt{x}}&=\frac{3}{\sqrt{10}-\sqrt{x}}\cdot\frac{\sqrt{10}+\sqrt{x}}{\sqrt{10}+\sqrt{x}}\\&=\frac{3\left(\sqrt{10}+\sqrt{x}\right)}{\left(\sqrt{10}\right)^2-\left(\sqrt{x}\right)^2}\\&=\frac{3\left(\sqrt{10}+\sqrt{x}\right)}{10-x}\end{aligned}$$

32.
$$\begin{aligned}\sqrt{x-5}-6&=0\\\sqrt{x-5}&=6\\\left(\sqrt{x-5}\right)^2&=6^2\\x-5&=36\\x&=41\end{aligned}$$

Check:
$$\begin{aligned}\sqrt{41-5}-6&\stackrel{?}{=}0\\\sqrt{36}-6&\stackrel{?}{=}0\\6-6&\stackrel{?}{=}0\\0&=0\end{aligned}$$

33.
$$\begin{aligned}\sqrt{3-x}+10&=11\\\sqrt{3-x}&=1\\\left(\sqrt{3-x}\right)^2&=1^2\\3-x&=1\\2&=x\end{aligned}$$

Check:
$$\begin{aligned}\sqrt{3-2}+10&\stackrel{?}{=}11\\\sqrt{1}+10&\stackrel{?}{=}11\\1+10&\stackrel{?}{=}11\\11&=11\end{aligned}$$

34.
$$\begin{aligned}\sqrt{x+5}-\sqrt{x-7}&=2\\\sqrt{x+5}&=2+\sqrt{x-7}\\\left(\sqrt{x+5}\right)^2&=\left(2+\sqrt{x-7}\right)^2\\x+5&=4+4\sqrt{x-7}+x-7\\5&=-3+4\sqrt{x-7}\\8&=4\sqrt{x-7}\\2&=\sqrt{x-7}\\2^2&=\left(\sqrt{x-7}\right)^2\\4&=x-7\\11&=x\end{aligned}$$

Check:
$$\begin{aligned}\sqrt{11+5}-\sqrt{11-7}&\stackrel{?}{=}2\\\sqrt{16}-\sqrt{4}&\stackrel{?}{=}2\\4-2&\stackrel{?}{=}2\\2&=2\end{aligned}$$

35. $\sqrt{x-4} = \sqrt{x+7} - 1$

$$\left(\sqrt{x-4}\right)^2 = \left(\sqrt{x+7} - 1\right)^2$$
$$x - 4 = x + 7 - 2\sqrt{x+7} + 1$$
$$-4 = 8 - 2\sqrt{x+7}$$
$$-12 = -2\sqrt{x+7}$$
$$6 = \sqrt{x+7}$$
$$6^2 = \left(\sqrt{x+7}\right)^2$$
$$36 = x + 7$$
$$29 = x$$

Check: $\sqrt{29-4} \stackrel{?}{=} \sqrt{29+7} - 1$
$$\sqrt{25} \stackrel{?}{=} \sqrt{36} - 1$$
$$5 \stackrel{?}{=} 6 - 1$$
$$5 = 5$$

36. $\sqrt{-2}\left(\sqrt{-8} + 3\right) = i\sqrt{2}\left(2i\sqrt{2} + 3\right)$
$$= 2i^2 \cdot 2 + 3i\sqrt{2}$$
$$= -4 + 3\sqrt{2}i$$

37. $(-4 + 11i) - (3 - 5i) = -4 + 11i - 3 + 5i$
$$= \left[-4 + (-3)\right] + (11i + 5i)$$
$$= -7 + 16i$$

38. $(5 + 2i)^2 = 5^2 + 2(5)(2i) + (2i)^2$
$$= 25 + 20i + 4i^2$$
$$= 25 + 20i + 4(-1)$$
$$= 21 + 20i$$

39. $\dfrac{2+3i}{6-2i} = \dfrac{2+3i}{6-2i} \cdot \dfrac{6+2i}{6+2i} = \dfrac{12 + 4i + 18i + 6i^2}{6^2 - (2i)^2}$
$$= \frac{12 + 22i + 6(-1)}{36 - 4(-1)} = \frac{6 + 22i}{40}$$
$$= \frac{6}{40} + \frac{22}{40}i = \frac{3}{20} + \frac{11}{20}i$$

40. $P = (x+1) + (x+5) + x + x + (2x+1) + (2x+5)$
$$= 8x + 12$$
$$= 4(2x + 3)$$

41. The fractional part of the task completed is

$$\frac{t}{2} + \frac{2t}{7} = \frac{7t}{14} + \frac{4t}{14} = \frac{11t}{14}.$$

42. (a)
$$F = kx$$
$$50 = k(7)$$
$$\tfrac{50}{7} = k$$
$$F = \tfrac{50}{7}x$$

For $F = 20$;
$$20 = \tfrac{50}{7}x$$
$$x = 2.8$$

The spring will stretch 2.8 centimeters with a 20-pound force.

(b) For $x = 4.2$:
$$F = \left(\tfrac{50}{7}\right)(4.2) = 30$$

The force needed to stretch the spring 4.2 centimeters is 30 pounds.

43. $c^2 = a^2 + b^2$
$$c = \sqrt{180^2 + 90^2}$$
$$c = \sqrt{32{,}400 + 8100}$$
$$c = \sqrt{40{,}500}$$
$$c = \sqrt{81 \cdot 100 \cdot 5}$$
$$c = 90\sqrt{5} \approx 201.25 \text{ feet}$$

44.
$$t = \sqrt{\frac{d}{16}}$$
$$5 = \sqrt{\frac{d}{16}}$$
$$(5)^2 = \left(\sqrt{\frac{d}{16}}\right)^2$$
$$25 = \frac{d}{16}$$
$$25(16) = d$$
$$400 \text{ feet} = d$$

CHAPTER 8
Quadratic Equations, Functions, and Inequalities

Section 8.1 Solving Quadratic Equations 244

Section 8.2 Completing the Square 249

Section 8.3 The Quadratic Formula 254

Mid-Chapter Quiz 259

Section 8.4 Graphs of Quadratic Functions 261

Section 8.5 Applications of Quadratic Equations 265

Section 8.6 Quadratic and Rational Inequalities 269

Review Exercises 275

Chapter Test 282

CHAPTER 8
Quadratic Equations, Functions, and Inequalities

Section 8.1 Solving Quadratic Equations

1. $x^2 - 15x + 54 = 0$

$(x - 6)(x - 9) = 0$

$x - 6 = 0 \quad x - 9 = 0$

$x = 6 \quad x = 9$

3. $x^2 - x - 30 = 0$

$(x - 6)(x + 5) = 0$

$x = 6, \quad x = -5$

5. $x^2 + 4x = 45$

$x^2 + 4x - 45 = 0$

$(x + 9)(x - 5) = 0$

$x = -9, \quad x = 5$

7. $x^2 - 16x + 64 = 0$

$(x - 8)(x - 8) = 0$

$x - 8 = 0 \quad x - 8 = 0$

$x = 8 \quad x = 8$

9. $9x^2 - 10x - 16 = 0$

$(9x + 8)(x - 2) = 0$

$9x + 8 = 0 \quad x - 2 = 0$

$9x = -8 \quad x = 2$

$x = -\frac{8}{9}$

11. $u(u - 9) - 12(u - 9) = 0$

$(u - 9)(u - 12) = 0$

$u - 9 = 0 \quad x - 12 = 0$

$u = 9 \quad u = 12$

13. $6x^2 = 54$

$x^2 = 9$

$x = \pm\sqrt{9}$

$x = \pm 3$

15. $25x^2 = 16$

$x^2 = \frac{16}{25}$

$x = \pm\sqrt{\frac{16}{25}}$

$x = \pm\frac{4}{5}$

17. $\frac{w^2}{4} = 49$

$w^2 = 196$

$w = \pm\sqrt{196}$

$w = \pm 14$

19. $4x^2 - 25 = 0$

$4x^2 = 25$

$x^2 = \frac{25}{4}$

$x = \pm\sqrt{\frac{25}{4}}$

$x = \pm\frac{5}{2}$

21. $(x + 4)^2 = 64$

$x + 4 = \pm\sqrt{64}$

$x = -4 \pm 8$

$x = 4, -12$

23. $(x - 3)^2 = 0.25$

$x - 3 = \pm\sqrt{0.25}$

$x = 3 \pm 0.5$

$x = 3.5, 2.5$

25. $(x - 2)^2 = 7$

$x - 2 = \pm\sqrt{7}$

$x = 2 \pm \sqrt{7}$

27. $(2x + 1)^2 = 50$

$2x + 1 = \pm\sqrt{50}$

$2x = -1 \pm 5\sqrt{2}$

$x = \dfrac{-1 \pm 5\sqrt{2}}{2}$

29. $z^2 = -36$

$z = \pm\sqrt{-36}$

$z = \pm 6i$

31. $x^2 + 4 = 0$

$x^2 = -4$

$x = \pm\sqrt{-4}$

$x = \pm 2i$

33. $9u^2 + 17 = 0$

$$9u^2 = -17$$
$$u = \pm\sqrt{-\frac{17}{9}}$$
$$u = \pm\frac{\sqrt{17}}{3}i$$

35. $(t - 3)^2 = -25$

$$t - 3 = \pm\sqrt{-25}$$
$$t = 3 \pm 5i$$

37. $(3z + 4)^2 + 144 = 0$

$$(3z + 4)^2 = -144$$
$$3z + 4 = \pm\sqrt{-144}$$
$$3z + 4 = \pm 12i$$
$$3z = -4 \pm 12i$$
$$z = \frac{-4 \pm 12i}{3}$$
$$z = -\frac{4}{3} \pm 4i$$

39. $(4m + 1)^2 = -80$

$$4m + 1 = \pm\sqrt{-80}$$
$$4m = -1 \pm 4\sqrt{5}i$$
$$m = \frac{-1 \pm 4\sqrt{5}i}{4}$$
$$m = -\frac{1}{4} \pm \sqrt{5}i$$

41. $36(t + 3)^2 = -100$

$$(t + 3)^2 = -\tfrac{100}{36}$$
$$t + 3 = \pm\sqrt{-\tfrac{100}{36}}$$
$$t = -3 \pm \tfrac{10}{6}i$$
$$t = -3 \pm \tfrac{5}{3}i$$

43. $S = 4\pi r^2$

$$\frac{900}{\pi} = 4\pi r^2$$
$$\frac{900}{\pi(4\pi)} = r^2$$
$$\frac{225}{\pi^2} = r^2$$
$$\sqrt{\frac{225}{\pi^2}} = r$$
$$\frac{15}{\pi} = r \approx 4.77 \text{ inches}$$

45. $x^4 - 5x^2 + 4 = 0$

Let $u = x^2$.

$$(x^2)^2 - 5x^2 + 4 = 0$$
$$u^2 - 5u + 4 = 0$$
$$(u - 4)(u - 1) = 0$$

$u - 4 = 0$ $\quad$ $u - 1 = 0$

$u = 4$ $\quad$ $u = 1$

$x^2 = 4$ $\quad$ $x^2 = 1$

$x = \pm 2$ $\quad$ $x = \pm 1$

47. $x^4 - 5x^2 + 6 = 0$

Let $u = x^2$.

$$(x^2)^2 - 5x^2 + 6 = 0$$
$$u^2 - 5u + 6 = 0$$
$$(u - 3)(u - 2) = 0$$

$u - 3 = 0$ $\quad$ $u - 2 = 0$

$u = 3$ $\quad$ $u = 2$

$x^2 = 3$ $\quad$ $x^2 = 2$

$x = \pm\sqrt{3}$ $\quad$ $x = \pm\sqrt{2}$

49. $(x^2 - 4)^2 + 2(x^2 - 4) - 3 = 0$

Let $u = x^2 - 4$.

$$(x^2 - 4)^2 + 2(x^2 - 4) - 3 = 0$$
$$u^2 + 2u - 3 = 0$$
$$(u + 3)(u - 1) = 0$$

$u + 3 = 0$ $\quad$ $u - 1 = 0$

$u = -3$ $\quad$ $u = 1$

$x^2 - 4 = -3$ $\quad$ $x^2 - 4 = 1$

$x^2 = 1$ $\quad$ $x^2 = 5$

$x = \pm 1$ $\quad$ $x = \pm\sqrt{5}$

51. $x - 3\sqrt{x} - 4 = 0$

Let $u = \sqrt{x}$.

$$\left(\sqrt{x}\right)^2 - 3\sqrt{x} - 4 = 0$$
$$u^2 - 3u - 4 = 0$$
$$(u - 4)(u + 1) = 0$$

$u = 4$ $u = -1$

$\sqrt{x} = 4$ $\sqrt{x} = -1$

$\left(\sqrt{x}\right)^2 = 4^2$ $\left(\sqrt{x}\right)^2 = (-1)^2$

$x = 16$ $x = 1$

Check: $16 - 3\sqrt{16} - 4 \stackrel{?}{=} 0$
$$16 - 12 - 4 \stackrel{?}{=} 0$$
$$0 = 0$$

Check: $1 - 3\sqrt{1} - 4 \stackrel{?}{=} 0$
$$1 - 3 - 4 \stackrel{?}{=} 0$$
$$-6 \neq 0$$

53. $x - 7\sqrt{x} + 10 = 0$

Let $u = \sqrt{x}$.

$$\left(\sqrt{x}\right)^2 - 7\left(\sqrt{x}\right) + 10 = 0$$
$$u^2 - 7u + 10 = 0$$
$$(u - 5)(u - 2) = 0$$

$u = 5$ $u = 2$

$\sqrt{x} = 5$ $\sqrt{x} = 2$

$x = 25$ $x = 4$

Check: $25 - 7\sqrt{25} + 10 \stackrel{?}{=} 0$
$$25 - 35 + 10 \stackrel{?}{=} 0$$
$$0 = 0$$

Check: $4 - 7\sqrt{4} + 10 \stackrel{?}{=} 0$
$$4 - 14 + 10 \stackrel{?}{=} 0$$
$$0 = 0$$

55. $x^{2/3} - x^{1/3} - 6 = 0$

Let $u = x^{1/3}$.

$$\left(x^{1/3}\right)^2 - x^{1/3} - 6 = 0$$
$$u^2 - u - 6 = 0$$
$$(u - 3)(u + 2) = 0$$

$u - 3 = 0$ $u + 2 = 0$

$u = 3$ $u = -2$

$x^{1/3} = 3$ $x^{1/3} = -2$

$\left(x^{1/3}\right)^3 = 3^3$ $\left(x^{1/3}\right)^3 = (-2)^3$

$x = 27$ $x = -8$

57. $2x^{2/3} - 7x^{1/3} + 5 = 0$

Let $u = x^{1/3}$.

$$2\left(x^{1/3}\right)^2 - 7x^{1/3} + 5 = 0$$
$$2u^2 - 7u + 5 = 0$$
$$(2u - 5)(u - 1) = 0$$

$2u - 5 = 0$ $u - 1 = 0$

$2u = 5$ $u = 1$

$u = \frac{5}{2}$ $x^{1/3} = 1$

$x^{1/3} = \frac{5}{2}$ $\left(x^{1/3}\right) = 1^3$

$\left(x^{1/3}\right) = \left(\frac{5}{2}\right)^3$ $x = 1$

$x = \frac{125}{8}$

59. $x^{2/5} - 3x^{1/5} + 2 = 0$

Let $u = x^{1/5}$.

$$\left(x^{1/5}\right)^2 - 3x^{1/5} + 2 = 0$$
$$u^2 - 3u + 2 = 0$$
$$(u - 2)(u - 1) = 0$$

$u - 2 = 0$ $u - 1 = 0$

$u = 2$ $u = 1$

$x^{1/5} = 2$ $x^{1/5} = 1$

$\left(x^{1/5}\right)^5 = 2^5$ $\left(x^{1/5}\right)^5 = 1^5$

$x = 32$ $x = 1$

61. $2x^{2/5} - 7x^{1/5} + 3 = 0$

Let $u = x^{1/5}$.

$$2(x^{1/5})^2 - 7x^{1/5} + 3 = 0$$
$$2u^2 - 7u + 3 = 0$$
$$(2u - 1)(u - 3) = 0$$

$2u - 1 = 0$ $u - 3 = 0$

$2u = 1$ $u = 3$

$u = \frac{1}{2}$ $x^{1/5} = 3$

$x^{1/5} = \frac{1}{2}$ $(x^{1/5})^5 = 3^5$

$(x^{1/5})^5 = \left(\frac{1}{2}\right)^5$ $x = 243$

$x = \frac{1}{32}$

63. $x^{1/3} - x^{1/6} - 6 = 0$

Let $u = x^{1/6}$.

$$(x^{1/6})^2 - x^{1/6} - 6 = 0$$
$$u^2 - u - 6 = 0$$
$$(u - 3)(u + 2) = 0$$

$u - 3 = 0$ $u + 2 = 0$

$u = 3$ $u = -2$

$x^{1/6} = 3$ $x^{1/6} = -2$

$(x^{1/6})^6 = 3^6$ $(x^{1/6})^6 = (-2)^6$

$x = 729$ $x = 64$

Check: $729^{1/3} - 729^{1/6} - 6 \stackrel{?}{=} 0$

$$9 - 3 - 6 \stackrel{?}{=} 0$$
$$0 = 0$$

Check: $64^{1/3} - 64^{1/6} - 6 \stackrel{?}{=} 0$

$$4 - 2 - 6 \stackrel{?}{=} 0$$
$$-4 \neq 0$$

65. Write the equation in the form $u^2 = d$, where u is an algebraic expression and d is a positive constant. Take the square roots of each side to obtain the solutions $u = \pm\sqrt{d}$.

67. Two complex solutions. When the squared expression is isolated on one side of the equation, the other side is negative. When the square root of each side is taken, the square root of the negative number is imaginary.

69. $x^2 + 900 = 0$

$$x^2 = -900$$
$$x = \pm\sqrt{-900}$$
$$x = \pm 30i$$

71. $\frac{2}{3}x^2 = 6$

$$\frac{3}{2} \cdot \frac{2}{3}x^2 = 6 \cdot \frac{3}{2}$$
$$x^2 = 9$$
$$x = \pm 3$$

73. $(p - 2)^2 - 108 = 0$

$$(p - 2)^2 = 108$$
$$p - 2 = \pm\sqrt{108}$$
$$p = 2 \pm 6\sqrt{3}$$

75. $(p - 2)^2 + 108 = 0$

$$(p - 2)^2 = -108$$
$$p - 2 = \pm\sqrt{-108}$$
$$p = 2 \pm 6\sqrt{3}i$$

77. $(x + 2)^2 + 18 = 0$

$$(x + 2)^2 = -18$$
$$x + 2 = \pm\sqrt{-18}$$
$$x = -2 \pm 3\sqrt{2}i$$

79. $x^{1/2} - 3x^{1/4} + 2 = 0$

Let $u = x^{1/4}$.

$$(x^{1/4})^2 - 3x^{1/4} + 2 = 0$$
$$u^2 - 3u + 2 = 0$$
$$(u - 2)(u - 1) = 0$$

$u - 2 = 0$ $u - 1 = 0$

$u = 2$ $u = 1$

$x^{1/4} = 2$ $x^{1/4} = 1$

$(x^{1/4})^4 = 2^4$ $(x^{1/4})^4 = 1^4$

$x = 16$ $x = 1$

Check: $16^{1/2} - 3(16)^{1/4} + 2 \stackrel{?}{=} 0$

$$4 - 3(2) + 2 \stackrel{?}{=} 0$$
$$0 = 0$$

Check: $1^{1/2} - 3(1)^{1/4} + 2 \stackrel{?}{=} 0$

$$1 - 3(1) + 2 \stackrel{?}{=} 0$$
$$0 = 0$$

81. $\frac{1}{x^2} - \frac{3}{x} + 2 = 0$

Let $u = \frac{1}{x}$.

$$\left(\frac{1}{x}\right)^2 - 3\left(\frac{1}{x}\right) + 2 = 0$$
$$u^2 - 3u + 2 = 0$$
$$(u - 2)(u - 1) = 0$$

$u - 2 = 0$ $u - 1 = 0$

$u = 2$ $u = 1$

$\frac{1}{x} = 2$ $\frac{1}{x} = 1$

$1 = 2x$ $1 = x$

$\frac{1}{2} = x$

83. $4x^{-2} - x^{-1} - 5 = 0$

Let $u = x^{-1}$.

$$4u^2 - u - 5 = 0$$
$$(u + 1)(4u - 5) = 0$$

$u = -1$ $u = \frac{5}{4}$

$x^{-1} = -1$ $x^{-1} = \frac{5}{4}$

$x = -1$ $x = \frac{4}{5}$

85. $(x^2 - 3x)^2 - 2(x^2 - 3x) - 8 = 0$

Let $u = x^2 - 3x$.

$$u^2 - 2u - 8 = 0$$
$$(u - 4)(u + 2) = 0$$

$u - 4 = 0$ $u + 2 = 0$

$u = 4$ $u = -2$

$$x^2 - 3x = 4$$
$$x^2 - 3x - 4 = 0$$
$$(x - 4)(x + 1) = 0$$

$x - 4 = 0$ $x + 1 = 0$

$x = 4$ $x = -1$

$$x^2 - 3x = -2$$
$$x^2 - 3x + 2 = 0$$
$$(x - 2)(x - 1) = 0$$

$x - 2 = 0$ $x - 1 = 0$

$x = 2$ $x = 1$

87. $16\left(\frac{x - 1}{x - 8}\right)^2 + 8\left(\frac{x - 1}{x - 8}\right) + 1 = 0$

Let $u = \left(\frac{x - 1}{x - 8}\right)$.

$$16u^2 + 8u + 1 = 0$$
$$(4u + 1)(4u + 1) = 0$$
$$u = -\frac{1}{4}$$
$$\frac{x - 1}{x - 8} = -\frac{1}{4}$$
$$4x - 4 = -x + 8$$
$$5x = 12$$
$$x = \frac{12}{5}$$

89.
$$0 = -16t^2 + 256$$
$$16t^2 = 256$$
$$t^2 = 16$$
$$t = 4 \text{ seconds}$$

91.
$$0 = -16t^2 + 128$$
$$16t^2 = 128$$
$$t^2 = 8$$
$$t = \pm\sqrt{8}$$
$$t = \pm 2\sqrt{2}$$
$$t = 2\sqrt{2} \approx 2.83 \text{ seconds}$$

93.
$$0 = 144 + 128 - 16^2$$
$$0 = -16t^2 + 128t + 144$$
$$0 = -16(t^2 - 8t - 9)$$
$$0 = -16(t - 9)(t + 1)$$

$t - 9 = 0$ $t + 1 = 0$

$t = 9$ seconds $t = -1$

95. $P = \$1500$, $A = \$1685.40$

$$A = P(1 + r)^2$$
$$1685.40 = 1500(1 + r)^2$$
$$1.1236 = (1 + r)^2$$
$$1.06 = 1 + r$$
$$0.06 = r$$
$$6\% = r$$

97. If $a = 0$, the equation would not be quadratic because it would be of degree 1, not 2.

99. In the equation $x^2 = m$, it is not possible to have one real solution and one complex solution because complex solutions always occur in complex conjugate pairs.

101. $3x - 8 > 4$

$3x > 12$

$x > 4$

103. $2x - 6 \le 9 - x$

$3x \le 15$

$x \le 5$

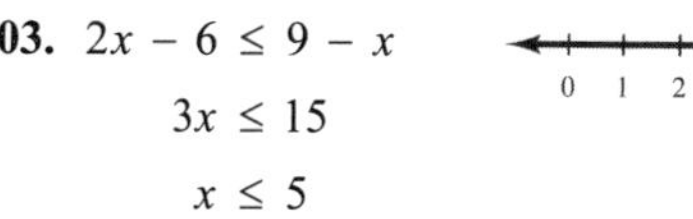

105. $\begin{aligned} x + y - z &= 4 \\ 2x + y + 2z &= 10 \\ x - 3y - 4z &= -7 \end{aligned} \Rightarrow \left[\begin{array}{ccc:c} 1 & 1 & -1 & 4 \\ 2 & 1 & 2 & 10 \\ 1 & -3 & -4 & -7 \end{array}\right] \begin{matrix} \\ -2R_1 + R_2 \\ -R_1 + R_3 \end{matrix} \left[\begin{array}{ccc:c} 1 & 1 & -1 & 4 \\ 0 & -1 & 4 & 2 \\ 0 & -4 & -3 & -11 \end{array}\right]$

$$-R_2 \left[\begin{array}{ccc:c} 1 & 1 & -1 & 4 \\ 0 & 1 & -4 & -2 \\ 0 & -4 & -3 & -11 \end{array}\right] \begin{matrix} -R_2 + R_1 \\ \\ 4R_2 + R_3 \end{matrix} \left[\begin{array}{ccc:c} 1 & 0 & 3 & 6 \\ 0 & 1 & -4 & -2 \\ 0 & 0 & -19 & -19 \end{array}\right] -\tfrac{1}{19}R_3$$

$$\left[\begin{array}{ccc:c} 1 & 0 & 3 & 6 \\ 0 & 1 & -4 & -2 \\ 0 & 0 & 1 & 1 \end{array}\right] \begin{matrix} -3R_3 + R_1 \\ 4R_3 + R_2 \end{matrix} \left[\begin{array}{ccc:c} 1 & 0 & 0 & 3 \\ 0 & 1 & 0 & 2 \\ 0 & 0 & 1 & 1 \end{array}\right] \quad \begin{aligned} x &= 3 \\ y &= 2 \\ z &= 1 \end{aligned} \quad (3, 2, 1)$$

107. $5\sqrt{3} - 2\sqrt{3} = (5 - 2)\sqrt{3} = 3\sqrt{3}$

109. $16\sqrt[3]{y} - 9\sqrt[3]{x} = 16\sqrt[3]{y} - 9\sqrt[3]{x}$

Radical expressions are not alike so cannot be combined.

111. $\sqrt{16m^4n^3} + m\sqrt{m^2n} = 4m^2n\sqrt{n} + m\sqrt[2]{n} = \left(4m^2n + m^2\right)\sqrt{n} = (4n + 1)m^2\sqrt{n}$

Section 8.2 Completing the Square

1. $x^2 + 8x + 16$

$\left[16 = \left(\frac{8}{2}\right)^2\right]$

3. $y^2 - 20y + 100$

$\left[100 = \left(-\frac{20}{2}\right)^2\right]$

5. $x^2 + 14x + 49$

$\left[49 = \left(\frac{14}{2}\right)^2\right]$

7. $t^2 + 5t + \frac{25}{4}$

$\left[\frac{25}{4} = \left(\frac{5}{2}\right)^2\right]$

9. $x^2 - 9x + \frac{81}{4}$

$\left[\frac{81}{4} = \left(-\frac{9}{2}\right)^2\right]$

11. $a^2 - \frac{1}{3}a + \frac{1}{36}$

$\left[\frac{1}{36} = \left(-\frac{1}{3}\right)\left(\frac{1}{2}\right)\right)^2\right]$

13. $y^2 + \frac{8}{5}y + \frac{16}{25}$

$\left[\frac{16}{25} = \left(\frac{1}{2} \cdot \frac{8}{5}\right)^2\right]$

15. $r^2 - 0.4r + 0.04$

$\left[0.04 = \left(-\frac{0.4}{2}\right)^2\right]$

17. (a) $x^2 - 20x + 100 = 100$

$(x - 10)^2 = 100$

$x - 10 = \pm 10$

$x = 10 \pm 10$

$x = 20, 0$

(b) $x^2 - 20x = 0$

$x(x - 20) = 0$

$x = 0, 20$

19. (a) $x^2 + 6x + 9 = 0 + 9$

$(x + 3)^2 = 9$

$x + 3 = \pm 3$

$x = -3 \pm 3$

$x = -6, 0$

(b) $x^2 + 6x = 0$

$x(x + 6) = 0$

$x = 0, \quad x + 6 = 0$

$x = -6, 0$

21. (a) $y^2 - 5y = 0$

$y^2 - 5y + \frac{25}{4} = \frac{25}{4}$

$\left(y - \frac{5}{2}\right)^2 = \frac{25}{4}$

$y - \frac{5}{2} = \pm\frac{5}{2}$

$y = \frac{5}{2} \pm \frac{5}{2}$

$= 0, 5$

(b) $y^2 - 5y = 0$

$y(y - 5) = 0$

$y = 0, \quad y - 5 = 0$

$y = 5$

23. (a) $t^2 - 8t + 16 = -7 + 16$

$(t - 4)^2 = 9$

$t - 4 = \pm 3$

$t = 4 \pm 3$

$t = 7, 1$

(b) $t^2 - 8t + 7 = 0$

$(t - 7)(t - 1) = 0$

$t = 7, \quad t = 1$

25. (a) $x^2 + 7x + \frac{49}{4} = -12 + \frac{49}{4}$

$\left(x + \frac{7}{2}\right)^2 = \frac{1}{4}$

$x + \frac{7}{2} = \pm\frac{1}{2}$

$x = -\frac{7}{2} \pm \frac{1}{2}$

$x = -\frac{6}{2}, -\frac{8}{2}$

$x = -3, -4$

(b) $x^2 + 7x + 12 = 0$

$(x + 4)(x + 3) = 0$

$x = -4, \quad x = -3$

27. (a) $x^2 - 3x + \frac{9}{4} = 18 + \frac{9}{4}$

$\left(x - \frac{3}{2}\right)^2 = \frac{81}{4}$

$x - \frac{3}{2} = \pm\frac{9}{2}$

$x = \frac{3}{2} \pm \frac{9}{2}$

$x = \frac{12}{2}, -\frac{6}{2}$

$x = 6, -3$

(b) $x^2 - 3x - 18 = 0$

$(x - 6)(x + 3) = 0$

$x = 6, -3$

29. (a) $x^2 + 8x + 7 = 0$

$x^2 + 8x + 16 = -7 + 16$

$(x + 4)^2 = 9$

$x + 4 = \pm 3$

$x = -4 \pm 3$

$x = -1, -7$

(b) $x^2 + 8x + 7 = 0$

$(x + 1)(x + 7) = 0$

$x + 1 = 0 \qquad x + 7 = 0$

$x = -1 \qquad x = -7$

31. (a) $x^2 - 10x + 21 = 0$

$x^2 - 10x + 25 = -21 + 25$

$(x - 5)^2 = 4$

$x - 5 = \pm 2$

$x = 5 \pm 2$

$x = 7, 3$

(b) $x^2 - 10x + 21 = 0$

$(x - 7)(x - 3) = 0$

$x - 7 = 0 \qquad x - 3 = 0$

$x = 7 \qquad x = 3$

33. $x^2 - 4x - 3 = 0$

$x^2 - 4x + 4 = 3 + 4$

$(x - 2)^2 = 7$

$x - 2 = \sqrt{7}$

$x = 2 \pm \sqrt{7}$

$x \approx 4.65, -0.65$

35. $x^2 + 4x - 3 = 0$
$$x^2 + 4x + 4 = 3 + 4$$
$$(x + 2)^2 = 7$$
$$x + 2 = \pm\sqrt{7}$$
$$x = -2 \pm \sqrt{7}$$
$$x \approx 0.65, -4.65$$

37. $3x^2 + 9x + 5 = 0$
$$x^2 + 3x + \frac{9}{4} = -\frac{5}{3} + \frac{9}{4}$$
$$\left(x + \frac{3}{2}\right)^2 = \frac{-20 + 27}{12}$$
$$\left(x + \frac{3}{2}\right)^2 = \frac{7}{12}$$
$$x + \frac{3}{2} = \pm\sqrt{\frac{7}{12}} \cdot \frac{\sqrt{3}}{\sqrt{3}}$$
$$x = -\frac{3}{2} \pm \frac{\sqrt{21}}{6}$$
$$x = \frac{-9 \pm \sqrt{21}}{6}$$
$$x \approx -0.74, -2.26$$

39. $2x^2 + 8x + 3 = 0$
$$x^2 + 4x + 4 = -\frac{3}{2} + 4$$
$$(x + 2)^2 = \frac{5}{2}$$
$$x + 2 = \pm\sqrt{\frac{5}{2}} \cdot \frac{\sqrt{2}}{\sqrt{2}}$$
$$x = -2 \pm \frac{\sqrt{10}}{2}$$
$$x \approx -0.42, -3.58$$

41. $4y^2 + 4y - 9 = 0$
$$y^2 + y + \frac{1}{4} = \frac{9}{4} + \frac{1}{4}$$
$$\left(y + \frac{1}{2}\right)^2 = \frac{10}{4}$$
$$y + \frac{1}{2} = \pm\sqrt{\frac{10}{4}}$$
$$y = -\frac{1}{2} \pm \frac{\sqrt{10}}{2}$$
$$y = \frac{-1 \pm \sqrt{10}}{2}$$
$$y \approx 1.08, -2.08$$

43. $x\left(x - \frac{2}{3}\right) = 14$
$$x^2 - \frac{2}{3}x + \frac{1}{9} = 14 + \frac{1}{9}$$
$$\left(x - \frac{1}{3}\right)^2 = \frac{127}{9}$$
$$x - \frac{1}{3} = \pm\sqrt{\frac{127}{9}}$$
$$x = \frac{1}{3} \pm \frac{\sqrt{127}}{3}$$
$$x = \frac{1 + \sqrt{127}}{3} \approx 4.09$$
$$x = \frac{1 - \sqrt{127}}{3} \approx -3.42$$

45. $0.1x^2 + 0.5x = -0.2$
$$0.1x^2 + 0.5x + 0.2 = 0$$
$$x^2 + 5x + 2 = 0$$
$$x^2 + 5x + \frac{25}{4} = -2 + \frac{25}{4}$$
$$\left(x + \frac{5}{2}\right)^2 = \frac{-8 + 25}{4}$$
$$\left(x + \frac{5}{2}\right)^2 = \frac{17}{4}$$
$$x + \frac{5}{2} = \pm\sqrt{\frac{17}{4}}$$
$$x = -\frac{5}{2} \pm \frac{\sqrt{17}}{2}$$
$$x = \frac{-5 \pm \sqrt{17}}{2}$$
$$x \approx -0.44, -4.56$$

47. $z^2 + 4z + 13 = 0$
$$z^2 + 4z + 4 = -13 + 4$$
$$(z + 2)^2 = -9$$
$$z + 2 = \pm\sqrt{-9}$$
$$z = -2 \pm 3i$$

49. $x^2 - 4x = -9$
$$x^2 - 4x + 4 = -9 + 4$$
$$(x - 2)^2 = -5$$
$$x - 2 = \pm\sqrt{-5}$$
$$x = 2 \pm \sqrt{5}i$$
$$x \approx 2 + 2.24i, 2 - 2.24i$$

51. $-x^2 + x - 1 = 0$

$$x^2 - x + 1 = 0$$
$$x^2 - x + \frac{1}{4} = -1 + \frac{1}{4}$$
$$\left(x - \frac{1}{2}\right)^2 = -\frac{3}{4}$$
$$x - \frac{1}{2} = \pm\sqrt{-\frac{3}{4}}$$
$$x = \frac{1}{2} \pm \frac{\sqrt{3}}{2}i$$
$$x = \frac{1 \pm \sqrt{3}i}{2}$$
$$x \approx 0.5 + 0.87i$$
$$x \approx 0.5 - 0.87i$$

53. $4z^2 - 3z + 2 = 0$

$$z^2 - \frac{3}{4}z + \frac{9}{64} = \frac{-1}{2} + \frac{9}{64}$$
$$\left(z - \frac{3}{8}\right)^2 = -\frac{23}{64}$$
$$z - \frac{3}{8} = \pm\sqrt{-\frac{23}{64}}$$
$$z = \frac{3}{8} \pm \frac{\sqrt{23}}{8}i$$
$$z = \frac{3}{8} + \frac{\sqrt{23}}{8}i \approx 0.38 + 0.60i$$
$$z = \frac{3}{8} - \frac{\sqrt{23}}{8}i \approx 0.38 - 0.60i$$

55. (a) Area of square $= x \cdot x = x^2$

Area of vertical rectangle $= 4 \cdot x = 4x$

Area of horizontal rectangle $= 4 \cdot x = 4x$

Total area $= x^2 + 4x + 4x = x^2 + 8x$

(b) Area of small square $= 4 \cdot 4 = 16$

Total area $= x^2 + 8x + 16$

(c) $(x + 4)(x + 4) = x^2 + 8x + 16$

57. *Verbal Model:*

Volume = Length · Width · Height

Labels: Length $= x + 4$

Width $= x$

Height $= 6$

Equation:
$$840 = (x + 4)(x)(6)$$
$$140 = x(x + 4)$$
$$140 = x^2 + 4x$$
$$0 = x^2 + 4x - 140$$
$$0 = (x + 14)(x - 10)$$

$x + 14 = 0 \qquad x - 10 = 0$

$x = -14 \qquad x = 10$

Not a solution

Thus, the dimensions are:
length $= 10 + 4 = 14$ inches, width $= x = 10$ inches, height $= 6$ inches

59. A perfect square trinomial is one that can be written as $(x + k)^2$.

61. Each side of the equation must be divided by 2 to obtain a leading coefficient of 1. The resulting equation is the same by the Multiplication Property of Equality.

63. $0.2x^2 + 0.1x = -0.5$

$$x^2 + \frac{1}{2}x + \frac{1}{16} = \frac{-5}{2} + \frac{1}{16}$$
$$\left(x + \frac{1}{4}\right)^2 = -\frac{39}{16}$$
$$x + \frac{1}{4} = \pm\sqrt{-\frac{39}{16}}$$
$$x = -\frac{1}{4} \pm \frac{\sqrt{39}}{4}i$$
$$x = -\frac{1}{4} + \frac{\sqrt{39}}{4}i \approx -0.25 + 1.56i$$
$$x = -\frac{1}{4} - \frac{\sqrt{39}}{4}i \approx -0.25 - 1.56i$$

65. $\frac{x}{2} - \frac{1}{x} = 1$

$$2x\left(\frac{x}{2} - \frac{1}{x}\right) = (1)2x$$
$$x^2 - 2 = 2x$$
$$x^2 - 2x + 1 = 2 + 1$$
$$(x - 1)^2 = 3$$
$$x - 1 = \pm\sqrt{3}$$
$$x = 1 \pm \sqrt{3}$$

67.
$$\frac{x^2}{8} = \frac{x+3}{2}$$
$$8\left(\frac{x^2}{8}\right) = \left(\frac{x+3}{2}\right)8$$
$$x^2 = 4(x+3)$$
$$x^2 = 4x + 12$$
$$x^2 - 4x + 4 = 12 + 4$$
$$(x-2)^2 = 16$$
$$x - 2 = \pm 4$$
$$x = 6, -2$$

69.
$$\sqrt{2x+1} = x - 3$$
$$\left(\sqrt{2x+1}\right)^2 = (x-3)^2$$
$$2x + 1 = x^2 - 6x + 9$$
$$0 = x^2 - 8x + 8$$
$$16 - 8 = x^2 - 8x + 16$$
$$8 = (x-4)^2$$
$$\pm\sqrt{8} = x - 4$$
$$4 \pm \sqrt{8} = x$$
$$4 \pm 2\sqrt{2} = x$$

71. *Verbal Model:* $\boxed{\text{Area}} = \boxed{\text{Length}} \cdot \boxed{\text{Width}}$

Labels: Length $= x$

Width $= \dfrac{200 - 4x}{3}$

Equation:
$$1400 = 2\left[x \cdot \left(\frac{200 - 4x}{3}\right)\right]$$
$$1400 = 2\left[\frac{200}{3}x - \frac{4x^2}{3}\right]$$
$$1400 = \frac{400x}{3} - \frac{8x^2}{3}$$
$$4200 = 400x - 8x^2$$
$$8x^2 - 400x + 4200 = 0$$
$$x^2 - 50x + 525 = 0$$
$$(x - 35)(x - 15) = 0$$

$x - 35 = 0$ $\qquad$ $x - 15 = 0$

$x = 35$ meters $\qquad$ $x = 15$ meters

$\dfrac{200 - 4x}{3} = 20$ meters $\qquad$ $\dfrac{200 - 4x}{3} = 46\dfrac{2}{3}$ meters

73.
$$R = x\left(80 - \tfrac{1}{2}x\right)$$
$$2750 = 80x - \tfrac{1}{2}x^2$$
$$0 = -\tfrac{1}{2}x^2 + 80x - 2750$$
$$0 = x^2 - 160x + 5500$$
$$0 = (x - 50)(x - 110)$$

$0 = x - 50$ $\qquad$ $x - 110 = 0$

$50 = x$ $\qquad$ $x = 110$

50 pairs, 110 pairs

75. Use the method of completing the square to write the quadratic equation in the form $u^2 = d$. Then use the Square Root Property to simplify.

77. (a) $d = 0$

(b) $d > 0$, and d is a perfect square.

(c) $d > 0$, and d is not a perfect square.

(d) $d < 0$

79. $3\sqrt{5}\sqrt{500} = 3 \cdot \sqrt{5 \cdot 500}$
$$= 3 \cdot \sqrt{25 \cdot 100} = 3 \cdot 5 \cdot 10 = 150$$

81. $\left(3 + \sqrt{2}\right)\left(3 - \sqrt{2}\right) = 3^2 - \left(\sqrt{2}\right)^2 = 9 - 2 = 7$

83. $\left(3 + \sqrt{2}\right)^2 = 3^2 + 2(3)\left(\sqrt{2}\right) + \left(\sqrt{2}\right)^2$
$$= 9 + 6\sqrt{2} + 2$$
$$= 11 + 6\sqrt{2}$$

85. $\dfrac{8}{\sqrt{10}} = \dfrac{8}{\sqrt{10}} \cdot \dfrac{\sqrt{10}}{\sqrt{10}} = \dfrac{8\sqrt{10}}{10} = \dfrac{4\sqrt{10}}{5}$

87. Product Rule: $\sqrt{ab} = \sqrt{a} \cdot \sqrt{b}$

Section 8.3 The Quadratic Formula

1. $2x^2 = 7 - 2x$

$2x^2 + 2x - 7 = 0$

3. $x(10 - x) = 5$

$10x - x^2 = 5$

$-x^2 + 10x - 5 = 0$

$x^2 - 10x + 5 = 0$

5. (a) $x^2 - 11x + 28 = 0$

$x = \frac{11 \pm \sqrt{(-11)^2 - 4(1)(28)}}{2(1)}$

$x = \frac{11 \pm \sqrt{121 - 112}}{2}$

$x = \frac{11 \pm \sqrt{9}}{2}$

$x = \frac{11 \pm 3}{2}$

$x = 7, 4$

(b) $(x - 7)(x - 4) = 0$

$x - 7 = 0 \quad x - 4 = 0$

$x = 7 \quad x = 4$

7. (a) $x^2 + 6x + 8 = 0$

$x = \frac{-6 \pm \sqrt{6^2 - 4(1)(8)}}{2(1)}$

$x = \frac{6 \pm \sqrt{36 - 32}}{2}$

$x = \frac{-6 \pm \sqrt{4}}{2}$

$x = \frac{-6 \pm 2}{2}$

$x = -2, -4$

(b) $(x + 4)(x + 2) = 0$

$x + 4 = 0 \quad x + 2 = 0$

$x = -4 \quad x = -2$

9. $x^2 - 2x - 4 = 0$

$x = \frac{-(-2) \pm \sqrt{(-2)^2 - 4(1)(-4)}}{2(1)}$

$x = \frac{2 \pm \sqrt{4 + 16}}{2}$

$x = \frac{2 \pm \sqrt{20}}{2}$

$x = \frac{2 \pm 2\sqrt{5}}{2}$

$x = \frac{2(1 \pm \sqrt{5})}{2}$

$x = 1 \pm \sqrt{5}$

11. $t^2 + 4t + 1 = 0$

$t = \frac{-4 \pm \sqrt{4^2 - 4(1)(1)}}{2(1)}$

$t = \frac{-4 \pm \sqrt{16 - 4}}{2}$

$t = \frac{4 \pm \sqrt{12}}{2}$

$t = \frac{-4 \pm 2\sqrt{3}}{2}$

$t = \frac{2(-2 \pm \sqrt{3})}{2}$

$t = -2 \pm \sqrt{3}$

13. $-x^2 + 10x - 23 = 0$

$x = \frac{-10 \pm \sqrt{10^2 - 4(-1)(-23)}}{2(-1)}$

$x = \frac{-10 \pm \sqrt{100 - 92}}{-2}$

$x = \frac{-10 \pm \sqrt{8}}{-2}$

$x = \frac{-10 \pm 2\sqrt{2}}{-2}$

$x = \frac{-2(5 \pm \sqrt{2})}{-2}$

$x = 5 \pm \sqrt{2}$

15. (a) $16x^2 + 8x + 1 = 0$

$$x = \frac{-8 \pm \sqrt{8^2 - 4(16)(1)}}{2(16)}$$
$$x = \frac{-8 \pm \sqrt{64 - 64}}{32}$$
$$x = \frac{-8 \pm \sqrt{0}}{32}$$
$$x = -\frac{8}{32} = -\frac{1}{4}$$

(b) $(4x + 1)(4x + 1) = 0$

$$4x + 1 = 0 \qquad 4x + 1 = 0$$
$$4x = -1 \qquad 4x = -1$$
$$x = -\frac{1}{4} \qquad x = -\frac{1}{4}$$

17. $2x^2 + 3x + 3 = 0$

$$x = \frac{-3 \pm \sqrt{3^2 - 4(2)(3)}}{2(2)}$$
$$x = \frac{-3 \pm \sqrt{9 - 24}}{4}$$
$$x = \frac{-3 \pm \sqrt{-15}}{4}$$
$$x = \frac{-3 \pm i\sqrt{15}}{4}$$
$$x = -\frac{3}{4} \pm \frac{\sqrt{15}}{4}i$$

19. $b^2 - 4ac = 1^2 - 4(1)(1) = 1 - 4 = -3$

2 distinct complex solutions

21. $b^2 - 4ac = (-2)^2 - 4(8)(-5) = 4 + 160 = 164$

Two distinct rational solutions

23. $b^2 - 4ac = (-24)^2 - 4(9)(16) = 576 - 576 = 0$

1 repeated rational solution

25. $b^2 - 4ac = (-1)^2 - 4(3)(2) = 1 - 24 = -23$

2 distinct complex solutions

27. $z^2 - 169 = 0$

$$z^2 = 169$$
$$z = \pm 13$$

29. $5y^2 + 15y = 0$

$$5y(y + 3) = 0$$
$$5y = 0 \qquad y + 3 = 0$$
$$y = 0 \qquad y = -3$$

31. $25(x - 3)^2 - 36 = 0$

$$(x - 3)^2 = \tfrac{36}{25}$$
$$x - 3 = \pm\sqrt{\tfrac{36}{25}}$$
$$x = 3 \pm \tfrac{6}{5}$$
$$x = \tfrac{15}{5} \pm \tfrac{6}{5}$$
$$x = \tfrac{21}{5}, \tfrac{9}{5}$$

33. $2y(y - 18) + 3(y - 18) = 0$

$$(y - 18)(2y + 3) = 0$$
$$y - 18 = 0 \qquad 2y + 3 = 0$$
$$y = 18 \qquad 2y = -3$$
$$y = -\tfrac{3}{2}$$

35. $x^2 + 8x + 25 = 0$

$$x^2 + 8x + 16 = -25 + 16$$
$$(x + 4)^2 = -9$$
$$x + 4 = \pm\sqrt{-9}$$
$$x = -4 \pm 3i$$

37. $3x^2 - 13x - 169 = 0$

$$x = \frac{-(-13) \pm \sqrt{(-13)^2 - 4(3)(169)}}{2(3)}$$
$$x = \frac{13 \pm \sqrt{169 - 2028}}{6}$$
$$x = \frac{13 \pm \sqrt{-1859}}{6}$$
$$x = \frac{13}{6} \pm \frac{13\sqrt{11}}{6}i$$

39. $25x^2 + 80x + 61 = 0$

$$x = \frac{-80 \pm \sqrt{80^2 - 4(25)(61)}}{2(25)}$$
$$x = \frac{-80 \pm \sqrt{6400 - 6100}}{50}$$
$$x = \frac{-80 \pm \sqrt{300}}{50}$$
$$x = \frac{-80 \pm 10\sqrt{3}}{50}$$
$$x = -\frac{8}{5} \pm \frac{\sqrt{3}}{5}$$

41. $7x(x + 2) + 5 = 3x(x + 1)$

$7x^2 + 14x + 5 = 3x^2 + 3x$

$4x^2 + 11x + 5 = 0$

$x = \dfrac{-11 \pm \sqrt{11^2 - 4(4)(5)}}{2(4)}$

$x = \dfrac{-11 \pm \sqrt{121 - 80}}{8}$

$x = \dfrac{-11 \pm \sqrt{41}}{8}$

43. $y = x^2 - 4x + 3$

Keystrokes:

[Y=] [X,T,θ] [x^2] [−] 4 [X,T,θ] [+] 3 [GRAPH]

$0 = x^2 - 4x + 3$

$0 = (x - 3)(x - 1)$

$x - 3 = 0 \qquad x - 1 = 0$

$x = 3 \qquad x = 1$

$(3, 0), (1, 0)$

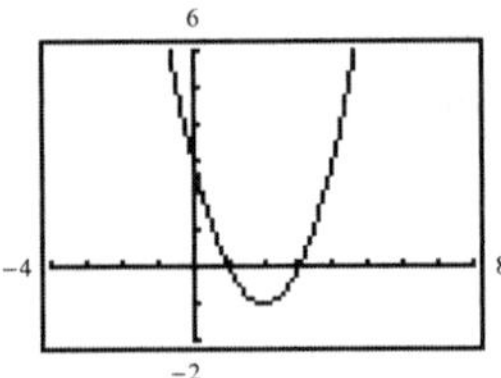

45. $-0.03x^2 + 2x - 0.4 = 0$

Keystrokes:

[Y=] [(−)] 0.03 [X,T,θ] [x^2] [+] 2 [X,T,θ] [−] 0.4 [GRAPH]

$x = \dfrac{-2 \pm \sqrt{2^2 - 4(-0.03)(-0.4)}}{2(-0.03)}$

$x = \dfrac{-2 \pm \sqrt{4 - 0.048}}{-0.06}$

$x = \dfrac{-2 \pm \sqrt{3.952}}{-0.06}$

$x \approx 0.20, 66.47$

$(0.20, 0), (66.47, 0)$

47. $x = 5 \qquad x = -2$

$x - 5 = 0 \qquad x + 2 = 0$

$(x - 5)(x + 2) = 0$

$x^2 - 3x - 10 = 0$

49. $x = 1 \qquad x = 7$

$x - 1 = 0 \qquad x - 7 = 0$

$(x - 1)(x - 7) = 0$

$x^2 - 8x + 7 = 0$

51. $x = 1 + \sqrt{2} \qquad x = 1 - \sqrt{2}$

$x - (1 + \sqrt{2}) = 0 \qquad x - (1 - \sqrt{2}) = 0$

$[x - (1 + \sqrt{2})][x - (1 - \sqrt{2})] = 0$

$[(x - 1) - \sqrt{2}][(x - 1) + \sqrt{2}] = 0$

$(x - 1)^2 - (\sqrt{2})^2 = 0$

$x^2 - 2x + 1 - 2 = 0$

$x^2 - 2x - 1 = 0$

53. $x = 5i \qquad x = -5i$

$x - 5i = 0 \qquad x + 5i = 0$

$(x - 5i)(x + 5i) = 0$

$x^2 - 25i^2 = 0$

$x^2 + 25 = 0$

55. $x = 12 \qquad x = 12$

$x - 12 = 0 \qquad x - 12 = 0$

$(x - 12)(x - 12) = 0$

$x^2 - 24x + 144 = 0$

57. $x = \frac{1}{2} \qquad x = \frac{1}{2}$

$x - \frac{1}{2} = 0 \qquad x - \frac{1}{2} = 0$

$\left(x - \frac{1}{2}\right)\left(x - \frac{1}{2}\right) = 0$

$x^2 - x + \frac{1}{4} = 0$

59. The Quadratic Formula states: The opposite of *b*, plus or minus the square root of *b* squared minus 4*ac*, all divided by 2*a*.

61. The equation in general form is $3x^2 + x - 3 = 0$, so $a = 3, b = 1$, and $c = -3$.

63. $\dfrac{x^2}{4} - \dfrac{2x}{3} = 1$

$12\left(\dfrac{x^2}{4} - \dfrac{2x}{3}\right) = (1)12$

$3x^2 - 8x - 12 = 0$

$x = \dfrac{-(-8) \pm \sqrt{(-8)^2 - 4(3)(-12)}}{2(3)}$

$x = \dfrac{8 \pm \sqrt{64 + 144}}{6}$

$x = \dfrac{8 \pm \sqrt{208}}{6}$

$x = \dfrac{8 \pm 4\sqrt{13}}{6}$

$x = \dfrac{4}{3} \pm \dfrac{2\sqrt{13}}{3}$

65. $\sqrt{x+3} = x - 1$

$$\left(\sqrt{x+3}\right)^2 = (x-1)^2$$
$$x + 3 = x^2 - 2x + 1$$
$$0 = x^2 - 3x - 2$$
$$x = \frac{-(-3) \pm \sqrt{(-3)^2 - 4(1)(-2)}}{2(1)}$$
$$x = \frac{3 \pm \sqrt{9+8}}{2}$$
$$x = \frac{3 \pm \sqrt{17}}{2}$$
$$x = \frac{3 + \sqrt{17}}{2}$$
$$x = \frac{3 - \sqrt{17}}{2} \text{ does not check.}$$

Check:
$$\sqrt{\frac{3+\sqrt{17}}{2} + 3} \stackrel{?}{=} \frac{3+\sqrt{17}}{2} - 1$$
$$\sqrt{\frac{3+\sqrt{17}+6}{2}} \stackrel{?}{=} \frac{3+\sqrt{17}-2}{2}$$
$$\sqrt{\frac{9+\sqrt{17}}{2}} \stackrel{?}{=} \frac{1+\sqrt{17}}{2}$$
$$\frac{\sqrt{18+2\sqrt{17}}}{2} \stackrel{?}{=} \frac{1+\sqrt{17}}{2}$$
$$2.5616 = 2.5616$$

Check:
$$\sqrt{\frac{3-\sqrt{17}}{2} + 3} \stackrel{?}{=} \frac{3-\sqrt{17}}{2} - 1$$
$$\sqrt{\frac{3-\sqrt{17}+6}{2}} \stackrel{?}{=} \frac{3-\sqrt{17}-2}{2}$$
$$\sqrt{\frac{9-\sqrt{17}}{2}} \stackrel{?}{=} \frac{1-\sqrt{17}}{2}$$
$$\frac{\sqrt{18-2\sqrt{17}}}{2} \stackrel{?}{=} \frac{1-\sqrt{17}}{2}$$
$$1.5616 \neq -1.5616$$

67. *Verbal Model:* Area = Length · Width

Labels: Length $= x + 6.3$

Width $= x$

Equation: $58.14 = (x + 6.3) \cdot x$

$$58.14 = x^2 + 6.3x$$
$$0 = x^2 + 6.3x - 58.14$$
$$x = \frac{-6.3 \pm \sqrt{6.3^2 - 4(1)(-58.14)}}{2(1)}$$
$$x = \frac{-6.3 \pm \sqrt{39.69 + 232.56}}{2}$$
$$x = \frac{-6.3 \pm \sqrt{272.25}}{2}$$

$x \approx 5.1$ inches

$x + 6.3 \approx 11.4$ inches

69. $d = -0.25t^2 + 1.7t + 3.5, \quad 0 \le t \le 7$

$$6 = -0.25t^2 + 1.7t + 3.5$$
$$0 = -0.25t^2 + 1.7t - 2.5$$
$$0 = t^2 - 6.8t + 10$$
$$t = \frac{-(-6.8) \pm \sqrt{6.8^2 - 4(1)(10)}}{2(1)}$$
$$t = \frac{6.8 \pm \sqrt{46.24 - 40}}{2}$$
$$t = \frac{6.8 \pm \sqrt{6.24}}{2}$$

$t = \dfrac{6.8 + \sqrt{6.24}}{2} \approx 4.65$ hours after a heavy rain

$t = \dfrac{6.8 - \sqrt{6.24}}{2} \approx 2.15$ hours after a heavy rain

71. (a) *Keystrokes:* [Y=] −0.013 [X,T,θ] [x^2] [+] 1.25 [X,T,θ] [+] 5.6 [GRAPH]

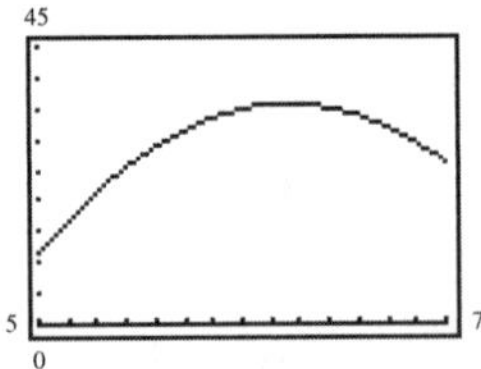

(b)
$$32 = -0.013x^2 + 1.25x + 5.6$$
$$0 = -0.013x^2 + 1.25x - 26.4$$
$$0 = 1.3x^2 - 125x + 2640$$
$$x = \frac{-(-125) \pm \sqrt{(-125)^2 - 4(1.3)(2640)}}{2(1.3)}$$
$$x = \frac{125 \pm \sqrt{15{,}625 - 13{,}728}}{2.6}$$
$$x = \frac{125 + \sqrt{1897}}{2.6}$$
$x \approx 64.8$ miles per hour
$$x = \frac{125 - \sqrt{1897}}{2.6}$$
$x \approx 31.3$ miles per hour

Using the graph, you would have to travel approximately 65 miles per hour or 31 miles per hour to obtain a fuel economy of 32 miles per gallon.

73. The Square Root Property would be convenient because the equation is of the form $u^2 = d$.

75. The Quadratic Formula would be convenient because the equation is already in general form, the expression cannot be factored, and the leading coefficient is not equal to 1.

77. When the Quadratic Formula is applied to $ax^2 + bx + c = 0$, the square root of the discriminant is evaluated. When the discriminant is positive, the square root of the discriminant is positive and will yield two real solutions (or x-intercepts). When the discriminant is zero, the equation has one real solution (or x-intercept). When the discriminant is negative, the square root of the discriminant is negative and will yield two complex solutions (or no x-intercepts).

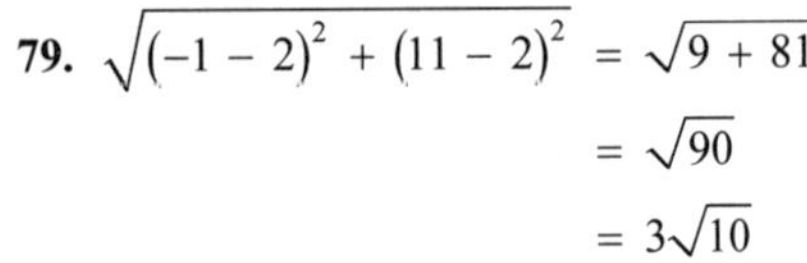

79.
$$\sqrt{(-1 - 2)^2 + (11 - 2)^2} = \sqrt{9 + 81} = \sqrt{90} = 3\sqrt{10}$$

81.
$$\sqrt{(-6 - (-3))^2 + (-2 - (-4))^2} = \sqrt{9 + 4} = \sqrt{13}$$

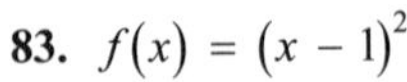

83. $f(x) = (x - 1)^2$

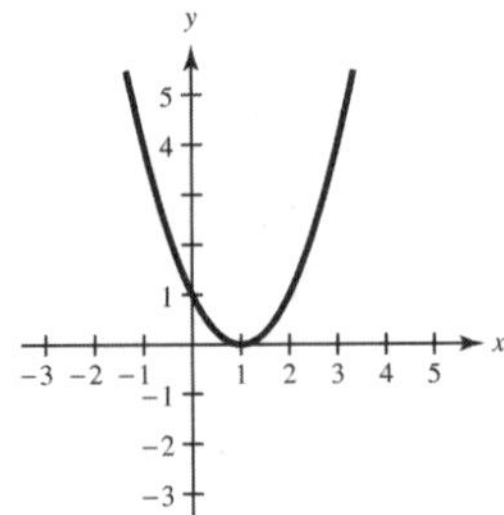

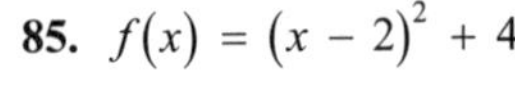

85. $f(x) = (x - 2)^2 + 4$

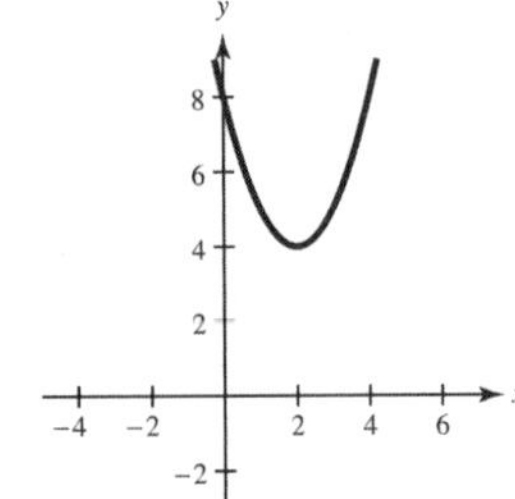

Mid-Chapter Quiz for Chapter 8

1.
$$2x^2 - 72 = 0$$
$$2(x^2 - 36) = 0$$
$$2(x - 6)(x + 6) = 0$$
$$x - 6 = 0 \qquad x + 6 = 0$$
$$x = 6 \qquad x = -6$$

2.
$$2x^2 + 3x - 20 = 0$$
$$(2x - 5)(x + 4) = 0$$
$$2x - 5 = 0 \qquad x + 4 = 0$$
$$x = \tfrac{5}{2} \qquad x = -4$$

3.
$$3x^2 = 36$$
$$x^2 = 12$$
$$x = \pm\sqrt{12}$$
$$x = \pm 2\sqrt{3}$$

4.
$$(u - 3)^2 - 16 = 0$$
$$(u - 3)^2 = 16$$
$$u - 3 = \pm 4$$
$$u = 3 \pm 4 = 7, -1$$

5.
$$m^2 + 7m + 2 = 0$$
$$m^2 + 7m + \frac{49}{4} = -2 + \frac{49}{4}$$
$$\left(m + \frac{7}{2}\right)^2 = \frac{41}{4}$$
$$m + \frac{7}{2} = \pm\sqrt{\frac{41}{4}}$$
$$m = -\frac{7}{2} \pm \frac{\sqrt{41}}{2}$$

6.
$$2y^2 + 6y - 5 = 0$$
$$y^2 + 3y = \frac{5}{2}$$
$$y^2 + 3y + \frac{9}{4} = \frac{5}{2} + \frac{9}{4}$$
$$\left(y + \frac{3}{2}\right)^2 = \frac{10}{4} + \frac{9}{4}$$
$$\left(y + \frac{3}{2}\right)^2 = \frac{19}{4}$$
$$y + \frac{3}{2} = \pm\frac{\sqrt{19}}{2}$$
$$y = -\frac{3}{2} \pm \frac{\sqrt{19}}{2}$$

7.
$$x^2 + 4x - 6 = 0$$
$$x = \frac{-4 \pm \sqrt{4^2 - 4(1)(-6)}}{2(1)}$$
$$x = \frac{-4 \pm \sqrt{16 + 24}}{2}$$
$$x = \frac{-4 \pm \sqrt{40}}{2}$$
$$x = \frac{-4 \pm 2\sqrt{10}}{2} = -2 \pm \sqrt{10}$$

8.
$$6v^2 - 3v - 4 = 0$$
$$v = \frac{-(-3) \pm \sqrt{(-3)^2 - 4(6)(-4)}}{2(6)}$$
$$v = \frac{3 \pm \sqrt{9 + 96}}{12}$$
$$v = \frac{3 \pm \sqrt{105}}{12}$$

9.
$$x^2 + 5x + 7 = 0$$
$$x = \frac{-5 \pm \sqrt{5^2 - 4(1)(7)}}{2(1)}$$
$$x = \frac{-5 \pm \sqrt{25 - 28}}{2}$$
$$x = \frac{-5 \pm \sqrt{-3}}{2}$$
$$x = \frac{-5 \pm \sqrt{3}i}{2} = -\frac{5}{2} \pm \frac{\sqrt{3}}{2}i$$

10.
$$36 = (t - 4)^2$$
$$\pm 6 = t - 4$$
$$4 \pm 6 = t$$
$$10, -2 = t$$

11.
$$(x - 10)(x + 3) = 0$$
$$(x - 10) = 0 \qquad x + 3 = 0$$
$$x = 10 \qquad x = -3$$

12.
$$x^2 - 3x - 10 = 0$$
$$(x - 5)(x + 2) = 0$$
$$x - 5 = 0 \qquad x + 2 = 0$$
$$x = 5 \qquad x = -2$$

13. $4b^2 - 12b + 9 = 0$

$(2b - 3)(2b - 3) = 0$

$2b - 3 = 0 \qquad 2b - 3 = 0$

$b = \frac{3}{2} \qquad b = \frac{3}{2}$

14. $3m^2 + 10m + 5 = 0$

$m = \dfrac{-10 \pm \sqrt{10^2 - 4(3)(5)}}{2(3)}$

$m = \dfrac{-10 \pm \sqrt{100 - 60}}{6}$

$m = \dfrac{10 \pm \sqrt{40}}{6}$

$m = \dfrac{-10 \pm 2\sqrt{10}}{6}$

$m = \dfrac{-5 \pm \sqrt{10}}{3}$

15. $x - 4\sqrt{x} - 21 = 0$

$x - 21 = 4\sqrt{x}$

$(x - 21)^2 = (4\sqrt{x})^2$

$x^2 - 42x + 441 = 16x$

$x^2 - 58x + 441 = 0$

$(x - 9)(x - 49) = 0$

$x - 9 = 0 \qquad x - 49 = 0$

$x = 9 \qquad x = 49$

Check: $9 - 4\sqrt{9} - 21 \stackrel{?}{=} 0$

$9 - 12 - 21 \stackrel{?}{=} 0$

$-24 \neq 0$

Not a solution

Check: $49 - 4\sqrt{49} - 21 \stackrel{?}{=} 0$

$49 - 28 - 21 \stackrel{?}{=} 0$

$0 = 0$

Solution

16. $x^4 + 7x^2 + 12 = 0$

$(x^2 + 4)(x^2 + 3) = 0$

$x^2 = -4 \qquad x^2 = -3$

$x = \pm\sqrt{-4} \qquad x = \pm\sqrt{-3}$

$x = \pm 2i \qquad x = \pm\sqrt{3}i$

17. $x - 4\sqrt{x} + 3 = 0$

Let $u = \sqrt{x}, u^2 = x$.

$u^2 - 4u + 3 = 0$

$(u - 3)(u - 1) = 0$

$u - 3 = 0 \qquad u - 1 = 0$

$u = 3 \qquad u = 1$

$\sqrt{x} = 3 \qquad \sqrt{x} = 1$

$x = 9 \qquad x = 1$

Check: $9 - 4\sqrt{9} + 3 \stackrel{?}{=} 0$

$9 - 12 + 3 \stackrel{?}{=} 0$

$0 = 0$

Solution

Check: $1 - 4\sqrt{1} + 3 \stackrel{?}{=} 0$

$1 - 4 + 3 \stackrel{?}{=} 0$

$0 = 0$

Solution

18. $x^4 - 14x^2 + 24 = 0$

Let $u = x^2, u^2 = x^4$.

$u^2 - 14u + 24 = 0$

$u^2 - 14u + 49 = -24 + 49$

$(u - 7)^2 = 25$

$u - 7 = \pm 5$

$u = 7 \pm 5$

$u = 12 \qquad u = 2$

$x^2 = 12 \qquad x^2 = 2$

$x = \pm\sqrt{12} \qquad x = \pm\sqrt{2}$

$x = \pm 2\sqrt{3}$

19. $R = x(180 - 1.5x)$

$5400 = 180x - 1.5x^2$

$0 = 1.5x^2 - 180x + 5400$

$0 = x^2 - 120x + 3600$

$0 = (x - 60)(x - 60)$

$0 = x - 60 \qquad x - 60 = 0$

$60 = x \qquad x = 60$

60 video games

20. *Verbal Model:* $\boxed{\text{Area}} = \boxed{\text{Length}} \cdot \boxed{\text{Width}}$

Equation: $2275 = x \cdot (100 - x)$

$2275 = 100x - x^2$

$0 = x^2 - 100x + 2275$

$0 = (x - 35)(x - 65)$

$x - 35 = 0 \qquad x - 65 = 0$

$x = 35 \text{ meters} \qquad x = 65 \text{ meters}$

35 meters × 65 meters

21. $h = -0.003x^2 + 1.19x + 5.2$

$10 = -0.003x^2 + 1.19x + 5.2$

$0 = -0.003x^2 + 1.19x - 4.8$

$0 = 3x^2 - 1190x + 4800$

$$x = \frac{-(-1190) \pm \sqrt{(-1190)^2 - 4(3)(4800)}}{2(3)}$$

$$x = \frac{1190 \pm \sqrt{1{,}416{,}100 - 57600}}{6}$$

$$x = \frac{1190 + \sqrt{1{,}358{,}500}}{6} \approx 392.6 \text{ feet}$$

$$x = \frac{1190 - \sqrt{1{,}358{,}500}}{6} \approx 4.07; \text{ not reasonable}$$

Section 8.4 Graphs of Quadratic Functions

1. $y = x^2 - 2x = (x^2 - 2x + 1) - 1 = (x - 1)^2 - 1$

vertex $= (1, -1)$

3. $y = x^2 - 4x + 7$

$= (x^2 - 4x + 4) + 7 - 4$

$= (x - 2)^2 + 3$

vertex $= (2, 3)$

5. $y = x^2 + 6x + 5$

$y = (x^2 + 6x + 9) + 5 - 9$

$y = (x + 3)^2 - 4$

vertex $= (-3, -4)$

7. $y = -x^2 + 6x - 10$

$y = -1(x^2 - 6x) - 10$

$y = -1(x^2 - 6x + 9) - 10 + 9$

$y = -1(x - 3)^2 - 1$

vertex $= (3, -1)$

9. $y = 2x^2 + 6x + 2$

$= 2\left(x^2 + 3x + \frac{9}{4}\right) + 2 - \frac{9}{2}$

$= 2\left(x + \frac{3}{2}\right)^2 - \frac{5}{2}$

vertex $= \left(-\frac{3}{2}, -\frac{5}{2}\right)$

11. $f(x) = x^2 - 8x + 15$

$a = 1, \quad b = -8$

$$x = \frac{-b}{2a} = \frac{-(-8)}{2(1)} = 4$$

$$f\left(\frac{-b}{2a}\right) = 4^2 - 8(4) + 15 = 16 - 32 + 15 = -1$$

vertex $= (4, -1)$

13. $g(x) = -x^2 - 2x + 1$

$a = -1, \quad b = -2$

$$x = \frac{-b}{2a} = \frac{-(-2)}{2(-1)} = -1$$

$$g\left(\frac{-b}{2a}\right) = -(-1)^2 - 2(-1) + 1$$

$= -1 + 2 + 1$

$= 2$

vertex $= (-1, 2)$

15. $y = 4x^2 + 4x + 4$

$a = 4, \quad b = 4$

$$x = \frac{-b}{2a} = \frac{-4}{2(4)} = -\frac{1}{2}$$

$$y = 4\left(-\frac{1}{2}\right)^2 + 4\left(-\frac{1}{2}\right) + 4$$

$= 4\left(\frac{1}{4}\right) - 2 + 4$

$= 1 - 2 + 4$

$= 3$

vertex $= \left(-\frac{1}{2}, 3\right)$

17. $g(x) = x^2 - 4$

x-intercepts:

$0 = x^2 - 4$

$0 = (x - 2)(x + 2)$

$x = 2, \quad x = -2$

vertex: $g(x) = (x - 0)^2 - 4$

$(0, -4)$

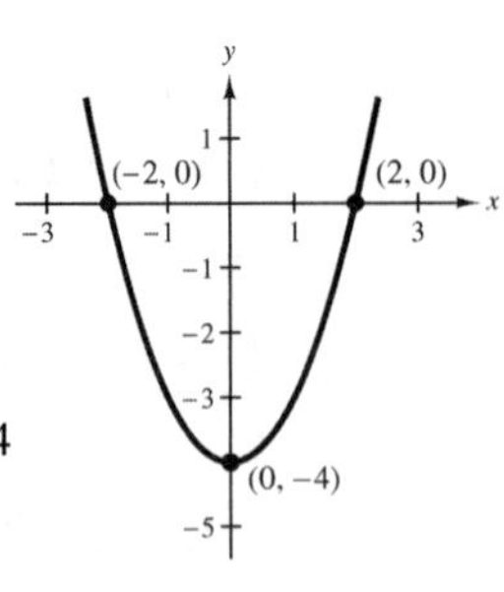

19. $f(x) = -x^2 + 4$

x-intercepts:

$0 = -x^2 + 4$

$x^2 = 4$

$x = \pm 2$

vertex:

$f(x) = -(x - 0)^2 + 4$

$(0, 4)$

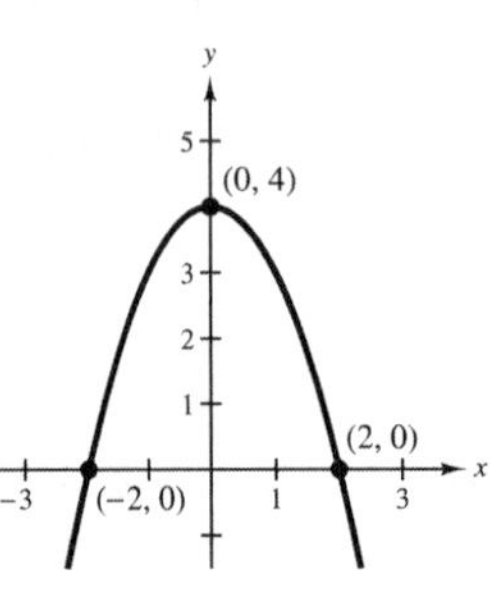

21. $y = (x - 4)^2$

x-intercepts:

$0 = (x - 4)^2$

$0 = x - 4$

$4 = x$

vertex:

$y = (x - 4)^2 + 0$

$(4, 0)$

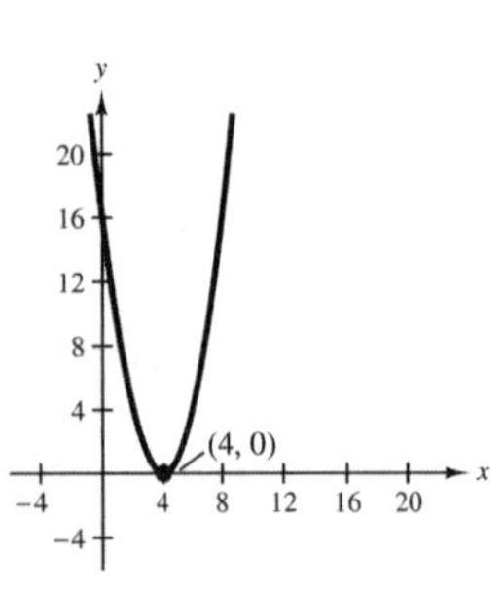

23. $y = x^2 - 9x - 18$

x-intercepts:

$0 = x^2 - 9x - 18$

$$x = \frac{-(-9) \pm \sqrt{(-9)^2 - 4(1)(-18)}}{2(1)}$$

$$x = \frac{9 \pm \sqrt{81 + 72}}{2} = \frac{9 \pm \sqrt{153}}{2}$$

$$= \frac{9 \pm 3\sqrt{17}}{2} = \frac{9}{2} \pm \frac{3\sqrt{17}}{2}$$

$$\approx (10.68, 0), (-1.68, 0)$$

vertex:

$$y = \left(x^2 - 9x + \frac{81}{4}\right) - 18 - \frac{81}{4}$$

$$= \left(x - \frac{9}{2}\right)^2 - \frac{72}{4} - \frac{81}{4}$$

$$= \left(x - \frac{9}{2}\right)^2 - \frac{153}{4}$$

$$\left(\frac{9}{2}, -\frac{153}{4}\right) = (4.5, -38.25)$$

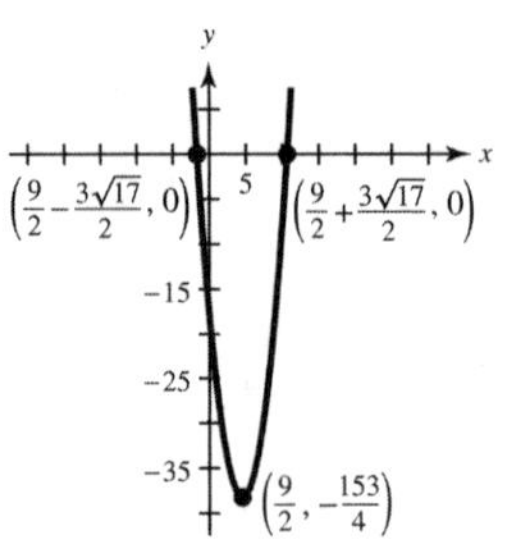

25. vertex $= (2, 0)$, point $= (0, 4)$

$4 = a(0 - 2)^2 + 0$	$y = 1(x - 2)^2 + 0$
$4 = a(4)$	$y = (x - 2)^2$
$1 = a$	$y = x^2 - 4x + 4$

27. vertex $= (2, 6)$, point $= (0, 4)$

$y = a(x - 2)^2 + 6$	$y = -\frac{1}{2}(x - 2)^2 + 6$
$4 = a(0 - 2)^2 + 6$	$y = -\frac{1}{2}(x^2 - 4x + 4) + 6$
$4 = a(4) + 6$	$y = -\frac{1}{2}x^2 + 2x - 2 + 6$
$-2 = a(4)$	$y = -\frac{1}{2}x^2 + 2x + 4$
$-\frac{2}{4} = a$	

29. vertex $= (-1, -3)$, point $= (0, -2)$

$$y = a(x - (-1))^2 - 3 \qquad y = 1(x + 1)^2 - 3$$
$$y = a(x + 1)^2 - 3 \qquad y = (x + 1)^2 - 3$$
$$-2 = a(0 + 1)^2 - 3$$
$$1 = a(0 + 1)^2$$
$$1 = a$$

31. vertex $= (2, 1)$, $a = 1$

$$y = 1(x - 2)^2 + 1 = x^2 - 4x + 5$$

33. vertex $= (2, -4)$, point $= (0, 0)$

$$0 = a(0 - 2)^2 - 4$$
$$4 = a(4)$$
$$1 = a$$
$$y = 1(x - 2)^2 - 4 = x^2 - 4x$$

35. vertex $= (-2, -1)$, point $= (1, 8)$

$$y = a(x - (-2))^2 - 1$$
$$8 = a(1 + 2)^2 - 1$$
$$9 = a(9)$$
$$1 = a$$
$$y = 1(x + 2)^2 - 1$$
$$y = (x + 2)^2 - 1$$

37. vertex $= (-1, 1)$, point $= (-4, 7)$

$$y = a(x + 1)^2 + 1$$
$$7 = a(-4 + 1)^2 + 1$$
$$6 = a(9)$$
$$\tfrac{2}{3} = \tfrac{6}{9} = a$$
$$y = \tfrac{2}{3}(x + 1)^2 + 1$$

39. vertex $= (2, -2)$, point $= (7, 8)$

$$y = a(x - 2)^2 - 2 \qquad y = \tfrac{2}{5}(x - 2)^2 - 2$$
$$8 = a(7 - 2)^2 - 2$$
$$10 = a(25)$$
$$\tfrac{2}{5} = a$$

41.
$$y = a(x - 0)^2 + 0$$
$$15 = a(30 - 0)^2$$
$$15 = a(900)$$
$$\frac{15}{900} = a$$
$$\frac{1}{60} = a$$
$$y = \frac{1}{60}x^2$$

43. The graph is the shape of a parabola that opens upward when $a > 0$ and downward when $a < 0$.

45. To find any x-intercepts, set $y = 0$ and solve the resulting equation for x. To find the y-intercept, set $x = 0$ and solve the resulting equation for y.

47. $h(x) = x^2 - 1$

Vertical shift 1 unit down

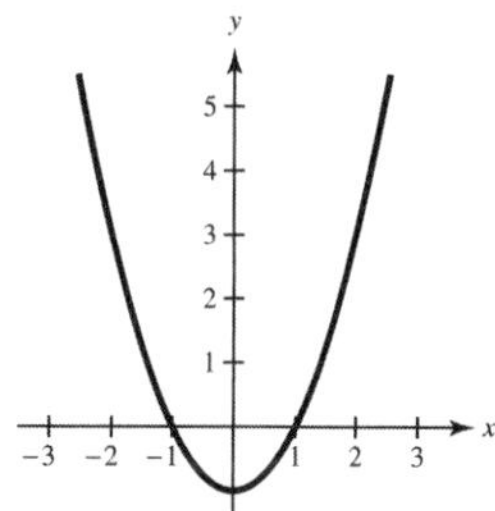

49. $h(x) = (x + 2)^2$

Horizontal shift 2 units left

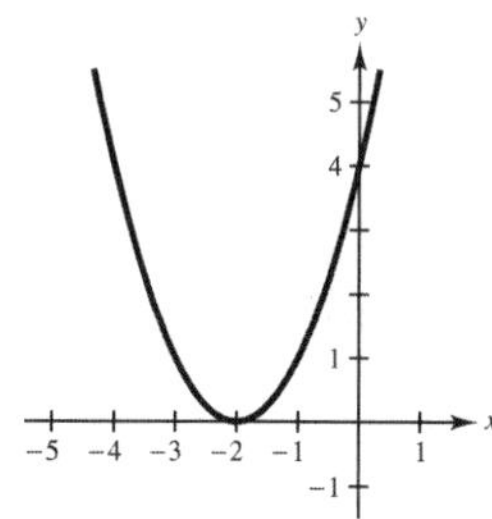

51. $h(x) = -(x + 5)^2$

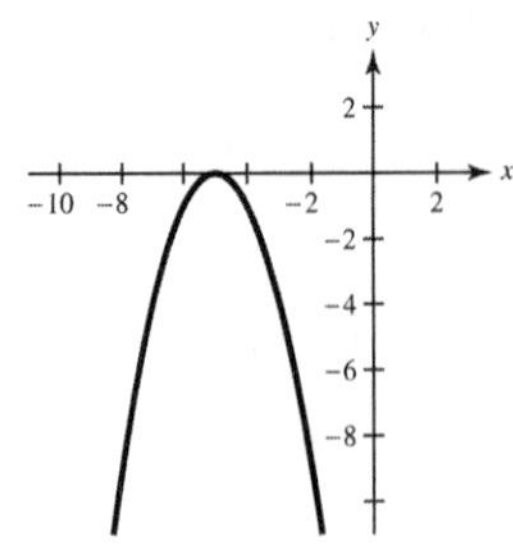

Horizontal shift 5 units left

Reflection in the x-axis

53. $h(x) = -(x - 2)^2 - 3$

Horizontal shift 2 units right

Vertical shift 3 units down

Reflection in the x-axis

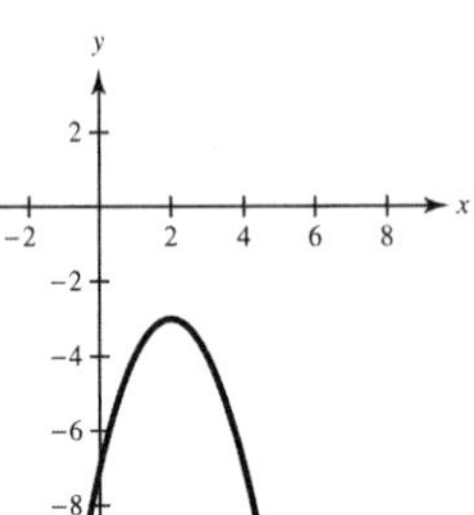

55. $y = -\frac{1}{12}x^2 + 2x + 4$

(a) $y = -\frac{1}{12}(0)^2 + 2(0) + 4$

$y = 4$ feet

(b) $y = -\frac{1}{12}x^2 + 2x + 4$

$y = -\frac{1}{12}(x^2 - 24x + 144) + 4 + 12$

$y = -\frac{1}{12}(x - 12)^2 + 16$

Maximum height $=$ 16 feet

(c) $0 = -\frac{1}{12}x^2 + 2x + 4$

$0 = x^2 - 24x - 48$

$x = \frac{24 \pm \sqrt{576 + 192}}{2} \approx 25.9$ feet

57. $y = -\frac{1}{480}x^2 + \frac{1}{2}x$

(a) $y = -\frac{1}{480}(0)^2 + \frac{1}{2}(0) = 0$ yards

(b) $y = -\frac{1}{480}(x^2 - 240x + 14{,}400) + 30$

$= -\frac{1}{480}(x - 120)^2 + 30$

Maximum height $=$ 30 yards

(c) $0 = -\frac{1}{480}x^2 + \frac{1}{2}x$

$0 = x^2 - 240x$

$0 = x(x - 240)$

$0 = x \qquad x - 240 = 0$

$x = 240$

240 yards

59. $y = -\frac{4}{9}x^2 + \frac{24}{9}x + 10$

$y = -\frac{4}{9}(x^2 - 6x) + 10$

$y = -\frac{4}{9}(x^2 - 6x + 9 - 9) + 10$

$y = -\frac{4}{9}(x^2 - 6x + 9) + 4 + 10$

$y = -\frac{4}{9}(x - 3)^2 + 14$

The maximum height of the diver is 14 feet.

61. If the discriminant is positive, the parabola has two x-intercepts; if it is zero, the parabola has one x-intercept; and if it is negative, the parabola has no x-intercepts.

63. Find the y-coordinate of the vertex of the graph of the function.

65. $(0, 0), (4, -2)$

$m = \frac{-2 - 0}{4 - 0} = -\frac{2}{4} = -\frac{1}{2}$

$y = mx + b$

$y = -\frac{1}{2}x + 0$

$y = -\frac{1}{2}x$

67. $(-1, -2), (3, 6)$

$m = \frac{6 - (-2)}{3 - (-1)} = \frac{8}{4} = 2$

$y - y_1 = m(x - x_1)$

$y - 6 = 2(x - 3)$

$y - 6 = 2x - 6$

$y = 2x$

69. $\left(\frac{3}{2}, 8\right), \left(\frac{11}{2}, \frac{5}{2}\right)$

$m = \frac{\frac{5}{2} - 8}{\frac{11}{2} - \frac{3}{2}} = \frac{\frac{5}{2} - \frac{16}{2}}{\frac{8}{2}} = \frac{-\frac{11}{2}}{4} = -\frac{11}{8}$

$y - y_1 = m(x - x_1)$

$y - 8 = -\frac{11}{8}\left(x - \frac{3}{2}\right)$

$y = -\frac{11}{8}x + \frac{33}{16} + 8$

$y = -\frac{11}{8}x + \frac{33}{16} + \frac{128}{16}$

$y = -\frac{11}{8}x + \frac{161}{16}$

71. $(0, 8), (5, 8)$

$m = \frac{8 - 8}{5 - 0} = \frac{0}{5} = 0$

$y = 8$

Horizontal line

73. $\sqrt{-64} = \sqrt{-1 \cdot 64} = 8i$

75. $\sqrt{-0.0081} = \sqrt{-1 \cdot 0.0081} = 0.09i$

Section 8.5 Applications of Quadratic Equations

1. *Verbal Model:* $\boxed{\text{Selling price per dozen eggs}} = \boxed{\text{Cost per dozen eggs}} + \boxed{\text{Profit per dozen eggs}}$

Equation: $\frac{21.60}{x} = \frac{21.60}{x + 6} + 0.30$

Labels: Number of eggs sold $= x$

Number of eggs purchased $= x + 6$

$$21.60(x + 6) = 21.60x + 0.30x(x + 6)$$
$$21.6x + 129.6 = 21.6x + 0.3x^2 + 1.8x$$
$$0 = 0.3x^2 + 1.8x - 129.6$$
$$0 = 3x^2 + 18x - 1296$$
$$0 = x^2 + 6x - 432$$
$$0 = (x + 24)(x - 18)$$
$$x = -24, \quad x = 18 \text{ dozen}$$

$$\text{Selling price} = \frac{21.60}{18} = \$1.20 \text{ per dozen}$$

3. *Verbal Model:* $\boxed{\text{Area}} = \boxed{\text{Length}} \cdot \boxed{\text{Width}}$

Labels: Length $= x + 4$

Width $= x$

Equation: $192 = (x + 4)x$

$$192 = x^2 + 4x$$
$$0 = x^2 + 4x - 192$$
$$0 = (x + 16)(x - 12)$$
$$x = -16, \quad x = 12 \text{ inches}$$
$$x + 4 = 16 \text{ inches}$$

5.
$$A = P(1 + r)^2$$
$$11{,}990.25 = 10{,}000(1 + r)^2$$
$$1.199025 = (1 + r)^2$$
$$1.095 = 1 + r$$
$$0.095 = r \text{ or } 9.5\%$$

7.
$$A = P(1 + r)^2$$
$$572.45 = 500(1 + r)^2$$
$$\frac{572.45}{500} = (1 + r)^2$$
$$1.1449 = (1 + r)^2$$
$$1.07 = 1 + r$$
$$0.07 = r \text{ or } 7\%$$

9. *Verbal Model:* [Cost per ticket] · [Number of people going] = 210

Equation:

$$\left(\frac{210}{x} - 3.50\right) \cdot (x + 3) = 210$$

$$210 + \frac{630}{x} - 3.5x - 10.50 = 210$$

$$210x + 630 - 3.5x^2 - 10.5x = 210x$$

$$-3.5x^2 - 10.5x + 630 = 0$$

$$0 = 3.5x^2 + 10.5x - 630$$

$$0 = 3.5(x^2 + 3x - 180)$$

$$0 = 3.5(x - 12)(x + 15)$$

$$x - 12 = 0 \qquad x + 15 = 0$$

$$x = 12 \qquad x = -15$$

There are $12 + 3 = 15$ people going to the game.

11. *Verbal Model:* [Work done by Machine A] + [Work done by Machine B] = [One complete job]

Equation:

$$\frac{1}{x}(6) + \frac{1}{x + 3}(6) = 1$$

$$x(x + 3)\left[\frac{1}{x}(6) + \frac{1}{x + 3}(6)\right] = (1)x(x + 3)$$

$$6(x + 3) + 6x = x(x + 3)$$

$$6x + 18 + 6x = x^2 + 3x$$

$$-x^2 + 9x + 18 = 0$$

$$x^2 - 9x - 18 = 0$$

$$x = \frac{9 \pm \sqrt{(-9)^2 - 4(1)(-18)}}{2(1)} = \frac{9 \pm \sqrt{81 + 72}}{2} = \frac{9 \pm \sqrt{153}}{2}$$

$$x = 10.684658, -1.6846584 \qquad x \approx 10.7 \text{ minutes}$$

$$x + 3 = 13.684658 \qquad x + 3 \approx 13.7 \text{ minutes}$$

13. *Common formula:* $a^2 + b^2 + c^2$

Equation:

$$x^2 + (100 - x)^2 = 80^2$$

$$x^2 + 10{,}000 - 200x + x^2 = 6400$$

$$2x^2 - 200x + 3600 = 0$$

$$x^2 - 100x + 1800 = 0$$

$$x = \frac{-(-100) \pm \sqrt{(-100)^2 - 4(1)(1800)}}{2(1)}$$

$$x = \frac{100 \pm \sqrt{10{,}000 - 7200}}{2} = \frac{100 \pm \sqrt{2800}}{2} \approx 76.5 \text{ yards}, 23.5 \text{ yards}$$

15. $3.5 = 5 + 18t - 16t^2$

$16t^2 - 18t - 1.5 = 0$

$$t = \frac{-(-18) \pm \sqrt{(-18)^2 - 4(16)(-1.5)}}{2(16)}$$

$$t = \frac{18 \pm \sqrt{324 + 96}}{32}$$

$$t = \frac{18 \pm \sqrt{400}}{32}$$

$t \approx 1.2, -0.1$

1.2 seconds will pass before you hit the ball.

17. The Pythagorean Theorem can be used to set the sum of the square of a missing length, and the square of the difference of the sum of the lengths and the missing length, equal to the square of the length of the hypotenuse.

19. The quotient of the investment amount and the number of units is equal to the quotient of the investment amount and the number of units sold, plus the profit per unit sold.

21. *Verbal Model:* 2 [Length] + 2 [Width] = [Perimeter]

Labels: Length $= l$

Width $= 1.4l$

Equation: $2l + 2(1.4l) = 54$

$2l + 2.8l = 54$

$4.8l = 54$

$l = 11.25$ inches

$w = 1.4l = 15.75$ inches

Verbal Model: [Length] · [Width] = [Area]

Equation: $11.25 \cdot 15.75 = A$

$177.1875 \text{ in.}^2 = A$

27. *Verbal Model:* [Length] · [Width] = [Area]

Labels: Length $= l$

Width $= l - 20$

Equation: $l \cdot (l - 20) = 12{,}000$

$l^2 - 20l = 12{,}000$

$l^2 - 20l + 100 = 12{,}000 + 100$

$(l - 10)^2 = 12{,}100$

$l - 10 = \pm\sqrt{12{,}100}$

$l = 10 + 110 = 120$ meters

$w = l - 20 = 100$ meters

Verbal Model: 2 [Length] + 2 [Width] = [Perimeter]

Equation: $2(120) + 2(100) = 440 \text{ meters} = P$

23. *Verbal Model:* [Area] = [Length] · [Width]

Labels: Length $= 2.5w$

Width $= w$

Equation: $250 = 2.5w \cdot w$

$250 = 2.5w^2$

$100 = w^2$

$10 = w$

$25 = 2.5w$

Verbal Model: 2 [Length] + 2 [Width] = [Perimeter]

Equation: $2(25) + 2(10) = P$

$70 \text{ feet} = P$

25. *Verbal Model:* 2 [Length] + 2 [Width] = [Perimeter]

Labels: Length $= w + 3$

Width $= w$

Equation: $2(w + 3) + 2w = 54$

$2w + 6 + 2w = 54$

$4w = 48$

$w = 12$ km

$l = w + 3 = 15$ km

Verbal Model: [Length] · [Width] = [Area]

Equation: $15 \cdot 12 = 180 \text{ km}^2 = A$

29. *Verbal Model:* [Selling price per DVD] = [Cost per DVD] + [Profit per DVD]

Labels: Number of DVDs sold $= x$

Number of DVDs purchased $= x + 15$

Equation:

$$\frac{50}{x} = \frac{50}{x+15} + 3$$

$$50(x+15) = 50x + 3x(x+15)$$

$$50x + 750 = 50x + 3x^2 + 45x$$

$$0 = 3x^2 + 45x - 750$$

$$0 = x^2 + 15x - 250$$

$$0 = (x+25)(x-10)$$

$x = -25,\quad x = 10$ DVDs

Selling price $= \frac{50}{10} = \$5$

31. *Verbal Model:* [Integer] · [Integer] = [Product]

Labels: First integer $= n$

Second integer $= n + 1$

Equation:

$$n \cdot (n+1) = 182$$

$$n^2 + n - 182 = 0$$

$$(n+14)(n-13) = 0$$

$$n + 14 = 0 \qquad n - 13 = 0$$

reject $\begin{cases} n = -14 \\ n + 1 = -13 \end{cases}$ $\qquad n = 13,\quad n + 1 = 14$

33. *Verbal Model:* [Height] · [Width] = [Area]

Labels: Height $= x$

Width $= 48 - 2x$

Equation:

$$x \cdot (48 - 2x) = 288$$

$$2x^2 - 48x + 288 = 0$$

$$x^2 - 24x + 144 = 0$$

$$(x-12)(x-12) = 0$$

$$x = 12$$

height $= 12$ inches

width $= 48 - 2(12)$

$= 48 - 24 = 24$ inches

35. *Verbal Model:* [Time for part 1] + [Time for part 2] = 5

Equation:

$$\frac{100}{r} + \frac{135}{r-5} = 5$$

$$r(r-5)\left[\frac{100}{r} + \frac{135}{r-5}\right] = 5r(r-5)$$

$$100(r-5) + 135r = 5r^2 - 25r$$

$$100r - 500 + 135r = 5r^2 - 25r$$

$$0 = 5r^2 - 260r + 500$$

$$0 = 5(r^2 - 52r + 100)$$

$$0 = 5(r-50)(r-2)$$

$$r - 50 = 0 \qquad r - 2 = 0$$

$$r = 50 \qquad r = 2, \text{ reject}$$

Thus, the average speed for the first part of the trip was 50 miles per hour.

37.
$$\begin{aligned} d &= \sqrt{(x_1 - x_2)^2 + (y_1 - y_2)^2} \\ 10 &= \sqrt{(x_1 - 2)^2 + (9 - 3)^2} \\ 100 &= (x_1 - 2)^2 + 36 \\ 64 &= (x_1 - 2)^2 \\ \pm 8 &= x_1 - 2 \\ 2 \pm 8 &= x_1 \end{aligned}$$

$x_1 = 10 \qquad x_1 = -6$

$(10, 9) \qquad (-6, 9)$

39. To solve a rational equation, each side of the equation is multiplied by the LCD. The resulting equations in this section are quadratic equations.

41. No. For each additional person, the cost-per-person decrease gets smaller because the discount is distributed to more people.

43.
$$\begin{aligned} 5 - 3x &> 17 \\ -3x &> 12 \\ x &< -4 \end{aligned}$$

45.
$$\begin{aligned} x^2 - 8x &= 0 \\ x^2 - 8x + 16 &= 16 \\ (x - 4)^2 &= 16 \\ x - 4 &= \pm 4 \\ x &= 4 \pm 4 \\ x &= 8, 0 \end{aligned}$$

Section 8.6 Quadratic and Rational Inequalities

1. $x - 4$

Negative: $(-\infty, 4)$

Positive: $(4, \infty)$

Choose a test value from each interval.

$(-\infty, 4) \Rightarrow x = 0 \Rightarrow 0 - 4 = -4 < 0$

$(4, \infty) \Rightarrow x = 5 \Rightarrow 5 - 4 = 1 > 0$

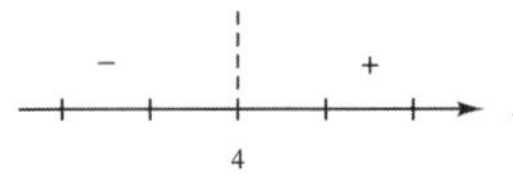

3. $3 - \frac{1}{2}x$

Negative: $(6, \infty)$

Positive: $(-\infty, 6)$

Choose a test value from each interval.

$(-\infty, 6) \Rightarrow x = 0 \Rightarrow 3 - \frac{1}{2}(0) = 3 > 0$

$(6, \infty) \Rightarrow x = 8 \Rightarrow 3 - \frac{1}{2}(8) = -1 < 0$

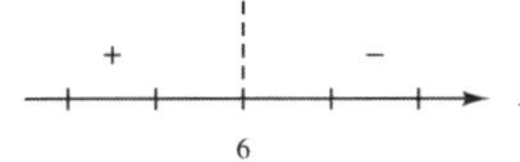

5. $4x(x - 5)$

$4x = 0 \qquad x - 5 = 0$

$x = 0 \qquad x = 5$ Critical numbers

Negative: $(0, 5)$

Positive: $(-\infty, 0) \cup (5, \infty)$

Choose a test value from each interval.

$(-\infty, 0) \Rightarrow x = -1 \Rightarrow 4(-1)(-1 - 5) = 24 > 0$

$(0, 5) \Rightarrow x = 1 \Rightarrow 4(1)(1 - 5) = -16 < 0$

$(5, \infty) \Rightarrow x = 6 \Rightarrow 4(6)(6 - 5) = 24 > 0$

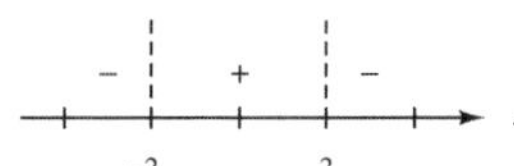

7. $4 - x^2 = (2 - x)(2 + x)$

Negative: $(-\infty, -2) \cup (2, \infty)$

Positive: $(-2, 2)$

Choose a test value from each interval.

$(-\infty, -2) \Rightarrow x = -3 \Rightarrow (2 - (-3))(2 + (-3)) = -5 < 0$

$(-2, 2) \Rightarrow x = 0 \Rightarrow (2 - 0)(2 + 0) = 4 > 0$

$(2, \infty) \Rightarrow x = 3 \Rightarrow (2 - 3)(2 + 3) = -5 < 0$

9. $3x(x - 2) < 0$

Critical numbers: $x = 0, 2$

Test intervals:

Positive: $(-\infty, 0)$

Negative: $(0, 2)$

Positive: $(2, \infty)$

Solution: $(0, 2)$

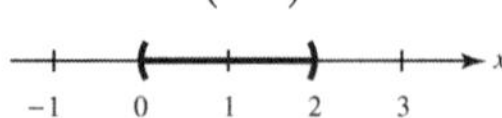

11. $3x(2 - x) \geq 0$

Critical numbers: $x = 0, 2$

Test intervals:

Negative: $(-\infty, 0]$

Positive: $[0, 2]$

Negative: $[2, \infty)$

Solution: $[0, 2]$

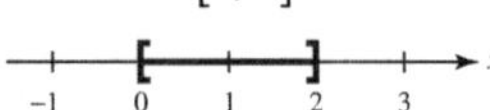

13. $x^2 + 4x > 0$

$x(x + 4) > 0$

Critical numbers: $x = -4, 0$

Test intervals:

Positive: $(-\infty, -4)$

Negative: $(-4, 0)$

Positive: $(0, \infty)$

Solution: $(-\infty, -4) \cup (0, \infty)$

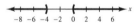

15. $x^2 - 3x - 10 \geq 0$

$(x - 5)(x + 2) \geq 0$

Critical numbers: $x = -2, 5$

Test intervals:

Positive: $(-\infty, -2)$

Negative: $(-2, 5)$

Positive: $(5, \infty)$

Solution: $(-\infty, -2] \cup [5, \infty)$

17. $x^2 > 4$

$x^2 - 4 > 0$

$(x - 2)(x + 2) > 0$

Critical numbers: $x = 2, -2$

Test intervals:

Positive: $(-\infty, 2)$

Negative: $(-2, 2)$

Positive: $(2, \infty)$

Solution: $(-\infty, -2) \cup (2, \infty)$

19. $x^2 + 5x \leq 36$

$x^2 + 5x - 36 \leq 0$

$(x + 9)(x - 4) \leq 0$

Critical numbers: $-9, 4$

Test intervals:

Positive: $(-\infty, -9)$

Negative: $(-9, 4)$

Positive: $(4, \infty)$

Solution: $[-9, 4]$

21. $u^2 + 2u - 2 > 1$

$u^2 + 2u - 3 > 0$

$(u + 3)(u - 1) > 0$

Critical numbers: $u = -3, 1$

Test intervals:

Positive: $(-\infty, -3)$

Negative: $(-3, 1)$

Positive: $(1, \infty)$

Solution: $(-\infty, -3) \cup (1, \infty)$

23. $x^2 + 4x + 5 < 0$

$$x = \frac{-4 \pm \sqrt{16 - 20}}{2}$$

No critical numbers

$x^2 + 4x + 5$ is not less than zero for any value of x.

Solution: none

25. $x^2 + 6x + 10 > 0$

$$x = \frac{-6 \pm \sqrt{36 - 4(10)}}{2}$$

$$x = \frac{-6 \pm \sqrt{-4}}{2}$$

$$x = \frac{-6 \pm 2i}{2}$$

$$x = -3 \pm i$$

No critical numbers

$x^2 + 6x + 10$ is greater than 0 for all values of x.

Solution: $(-\infty, \infty)$

−3 −2 −1 0 1 2 3 x

27. $y^2 + 16y + 64 \le 0$

$$(y + 8)^2 \le 0$$

Critical number: $y = -8$

Test intervals:

Positive: $(-\infty, -8)$

Positive: $(-8, \infty)$

Solution: -8

−10 −8 −6 −4 −2 0 2 y

29. $\dfrac{5}{x - 3} > 0$

Critical number: $x = 3$

Test intervals:

Negative: $(-\infty, 3)$

Positive: $(3, \infty)$

Solution: $(3, \infty)$

0 1 2 3 4 x

31. $\dfrac{-5}{x - 3} > 0$

Critical number: $x = 3$

Test intervals:

Positive: $(-\infty, 3)$

Negative: $(3, \infty)$

Solution: $(-\infty, 3)$

0 1 2 3 4 5 6 x

33. $\dfrac{x + 4}{x - 2} > 0$

Critical numbers: $x = -4, 2$

Test intervals:

Positive: $(-\infty, -4)$

Negative: $(-4, 2)$

Positive: $(2, \infty)$

Solution: $(-\infty, -4) \cup (2, \infty)$

−6 −4 −2 0 2 4 x

35.

$$\frac{3}{y - 1} \le -1$$

$$\frac{3}{y - 1} + 1 \le 0$$

$$\frac{3 + (y - 1)}{y - 1} \le 0$$

$$\frac{y + 2}{y - 1} \le 0$$

Critical numbers: $x = -2, 1$

Test intervals:

Positive: $(-\infty, -2)$

Negative: $(-2, 1)$

Positive: $(1, \infty)$

Solution: $[-2, 1)$

−3 −2 −1 0 1 2 y

37. $h = -16t^2 + 128t$

$$\begin{aligned} \text{height} &> 240 \\ -16t^2 + 128t &> 240 \\ -16t^2 + 128t - 240 &> 0 \\ t^2 - 8t + 15 &< 0 \\ (t - 3)(t - 5) &< 0 \end{aligned}$$

Critical numbers: $x = 3, 5$

Test intervals:

Positive: $(-\infty, 3)$

Negative: $(3, 5)$

Positive: $(5, \infty)$

Solution: $(3, 5)$

39. The critical numbers are -1 and 3.

41. $x = 4$ is not a solution of the inequality $x(x - 4) < 0$.

It does not satisfy the inequality:

$$4(4 - 4) \stackrel{?}{<} 0$$

$$4(0) \stackrel{?}{<} 0$$

$$0 \not< 0$$

43. $x(2x - 5) = 0$

$x = 0, \qquad 2x - 5 = 0$

$x = \frac{5}{2}$

Critical numbers: $0, \frac{5}{2}$

45. $4x^2 - 81 = 0$

$x^2 = \frac{81}{4}$

$x = \pm\frac{9}{2}$

Critical numbers: $\frac{9}{2}, -\frac{9}{2}$

47.

$$\begin{aligned} 6 - (x - 2)^2 &< 0 \\ \left[\sqrt{6} - (x - 2)\right]\left[\sqrt{6} + (x - 2)\right] &< 0 \\ -\left(x - 2 - \sqrt{6}\right)\left(x - 2 + \sqrt{6}\right) &< 0 \end{aligned}$$

Critical numbers: $2 - \sqrt{6}, 2 + \sqrt{6}$

Test intervals:

Negative: $\left(-\infty, 2 - \sqrt{6}\right)$

Positive: $\left(2 - \sqrt{6}, 2 + \sqrt{6}\right)$

Negative: $\left(2 + \sqrt{6}, \infty\right)$

Solution: $\left(-\infty, 2 - \sqrt{6}\right) \cup \left(2 + \sqrt{6}, \infty\right)$

49.

$$\begin{aligned} 16 &\le (u + 5)^2 \\ (u + 5)^2 &\ge 16 \\ u^2 + 10u + 25 - 16 &\ge 0 \\ u^2 + 10u + 9 &\ge 0 \\ (u + 9)(u + 1) &\ge 0 \end{aligned}$$

Critical numbers: $u = -9, -1$

Test intervals:

Positive: $(-\infty, -9]$

Negative: $(-9, -1]$

Positive: $[-1, \infty)$

Solution: $(-\infty, -9] \cup [-1, \infty)$

51. $\dfrac{u - 6}{3u - 5} \le 0$

Critical numbers: $u = 6, \dfrac{5}{3}$

Test intervals:

Positive: $\left(-\infty, \dfrac{5}{3}\right)$

Negative: $\left(\dfrac{5}{3}, 6\right]$

Positive: $[6, \infty)$

Solution: $\left(\dfrac{5}{3}, 6\right]$

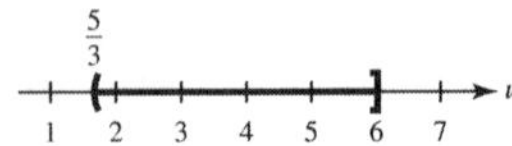

53. $\dfrac{2(4 - t)}{4 + t} > 0$

Critical numbers: $t = -4, 4$

Test intervals:

Negative: $(-\infty, 4)$

Positive: $(-4, 4)$

Negative: $(4, \infty)$

Solution: $(-4, 4)$

55.

$$\frac{1}{x + 2} > -3$$

$$\frac{1}{x + 2} + 3 > 0$$

$$\frac{1 + 3(x + 2)}{x + 2} > 0$$

$$\frac{1 + 3x + 6}{x + 2} > 0$$

$$\frac{3x + 7}{x + 2} > 0$$

Critical numbers: $x = -\frac{7}{3}, -2$

Test intervals:

Positive: $\left(-\infty, -\frac{7}{3}\right)$

Negative: $\left(-\frac{7}{3}, -2\right)$

Positive: $(-2, \infty)$

Solution: $\left(-\infty, -\frac{7}{3}\right) \cup (-2, \infty)$

57.

$$\frac{6x}{x - 4} < 5$$

$$\frac{6x}{x - 4} - 5 < 0$$

$$\frac{6x - 5(x - 4)}{x - 4} < 0$$

$$\frac{6x - 5x + 20}{x - 4} < 0$$

$$\frac{x + 20}{x - 4} < 0$$

Critical numbers: $x = -20, 4$

Test intervals:

Positive: $(-\infty, -20)$

Negative: $(-20, 4)$

Positive: $(4, \infty)$

Solution: $(-20, 4)$

59.

$$1000(1 + r)^2 > 1150$$

$$1000(1 + 2r + r^2) > 1150$$

$$1000 + 2000r + 1000r^2 > 1150$$

$$1000r^2 + 2000r - 150 > 0$$

$$20r^2 + 40r - 3 > 0$$

Critical numbers: $r = \dfrac{-40 + \sqrt{1840}}{40}, \dfrac{-40 - \sqrt{1840}}{40}$

r cannot be negative.

Test intervals:

Negative: $\left(0, \dfrac{-40 + \sqrt{1840}}{40}\right)$

Positive: $\left(\dfrac{-40 + \sqrt{1840}}{40}, \infty\right)$

Solution: $\left(\dfrac{-40 + \sqrt{1840}}{40}, \infty\right)$

$(0.0724, \infty)$, $r > 7.24\%$

61.

$$\begin{aligned} \text{Area} &> 240 \\ l(32 - l) &> 240 \\ 32l - l^2 &> 240 \\ -l^2 + 32l - 240 &> 0 \\ l^2 - 32l + 240 &< 0 \\ (l - 20)(l - 12) &< 0 \end{aligned}$$

Critical numbers: $l = 20, 12$

Test intervals:

Positive: $(-\infty, 12)$

Negative: $(12, 20)$

Positive: $(20, \infty)$

Solution: $(12, 20)$

63. *Verbal Model:* $\boxed{\text{Profit}} = \boxed{\text{Revenue}} - \boxed{\text{Cost}}$

$$\begin{aligned} P(x) &= x(125 - 0.0005x) - (3.5x + 185{,}000) \\ &= 125x - 0.0005x^2 - 3.5x - 185{,}000 \\ &= -0.0005x^2 + 121.5x - 185{,}000 \end{aligned}$$

Inequality:

$$-0.0005x^2 + 121.5x - 185{,}000 \ge 6{,}000{,}000$$

$$-0.0005x^2 + 121.5x - 6{,}185{,}000 \ge 0$$

$$x = \frac{-121.5 \pm \sqrt{(121.5)^2 - 4(-0.0005)(-6{,}185{,}000)}}{2(-0.0005)}$$

$$x = \frac{121.5 \pm \sqrt{1476.25 - 12{,}370}}{0.001} = \frac{121.5 \pm \sqrt{2392.25}}{0.001}$$

$$x \approx 72{,}589; 170{,}411$$

Critical numbers: $x \approx 72{,}589; 170{,}411$

Test intervals:

Positive: $(-\infty, 72{,}589)$

Negative: $(72{,}589, 170{,}411)$

Positive: $(170{,}411, \infty)$

Solution: $(72{,}589, 170{,}411)$

65. The critical numbers of a polynomial are its zeros, so the value of the polynomial is zero at its critical numbers.

67. No solution. The value of the polynomial is positive for every real value of x, so there are no values that would make the polynomial negative.

69. $\dfrac{4xy^3}{x^2y} \cdot \dfrac{y}{8x} = \dfrac{4xy^4}{8x^3y} = \dfrac{y^3}{2x^2}, \quad y \ne 0$

71. $\dfrac{x^2 - x - 6}{4x^3} \cdot \dfrac{x + 1}{x^2 + 5x + 6} = \dfrac{(x - 3)(x + 2)(x + 1)}{4x^3(x + 3)(x + 2)} = \dfrac{(x - 3)(x + 1)}{4x^3(x + 3)}, \quad x \ne -2$

73. $\dfrac{x^2 + 8x + 16}{x^2 - 6x} \div (3x - 24) = \dfrac{(x + 4)(x + 4)}{x(x - 6)} \cdot \dfrac{1}{3(x - 8)} = \dfrac{(x + 4)^2}{3x(x - 6)(x - 8)}$

75. $x = -\frac{1}{3}$

$x^2 = \left(-\frac{1}{3}\right)^2 = \frac{1}{9}$

77. $x = 1.06$

$\frac{100}{x^4} = \frac{100}{(1.06)^4} = 79.21$

Review Exercises for Chapter 8

1. $x^2 + 12x = 0$

$x(x + 12) = 0$

$x = 0 \qquad x + 12 = 0$

$x = 0 \qquad x = -12$

3. $3y^2 - 27 = 0$

$3(y^2 - 9) = 0$

$3(y - 3)(y + 3) = 0$

$y - 3 = 0 \qquad y + 3 = 0$

$y = 3 \qquad y = -3$

5. $4y^2 + 20y + 25 = 0$

$(2y + 5)(2y + 5) = 0$

$2y + 5 = 0 \qquad 2y + 5 = 0$

$2y = -5 \qquad 2y = -5$

$y = -\frac{5}{2} \qquad y = -\frac{5}{2}$

7. $2x^2 - 2x - 180 = 0$

$2(x^2 - x - 90) = 0$

$2(x - 10)(x + 9) = 0$

$x - 10 = 0 \qquad x + 9 = 0$

$x = 10 \qquad x = -9$

9. $2x^2 - 9x - 18 = 0$

$(2x + 3)(x - 6) = 0$

$2x + 3 = 0 \qquad x - 6 = 0$

$x = -\frac{3}{2} \qquad x = 6$

11. $z^2 = 144$

$z = \pm\sqrt{144}$

$z = \pm 12$

13. $y^2 - 12 = 0$

$y^2 = 12$

$y = \pm\sqrt{12}$

$y = \pm 2\sqrt{3}$

15. $(x - 16)^2 = 400$

$x - 16 = \pm\sqrt{400}$

$x = 16 \pm 20$

$x = 36, -4$

17. $z^2 = -121$

$z = \pm\sqrt{-121}$

$z = \pm 11i$

19. $y^2 + 50 = 0$

$y^2 = -50$

$y = \pm\sqrt{-50}$

$y = \pm 5\sqrt{2}i$

21. $(y + 4)^2 + 18 = 0$

$(y + 4)^2 = -18$

$y + 4 = \pm\sqrt{-18}$

$y = -4 \pm 3\sqrt{2}i$

23. $x^4 - 4x^2 - 5 = 0$

$(x^2 - 5)(x^2 + 1) = 0$

$x^2 + 1 = 0$

$x^2 - 5 = 0 \qquad x^2 = -1$

$x^2 = 5 \qquad x = \pm\sqrt{-1}$

$x = \pm\sqrt{5} \qquad x = \pm i$

25. $x - 4\sqrt{x} + 3 = 0$

$(\sqrt{x} - 3)(\sqrt{x} - 1) = 0$

$(\sqrt{x} - 3) = 0 \qquad (\sqrt{x} - 1) = 0$

$\sqrt{x} = 3 \qquad \sqrt{x} = 1$

$(\sqrt{x})^2 = 3^2 \qquad (\sqrt{x})^2 = 1^2$

$x = 9 \qquad x = 1$

Check: $9 - 4\sqrt{9} + 3 \stackrel{?}{=} 0$

$9 - 12 + 3 \stackrel{?}{=} 0$

$0 = 0$

Check: $1 - 4\sqrt{1} + 3 \stackrel{?}{=} 0$

$1 - 4 + 3 \stackrel{?}{=} 0$

$0 = 0$

27. $(x^2 - 2x)^2 - 4(x^2 - 2x) - 5 = 0$

$[(x^2 - 2x) - 5][(x^2 - 2x) + 1] = 0$

$(x^2 - 2x - 5)(x^2 - 2x + 1) = 0$

$x = \dfrac{-(-2) \pm \sqrt{(-2)^2 - 4(1)(-5)}}{2(1)}$

$x = \dfrac{2 \pm \sqrt{4 + 20}}{2}$

$x = \dfrac{2 \pm \sqrt{24}}{2}$

$x = \dfrac{2 \pm 2\sqrt{6}}{2}$ $\qquad (x - 1)^2 = 0$

$x = 1 \pm \sqrt{6}$ $\qquad x = 1$

29. $x^{2/3} + 3x^{1/3} - 28 = 0$

$(x^{1/3} + 7)(x^{1/3} - 4) = 0$

$x^{1/3} + 7 = 0$ $\qquad x^{1/3} - 4 = 0$

$x^{1/3} = -7$ $\qquad x^{1/3} = 4$

$\sqrt[3]{x} = -7$ $\qquad \sqrt[3]{x} = 4$

$(\sqrt[3]{x})^3 = (-7)^3$ $\qquad (\sqrt[3]{x})^3 = 4^3$

$x = -343$ $\qquad x = 64$

31. $z^2 + 18z + 81$

$\left[81 = \left(\frac{18}{2}\right)^2\right]$

33. $x^2 - 15x + \frac{225}{4}$

$\left[\frac{225}{4} = \left(-\frac{15}{2}\right)^2\right]$

35. $y^2 + \dfrac{2}{5}y + \dfrac{1}{25}$

$\left[\dfrac{1}{25} = \left(\dfrac{2/5}{2}\right)^2\right]$

$\left[\dfrac{1}{25} = \left(\dfrac{1}{5}\right)^2\right]$

37. $x^2 - 6x - 3 = 0$

$x^2 - 6x + 9 = 3 + 9$

$(x - 3)^2 = 12$

$x - 3 = \pm\sqrt{12}$

$x = 3 \pm 2\sqrt{3}$

$x \approx 6.46, -0.46$

39. $v^2 + 5v + 4 = 0$

$v^2 + 5v + \frac{25}{4} = -4 + \frac{25}{4}$

$\left(v + \frac{5}{2}\right)^2 = -\frac{16}{4} + \frac{25}{4}$

$\left(v + \frac{5}{2}\right)^2 = \frac{9}{4}$

$v + \frac{5}{2} = \pm\frac{3}{2}$

$v = -\frac{5}{2} \pm \frac{3}{2}$

$v = -\frac{2}{2}, -\frac{8}{2}$

$v = -1, -4$

41. $y^2 - \dfrac{2}{3}y + 2 = 0$

$y^2 - \dfrac{2}{3}y = -2$

$y^2 - \dfrac{2}{3}y + \dfrac{1}{9} = -2 + \dfrac{1}{9}$

$\left(y - \dfrac{1}{3}\right)^2 = \dfrac{-17}{9}$

$y - \dfrac{1}{3} = \pm\sqrt{\dfrac{-17}{9}}$

$y = \dfrac{1}{3} \pm \dfrac{\sqrt{17}i}{3}$

$y \approx 0.33 + 1.37i,\ 0.33 - 1.37i$

43. $v^2 + v - 42 = 0$

$v = \dfrac{-1 \pm \sqrt{1^2 - 4(1)(-42)}}{2(1)}$

$v = \dfrac{-1 \pm \sqrt{1 + 168}}{2}$

$v = \dfrac{-1 \pm \sqrt{169}}{2}$

$v = \dfrac{-1 \pm 13}{2}$

$v = \dfrac{12}{2}, \dfrac{-14}{2}$

$v = 6, -7$

45. $5x^2 - 16x + 2 = 0$

$$x = \frac{-(-16) \pm \sqrt{(-16)^2 - 4(5)(2)}}{2(5)}$$

$$x = \frac{16 \pm \sqrt{256 - 40}}{10}$$

$$x = \frac{16 \pm \sqrt{216}}{10}$$

$$x = \frac{16 \pm 6\sqrt{6}}{10}$$

$$x = \frac{8 \pm 3\sqrt{6}}{5}$$

47. $8x^2 - 6x + 2 = 0$ (Divide by 2.)

$4x^2 - 3x + 1 = 0$

$$x = \frac{-(-3) \pm \sqrt{(-3)^2 - 4(4)(1)}}{2(4)}$$

$$x = \frac{3 \pm \sqrt{9 - 16}}{8}$$

$$x = \frac{3 \pm \sqrt{-7}}{8} = \frac{3}{8} \pm \frac{\sqrt{7}}{8}i$$

49. $x^2 + 4x + 4 = 0$

$$\begin{aligned} b^2 - 4ac &= 4^2 - 4(1)(4) \\ &= 16 - 16 \\ &= 0 \end{aligned}$$

One repeated rational solution

51. $s^2 - s - 20 = 0$

$$\begin{aligned} b^2 - 4ac &= (-1)^2 - 4(1)(-20) \\ &= 1 + 80 \\ &= 81 \end{aligned}$$

Two distinct rational solutions

53. $4t^2 + 16t + 10 = 0$

$$\begin{aligned} b^2 - 4ac &= 16^2 - 4(4)(10) \\ &= 256 - 160 \\ &= 96 \end{aligned}$$

Two distinct irrational solutions

55. $v^2 - 6v + 21 = 0$

$$\begin{aligned} b^2 - 4ac &= (-6)^2 - 4(1)(21) \\ &= 36 - 84 \\ &= -48 \end{aligned}$$

Two distinct complex solutions

57. $x = 3 \qquad x = -7$

$x - 3 = 0 \qquad x + 7 = 0$

$(x - 3)(x + 7) = 0$

$x^2 + 4x - 21 = 0$

59. $x = 5 + \sqrt{7} \qquad x = 5 - \sqrt{7}$

$x - (5 + \sqrt{7}) = 0 \qquad x - (5 - \sqrt{7}) = 0$

$(x - 5) - \sqrt{7} = 0 \qquad (x - 5) + \sqrt{7} = 0$

$$\left[(x - 5) - \sqrt{7}\right]\left[(x - 5) + \sqrt{7}\right] = 0$$

$$(x - 5)^2 - (\sqrt{7})^2 = 0$$

$$x^2 - 10x + 25 - 7 = 0$$

$$x^2 - 10x + 18 = 0$$

61. $x = 6 + 2i \qquad x = 6 - 2i$

$x - (6 + 2i) = 0 \qquad x - (6 - 2i) = 0$

$(x - 6) - 2i = 0 \qquad (x - 6) + 2i = 0$

$$\left[(x - 6) - 2i\right]\left[(x - 6) + 2i\right] = 0$$

$$(x - 6)^2 - (2i)^2 = 0$$

$$x^2 - 12x + 36 + 4 = 0$$

$$x^2 - 12x + 40 = 0$$

63.
$$\begin{aligned} y &= x^2 - 8x + 3 \\ &= (x^2 - 8x + 16) + 3 - 16 \\ &= (x - 4)^2 - 13 \end{aligned}$$

Vertex: $(4, -13)$

65.
$$\begin{aligned} y &= 2x^2 - x + 3 \\ &= 2\left(x^2 - \tfrac{1}{2}x\right) + 3 \\ &= 2\left(x^2 - \tfrac{1}{2}x + \tfrac{1}{16}\right) + 3 - \tfrac{1}{8} \\ &= 2\left(x - \tfrac{1}{4}\right)^2 + \tfrac{23}{8} \end{aligned}$$

Vertex: $\left(\frac{1}{4}, \frac{23}{8}\right)$

67. $y = x^2 + 8x$

x-intercepts:

$0 = x^2 + 8x$

$0 = x(x + 8)$

$x = 0, \quad x = -8$

Vertex:

$y = x^2 + 8x + 16 - 16$

$y = (x + 4)^2 - 16$

$(-4, -16)$

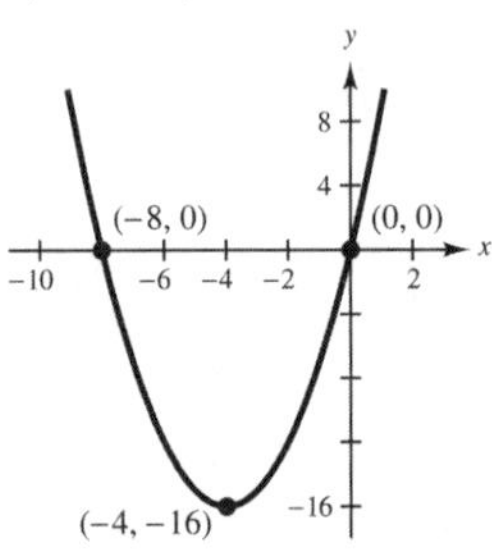

69. $f(x) = -x^2 - 2x + 4$

x-intercepts:

$0 = x^2 + 2x - 4$

$x = \dfrac{-2 \pm \sqrt{4 + 16}}{2}$

$x = \dfrac{-2 \pm \sqrt{20}}{2}$

$x = \dfrac{-2 \pm 2\sqrt{5}}{2}$

$x = -1 \pm \sqrt{5}$

Vertex:

$f(x) = -1(x^2 + 2x) + 4$

$= -(x^2 + 2x + 1) + 4 + 1$

$= -(x + 1)^2 + 5$

$(-1, 5)$

71. Vertex: $(2, -5)$; y-intercept: $(0, 3)$

$y = a(x - h)^2 + k$

$y = a(x - 2)^2 - 5$

$3 = a(0 - 2)^2 - 5$

$3 = a(4) - 5$

$8 = a(4)$

$2 = a$

$y = 2(x - 2)^2 - 5$ or $y = 2x^2 - 8x + 3$

73. Vertex: $(5, 0)$; passes through the point $(1, 1)$

$y = a(x - h)^2 + k$

$1 = a(1 - 5)^2 + 0$

$1 = a(16)$

$\frac{1}{16} = a$

$y = \frac{1}{16}(x - 5)^2 + 0$ or $y = \frac{1}{16}x^2 - \frac{5}{8}x + \frac{25}{16}$

75. (a) $y = -\dfrac{1}{10}(0)^2 + 3(0) + 6$

$y = 0 + 0 + 6$

$y = 6$ feet

(b) $x = -\dfrac{b}{2a} = -\dfrac{3}{2\left(-\dfrac{1}{10}\right)} = \dfrac{-3}{-\dfrac{1}{5}} = 15$

$y = \dfrac{1}{10}(15)^2 + 3(15) + 6$

$= -\dfrac{1}{10}(225) + 45 + 6$

$= -22.5 + 45 + 6 = 28.5$ feet

(c) $0 = -\dfrac{1}{10}x^2 + 3x + 6$

$x = \dfrac{-3 \pm \sqrt{3^2 - 4\left(-\dfrac{1}{10}\right)(6)}}{2\left(-\dfrac{1}{10}\right)}$

$x = \dfrac{-3 \pm \sqrt{9 + 2.4}}{-\dfrac{1}{5}}$

$x = \dfrac{-3 \pm \sqrt{11.4}}{-\dfrac{1}{5}}$

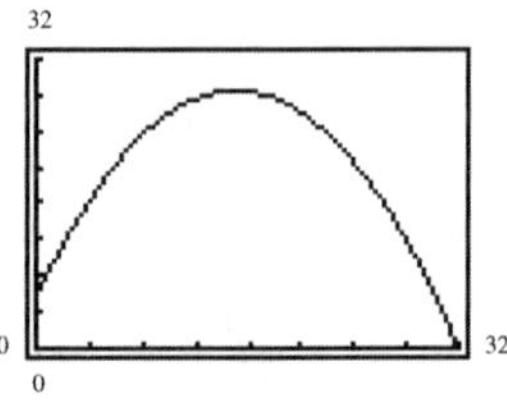

$x = -5(-3 \pm \sqrt{11.4}) = 15 \pm 5\sqrt{11.4} = 31.9$

The ball is 31.9 feet from the child when it hits the ground.

77. *Verbal Model:* $\boxed{\text{Selling price per car}} = \boxed{\text{Cost per car}} + \boxed{\text{Profit per car}}$

Labels: Number of cars sold $= x$

Number of cars bought $= x + 4$

Selling price per car $= 80{,}000/x$

Cost per car: $80{,}000/(x + 4)$

Profit per car: 1,000

Equation:

$$\frac{80{,}000}{x} = \frac{80{,}000}{x + 4} + 1000$$

$$80{,}000(x + 4) = 80{,}000x + 1000x(x + 4),\ x \neq 0,\ x \neq -4$$

$$80{,}000x + 320{,}000 = 80{,}000x + 1000x^2 + 4000x$$

$$0 = 1000x^2 + 4000x - 320{,}000$$

$$0 = x^2 + 4x - 320$$

$$0 = (x - 16)(x + 20)$$

$x - 16 = 0$ $\qquad$ $x + 20 = 0$

$x = 16$ $\qquad$ $x = -20$

By choosing the positive value, it follows that the dealer sold 16 cars at an average price of $80{,}000/16 = \$5000$ per car.

79. *Verbal Model:* $\boxed{\text{Area}} = \boxed{\text{Length}} \cdot \boxed{\text{Width}}$

Labels: Width $= x$

Length $= x + 12$

Equation:

$$85 = (x + 12)x$$

$$0 = x^2 + 12x - 85$$

$$0 = (x + 17)(x - 5)$$

reject $x = -17,$ $\qquad$ $x = 5$ inches

$x + 12 = 17$ inches

81. *Verbal Model:* $\boxed{\text{Cost per ticket}} \cdot \boxed{\text{Number of tickets}} = \boxed{\$96}$

Labels: Number in team $= x$

Number going to game $= x + 3$

Equation:

$$\left(\frac{96}{x} - 1.60\right)(x + 3) = 96$$

$$\left(\frac{96 - 1.60x}{x}\right)(x + 3) = 96$$

$$(96 - 1.6x)(x + 3) = 96x$$

$$96x - 1.6x^2 - 4.8x + 288 = 96x$$

$$1.6x^2 + 4.8x - 288 = 0$$

$$x^2 + 3x - 180 = 0$$

$$(x - 12)(x + 15) = 0$$

$x - 12 = 0$ $\qquad$ $x + 15 = 0$

$x = 12$ $\qquad$ $x = -15$ reject

$x + 3 = 15$

15 people are going to the game.

83. *Formula:* $c^2 = a^2 + b^2$

Labels: $c = 16$ $\quad a + b = 20$

$a = x$ $\quad x + b = 20$

$b = 20 - x$

Equation:

$$16^2 = x^2 + (20 - x)^2$$
$$256 = x^2 + 400 - 40x + x^2$$
$$0 = 2x^2 - 40x + 144$$
$$0 = x^2 - 20x + 72$$
$$x = \frac{20 \pm \sqrt{(-20)^2 - 4(1)(72)}}{2(1)}$$
$$x = \frac{20 \pm \sqrt{112}}{2}$$
$$x = \frac{20 \pm 4\sqrt{7}}{2}$$
$$x = 10 \pm 2\sqrt{7}$$
$$x \approx 15.3,\ 4.7$$

So, the possible distances from campus to the secretary's location are about 4.7 miles or 15.3 miles.

85. *Verbal Model:* [Work done by Person 1] + [Work done by Person 2] = [One complete job]

Labels: Time Person 1 $= x$

Time Person 2 $= x + 2$

Equation:

$$\frac{1}{x}(10) + \frac{1}{x+2}(10) = 1$$
$$x(x+2)\left[10\left(\frac{1}{x} + \frac{1}{x+2}\right)\right] = [1]x(x+2)$$
$$10(x+2) + 10x = x(x+2)$$
$$10x + 20 + 10x = x^2 + 2x$$
$$0 = x^2 - 18x - 20$$

$$x = \frac{-(-18) \pm \sqrt{(-18)^2 - 4(1)(-20)}}{2(1)}$$
$$x = \frac{18 \pm \sqrt{324 + 80}}{2}$$
$$x = \frac{18 \pm \sqrt{404}}{2}$$
$$x = \frac{18 \pm 2\sqrt{101}}{2}$$
$$x = 9 \pm \sqrt{101}$$
$$x \approx 19,\ x = -1,\ \text{reject}$$
$$x + 2 \approx 21$$

19 hours, 21 hours

87. $2x(x + 7)$

$2x = 0 \qquad x + 7 = 0$

$x = 0 \qquad x = -7$

Critical numbers: $x = 0, -7$

89. $x^2 - 6x - 27$

$(x - 9)(x + 3)$

$x - 9 = 0 \qquad x + 3 = 0$

$x = 9 \qquad x = -3$

Critical numbers: $x = 9, -3$

91. $5x(7 - x) > 0$

Critical numbers: $x = 0, 7$

Test intervals:

Negative: $(-\infty, 0)$

Positive: $(0, 7)$

Negative: $(7, \infty)$

Solution: $(0, 7)$

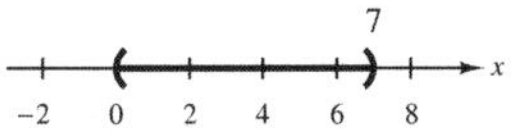

93. $16 - (x - 2)^2 \le 0$

$(4 - x + 2)(4 + x - 2) \le 0$

$(6 - x)(2 + x) \le 0$

Critical numbers: $x = -2, 6$

Test intervals:

Negative: $(-\infty, -2]$

Positive: $[-2, 6]$

Negative: $[6, \infty)$

Solution: $(-\infty, -2] \cup [6, \infty)$

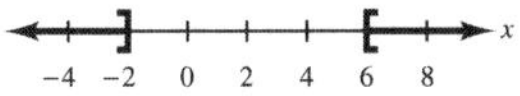

95. $2x^2 + 3x - 20 < 0$

$(2x - 5)(x + 4) < 0$

Critical numbers: $x = -4, \frac{5}{2}$

Test intervals:

Positive: $(-\infty, -4)$

Negative: $\left(-4, \frac{5}{2}\right)$

Positive: $\left(\frac{5}{2}, \infty\right)$

Solution: $\left(-4, \frac{5}{2}\right)$

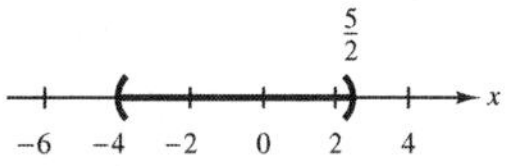

97. $\dfrac{x + 3}{2x - 7} \ge 0$

Critical numbers: $x = -3, \dfrac{7}{2}$

Test intervals:

Positive: $(-\infty, -3)$

Negative: $\left[-3, \dfrac{7}{2}\right]$

Positive: $\left(\dfrac{7}{2}, \infty\right)$

Solution: $(-\infty, -3] \cup \left(\dfrac{7}{2}, \infty\right)$

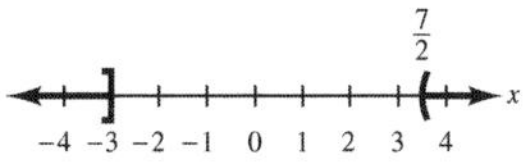

99. $\dfrac{x + 4}{x - 1} < 0$

Critical numbers: $x = -4, 1$

Test intervals:

Positive: $(-\infty, -4)$

Negative: $(-4, 1)$

Positive: $(1, \infty)$

Solution: $(-4, 1)$

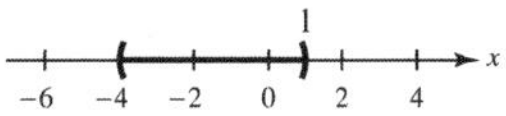

101. $h = -16t^2 + 312t$

$$-16t^2 + 312t > 1200$$

$$-16t^2 + 312t - 1200 > 0 \quad \text{(Divide by } -16.)$$

$$t^2 - 19.5t + 75 < 0$$

$$t = \frac{-(-19.5) \pm \sqrt{(-19.5)^2 - 4(1)(75)}}{2(1)}$$

$$t = \frac{19.5 \pm \sqrt{80.25}}{2}$$

$t \approx 14.2, 5.3$

Critical numbers: $t = 14.2, 5.3$

Test intervals:

Positive: $(-\infty, 5.3)$

Negative: $(5.3, 14.2)$

Positive: $(14.2, \infty)$

Solution: $(5.3, 14.2)$

$5.3 < t < 14.2$

Chapter Test for Chapter 8

1. $x(x - 3) - 10(x - 3) = 0$

$$(x - 3)(x - 10) = 0$$

$x - 3 = 0 \qquad x - 10 = 0$

$x = 3 \qquad x = 10$

2. $6x^2 - 34x - 12 = 0$

$$2(3x^2 - 17x - 6) = 0$$

$$2(3x + 1)(x - 6) = 0$$

$3x + 1 = 0 \qquad x - 6 = 0$

$x = -\frac{1}{3} \qquad x = 6$

3. $(x - 2)^2 = 0.09$

$$x - 2 = \pm 0.3$$

$$x = 2 \pm 0.3$$

$$x = 2.3, 1.7$$

4. $(x + 4)^2 + 100 = 0$

$$(x + 4)^2 = -100$$

$$x + 4 = \pm\sqrt{-100}$$

$$x = -4 \pm 10i$$

5. $2x^2 - 6x + 3 = 0$

$$x^2 - 3x + \frac{9}{4} = -\frac{3}{2} + \frac{9}{4}$$

$$\left(x - \frac{3}{2}\right)^2 = \frac{-6 + 9}{4}$$

$$\left(x - \frac{3}{2}\right)^2 = \frac{3}{4}$$

$$x - \frac{3}{2} = \pm\sqrt{\frac{3}{4}}$$

$$x = \frac{3}{2} \pm \frac{\sqrt{3}}{2}$$

6. $2y(y - 2) = 7$

$$2y^2 - 4y - 7 = 0$$

$$y = \frac{-(-4) \pm \sqrt{(-4)^2 - 4(2)(-7)}}{2(2)}$$

$$y = \frac{4 \pm \sqrt{16 + 56}}{4}$$

$$y = \frac{4 \pm \sqrt{72}}{4}$$

$$y = \frac{4 \pm 6\sqrt{2}}{4}$$

$$y = \frac{2 \pm 3\sqrt{2}}{2} \approx 7.41 \text{ and } -0.41$$

7. $\frac{1}{x^2} - \frac{6}{x} + 4 = 0$

$1 - 6x + 4x^2 = 0$

$4x^2 - 6x + 1 = 0$

$x = \frac{-(-6) \pm \sqrt{(-6)^2 - 4(4)(1)}}{2(4)}$

$x = \frac{6 \pm \sqrt{36 - 16}}{8}$

$x = \frac{6 \pm \sqrt{20}}{8}$

$x = \frac{6 \pm 2\sqrt{5}}{8}$

$x = \frac{3 \pm \sqrt{5}}{4}$

8. $x^{2/3} - 9x^{1/3} + 8 = 0$

$(x^{1/3} - 8)(x^{1/3} - 1) = 0$

$x^{1/3} - 8 = 0$ $\qquad$ $x^{1/3} - 1 = 0$

$x^{1/3} = 8$ $\qquad$ $x^{1/3} = 1$

$(x^{1/3})^3 = 8^3$ $\qquad$ $(x^{1/3})^3 = 1^3$

$x = 512$ $\qquad$ $x = 1$

9. $b^2 - 4ac = (-12)^2 - 4(5)(10)$

$= 144 - 200$

$= -56$

A negative discriminant tells us the equation has two complex solutions.

10. $x = -7$ $\qquad$ $x = -3$

$x + 7 = 0$ $\qquad$ $x + 3 = 0$

$(x + 7)(x + 3) = 0$

$x^2 + 10x + 21 = 0$

11. $y = -x^2 + 2x - 4$

x-intercepts:

$0 = x^2 - 2x + 4$

$x = \frac{-(-2) \pm \sqrt{(-2)^2 - 4(1)(4)}}{2(1)}$

$x = \frac{2 \pm \sqrt{4 - 16}}{2}$

$x = \frac{2 \pm \sqrt{-12}}{2}$

Not real

No x-intercepts

Vertex:

$y = -1(x^2 - 2x) - 4$

$= -(x^2 - 2x + 1) - 4 + 1$

$= -(x - 1)^2 - 3$

$(1, -3)$

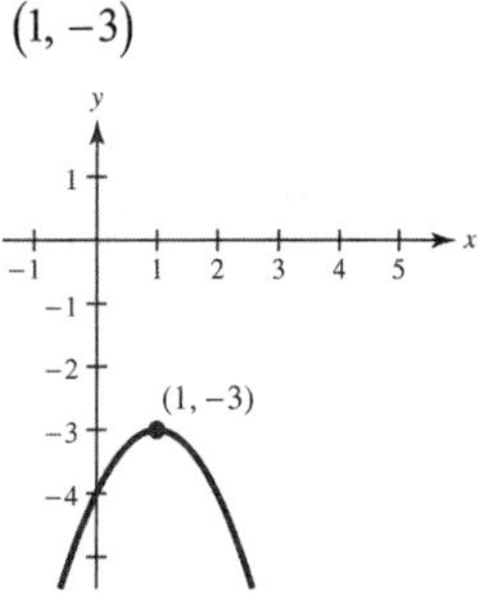

12. $y = x^2 - 2x - 15$

x-intercepts:

$0 = x^2 - 2x - 15$

$0 = (x - 5)(x + 3)$

$x = 5, \quad x = -3$

Vertex:

$y = (x^2 - 2x + 1) - 15 - 1$

$y = (x - 1)^2 - 16$

$(1, -16)$

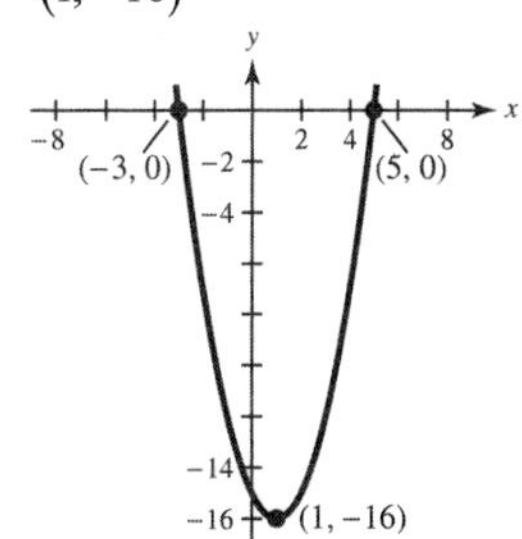

13. $16 \le (x - 2)^2$

$(x - 2)^2 \ge 16$

$x^2 - 4x + 4 \ge 16$

$x^2 - 4x - 12 \ge 0$

$(x - 6)(x + 2) \ge 0$

Critical numbers: $x = -2, 6$

Test intervals:

Positive: $(-\infty, -2]$

Negative: $[-2, 6]$

Positive: $[6, \infty)$

Solution: $(-\infty, -2] \cup [6, \infty)$

14. $2x(x - 3) < 0$

Critical numbers: $x = 0, 3$

Test intervals:

Positive: $(-\infty, 0)$

Negative: $(0, 3)$

Positive: $(3, \infty)$

Solution: $(0, 3)$

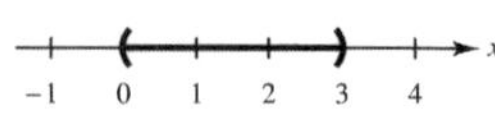

15. $\dfrac{x + 1}{x - 5} \le 0$

Critical numbers: $x = -1, 5$

Test intervals:

Positive: $(-\infty, -1)$

Negative: $(-1, 5)$

Positive: $(5, \infty)$

Solution: $[-1, 5)$

16. *Verbal Model:* $\boxed{\text{Area}} = \boxed{\text{Length}} \cdot \boxed{\text{Width}}$

Labels: Length $= x$

Width $= x - 22$

Equation: $240 = x(x - 22)$

$0 = x^2 - 22x - 240$

$0 = (x - 30)(x + 8)$

$0 = x - 30$ reject $x = -8$

$30 \text{ feet} = x$ $x - 22 = 8 \text{ feet}$

$8 \text{ feet} \times 30 \text{ feet}$

17. *Verbal Model:* $\boxed{\begin{array}{c}\text{Cost per person}\\ \text{Current Group}\end{array}} - \boxed{\begin{array}{c}\text{Cost per person}\\ \text{New Group}\end{array}} = 6.25$

Labels: Number Current Group $= x$

Number New Group $= x + 10$

Equation:

$$\frac{1250}{x} - \frac{1250}{x + 10} = 6.25$$

$$x(x + 10)\left(\frac{1250}{x} - \frac{1250}{x + 10}\right) = (6.25)x(x + 10)$$

$$1250(x + 10) - 1250x = 6.25x(x + 10)$$

$$1250x + 12500 - 1250x = 6.25x^2 + 62.5x$$

$$0 = 6.25x^2 + 62.5x - 12{,}500$$

$$0 = x^2 + 10x - 2000$$

$$0 = (x + 50)(x - 40)$$

reject $x = -50$, $x = 40$ club members

18.

$$\begin{aligned} 35 &= -16t^2 + 75 \\ 16t^2 &= 40 \\ t^2 &= \frac{40}{16} = \frac{5}{2} \\ t &= \sqrt{\frac{5}{2}} \\ t &= \frac{\sqrt{10}}{2} \approx 1.5811388 \\ t &\approx 1.58 \text{ seconds} \end{aligned}$$

19. *Formula*: $c^2 = a^2 + b^2$

Labels: $c = 125$ $\quad a + b = 155$

$a = x$ $\quad x + b = 155$

$b = 155 - x$ $\quad b = 155 - x$

Equation:

$$\begin{aligned} 125^2 &= x^2 + (155 - x)^2 \\ 15{,}625 &= x^2 + 24025 - 310x + x^2 \\ 0 &= 2x^2 - 310x + 8400 \\ 0 &= x^2 - 155x + 4200 \\ 0 &= (x - 35)(x - 120) \end{aligned}$$

$0 = x - 35 \qquad x - 120 = 0$

$35 \text{ feet} = x \qquad x = 120 \text{ feet}$

CHAPTER 9
Exponential and Logarithmic Functions

Section 9.1 Exponential Functions **287**

Section 9.2 Composite and Inverse Functions **291**

Section 9.3 Logarithmic Functions **294**

Mid-Chapter Quiz **298**

Section 9.4 Properties of Logarithms **299**

Section 9.5 Solving Exponential and Logarithmic Equations **302**

Section 9.6 Applications **307**

Review Exercises **310**

Chapter Test **316**

CHAPTER 9
Exponential and Logarithmic Functions

Section 9.1 Exponential Functions

1. $2^{-2.3} \approx 0.203$

Keystrokes:

Scientific: 2 $\boxed{g^x}$ 2.3 $\boxed{+/-}$ $\boxed{=}$

Graphing: 2 $\boxed{\wedge}$ $\boxed{-}$ 2.3 $\boxed{\text{ENTER}}$

3. $5^{\sqrt{2}} \approx 9.739$

Keystrokes:

Scientific: 5 $\boxed{g^x}$ 2 $\boxed{\sqrt{\ }}$ $\boxed{=}$

Graphing: 5 $\boxed{\wedge}$ $\boxed{\sqrt{\ }}$ 2 $\boxed{)}$ $\boxed{\text{ENTER}}$

5. $6^{1/3} \approx 1.817$

Keystrokes:

Scientific: 6 $\boxed{g^x}$ $\boxed{(}$ 1 ÷ 3 $\boxed{)}$ $\boxed{=}$

Graphing: 6 $\boxed{\wedge}$ $\boxed{(}$ 1/3 $\boxed{)}$ $\boxed{\text{Enter}}$

7. $f(x) = 3^x$

(a) $f(-2) = 3^{-2} = \frac{1}{9}$

(b) $f(0) = 3^0 = 1$

(c) $f(1) = 3^1 = 3$

9. $g(x) = 2.2^{-x}$

(a) $g(1) = 2 \cdot 2^{-1} \approx 0.455$

(b) $g(3) = 2 \cdot 2^{-3} \approx 0.094$

(c) $g(\sqrt{6}) = 2 \cdot 2^{-\sqrt{6}} \approx 0.145$

11. $f(t) = 500\left(\frac{1}{2}\right)^t$

(a) $f(0) = 500\left(\frac{1}{2}\right)^0 = 500$

(b) $f(1) = 500\left(\frac{1}{2}\right)^1 = 250$

(c) $f(\pi) = 500\left(\frac{1}{2}\right)^\pi \approx 56.657$

13. $f(x) = 1000(1.05)^{2x}$

(a) $f(0) = 1000(1.05)^{(2)(0)} = 1000$

(b) $f(5) = 1000(1.05)^{2(5)} \approx 1628.895$

(c) $f(10) = 1000(1.05)^{2(10)} \approx 2653.298$

15. $h(x) = \dfrac{5000}{(1.06)^{8x}}$

(a) $h(5) = \dfrac{5000}{(1.06)^{8(5)}} \approx 486.111$

(b) $h(10) = \dfrac{5000}{(1.06)^{8(10)}} \approx 47.261$

(c) $h(20) = \dfrac{5000}{(1.06)^{8(20)}} \approx 0.447$

17. $f(x) = 2^x; d$

18. $g(x) = 6^x; c$

19. $h(x) = 2^{-x}; a$

20. $k(x) = 6^{-x}; b$

21. $f(x) = 3^x$

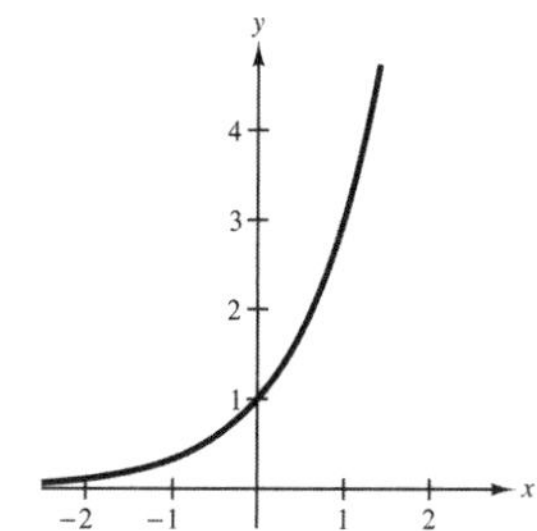

Domain: $-\infty < x < \infty$

Range: $y > 0$

Table of values:

x	-2	-1	0	1	2
$f(x)$	0.1	0.3	1	3	9

23. $f(x) = 3^{-x} = \left(\frac{1}{3}\right)^x$

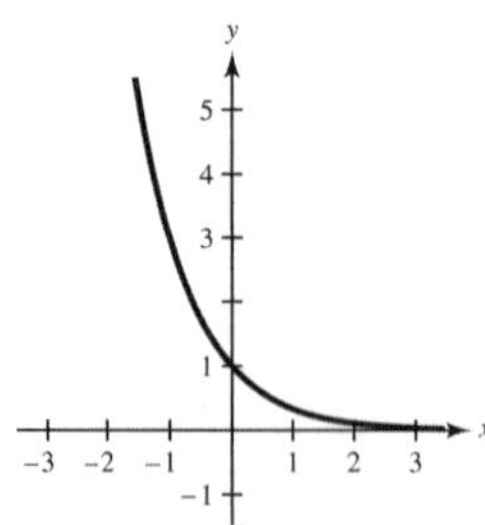

Domain: $-\infty < x < \infty$

Range: $y > 0$

25. The function g is related to $f(x) = 3^x$ by $g(x) = f(x) - 1$. To sketch g, shift the graph of f one unit downward.

y-intercept: $(0, 0)$

Asymptote: $y = -1$

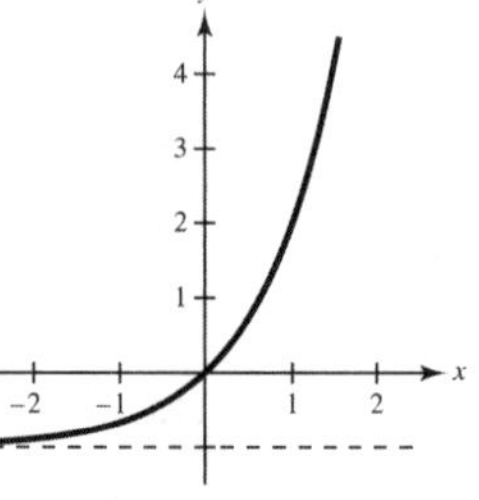

27. The function g is related to $f(x) = 5^x$ by $g(x) = f(x - 1)$. To sketch the graph of g, shift the graph of f one unit upward to the right.

y-intercept: $\left(0, \frac{1}{5}\right)$

Asymptote: x-axis

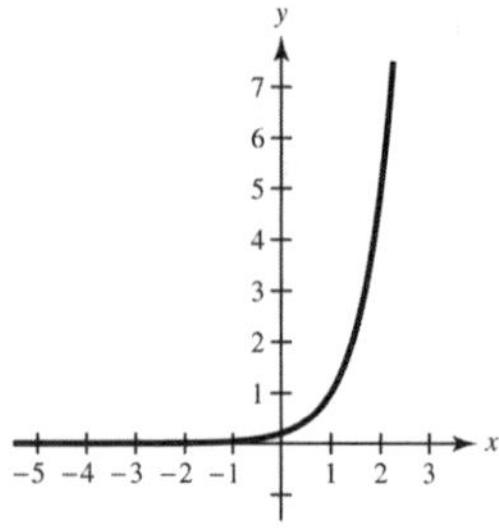

29. The function g is related to $f(x) = 2^x$ by $g(x) = f(x) + 3$. To sketch the graph of g, shift the graph of f three units upward.

y-intercept: $(0, 4)$

Asymptote: $y = 3$

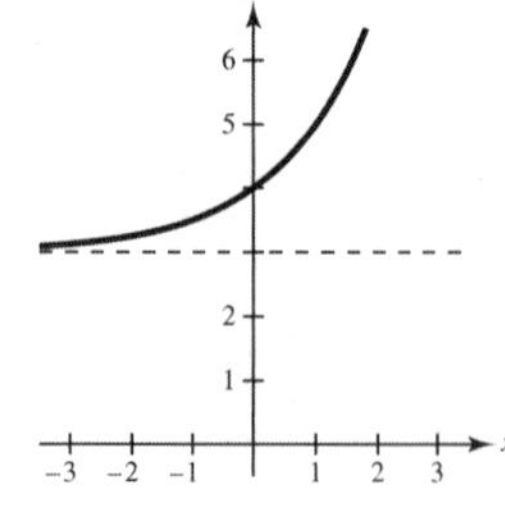

31. The function g is related to $f(x) = 2^x$ by $g(x) = f(x - 4)$. To sketch the graph of g, shift the graph of f four units to the right.

y-intercept: $\left(0, \frac{1}{16}\right)$

Asymptote: x-axis

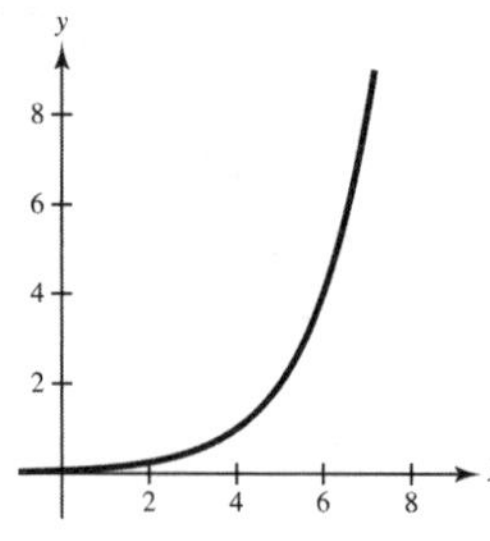

33. $g(x) = 2^{-x}$; b

34. $g(x) = 2^x - 1$; d

35. $g(x) = 2^{x-1}$; a

36. $g(x) = -2^x$; c

37. The function g is related to $f(x) = 4^x$ by $g(x) = -f(x)$. To sketch the graph of g, reflect the graph of f in the x-axis.

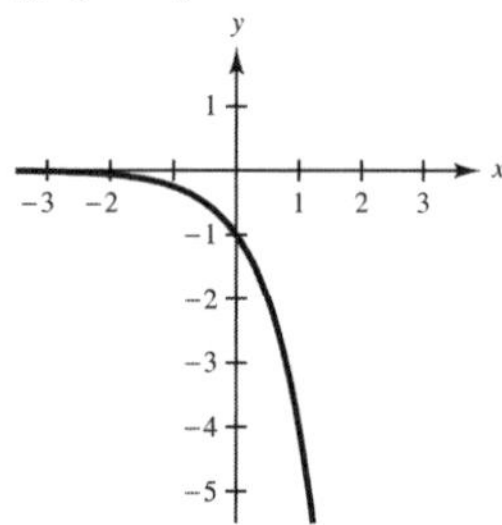

39. The function g is related to $f(x) = 5^x$ by $g(x) = f(-x)$. To sketch the graph of g, reflect the graph of f in the x-axis.

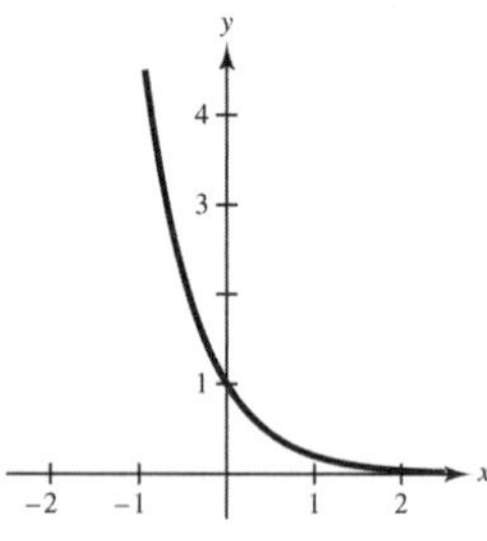

41. $e^{1/3} \approx 1.396$

Keystrokes:

Scientific: [(] [1] [÷] [3] [)] [INV] [ln x] [=]

Graphing: [e^x] [(] [1] [÷] [3] [)] [ENTER]

43. $3\left(2e^{1/2}\right)^3 = 3 \cdot 8 \cdot e^{3/2} = 24e^{1.5} \approx 107.561$

Keystrokes:

Scientific: 24 [×] 1.5 [e^x] [=]

Graphing: 24 [e^x] 1.5 [)] [ENTER]

45. $g(x) = 10e^{-0.5x}$

(a) $g(-4) = 10e^{-0.5(-4)} = 10e^2 \approx 73.891$

(b) $g(4) = 10e^{-0.5(4)} = 10e^{-2} \approx 1.353$

(c) $g(8) = 10e^{-0.5(8)} = 10e^{-4} \approx 0.183$

47. The function g is related to $f(x) = e^x$ by $g(x) = -f(x)$. To sketch the graph of g, reflect the graph of f in the x-axis.

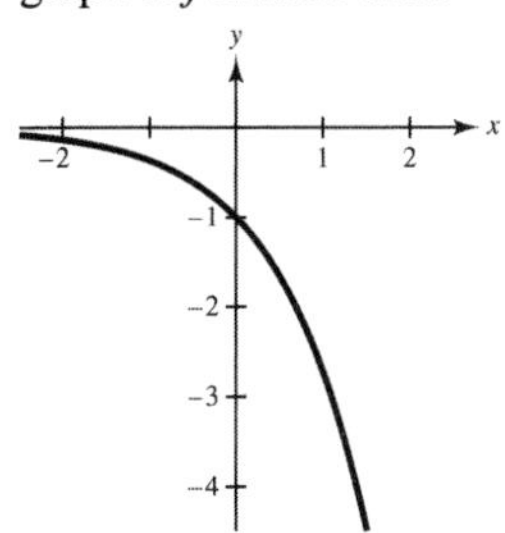

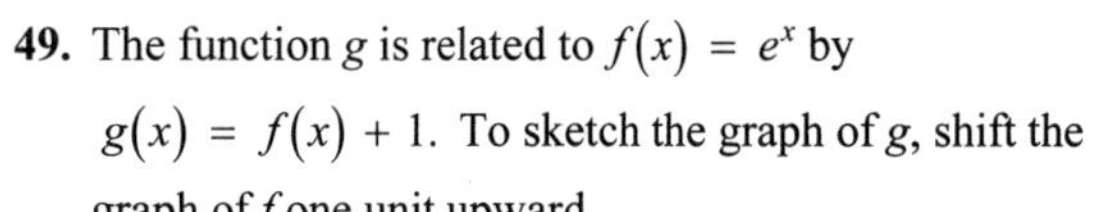

49. The function g is related to $f(x) = e^x$ by $g(x) = f(x) + 1$. To sketch the graph of g, shift the graph of f one unit upward.

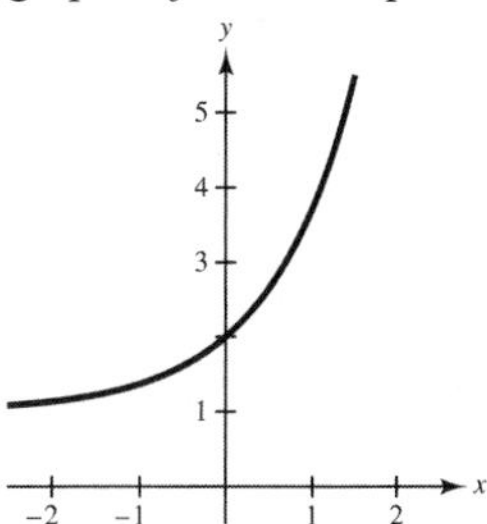

51. $y = 16\left(\frac{1}{2}\right)^{80/30} \approx 2.520$ grams

Keystrokes:

Scientific: 16 [×] 0.5 [y^x] [(] 8 [÷] 3 [)] [=]

Graphing: 16 [×] 0.5 [^] [(] 8 [÷] 3 [)] [ENTER]

53.

n	1	4	12	365	Continuous
A	\$275.90	\$283.18	\$284.89	\$285.74	\$285.77

Compounded 1 time: $A = 100\left(1 + \frac{0.07}{1}\right)^{1(15)} = \$275.90$

Compounded 4 times: $A = 100\left(1 + \frac{0.07}{4}\right)^{4(15)} = \$283.18$

Compounded 12 times: $A = 100\left(1 + \frac{0.07}{12}\right)^{12(15)} = \$284.89$

Compounded 365 times: $A = 100\left(1 + \frac{0.07}{365}\right)^{365(15)} = \$285.74$

Compounded continuously: $A = Pe^{rt} = 100e^{0.07(15)} = \$285.77$

55.

n	1	4	12	365	Continuous
A	\$4956.46	\$5114.30	\$5152.11	\$5170.78	\$5171.42

Compounded 1 time: $A = 2000\left(1 + \frac{0.095}{1}\right)^{1(10)} = \$4956.46$

Compounded 4 times: $A = 2000\left(1 + \frac{0.095}{4}\right)^{4(10)} = \$5114.30$

Compounded 12 times: $A = 2000\left(1 + \frac{0.095}{12}\right)^{12(10)} = \$5152.11$

Compounded 365 times: $A = 2000\left(1 + \frac{0.095}{365}\right)^{365(10)} = \$5170.78$

Compounded continuously: $A = 2000e^{0.095(10)} = \$5171.42$

57. To obtain the graph of $g(x) = -2^x$ from the graph of $f(x) = 2^x$, reflect the graph of f in the x-axis.

59. The behavior is reversed. When the graph of the original function is increasing, the reflection is decreasing. When the graph of the original function is decreasing, the reflected graph is increasing.

61. $3^x \cdot 3^{x+2} = 3^{x+(x+2)} = 3^{2x+2}$

63. $3(e^x)^{-2} = 3 \cdot \frac{1}{(e^x)^2} = \frac{3}{e^{2x}}$

65. $\frac{e^{x+2}}{e^x} = e^{x+2-x} = e^2$

67. $\sqrt[3]{-8e^{3x}} = -2e^x$ because

$-2 \cdot -2 \cdot -2 \cdot e^x \cdot e^x \cdot e^x = -8e^{3x}$.

69.

n	1	4	12	365	Continuous
P	\$2541.75	\$2498.00	\$2487.98	\$2483.09	\$2482.93

Compounded 1 time:

$$5000 = P\left(1 + \frac{0.07}{1}\right)^{1(10)}$$
$$\frac{5000}{(1.07)^{10}} = P$$
$$\$2541.75 = P$$

Compounded 4 times:

$$5000 = \left(1 + \frac{0.07}{4}\right)^{4(10)}$$
$$\frac{5000}{(1.0175)^{40}} = P$$
$$\$2498.00 = P$$

Compounded 12 times:

$$5000 = P\left(1 + \frac{0.07}{12}\right)^{12(10)}$$
$$\frac{5000}{(1.0058\overline{3})^{120}} = P$$
$$\$2487.98 = P$$

Compounded 365 times:

$$5000 = P\left(1 + \frac{0.07}{365}\right)^{365(10)}$$
$$\frac{5000}{(1.0001918)^{3650}} = P$$
$$\$2483.09 = P$$

Compounded continuously:

$$5000 = Pe^{0.07(10)}$$
$$\frac{5000}{e^{0.7}} = P$$
$$\$2482.93 = P$$

71. $p = 25 - 0.4^{0.02x}$

(a) $p = 25 - 0.4e^{0.02(100)} = 25 - 0.4e^2 \approx \$22.04$

(b) $p = 25 - 0.4e^{0.02(125)} = 25 - 0.4e^{2.5} \approx \$20.13$

73. $v(t) = 64{,}000(2)^{t/15}$

(a) $v(5) = 64{,}000(2)^{5/15} = 64{,}000(2)^{1/3} \approx \$80{,}634.95$

(b) $v(20) = 64{,}000(2)^{20/15}$

$= 64{,}000(2)^{4/3}$

$\approx \$161{,}269.89$

75. $f(t) = 2^{t-1}$

$f(30) = 2^{29} = 536{,}870{,}912$ pennies

77. By definition, the base of an exponential function must be positive and not equal to 1. If the base is 1, the function simplifies to the constant function $y = 1$.

79. False. e is an irrational number.

$\frac{271{,}801}{99{,}990}$ is rational because its equivalent decimal form is a repeating decimal.

81. Because $1 < \sqrt{2} < 2$ and $2 > 0$, $2^1 < 2^{\sqrt{2}} < 2^2$.

83. When $k > 1$, the values of f will increase. When $0 < k < 1$, the values of f will decrease. When $k = 1$, the values of f remain constant.

85. $g(s) = \sqrt{s - 4}$, Domain: $[4, \infty)$

$s - 4 \geq 0$

$s \geq 4$

87. $y^2 = x - 1$

y is not a function of x.

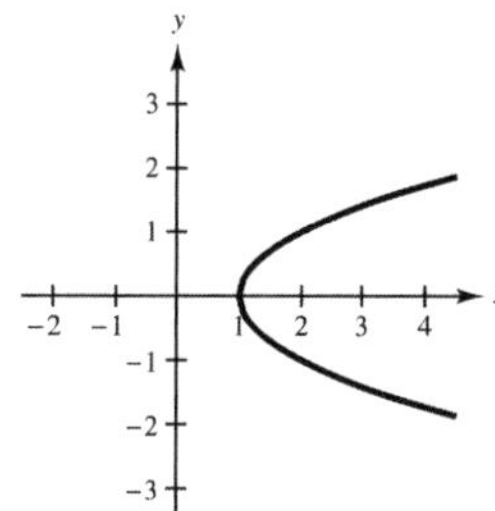

Section 9.2 Composite and Inverse Functions

1. $f(x) = 2x + 3, g(x) = x - 6$

(a) $(f \circ g)(x) = 2(x - 6) + 3$

$= 2x - 12 + 3 = 2x - 9$

(b) $(g \circ f)(x) = (2x + 3) - 6 = 2x - 3$

(c) $(f \circ g)(4) = 2(4) - 9 = 8 - 9 = -1$

(d) $(g \circ f)(7) = 2(7) - 3 = 14 - 3 = 11$

3. $f(x) = x^2 + 3, g(x) = x + 2$

(a) $(f \circ g)(x) = (x + 2)^2 + 3$

$= x^2 + 4x + 4 + 3 = x^2 + 4x + 7$

(b) $(g \circ f)(x) = (x^2 + 3) + 2 = x^2 + 5$

(c) $(f \circ g)(2) = 2^2 + 4(2) + 7 = 4 + 8 + 7 = 19$

(d) $(g \circ f)(-3) = (-3)^2 + 5 = 9 + 5 = 14$

5. $f(x) = \sqrt{x + 2}, g(x) = x - 4$

(a) $(f \circ g)(x) = \sqrt{(x - 4) + 2} = \sqrt{x - 2}$

Domain: $x - 2 \geq 0$

$x \geq 2, \quad [2, \infty)$

(b) $(g \circ f)(x) = \sqrt{x + 2} - 4$

Domain: $x + 2 \geq 0$

$x \geq -2, \quad [-2, \infty)$

7. $f(x) = x^2 - 2$

No, it does not have an inverse because it is possible to find a horizontal line that intersects the graph of f at more than one point.

9. $f(x) = x^2, x \geq 0$

Yes, it does have an inverse because no horizontal line intersects the graph of f at more than one point.

11. $g(x) = \sqrt{25 - x^2}$

No, it does not have an inverse because it is possible to find a horizontal line that intersects the graph of g at more than one point.

13. $f(x) = 3 - 4x$

$y = 3 - 4x$

$x = 3 - 4y$

$x - 3 = -4y$

$\dfrac{x - 3}{-4} = y$

$\dfrac{3 - x}{4}$ or $\dfrac{x - 3}{-4} = f^{-1}(x)$

15. $f(t) = t^3 - 1$

$y = t^3 - 1$

$t = y^3 - 1$

$t + 1 = y^3$

$\sqrt[3]{t + 1} = y$

$\sqrt[3]{t + 1} = f^{-1}(t)$

17. $f(x) = x + 4$, $f^{-1}(x) = x - 4$

$(0, 4)$ $\quad (4, 0)$

$(-4, 0)$ $\quad (0, -4)$

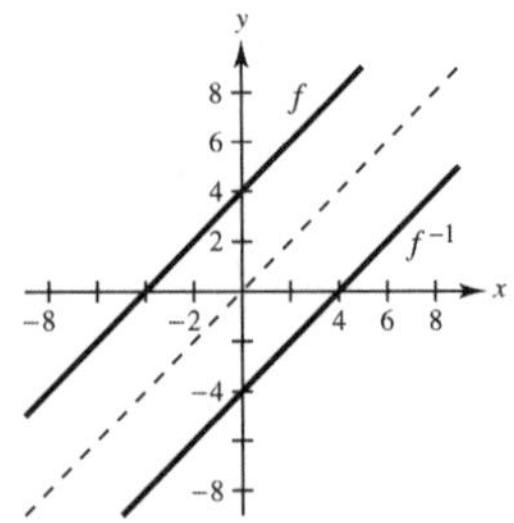

19. $f(x) = 3x - 1$, $f^{-1}(x) = \frac{1}{3}(x + 1)$

$(0, -1)$ $\quad (-1, 0)$

$\left(\frac{1}{3}, 0\right)$ $\quad \left(0, \frac{1}{3}\right)$

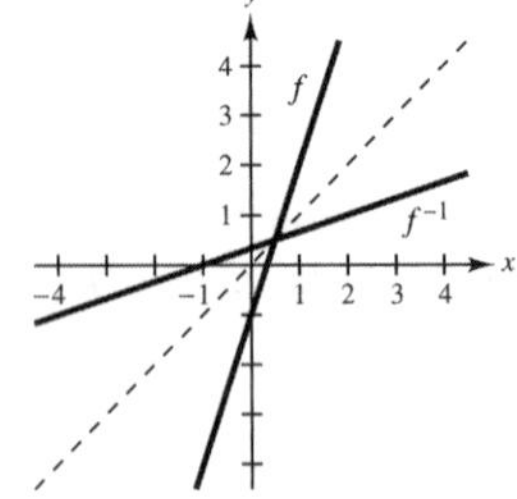

21. $f(x) = x^2 - 1, x \geq 0$ $\quad f^{-1}(x) = \sqrt{x + 1}$

$(0, -1)$ $\quad (-1, 0)$

$(1, 0)$ $\quad (0, 1)$

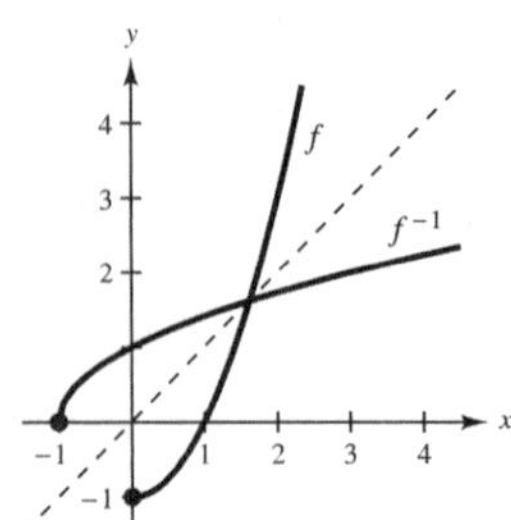

23. The graph of the inverse function is *b*.

24. The graph of the inverse function is *c*.

25. The graph of the inverse function is *d*.

26. The graph of the inverse function is *a*.

27.

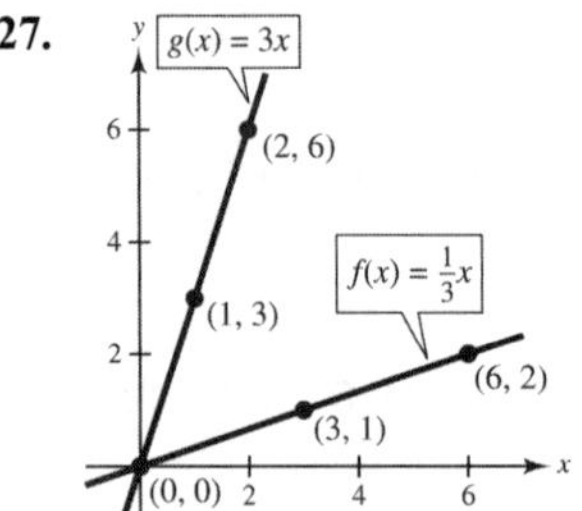

29.

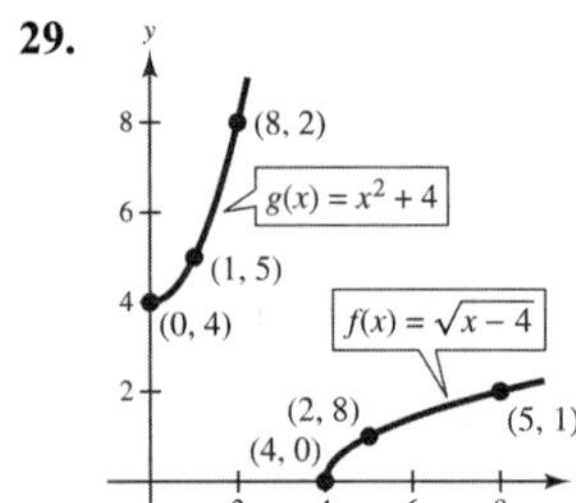

31. The equation obtained by replacing $f(x)$ with y is $y = 2x + 6$.

33. The inverse function of $f(x) = 2x + 6$ is

$f^{-1}(x) = \dfrac{x - 6}{2}$.

35. $(f \circ g)(-3) = f[g(-3)] = f[1] = -1$

37. $(g \circ f)(-2) = g[f(-2)] = g[3] = 1$

39. $f(x) = 1 - 2x,\ g(x) = \frac{1}{2}(1 - x)$

$$\begin{aligned} f(g(x)) &= f\left[\tfrac{1}{2}(1 - x)\right] \\ &= 1 - 2\left[\tfrac{1}{2}(1 - x)\right] \\ &= 1 - (1 - x) = 1 - 1 + x = x \\ g(f(x)) &= g(1 - 2x) \\ &= \tfrac{1}{2}\left[1 - (1 - 2x)\right] \\ &= \tfrac{1}{2}[1 - 1 + 2x] = \tfrac{1}{2}[2x] = x \end{aligned}$$

41. $g(x) = x^2 + 4$

$g(x)$ is not one-to-one so an inverse does not exist.

43. $h(x) = \sqrt{x}$

$$\begin{aligned} y &= \sqrt{x} \\ x &= \sqrt{y} \\ x^2 &= y \\ x^2 &= h^{-1}(x), \quad x \ge 0 \end{aligned}$$

45. $f(x) = (x - 2)^2, \quad x \ge 2$

$$\begin{aligned} y &= (x - 2)^2 \\ x &= (y - 2)^2 \\ \sqrt{x} &= y - 2 \\ \sqrt{x} + 2 &= y \\ \sqrt{x} + 2 &= f^{-1}(x), \quad x \ge 0 \end{aligned}$$

47. $r(t) = 0.6t,\ A(r) = \pi r^2$

$$A(r(t)) = A(0.6t) = \pi(0.6t)^2 = \pi(0.36t^2) = 0.36\pi t^2$$

Input: time
Output: area

$$\begin{aligned} A(r(3)) &= A[0.6(3)] \\ &= A(1.8) \\ &= \pi(1.8)^2 \\ &= \pi(3.24) \\ &\approx 10.2 \text{ square feet} \end{aligned}$$

49. (a)

$$\begin{aligned} y &= 9 + 0.65x \\ x &= 9 + 0.65y \\ \frac{x - 9}{0.65} &= y \\ \frac{100}{65}(x - 9) &= y \\ \frac{20}{13}(x - 9) &= y \end{aligned}$$

(b)

$$\begin{aligned} y &= \frac{20}{13}(14.20 - 9) \\ y &= \frac{20}{13}(5.20) \\ y &= 8 \text{ units} \end{aligned}$$

51. True. The x-coordinate of a point on the graph of f becomes the y-coordinate of a point on the graph of f^{-1}.

53. False: $f(x) = \sqrt{x - 1}$, Domain $[1, \infty)$
$f^{-1}(x) = x^2 + 1$, Domain $[0, \infty)$

55. Interchange the coordinates of each ordered pair. The inverse of the function defined by $\{(3, 6), (5, -2)\}$ is $\{(6, 3), (-2, 5)\}$.

57. $h(x) = -x^2$

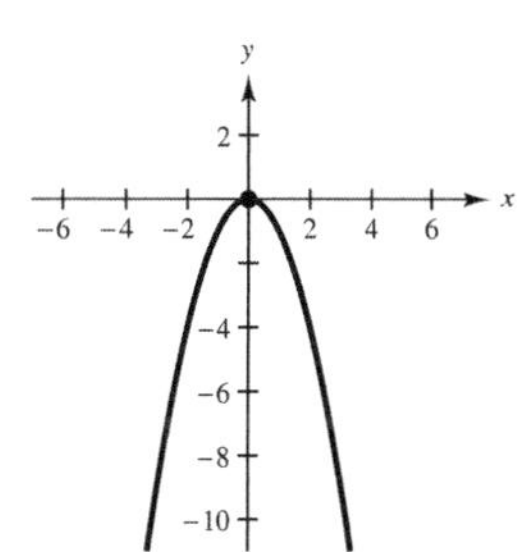

Reflection in the x-axis

59. $k(x) = (x + 3)^2 - 5$

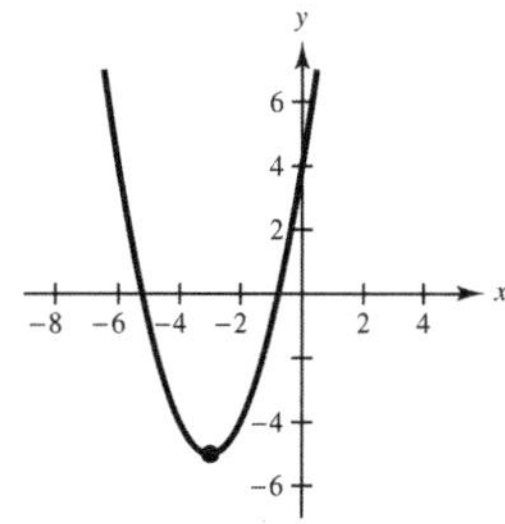

Horizontal shift 3 units left
Vertical shift 5 units down

61. $16 - (y + 2)^2 = [4 - (y + 2)][4 + (y + 2)]$
$= (4 - y - 2)(4 + y + 2)$
$= (2 - y)(6 + y)$

63. $5 - u + 5u^2 - u^3 = 1(5 - u) + u^2(5 - u)$
$= (5 - u)(1 + u^2)$
$= -1(u - 5)(u^2 + 1)$
$= -(u^2 + 1)(u - 5)$

65. $3x - 4y = 6$

Intercepts:

$3(0) - 4y = 6$

$y = -\frac{3}{2}$

$3x - 4(0) = 6$

$3x = 6$

$x = 2$

67. $y = -(x - 2)^2 + 1$

Intercepts: Vertex:

$y = -(0 - 2)^2 + 1 = -3, (0, -3)$ $(2, 1)$

$0 = -(x - 2)^2 + 1$

$(x - 2)^2 = 1$

$x - 2 = \pm 1$

$x = 2 \pm 1$

$x = 3, 1$

$(3, 0), (1, 0)$

Section 9.3 Logarithmic Functions

1. $\log_2 8 = 3$ because $2^3 = 8$.

3. $\log_9 3 = \frac{1}{2}$ because $9^{1/2} = 3$.

5. $\log_{16} 8 = \frac{3}{4}$ because $16^{3/4} = 8$.

7. $\log_4 1 = 0$ because $4^0 = 1$.

9. $\log_2 \frac{1}{16} = -4$ because $2^{-4} = \frac{1}{16}$.

11. $\log_2 (-3)$ is undefined because 2^x is never negative.

13. $\log_{10} 1000 = 3$ because $10^3 = 1000$.

15. $\log_{10} 0.1 = -1$ because $10^{-1} = \frac{1}{10}$.

17. $\log_{10} \frac{1}{10{,}000} = -4$ because $10^{-4} = \frac{1}{10{,}000}$.

19. $\log_{10} 42 \approx 1.6232$

21. $\log_{10} 0.023 \approx -1.6383$

23. $\log_{10} (\sqrt{5} + 3) \approx 0.7190$

25. $f(x) = 3^x$; $g(x) = \log_3 x$

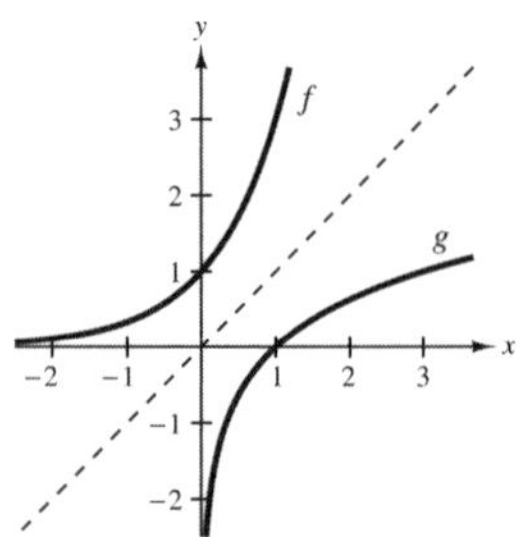

27. $f(x) = 6^x$; $g(x) = \log_6 x$

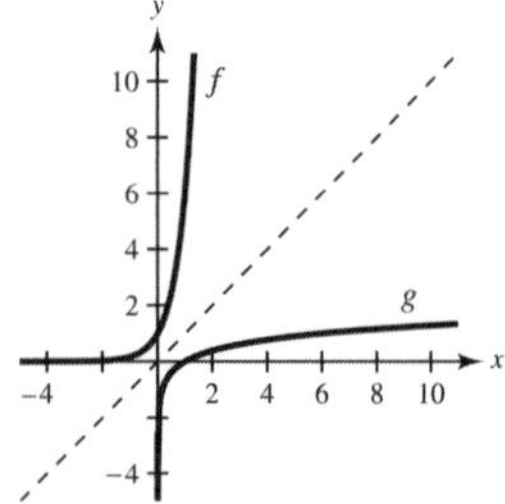

29. $f(x) = \left(\frac{1}{2}\right)^x$; $g(x) = \log_{1/2} x$

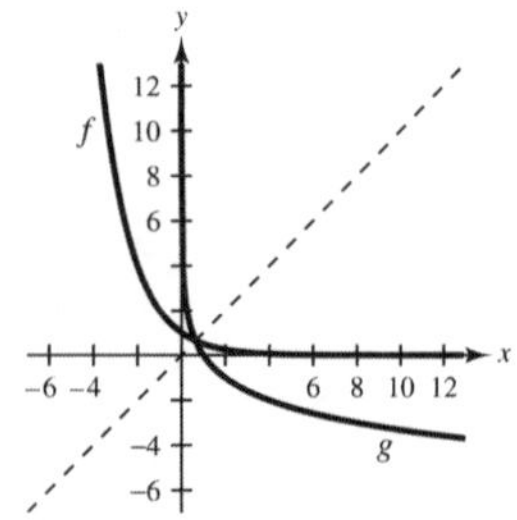

31. $f(x) = 4 + \log_3 x$ matches graph (c).

32. $f(x) = -\log_3 x$ matches graph (b).

33. $f(x) = \log_3(-x)$ matches graph (a).

34. $f(x) = \log_3(x + 2)$ matches graph (d).

35. $h(x) = 3 + \log_2 x$

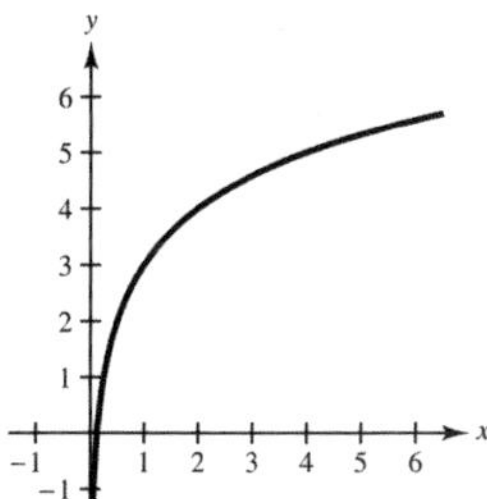

Vertical shift 3 units up

Vertical asymptote: $x = 0$

Domain: $(0, \infty)$

37. $h(x) = \log_2(x - 2)$

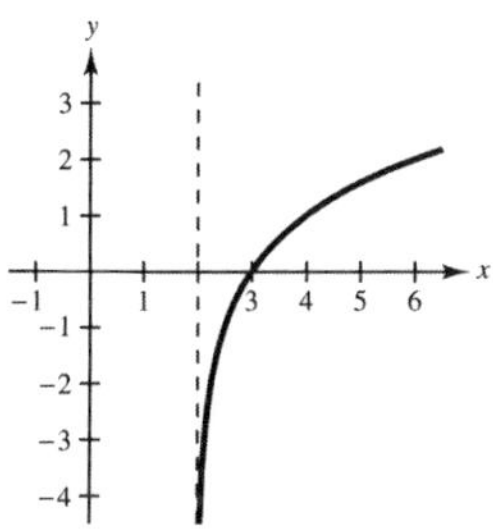

Horizontal shift 2 units right

Vertical asymptote: $x = 2$

Domain: $(2, \infty)$

39. $h(x) = \log_2(-x)$

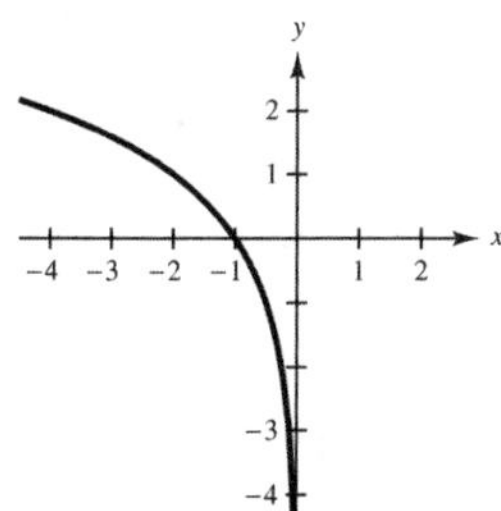

Reflection in the y-axis

Vertical asymptote: $x = 0$

Domain: $(-\infty, 0)$

41. $\ln e^3 = 3$

43. $\ln \dfrac{1}{e^2} = \ln e^{-2} = -2$

45. $f(x) = 3 + \ln x$

Vertical asymptote: $x = 0$

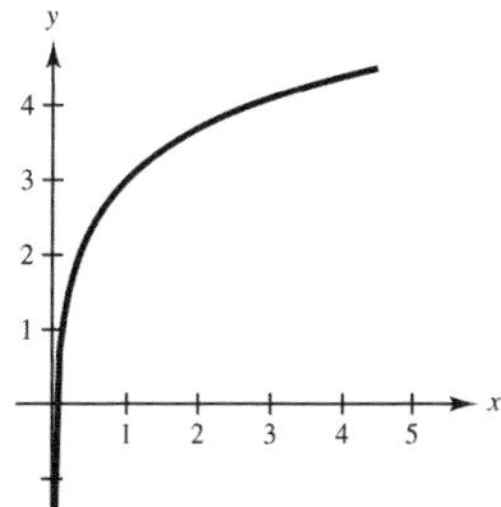

Table of values:

x	1	e
y	3	4

47. $f(x) = -\ln x$

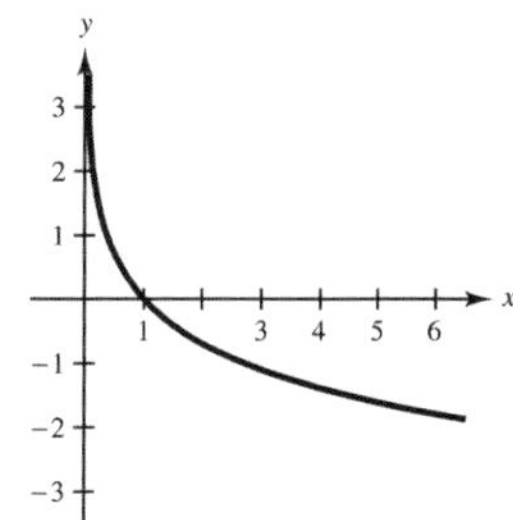

Vertical asymptote: $x = 0$

Table of values:

x	1	e
y	0	-1

49. $\log_9 36 = \dfrac{\log 36}{\log 9} \approx 1.6309$

$= \dfrac{\ln 36}{\ln 9} \approx 1.6309$

51. $\log_5 14 = \dfrac{\log 14}{\log 5} \approx 1.6397$

$= \dfrac{\ln 14}{\ln 5} \approx 1.6397$

53. $\log_2 0.72 = \dfrac{\log 0.72}{\log 2} \approx -0.4739$

$= \dfrac{\ln 0.72}{\ln 2} \approx -0.4739$

55. $\log_{15} 1250 = \dfrac{\log 1250}{\log 15} \approx 2.6332$

$= \dfrac{\ln 1250}{\ln 15} \approx 2.6332$

57. $\log_{1/4} 16 = \dfrac{\log 16}{\log 1/4} = -2$

$= \dfrac{\ln 16}{\ln 1/4} = -2$

59. $\log_4 \sqrt{42} = \dfrac{\log \sqrt{42}}{\log 4} \approx 1.3481$

$= \dfrac{\ln \sqrt{42}}{\ln 4} \approx 1.3481$

61. $\log_2(1 + e) = \dfrac{\log(1 + e)}{\log 2} \approx 1.8946$

$= \dfrac{\ln(1 + e)}{\ln 2} \approx 1.8946$

63. The inverse of the function $y = a^x$ is $y = \log_a x$ or $x = a^y$.

65. The base of the logarithm $\ln 5$ is e.

67. $\log_7 49 = 2$

$7^2 = 49$

69. $\log_2 \frac{1}{32} = -5$

$2^{-5} = \frac{1}{32}$

71. $\log_{36} 6 = \frac{1}{2}$

$36^{1/2} = 6$

73. $\log_8 4 = \frac{2}{3}$

$8^{2/3} = 4$

75. $6^2 = 36$

$\log_6 36 = 2$

77. $8^{2/3} = 4$

$\log_8 4 = \frac{2}{3}$

79. $h(s) = -2 \log_3 s$

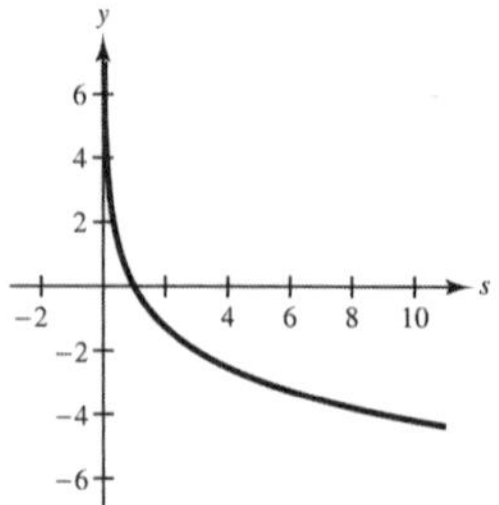

Vertical asymptote: $x = 0$

Table of values:

x	1	3
y	0	−2

81. $f(x) = \log_{10}(10x)$

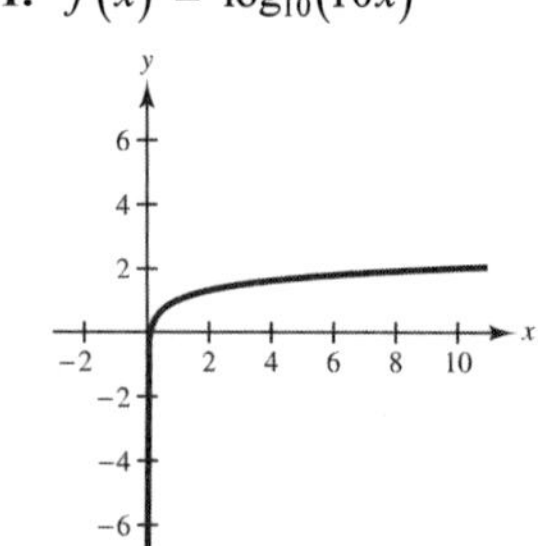

Vertical asymptote: $x = 0$

Table of values:

x	1	10
y	1	2

83. $y = \log_4(x - 1) - 2$

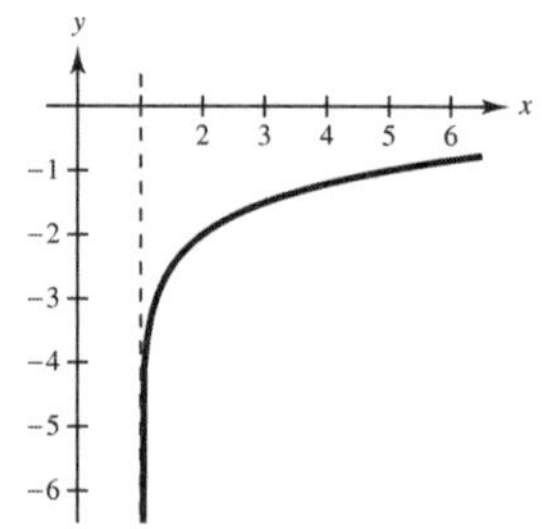

Vertical asymptote: $x = 1$

Table of values:

x	2	5
y	−2	−1

85. $y = -\log_3 x + 2$

Domain: $(0, \infty)$

Vertical asymptote: $x = 0$

Table of values:

x	1	3
y	2	1

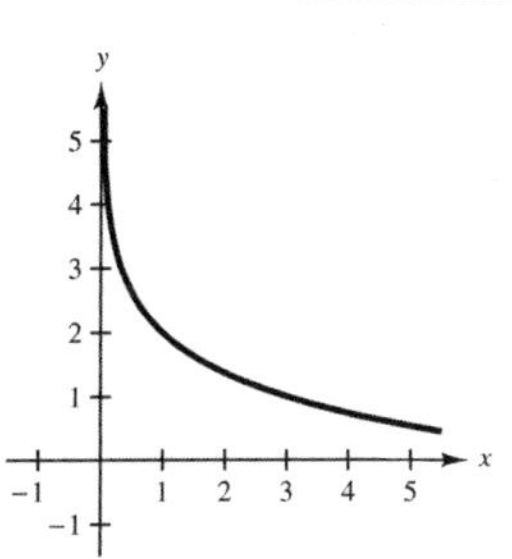

87. $h = 116 \log_{10}(55 + 40) - 176$

$= 116 \log_{10}(95) - 176$

≈ 53.4 inches

89. (a) *Keystrokes:*

[Y=] 10 [LN] [(] [(] 10 [+] [√] [(] 100 [−] [X,T,θ] [x^2] [)] [)] [÷] [X,T,θ] [)] [−] [√] [(] 100 [−] [X,T,θ] [x^2] [)] [GRAPH]

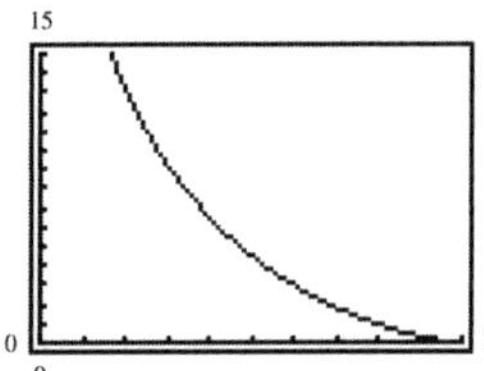

Domain: $(0, 10]$

(b) Vertical asymptote: $x = 0$

(c) $y = 13.126$ when $x = 2$. Trace to $x = 2$.

Position of boat is $(2, 13.126)$

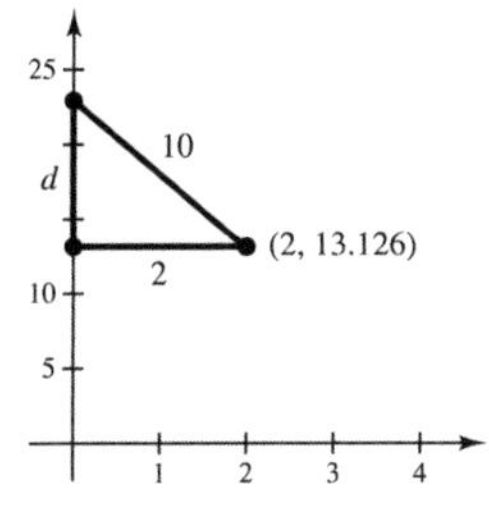

$10^2 - 2^2 = d^2$

$d \approx 9.798$

$9.798 + 13.126 = 22.924$

position of person is $(0, 22.924)$

91. Domain = positive real numbers, $(0, \infty)$

93. If $1000 \le x \le 10{,}000$, then $f(x) = \log_{10} x$ lies $3 \le f(x) \le 4$.

95. When $f(x)$ increases by 1 unit, x increases by a factor of 10.

97. False. $8 = 2^3$ is equivalent to $3 = \log_2 8$.

99. Logarithmic functions with base 10 are common logarithms. Logarithmic functions with base e are natural logarithms.

101. A vertical shift or reflection in the x-axis of a logarithmic graph does not affect the domain or range. A horizontal shift or reflection in the y-axis of a logarithmic graph affects the domain, but the range stays the same.

103. $(-m^6n)(m^4n^3) = -(m^{6+4}n^{1+3}) = -m^{10}n^4$

105. $\dfrac{36x^4y}{8xy^3} = \dfrac{36}{8} \cdot x^{4-1} \cdot y^{1-3} = \dfrac{9x^3}{2y^2}, \quad x \neq 0$

107. $$\begin{aligned}25\sqrt{3x} - 3\sqrt{12x} &= 25\sqrt{3x} - 3\sqrt{4 \cdot 3x}\\ &= 25\sqrt{3x} - 6\sqrt{3x}\\ &= (25 - 6)\sqrt{3x}\\ &= 19\sqrt{3x}\end{aligned}$$

109. $$\begin{aligned}\sqrt{u}\left(\sqrt{20} - \sqrt{5}\right) &= \sqrt{u}\left(\sqrt{4 \cdot 5} - \sqrt{5}\right)\\ &= \sqrt{u}\left(2\sqrt{5} - \sqrt{5}\right)\\ &= \sqrt{u} \cdot \sqrt{5}\\ &= \sqrt{5u}\end{aligned}$$

Mid-Chapter Quiz for Chapter 9

1. (a) $f(2) = \left(\dfrac{4}{3}\right)^2 = \dfrac{16}{9}$

(b) $f(0) = \left(\dfrac{4}{3}\right)^0 = 1$

(c) $f(-1) = \left(\dfrac{4}{3}\right)^{-1} = \dfrac{3}{4}$

(d) $f(1.5) = \left(\dfrac{4}{3}\right)^{1.5} \approx 1.54 = \dfrac{8\sqrt{3}}{9}$

2. $g(x) = 3^{x-1}$

Horizontal asymptote: $y = 0$

3. $f(x) = 2^x$; b

4. $g(x) = -2^x$; d

5. $h(x) = 2^{-x}$; a

6. $s(x) = 2^x - 2$; c

7. (a) $(f \circ g)(x) = f[g(x)] = 2x^3 - 3$

(b) $(g \circ f)(x) = g[f(x)] = (2x - 3)^3$

(c) $$\begin{aligned}(f \circ g)(-2) &= f[g(-2)]\\ &= f[-8] = 2(-8) - 3 = -19\end{aligned}$$

(d) $(g \circ f)(4) = g[f(4)] = g[5] = 5^3 = 125$

8. $$\begin{aligned}(f \circ g)(x) &= f\left[\tfrac{1}{2}(5 - x)\right] = 5 - 2\left[\tfrac{1}{2}(5 - x)\right]\\ &= 5 - (5 - x)\\ &= 5 - 5 + x = x\\ (g \circ f)(x) &= g[5 - 2x] = \tfrac{1}{2}[5 - (5 - 2x)]\\ &= \tfrac{1}{2}[5 - 5 + 2x]\\ &= \tfrac{1}{2}(2x) = x\end{aligned}$$

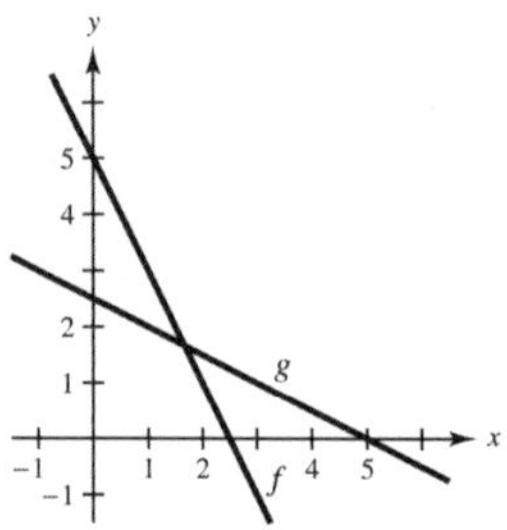

9. $$\begin{aligned}f(x) &= 10x + 3\\ y &= 10x + 3\\ x &= 10y + 3\\ x - 3 &= 10y\\ \frac{x - 3}{10} &= y\\ \frac{x - 3}{10} &= f^{-1}(x)\end{aligned}$$

10. $$\begin{aligned}f(t) &= \tfrac{1}{2}t^3 + 2\\ y &= \tfrac{1}{2}t^3 + 2\\ t &= \tfrac{1}{2}y^3 + 2\\ t - 2 &= \tfrac{1}{2}y^3\\ 2t - 4 &= y^3\\ \sqrt[3]{2t - 4} &= y\\ \sqrt[3]{2t - 4} &= f^{-1}(t)\end{aligned}$$

11. $\log_9\left(\frac{1}{81}\right) = -2$

$9^{-2} = \frac{1}{81}$

12. $2^6 = 64$

$\log_2 64 = 6$

13. $\log_5 125 = 3$ because $5^3 = 125$.

14. $f(t) = \ln(t + 3)$

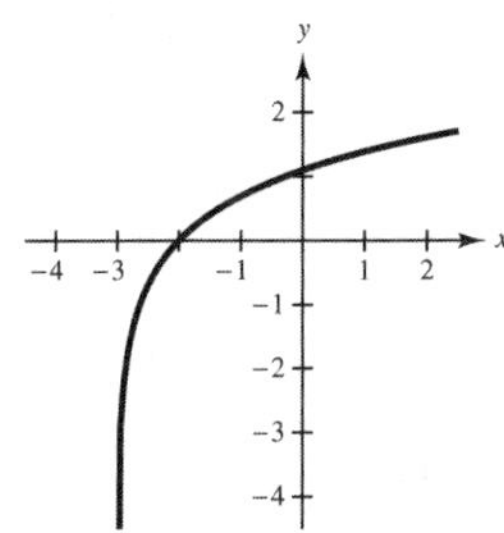

Vertical asymptote: $y = -3$

15. $h(x) = 1 + \ln(x)$

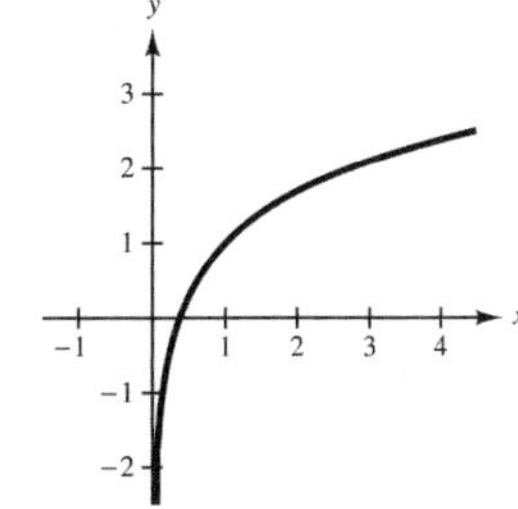

Vertical asymptote: $x = 0$

16. $f(x) = \log_5(x - 2) + 1$

The graph of $f(x) = \log_5 x$ has been shifted 2 units right and 1 unit up, so $h = 2, k = 1$.

17. $\log_3 782 = \dfrac{\log 782}{\log 3} \approx 6.0639$

18.

n	1	4	12	365	Continuous compounding
A	\$2979.31	\$3042.18	\$3056.86	\$3064.06	\$3064.31

Compounded 1 time per year: $A = 1200\left(1 + \dfrac{0.0625}{1}\right)^{1(15)} \approx \$2979.31$

Compounded 4 times per year: $A = 1200\left(1 + \dfrac{0.0625}{4}\right)^{4(15)} \approx \$3042.18$

Compounded 12 times per year: $A = 1200\left(1 + \dfrac{0.0625}{12}\right)^{12(15)} \approx \$3056.86$

Compounded 365 times per year: $A = 1200\left(1 + \dfrac{0.0625}{365}\right)^{365(15)} \approx \$3064.06$

Compounded continuously: $A = 1200e^{(0.0625)(15)} \approx \$3064.31$

19. $y = 14\left(\frac{1}{2}\right)^{t/40}$

$y = 14\left(\frac{1}{2}\right)^{125/40}$

$y = 14\left(\frac{1}{2}\right)^{3.125}$

$y \approx 1.60$ grams

Section 9.4 Properties of Logarithms

1. $\ln \frac{5}{3} = \ln 5 - \ln 3 \approx 1.6094 - 1.0986 \approx 0.5108$

3. $\ln 9 = \ln 3^2$

$= 2 \cdot \ln 3$

$\approx 2 \cdot 1.0986$

≈ 2.1972

5. $\ln 75 = \ln(3 \cdot 5^2)$
$= \ln 3 + 2 \ln 5$
$\approx 1.0986 + 2(1.6094)$
$\approx 1.0986 + 3.2188$
≈ 4.3174

7. $\ln \sqrt{45} = \frac{1}{2}\ln(3^2 \cdot 5) = \frac{1}{2}[2 \ln 3 + \ln 5] = \ln 3 + \frac{1}{2}\ln 5$
$\approx 1.0986 + \frac{1}{2}(1.6094)$
$\approx 1.0986 + 0.8047$
≈ 1.9033

9. $-3 \log_4 2 = \log_4 2^{-3} = \log_4 \frac{1}{2^3} = \log_4 \frac{1}{8}$

11. $-3 \log_{10} 3 + \log_{10} \frac{3}{2} = \log_{10} 3^{-3} + \log_{10} \frac{3}{2}$
$= \log_{10}\left(\frac{1}{3^3} \cdot \frac{3}{2}\right)$
$= \log_{10}\left(\frac{1}{3^2} \cdot \frac{1}{2}\right)$
$= \log_{10} \frac{1}{18}$

13. $-\ln \frac{1}{7} = \ln\left(\frac{1}{7}\right)^{-1} = \ln 7 = \ln \frac{56}{8} = \ln 56 - \ln 8$

15. $\log_{12} 12^3 = 3 \cdot \log_{12} 12 = 3 \cdot 1 = 3$

17. $\log_4\left(\frac{1}{16}\right)^2 = \log_4\left(4^{-2}\right)^2 = \log_4 4^{-4}$
$= -4 \log_4 4 = -4 \cdot 1 = -4$

19. $\log_5 \sqrt[3]{5} = \log_5 5^{1/3} = \frac{1}{3} \cdot \log_5 5 = \frac{1}{3} \cdot 1 = \frac{1}{3}$

21. $\ln 14^0 = 0 \cdot \ln 14 = 0$

23. $\ln e^{-9} = -9 \ln e = -9 \cdot 1 = -9$

25. $\log_8 4 + \log_8 16 = \log_8 4 \cdot 16 = \log_8 64 = 2$ because $8^2 = 64$.

27. $\log_3 54 - \log_3 2 = \log_3 \frac{54}{2} = \log_3 27 = \log_3 3^3 = 3$

29. $\log_2 5 - \log_2 40 = \log_2 \frac{5}{40} = \log_2 \frac{1}{8} = \log_2 2^{-3} = -3$ because $2^{-3} = 2^{-3}$.

31. $\ln e^8 + \ln e^4 = \ln e^8 \cdot e^4 = \ln e^{12}$
$= 12 \ln e = 12 \cdot 1 = 12$

33. $\ln \frac{e^3}{e^2} = \ln e = 1$

35. $\log_3 11x = \log_3 11 + \log_3 x$

37. $\ln 3y = \ln 3 + \ln y$

39. $\log_7 x^2 = 2 \log_7 x$

41. $\log_4 x^{-3} = -3 \log_4 x$

43. $\log_4 \sqrt{3x} = \log_4(3x)^{1/2} = \frac{1}{2}\log_4(3x)$
$= \frac{1}{2}(\log_4 3 + \log_4 x)$

45. $\log_2 \frac{z}{17} = \log_2 z - \log_2 17$

47. $\log_9 \frac{\sqrt{x}}{12} = \log_9 \sqrt{x} - \log_9 12$
$= \log_9 x^{1/2} - \log_9 12$
$= \frac{1}{2}\log_9 x - \log_9 12$

49. $\ln x^2(y + 2) = \ln x^2 + \ln(y + 2) = 2 \ln x + \ln(y + 2)$

51. $\log_4\left[x^6(x + 7)^2\right] = \log_4 x^6 + \log_4(x + 7)^2$
$= 6 \log_4 x + 2 \log_4(x + 7)$

53. $\log_3 \sqrt[3]{x + 1} = \frac{1}{3}\log_3(x + 1)$

55. $\ln\sqrt{x(x + 2)} = \frac{1}{2}[\ln x + \ln(x + 2)]$

57. $\ln \frac{xy^2}{z^3} = \ln(xy^2) - \ln z^3 = \ln x + \ln y^2 - \ln z^3$
$= \ln x + 2 \ln y - 3 \ln z$

59. $\log_{12} x - \log_{12} 3 = \log_{12} \frac{x}{3}$

61. $\log_3 5 + \log_3 x = \log_3 5x$

63. $7 \log_2 x + 3 \log_2 z = \log_2 x^7 + \log_2 z^3 = \log_2 x^7z^3$

65. $4(\ln x + \ln y) = \ln(xy)^4$ or $\ln x^4y^4$, $x > 0, y > 0$

67. $\log_4(x + 8) - 3 \log_4 x = \log_4(x + 8) - \log_4 x^3$
$= \log_4 \frac{(x + 8)}{x^3}, \quad x > 0$

69. $f(t) = 80 - \log_{10}(t + 1)^{12}$
$= 80 - 12 \log_{10}(t + 1)$
$f(2) = 80 - 12 \log_{10}(2 + 1) \approx 74.27$
$f(8) = 80 - 12 \log_{10}(8 + 1) \approx 68.55$

71. $B = 10 \log_{10}\left(\frac{I}{10^{-12}}\right)$

$= 10 \log_{10} I - 10 \log_{10}\left(10^{-12}\right)$

$= 10 \log_{10} I + 120 \log_{10}(10)$

$= 10 \log_{10} I + 120$

$B = 10 \log_{10}\left(10^{-1}\right) + 120$

$= -10 \log_{10}(10) + 120$

$= -10 + 120$

$= 110$ decibels

73. The expression is obtained by condensing.

75. The Power Property of Logarithms can be used to condense $2\left(\log_2 \frac{3x}{y}\right)$.

77. $3 \ln x + \ln y - 2 \ln z = \ln x^3 + \ln y - \ln z^2 = \ln \frac{x^3 y}{z^2}$

79. $2\left[\ln x - \ln(x + 1)\right] = 2 \ln \frac{x}{x + 1}$

$= \ln\left(\frac{x}{x + 1}\right)^2$

$= \ln \frac{x^2}{(x + 1)^2}, \quad x > 0$

81. $\frac{1}{3} \log_5(x + 3) - \log_5(x - 6) = \log_5(x + 3)^{1/3} - \log_5(x - 6) = \log_5 \frac{\sqrt[3]{x + 3}}{x - 6}$

83. $5 \log_6(c + d) - \frac{1}{2} \log_6(m - n) = \log_6(c + d)^5 - \log_6(m - n)^{1/2} = \log_6 \frac{(c + d)^5}{\sqrt{m - n}}$

85. $\frac{1}{5}(3 \log_2 x - 4 \log_2 y) = \frac{1}{5}\left(\log_2 x^3 - \log_2 y^4\right) = \frac{1}{5}\left(\log_2 \frac{x^3}{y^4}\right) = \log_2 \sqrt[5]{\frac{x^3}{y^4}}, \quad y > 0$

87. $\ln 3e^2 = \ln 3 + \ln e^2 = \ln 3 + 2 \ln e = \ln 3 + 2$

89. $\log_5 \sqrt{50} = \frac{1}{2}\left[\log_5\left(5^2 \cdot 2\right)\right] = \frac{1}{2}\left[2 \log_5 5 + \log_5 2\right] = \frac{1}{2}\left[2 + \log_5 2\right] = 1 + \frac{1}{2} \log_5 2$

91. $\log_8 \frac{8}{x^3} = \log_8 8 - \log_8 x^3 = 1 - 3 \log_8 x$

93. $E = 1.4(\log_{10} C_2 - \log_{10} C_1) = 1.4\left(\log_{10} \frac{C_2}{C_1}\right)$

$= \log_{10}\left(\frac{C_2}{C_1}\right)^{1.4}$

95. True; $\log_2 8x = \log_2 8 + \log_2 x = 3 + \log_2 x$

97. False; $\log_3(u + v)$ does not simplify.

99. True, $f(ax) = \log_a ax = \log_a a + \log_a x$

$= 1 + \log_a x$

$= 1 + f(x)$

101. False; 0 is not in the domain of f.

103. False; $f(x - 3) = \ln(x - 3) \neq \ln x - \ln 3$

105. Choose two values for x and y, such as $x = 3$ and $y = 5$, and show the two expressions are not equal.

$\frac{\ln 3}{\ln 5} \neq \ln \frac{3}{5} = \ln 3 - \ln 5$

$0.6826062 \neq -0.5108256 = -0.5108256$

or

$\frac{\ln e}{\ln e} \neq \ln \frac{e}{e}$

$1 \neq \ln 1$

$1 \neq 0$

107. $\frac{2}{3}x + \frac{2}{3} = 4x - 6$

$2x + 2 = 12x - 18$

$-10x = -20$

$x = \frac{20}{10}$

$x = 2$

109. $\frac{5}{2x} - \frac{4}{x} = 3$

$5 - 8 = 6x$

$-3 = 6x$

$-\frac{3}{6} = x$

$-\frac{1}{2} = x$

111. $|x - 4| = 3$

$x - 4 = 3$ or $x - 4 = -3$

$x = 7$ $\qquad x = 1$

113. $g(x) = -(x + 2)^2$

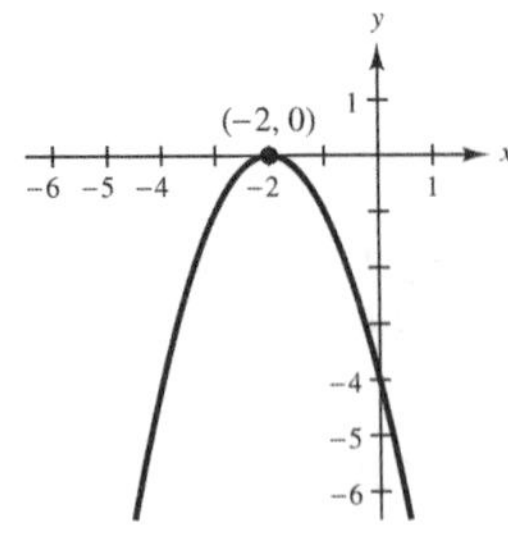

Vertex: $(-2, 0)$

x-intercept: $(-2, 0)$

115. $g(x) = -2x^2 + 4x - 7$

$$g(x) = -2(x^2 - 2x) - 7$$
$$= -2(x^2 - 2x + 1) - 7 + 2$$
$$= -2(x - 1)^2 - 5$$

Vertex: $(1, -5)$

x-intercepts: none

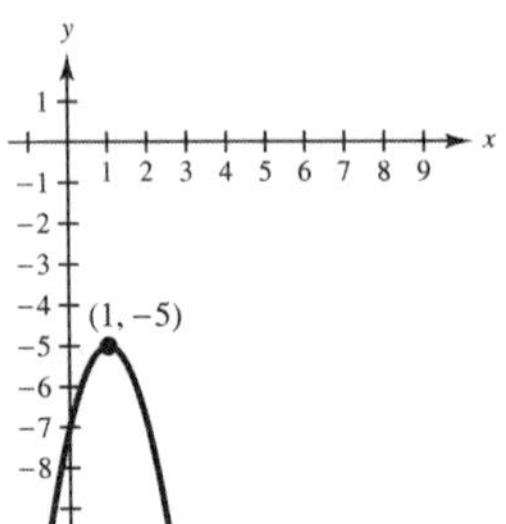

117. $f(x) = 4x + 9,\ g(x) = x - 5$

(a) $(f \circ g)(x) = f(x - 5) = 4(x - 5) + 9 = 4x - 20 + 9 = 4x - 11$

Domain: $(-\infty, \infty)$

(b) $(g \circ f)(x) = g(4x + 9) = (4x + 9) - 5 = 4x + 4$

Domain: $(-\infty, \infty)$

119. $f(x) = \dfrac{1}{x},\ g(x) = x + 2$

(a) $(f \circ g)(x) = f(x + 2) = \dfrac{1}{x + 2}$

Domain: $(-\infty, -2) \cup (-2, \infty)$

(b) $(g \circ f)(x) = g\left(\dfrac{1}{x}\right) = \dfrac{1}{x} + 2 = \dfrac{1 + 2x}{x}$

Domain: $(-\infty, 0) \cup (0, \infty)$

Section 9.5 Solving Exponential and Logarithmic Equations

1. $7^x = 7^3$

$x = 3$

3. $e^{1-x} = e^4$

$$1 - x = 4$$
$$-x = 3$$
$$x = -3$$

5. $3^{2-x} = 81$

$$3^{2-x} = 3^4$$
$$2 - x = 4$$
$$-x = 2$$
$$x = -2$$

7. $6^{2x} = 36$

$$6^{2x} = 6^2$$
$$2x = 2$$
$$x = 1$$

9. $\ln 5x = \ln 22$

$$5x = 22$$
$$x = \tfrac{22}{5}$$

11. $\ln(3 - x) = \ln 10$

$$3 - x = 10$$
$$-x = 7$$
$$x = -7$$

13. $\log_3(4 - 3x) = \log_3(2x + 9)$
$$4 - 3x = 2x + 9$$
$$-5 = 5x$$
$$-1 = x$$

15. $3^x = 91$
$$\log_3 3^x = \log_3 91$$
$$x = \log_3 91$$
$$x = \frac{\log 91}{\log 3}$$
$$x \approx 4.11$$

17. $5^x = 8.2$
$$\log_5 5^x = \log_5 8.2$$
$$x = \log_5 8.2$$
$$x = \frac{\log 8.2}{\log 5}$$
$$x \approx 1.31$$

19. $3^{2-x} = 8$
$$\log_3 3^{2-x} = \log_3 8$$
$$2 - x = \log_3 8$$
$$-x = \log_3 8 - 2$$
$$x = 2 - \frac{\log 8}{\log 3}$$
$$x \approx 0.11$$

21. $10^{x+6} = 250$
$$\log 10^{x+6} = \log 250$$
$$x + 6 = \log 250$$
$$x = \log 250 - 6$$
$$x \approx -3.60$$

23. $\frac{1}{4}e^x = 5$
$$e^x = 20$$
$$\ln e^x = \ln 20$$
$$x = \ln 20$$
$$x \approx 3.00$$

25. $4e^{-x} = 24$
$$e^{-x} = 6$$
$$\ln e^{-x} = \ln 6$$
$$-x = \ln 6$$
$$x = -\ln 6$$
$$x \approx -1.79$$

27. $7 + e^{2-x} = 28$
$$e^{2-x} = 21$$
$$\ln e^{2-x} = \ln 21$$
$$2 - x = \ln 21$$
$$-x = \ln 21 - 2$$
$$x = 2 - \ln 21$$
$$x \approx -1.04$$

29. $4 + e^{2x} = 10$
$$e^{2x} = 6$$
$$\ln e^{2x} = \ln 6$$
$$2x = \ln 6$$
$$x = \frac{\ln 6}{2}$$
$$x \approx 0.90$$

31. $17 - e^{x/4} = 14$
$$-e^{x/4} = -3$$
$$e^{x/4} = 3$$
$$\ln e^{x/4} = \ln 3$$
$$\frac{x}{4} = \ln 3$$
$$x = 4 \ln 3$$
$$x \approx 4.39$$

33. $8 - 12e^{-x} = 7$
$$-12e^{-x} = -1$$
$$e^{-x} = \frac{1}{12}$$
$$\ln e^{-x} = \ln \frac{1}{12}$$
$$-x = \ln \frac{1}{12}$$
$$x = -\ln \frac{1}{12} \approx 2.48$$

35. $4(1 + e^{x/3}) = 84$
$$1 + e^{x/3} = 21$$
$$e^{x/3} = 20$$
$$\ln e^{x/3} = \ln 20$$
$$\frac{x}{3} = \ln 20$$
$$x = 3 \ln 20$$
$$x \approx 8.99$$

37. $\log_{10} x = -1$
$$10^{\log_{10} x} = 10^{-1}$$
$$x = 10^{-1}$$
$$x = \frac{1}{10} = 0.1$$

39. $4 \log_3 x = 28$
$\log_3 x = 7$
$3^{\log_3 x} = 3^7$
$x = 3^7$
$x = 2187$

41. $\frac{1}{6} \log_3 x = \frac{1}{3}$
$\log_3 x = 2$
$x = 3^2$
$x = 9$

43. $\log_{10} 4x = 2$
$10^{\log_{10} 4x} = 10^2$
$4x = 10^2$
$x = \frac{10^2}{4}$
$x = \frac{100}{4} = 25$

45. $2 \log_4(x + 5) = 3$
$\log_4(x + 5) = \frac{3}{2}$
$4^{\log_4(x+5)} = 4^{1.5}$
$x + 5 = 4^{1.5}$
$x = 4^{1.5} - 5$
$x = 3$

47. $2 \log_{10} x = 10$
$\log_{10} x = 5$
$x = 10^5$
$x = 100{,}000$

49. $3 \ln 0.1x = 4$
$\ln 0.1x = \frac{4}{3}$
$0.1x = e^{4/3}$
$x = 10e^{4/3}$
$x \approx 37.94$

51. $\log_{10} x + \log_{10}(x - 3) = 1$
$\log_{10} x(x - 3) = 1$
$10^{\log_{10} x(x-3)} = 10^1$
$x(x - 3) = 10$
$x^2 - 3x - 10 = 0$
$(x - 5)(x + 2) = 0$
$x = 5, x = -2$ (which is extraneous)
$x = 5$

53. *Formula:* $A = Pe^{rt}$

Labels: Principal $= P = \$10{,}000$
Amount $= A = \$11{,}051.71$
Time $= t = 2$ years
Annual interest rate $= r$

Equation: $11{,}051.71 = 10{,}000e^{r(2)}$
$\frac{11{,}051.71}{10{,}000} = e^{2r}$
$1.105171 = e^{2r}$
$\ln 1.105171 = \ln e^{2r}$
$\ln 1.105171 = 2r$
$\frac{\ln 1.105171}{2} = r$
$0.05 \approx r \approx 5\%$

55. $F = 200e^{-0.5\pi\theta/180}$
$80 = 200e^{-0.5\pi\theta/180}$
$0.4 = e^{-0.5\pi\theta/180}$
$\ln 0.4 = \ln e^{-0.5\pi\theta/180}$
$\ln 0.4 = \frac{-0.5\pi\theta}{180}$
$(\ln 0.4)\left(-\frac{180}{0.5\pi}\right) = \theta$
$105° \approx \theta$

57. The one-to-one property can be applied to $3^{x-2} = 3^3$ because each side of the equation is in exponential form with the same base.

59. To apply an inverse property to the equation $\log_6 3x = 2$, exponentiate each side of the equation using the base 6, and then rewrite the left side of the equation as $3x$.

61. $5^x = \frac{1}{125}$
$5^x = 5^{-3}$
so $x = -3$.

63. $2^{x+2} = \frac{1}{16}$
$2^{x+2} = 2^{-4}$
so $x + 2 = -4$
$x = -6$.

65. $\log_6 3x = \log_6 18$
so $3x = 18$
$x = 6$.

67. $\log_4(x-8) = \log_4(-4)$

No solution

$\log_4(-4)$ does not exist.

69.
$$\begin{aligned}\frac{1}{5}4^{x+2} &= 300\\ 4^{x+2} &= 1500\\ \log_4 4^{x+2} &= \log_4 1500\\ x+2 &= \frac{\log 1500}{\log 4}\\ x &= \frac{\log 1500}{\log 4} - 2\\ x &\approx 3.28\end{aligned}$$

71.
$$\begin{aligned}6 + 2^{x-1} &= 1\\ 2^{x-1} &= -5\\ \log_2 2^{x-1} &= \log_2(-5)\end{aligned}$$

No solution

$\log_2(-5)$ is not possible.

73.
$$\begin{aligned}\log_5(x+3) - \log_5 x &= 1\\ \log_5\left(\frac{x+3}{x}\right) &= 1\\ 5^{\log_5[(x+3)/x]} &= 5^1\\ \frac{x+3}{x} &= 5\\ x+3 &= 5x\\ 3 &= 4x\\ \frac{3}{4} &= x\\ 0.75 &= x\end{aligned}$$

75.
$$\begin{aligned}\log_2(x-1) + \log_2(x+3) &= 3\\ \log_2(x-1)(x+3) &= 3\\ x^2 + 2x - 3 &= 2^3\\ x^2 + 2x - 11 &= 0\end{aligned}$$

$$x = \frac{-2 \pm \sqrt{4 - 4(1)(-11)}}{2(1)} = \frac{-2 \pm \sqrt{4+44}}{2} = \frac{-2 \pm \sqrt{48}}{2}$$

$x \approx 2.46$ and -4.46 (which is extraneous)

77.
$$\begin{aligned}\log_{10} 4x - \log_{10}(x-2) &= 1\\ \log_{10}\left(\frac{4x}{x-2}\right) &= 1\\ \frac{4x}{x-2} &= 10^1\\ 4x &= 10(x-2)\\ 4x &= 10x - 20\\ -6x &= -20\\ x &= \frac{20}{6}\\ x &= \frac{10}{3}\\ x &\approx 3.33\end{aligned}$$

79.
$$\begin{aligned}\log_2 x + \log_2(x+2) - \log_2 3 &= 4\\ \log_2 \frac{x(x+2)}{3} &= 4\\ 2^{\log_2[(x^2+2x)/3]} &= 2^4\\ \frac{x^2+2x}{3} &= 16\\ x^2 + 2x &= 48\\ x^2 + 2x - 48 &= 0\\ (x+8)(x-6) &= 0\end{aligned}$$

$x = -8$ (which is extraneous)

$x = 6.00$

81.
$$\begin{aligned}5000 &= 2500e^{0.09t}\\ \frac{5000}{2500} &= e^{0.09t}\\ 2 &= e^{0.09t}\\ \ln 2 &= \ln\left(e^{0.09t}\right)\\ \ln 2 &= 0.09t\\ \frac{\ln 2}{0.09} &= t\\ 7.70 \text{ years} &\approx t\end{aligned}$$

83.
$$B = 10\log_{10}\left(\frac{I}{10^{-16}}\right)$$
$$80 = 10\log_{10}\left(\frac{I}{10^{-16}}\right)$$
$$8 = \log_{10}\left(\frac{I}{10^{-16}}\right)$$
$$10^8 = 10^{\log_{10}\left(\frac{I}{10^{-16}}\right)}$$
$$10^8 = \frac{I}{10^{-16}}$$
$$10^8 \cdot 10^{-16} = I$$
$$10^{-8} = I$$

watts per square centimeter

85. (a)
$$kt = \ln\frac{T - S}{T_0 - S}$$
$$k(3) = \ln\frac{78 - 65}{85 - 65}$$
$$k = \frac{1}{3}\ln\frac{13}{20}$$
$$k \approx -0.144$$

(b)
$$-0.144t = \ln\frac{85 - 65}{98.6 - 65}$$
$$t = \frac{\ln\frac{20}{33.6}}{-0.144}$$
$$t \approx 3.6 \text{ hours}$$
$$\approx 3 \text{ hours } 36 \text{ minutes}$$

10 P.M. − 3 hr 36 min = 6:24 P.M.

(c)
$$-0.144(2) = \ln\frac{T - 65}{98.6 - 65}$$
$$-0.288 = \ln\frac{T - 65}{33.6}$$
$$e^{-0.288} = \frac{T - 65}{33.6}$$
$$33.6e^{-0.288} + 65 = T$$
$$90.2°\text{F} \approx T$$

87. The equation $2^{x-1} = 32$ can be solved without logarithms.

$$2^{x-1} = 32 \qquad 2^{x-1} = 30$$
$$2^{x-1} = 2^5 \qquad x - 1 = \log_2 30$$
$$x - 1 = 5 \qquad x = 1 + \log_2 30$$
$$x = 6$$

89. To solve an exponential equation, first isolate the exponential expression, then take the logarithms of both sides of the equation, and solve for the variable.

To solve a logarithmic equation, first isolate the logarithmic expression, then exponentiate both sides of the equation, and solve for the variable.

91.
$$x^2 = -25$$
$$x = \pm\sqrt{-25}$$
$$x = \pm 5i$$

93.
$$9n^2 - 16 = 0$$
$$n^2 = \tfrac{16}{9}$$
$$n = \pm\sqrt{\tfrac{16}{9}}$$
$$n = \pm\tfrac{4}{3}$$

95.
$$t^4 - 13t^2 + 36 = 0$$
$$(t^2 - 9)(t^2 - 4) = 0$$
$$t^2 = 9 \qquad t^2 = 4$$
$$t = \pm 3 \qquad t = \pm 2$$

97. *Verbal Model:* $\boxed{\text{Area}} = \boxed{\text{Length}} \cdot \boxed{\text{Width}}$

Labels: Length $= x$

Width $= 2.5x$

Equation:
$$2x + 2(2.5x) = 42$$
$$7x = 42$$
$$x = 6$$
$$A = 6 \cdot 2.5(6)$$
$$A = 90 \text{ in.}^2$$

99. *Verbal Model:* $\boxed{\text{Perimeter}} = 2 \cdot \boxed{\text{Length}} + 2 \cdot \boxed{\text{Width}}$

Labels: Length $= w + 4$

Width $= w$

Equation:
$$w(w + 4) = 192$$
$$w^2 + 4w - 192 = 0$$
$$(w - 12)(w + 16) = 0$$
$$w = 12, \quad w = -16$$
$$P = 2(12 + 4) + 2(12)$$
$$= 32 + 24$$
$$= 56 \text{ km}$$

Section 9.6 Applications

1.
$$A = P\left(1 + \frac{r}{n}\right)^{nt}$$
$$1004.83 = 500\left(1 + \frac{r}{12}\right)^{12(10)}$$
$$2.00966 = \left(1 + \frac{r}{12}\right)^{120}$$
$$(2.00966)^{1/120} = 1 + \frac{r}{12}$$
$$1.0058333 \approx 1 + \frac{r}{12}$$
$$0.0058333 \approx \frac{r}{12}$$
$$0.07 \approx r$$
$$7\% \approx r$$

3.
$$A = Pe^{rt}$$
$$2P = Pe^{0.08t}$$
$$2 = e^{0.08t}$$
$$\ln 2 = 0.08t$$
$$\frac{\ln 2}{0.08} = t$$
$$t \approx 8.66 \text{ years}$$

5.
$$A = Pe^{rt}$$
$$2P = Pe^{0.0675t}$$
$$2 = e^{0.0675t}$$
$$\ln 2 = 0.0675t$$
$$\frac{\ln 2}{0.0675} = t$$
$$t \approx 10.27 \text{ years}$$

7. $8954.24 = 5000\left(1 + \frac{0.06}{n}\right)^{n(10)}$

$$8954.24 = 5000\left(1 + \frac{0.06}{1}\right)^{1(10)}$$
$$8954.24 = 8954.24$$

Yearly compounding

9. $A = Pe^{rt}$

$$A = 1000e^{0.08(1)}$$
$$A = \$1083.29$$
$$\text{Effective yield} = \frac{83.29}{1000} = 0.08329 \approx 8.33\%$$

11. $A = P\left(1 + \frac{r}{n}\right)^{nt}$

$$A = 1000\left(1 + \frac{0.07}{12}\right)^{12(1)}$$
$$A = \$1072.29$$
$$\text{Effective yield} = \frac{72.29}{1000} = 0.07229 \approx 7.23\%$$

13. $A = P\left(1 + \frac{r}{n}\right)^{nt}$

$$A = 1000\left(1 + \frac{0.06}{4}\right)^{4(1)}$$
$$A = \$1061.36$$
$$\text{Effective yield} = \frac{61.36}{1000} = 0.06136 \approx 6.14\%$$

15. $A = 1000\left(1 + \frac{0.0525}{365}\right)^{365(1)}$

$$A = \$1053.90$$
$$\text{Effective yield} = \frac{53.90}{1000} = 0.0539 = 5.39\%$$

17. No. The effective yield is the ratio of the year's interest to the amount invested. This *ratio* will remain the same regardless of the amount invested.

19. (a)
$$y = Ce^{kt}$$
$$1 \times 10^6 = 73e^{k(20)}$$
$$\frac{1 \times 10^6}{73} = e^{20k}$$
$$\ln \frac{1 \times 10^6}{73} = 20k$$
$$\frac{1}{20} \ln \frac{1 \times 10^6}{73} = k$$
$$0.4763 \approx k$$
$$y = 73e^{0.4763t}$$

(b)
$$5300 = 73e^{0.4763t}$$
$$\frac{5300}{73} = e^{0.4763t}$$
$$\ln \frac{5300}{73} = 0.4763t$$
$$\frac{\ln \frac{5300}{73}}{0.4763} = t$$
$$9 \text{ hours} \approx t$$

21. $y = Ce^{kt}$

$5 = Ce^{k(0)}$

$5 = C$

$2.5 = 5e^{k(1620)}$

$0.5 = e^{1620k}$

$\ln 0.5 = \ln e^{1620k}$

$\ln 0.5 = 1620k$

$\frac{\ln 0.5}{1620} = k$

$-0.00043 \approx k$

$y = 5e^{-0.00043(1000)}$

$y \approx 3.3 \text{ grams}$

23. May 22, 1960: $R = \log_{10} I$

$9.5 = \log_{10} I$

$I = 10^{9.5}$

February 11, 2011: $R = \log_{10} I$

$6.8 = \log_{10} I$

$I = 10^{6.8}$

Ratio: $\frac{\text{May 22}}{\text{February 11}} = \frac{10^{9.5}}{10^{6.8}} = 10^{2.7} \approx 501$

The earthquake of 1960 was about 501 times as intense.

25. June 13, 2011: $R = \log_{10} I$

$6.0 = \log_{10} I$

$I = 10^{6.0}$

July 6, 2011: $R = \log_{10} I$

$7.6 = \log_{10} I$

$I = 10^{7.6}$

Ratio: $\frac{\text{July 6}}{\text{June 13}} = \frac{10^{7.6}}{10^{6.0}} = 10^{1.6} \approx 40$

The earthquake on July 6 was about 40 times as intense.

27. The formula $A = Pe^{rt}$ is for continuously compounded interest.

29. Substituting 0.06 for r is replacing the variable r, representing the interest rate, with the interest rate in decimal form.

31.

$$A = P\left(1 + \frac{r}{n}\right)^{nt}$$

$$5000 = 2500\left(1 + \frac{0.075}{12}\right)^{12t}$$

$$2 = (1.00625)^{12t}$$

$$\log_{1.00625} 2 = \log_{1.00625}(1.00625)^{12t}$$

$$\frac{\log 2}{\log 1.00625} = 12t$$

$$\frac{\log 2}{12 \log 1.00625} = t$$

$$t \approx 9.27 \text{ years}$$

33.

$$A = P\left(1 + \frac{r}{n}\right)^{nt}$$

$$1800 = 900\left(1 + \frac{0.0575}{4}\right)^{4t}$$

$$2 = (1.014375)^{4t}$$

$$\log_{1.014375} 2 = \log_{1.014375}(1.014375)^{4t}$$

$$\frac{\log 2}{\log 1.014375} = 4t$$

$$\frac{\log 2}{4 \log 1.014375} = t$$

$$t \approx 12.14 \text{ years}$$

35. $y = Ce^{kt}$

$3 = Ce^{k(0)}$

$3 = C$

$8 = 3e^{k(2)}$

$\frac{8}{3} = e^{2k}$

$\ln \frac{8}{3} = \ln e^{2k}$

$\ln \frac{8}{3} = 2k$

$\frac{\ln \frac{8}{3}}{2} = k \approx 0.4904$

37. $y = Ce^{kt}$

$400 = Ce^{k(0)}$

$400 = C$

$200 = 400e^{k(3)}$

$\frac{1}{2} = e^{3k}$

$\ln \frac{1}{2} = \ln e^{3k}$

$\ln \frac{1}{2} = 3k$

$\frac{\ln \frac{1}{2}}{3} = k \approx -0.2310$

39. $\text{pH} = -\log_{10}\left[H^+\right]$

$\text{pH} = -\log_{10}\left(9.2 \times 10^{-8}\right) \approx 7.04$

41. $\text{pH} = -\log_{10}[\text{H}^+]$

Fruit:

$$2.5 = -\log_{10}[\text{H}^+]$$
$$-2.5 = -\log_{10}[\text{H}^+]$$
$$10^{-2.5} = 10^{\log_{10}[\text{H}^+]}$$
$$0.0031623 \approx \text{H}^+$$

Tablet:

$$9.5 = -\log_{10}[\text{H}^+]$$
$$-9.5 = -\log_{10}[\text{H}^+]$$
$$10^{-9.5} = 10^{\log_{10}[\text{H}^+]}$$
$$3.1623 \times 10^{-10} \approx \text{H}^+$$

$$\frac{\text{H}^+ \text{ of fruit}}{\text{H}^+ \text{ of tablet}} \approx \frac{3.1623 \times 10^{-3}}{3.1623 \times 10^{-10}} = 10^7$$

The H^+ of fruit is 10^7 times as great.

43. $P = \dfrac{10.9}{1 + 0.80e^{-0.031t}}$

When $t = 22$: $P = \dfrac{10.9}{1 + 0.80e^{-0.031(22)}}$

≈ 7.761 billion people

45. (a) $p(0) = \dfrac{5000}{1 + 4e^{-0/6}} = \dfrac{5000}{5} = 1000$ rabbits

(b) $p(9) = \dfrac{5000}{1 + 4e^{-9/6}} \approx 2642$ rabbits

(c)
$$2000 = \frac{5000}{1 + 4e^{-t/6}}$$
$$1 + 4e^{-t/6} = 2.5$$
$$e^{-t/6} = 0.375$$
$$-\frac{t}{6} = \ln 0.375$$
$$t = -6 \ln 0.375$$
$$t \approx 5.88 \text{ years}$$

47. If $k > 0$, the model represents exponential growth and if $k < 0$, the model represents exponential decay.

49. When the investment is compounded more than once in a year (quarterly, monthly, daily, continuous), the effective yield is greater than the interest rate.

51. $x^2 - 7x - 5 = 0$

$$x = \frac{-(-7) \pm \sqrt{(-7)^2 - 4(1)(-5)}}{2(1)}$$
$$x = \frac{7 \pm \sqrt{49 + 20}}{2}$$
$$x = \frac{7 \pm \sqrt{69}}{2}$$
$$x = \frac{7}{2} \pm \frac{\sqrt{69}}{2}$$

53. $3x^2 + 9x + 4 = 0$

$$x = \frac{-9 \pm \sqrt{9^2 - 4(3)(4)}}{2(3)}$$
$$x = \frac{-9 \pm \sqrt{81 - 48}}{6}$$
$$x = \frac{-9 \pm \sqrt{33}}{6}$$
$$x = -\frac{3}{2} \pm \frac{\sqrt{33}}{6}$$

55. $\dfrac{4}{x - 4} > 0$

Critical number: 4

Test intervals:

Negative: $(-\infty, 4)$

Positive: $(4, \infty)$

Solution: $(4, \infty)$

57.
$$\frac{2x}{x - 3} > 1$$
$$\frac{2x}{x - 3} - 1 > 0$$
$$\frac{2x - (x - 3)}{x - 3} > 0$$
$$\frac{x + 3}{x - 3} > 0$$

Critical numbers: $-3, 3$

Test intervals:

Positive: $(-\infty, -3)$

Negative: $(-3, 3)$

Positive: $(3, \infty)$

Solution: $(-\infty, -3) \cup (3, \infty)$

Review Exercises for Chapter 9

1. $f(x) = 4^x$

(a) $f(-3) = 4^{-3} = \frac{1}{4^3} = \frac{1}{64}$

(b) $f(1) = 4^1 = 4$

(c) $f(2) = 4^2 = 16$

3. $g(t) = 5^{-t/3}$

(a) $g(-3) = 5^{-(-3)/3} = 5^1 = 5$

(b) $g(\pi) = 5^{-\pi/3} \approx 0.185$

(c) $g(6) = 5^{-6/3} = 5^{-2} = \frac{1}{25}$

5. $f(x) = 3^x$

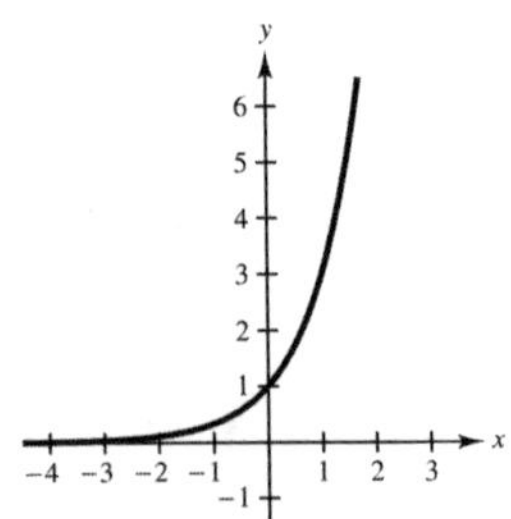

Table of values:

x	-1	0	1
y	$\frac{1}{3}$	1	3

7. $f(x) = 3^x - 3$

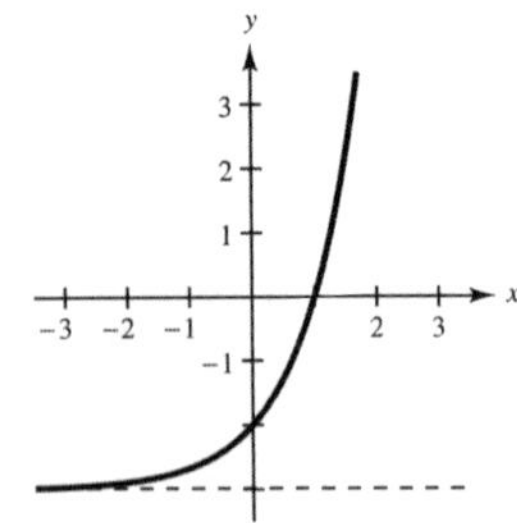

9. $f(x) = 3^{(x+1)}$

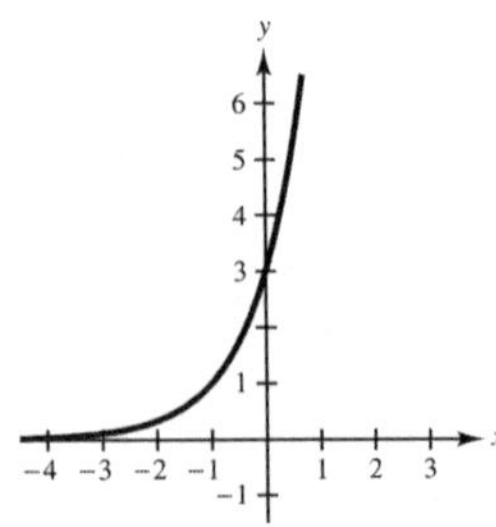

Table of values:

x	-1	0	1
y	1	3	9

11. $f(x) = 3e^{-2x}$

(a) $f(3) = 3e^{-2(3)} = 3e^{-6} \approx 0.007$

(b) $f(0) = 3e^{-2(0)} = 3e^0 = 3$

(c) $f(-19) = 3e^{-2(-19)} = 3e^{38} \approx 9.56 \times 10^{16}$

13. The function y is related to $f(x) = e^x$ by $y = f(-x) + 1$. To sketch y, reflect the graph of f in the y-axis, and shift the graph one unit upward.

y-intercept: $(0, 2)$

Asymptote: $y = 1$

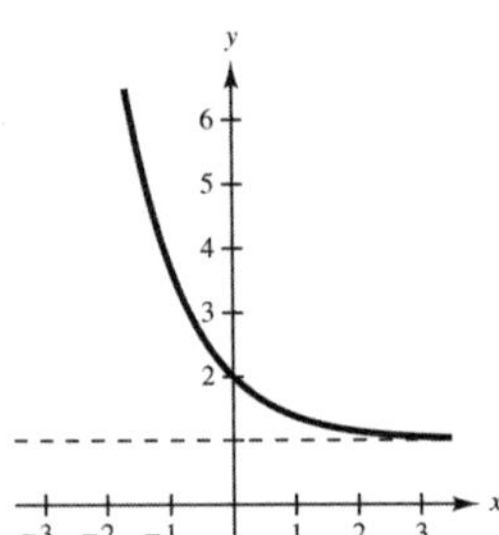

15. The function g is related to $f(x) = e^x$ by $y = f(x + 2)$. To sketch g, shift the graph of f two units to the left.

y-intercept: $(0, e^2)$

Asymptote: $y = 0$

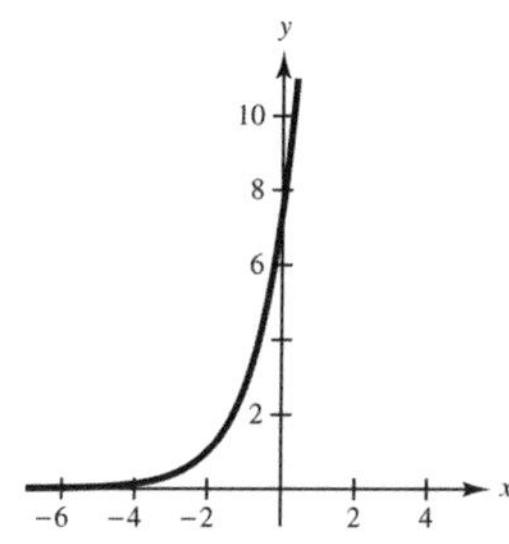

17. (a) $A = P\left(1 + \frac{r}{n}\right)^{nt}$

$= 5000\left(1 + \frac{0.10}{1}\right)^{1(40)}$

$= 5000(1.10)^{40}$

$\approx \$226{,}296.28$

(b) $A = P\left(1 + \frac{r}{n}\right)^{nt}$

$= 5000\left(1 + \frac{0.10}{4}\right)^{4(40)}$

$= 5000(1.025)^{160}$

$\approx \$259{,}889.34$

(c) $A = P\left(1 + \frac{r}{n}\right)^{nt}$

$= 5000\left(1 + \frac{0.10}{12}\right)^{12(40)}$

$= 5000(1.008\overline{3})^{480}$

$\approx \$268{,}503.32$

(d) $A = P\left(1 + \frac{r}{n}\right)^{nt}$

$= 5000\left(1 + \frac{0.10}{365}\right)^{365(40)}$

$\approx 5000(1.00027397)^{14600}$

$\approx \$272{,}841.23$

(e) $A = Pe^{rt}$

$= 5000e^{0.10(40)}$

$= 5000e^{4}$

$\approx \$272{,}990.75$

19. $y = 21\left(\frac{1}{2}\right)^{t/25}, \quad t \geq 0$

$y = 21\left(\frac{1}{2}\right)^{58/25}$

$y \approx 4.21$ grams

21. (a) $(f \circ g)(x) = x^2 + 2$

so $(f \circ g)(2) = 2^2 + 2 = 6.$

(b) $(g \circ f)(x) = (x + 2)^2 = x^2 + 4x + 4$

so $(g \circ f)(-1) = (-1)^2 + 4(-1) + 4$

$= 1 - 4 + 4 = 1.$

23. (a) $(f \circ g)(x) = \sqrt{x^2 - 1 + 1} = \sqrt{x^2} = |x|$

so $(f \circ g)(5) = |5| = 5.$

(b) $(g \circ f)(x) = \left(\sqrt{x + 1}\right)^2 - 1 = x + 1 - 1 = x$

so $(g \circ f)(-1) = -1.$

25. $f(x) = \sqrt{x + 6}, \quad g(x) = 2x$

(a) $(f \circ g)(x) = f(2x) = \sqrt{2x + 6}$

Domain: $[-3, \infty)$

$2x + 6 \geq 0$

$x \geq -3$

(b) $(g \circ f)(x) = g\left(\sqrt{x + 6}\right) = 2\sqrt{x + 6}$

Domain: $[-6, \infty)$

$x + 6 \geq 0$

$x \geq -6$

27. No, $f(x)$ does not have an inverse. f is not one-to-one.

29. $f(x) = 3x + 4$

$y = 3x + 4$

$x = 3y + 4$

$x - 4 = 3y$

$\frac{x - 4}{3} = y$

$\frac{x - 4}{3} = f^{-1}(x) = \frac{1}{3}(x - 4)$

31. $h(x) = \sqrt{5x}$

$y = \sqrt{5x}$

$x = \sqrt{5y}$

$x^2 = 5y$

$\frac{1}{5}x^2 = y$

$\frac{1}{5}x^2 = h^{-1}(x), \quad x \geq 0$

33. $f(t) = t^3 + 4$

$y = t^3 + 4$

$t = y^3 + 4$

$t - 4 = y^3$

$\sqrt[3]{t - 4} = y$

$\sqrt[3]{t - 4} = f^{-1}(t)$

35.

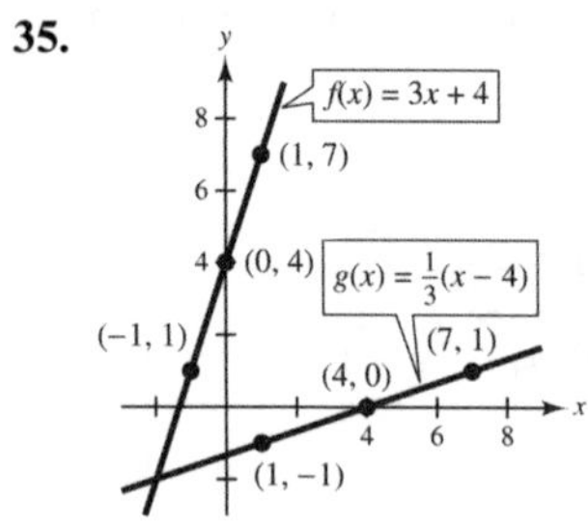

37. $\log_{10} 1000 = 3$ because $10^3 = 1000$.

39. $\log_3 \frac{1}{9} = -2$ because $3^{-2} = \frac{1}{9}$.

41. $\log_2 64 = 6$ because $2^6 = 64$.

43. $\log_3 1 = 0$ because $3^0 = 1$.

45. $f(x) = \log_3 x$

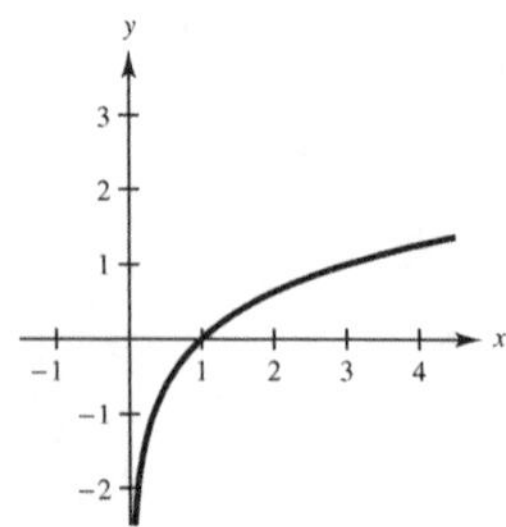

Vertical asymptote: $x = 0$

Table of values:

x	1	3
y	0	1

47. $f(x) = -1 + \log_3 x$

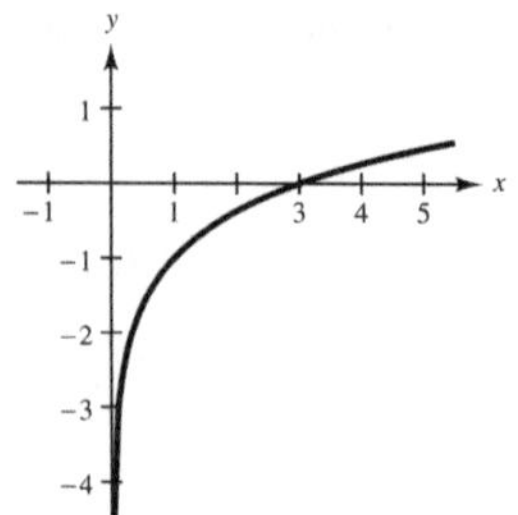

Vertical asymptote: $x = 0$

49. $f(x) = \log_2(x - 4)$

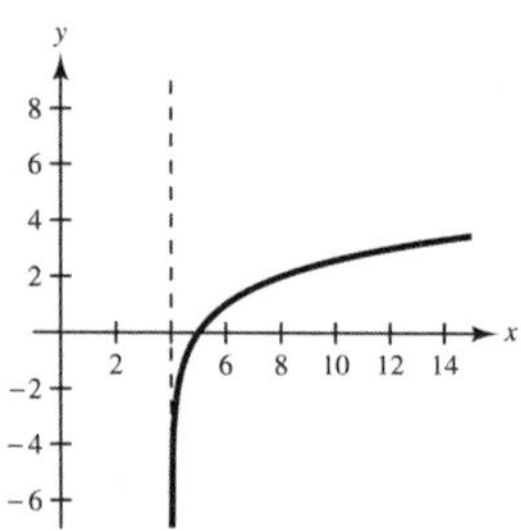

Vertical asymptote: $x = 4$

Table of values:

x	5	6
y	0	1

51. $\ln e^7 = 7$

53. $y = \ln(x - 3)$

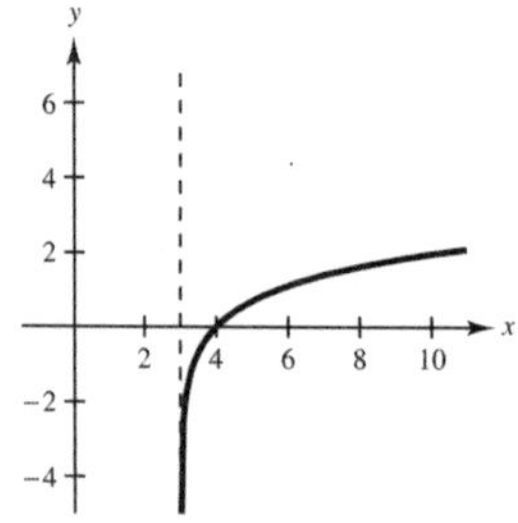

Table of values:

x	4	5
y	0	0.7

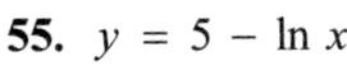

55. $y = 5 - \ln x$

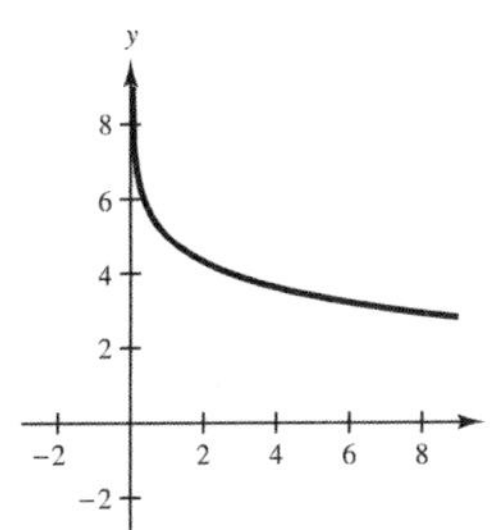

Table of values:

x	1	e
y	5	4

57. (a) $\log_4 9 = \dfrac{\log 9}{\log 4} \approx 1.5850$

(b) $\log_4 9 = \dfrac{\ln 9}{\ln 4} \approx 1.5850$

59. (a) $\log_8 160 = \dfrac{\log 160}{\log 8} \approx 2.4409$

(b) $\log_8 160 = \dfrac{\ln 160}{\ln 8} \approx 2.4409$

61. $\log_5 18 = \log_5 3^2 + \log_5 2$

$= 2\log_5 3 + \log 2$

$\approx 2(0.6826) + 0.4307$

≈ 1.7959

63. $\log_5 \frac{1}{2} = \log_5 1 - \log_5 2$

$\approx 0 - (0.4307)$

≈ -0.4307

65. $\log_5 (12)^{2/3} = \frac{2}{3}[2\log_5 2 + \log_5 3]$

$\approx \frac{2}{3}[2(0.4307) + 0.6826]$

≈ 1.0293

67. $\log_4 6x^4 = \log_4 6 + 4\log_4 x$

69. $\log_5 \sqrt{x+2} = \frac{1}{2}\log_5(x+2)$

71. $\ln \dfrac{x+2}{x+3} = \ln(x+2) - \ln(x+3)$

73. $5\log_2 y = \log_2 y^5$

75. $\log_8 16x + \log_8 2x^2 = \log_8(16x \cdot 2x^2) = \log_8(32x^3)$

77. $-2(\ln 2x - \ln 3) = \ln\left(\dfrac{2x}{3}\right)^{-2}$

$= \ln\left(\dfrac{3}{2x}\right)^2$

$= \ln \dfrac{9}{4x^2},\ x > 0$

79. $4[\log_2 k - \log_2(k-t)] = 4\left[\log_2\left(\dfrac{k}{k-t}\right)\right]$

$= \log_2\left(\dfrac{k}{k-t}\right)^4,\quad t < k, k > 0$

81. False

$\log_2 4x = \log_2 4 + \log_2 x = 2 + \log_2 x$

83. True

$\log_{10} 10^{2x} = 2x\log_{10} 10 = 2x$

85. True

$\log_4 \dfrac{16}{x} = \log_4 16 - \log_4 x = 2 - \log_4 x$

87. $B = 10\log_{10}\left(\dfrac{I}{10^{-12}}\right)$

$= 10\log_{10}\left(\dfrac{10^{-4}}{10^{-12}}\right)$

$= 10\log_{10}(10^8)$

$= 10 \cdot 8$

$= 80$ decibels

89. $2^x = 64$

$2^x = 2^6$

$x = 6$

91. $4^{x-3} = \frac{1}{16}$

$4^{x-3} = 4^{-2}$

$x - 3 = -2$

$x = 1$

93. $\log_7(x+6) = \log_7 12$

$x + 6 = 12$

$x = 6$

95. $3^x = 500$

$\log_3 3^x = \log_3 500$

$x = \dfrac{\log 500}{\log 3}$

$x \approx 5.66$

97.
$$\begin{aligned} 2e^{0.5x} &= 45 \\ e^{0.5x} &= 22.5 \\ \ln e^{0.5x} &= \ln 22.5 \\ 0.5x &= \ln 22.5 \\ x &= 2 \ln 22.5 \\ x &\approx 6.23 \end{aligned}$$

99.
$$\begin{aligned} 12\left(1 - 4^x\right) &= 18 \\ 1 - 4^x &= \tfrac{18}{12} \\ -4^x &= \tfrac{3}{2} - 1 \\ -4^x &= \tfrac{1}{2} \\ 4^x &= -\tfrac{1}{2} \end{aligned}$$

No solution; there is no power that will raise 4 to $-\frac{1}{2}$.

101.
$$\begin{aligned} \ln x &= 7.25 \\ e^{\ln x} &= e^{7.25} \\ x &= e^{7.25} \\ x &\approx 1408.10 \end{aligned}$$

103.
$$\begin{aligned} \log_{10} 4x &= 2.1 \\ 4x &= 10^{2.1} \\ x &= \frac{10^{2.1}}{4} \\ x &\approx 31.47 \end{aligned}$$

105.
$$\begin{aligned} \log_3(2x + 1) &= 2 \\ 3^{\log_3(2x+1)} &= 3^2 \\ 2x + 1 &= 9 \\ 2x &= 8 \\ x &= 4.00 \end{aligned}$$

107.
$$\begin{aligned} \tfrac{1}{3}\log_2 x + 5 &= 7 \\ \tfrac{1}{3}\log_2 x &= 2 \\ \log_2 x &= 6 \\ 2^{\log_2 x} &= 2^6 \\ x &= 2^6 = 64.00 \end{aligned}$$

109.
$$\begin{aligned} \log_3 x + \log_3 7 &= 4 \\ \log_3 7x &= 4 \\ 7x &= 3^4 \\ x &= \frac{3^4}{7} \\ x &\approx 11.57 \end{aligned}$$

111.
$$\begin{aligned} A &= Pe^{rt} \\ 5751.37 &= 5000e^{r(2)} \\ \frac{5751.37}{5000} &= e^{2r} \\ \ln 1.150274 &= \ln e^{2r} \\ \ln 1.150274 &= 2r \\ \frac{\ln 1.150274}{2} &= r \approx 7\% \end{aligned}$$

113.
$$\begin{aligned} A &= P\left(1 + \frac{r}{n}\right)^{nt} \\ 410.90 &= 250\left(1 + \frac{r}{4}\right)^{4(10)} \\ 1.6436 &= \left(1 + \frac{r}{4}\right)^{40} \\ (1.6436)^{1/40} &= 1 + \frac{r}{4} \\ 1.0124997 &\approx 1 + \frac{r}{4} \\ 0.0124997 &\approx \frac{r}{4} \\ 0.0499 &\approx r \\ 5\% &\approx r \end{aligned}$$

115.
$$\begin{aligned} A &= P\left(1 + \frac{r}{n}\right)^{nt} \\ 15{,}399.30 &= 5000\left(1 + \frac{r}{365}\right)^{365(15)} \\ 3.07986 &= \left(1 + \frac{r}{365}\right)^{5475} \\ (3.07986)^{1/5475} &= 1 + \frac{r}{365} \\ 1.000205479 &\approx 1 + \frac{r}{365} \\ 0.000205479 &\approx \frac{r}{365} \\ 0.074999 &\approx r \\ 7.5\% &\approx r \end{aligned}$$

117.
$$\begin{aligned} A &= Pe^{rt} \\ 46{,}422.61 &= 1800e^{r(50)} \\ \frac{46{,}422.61}{1800} &= e^{50r} \\ 25.79033\overline{8} &= e^{50r} \\ \ln 25.79033\overline{8} &= 50r \\ \frac{\ln 25.79033\overline{8}}{50} &= r \\ 0.065 &\approx r \\ 6.5\% &\approx r \end{aligned}$$

119. $A = P\left(1 + \frac{r}{n}\right)^{nt}$

$$A = 1000\left(1 + \frac{0.055}{365}\right)^{365(1)}$$

$$A = \$1056.54$$

$$\text{Effective yield} = \frac{56.54}{1000} = 0.05654 \approx 5.65\%$$

121. $A = P\left(1 + \frac{r}{n}\right)^{nt}$

$$A = 1000\left(1 + \frac{0.075}{4}\right)^{4(1)}$$

$$A = \$1077.14$$

$$\text{Effective yield} = \frac{77.14}{1000} = 0.07714 \approx 7.71\%$$

123. $A = Pe^{rt}$

$$A = 1000e^{0.075(1)}$$

$$A = \$1077.88$$

$$\text{Effective yield} = \frac{77.88}{1000} = 0.07788 \approx 7.79\%$$

125.
$$\begin{aligned} y &= Ce^{kt} \\ 3.5 &= Ce^{k(0)} \\ 3.5 &= C \end{aligned}$$

$$\begin{aligned} 1.75 &= 3.5e^{k(1620)} \\ 0.5 &= e^{1620k} \\ \ln 0.5 &= \ln e^{1620k} \\ \ln 0.5 &= 1620k \\ \frac{\ln 0.5}{1620} &= k \\ -0.00043 &\approx k \end{aligned}$$

$$\begin{aligned} y &= 3.5e^{-0.00043(1000)} \\ y &\approx 2.282 \text{ grams} \end{aligned}$$

127. April 18, 1906:
$$\begin{aligned} R &= \log_{10} I \\ 8.3 &= \log_{10} I \\ I &= 10^{8.3} \end{aligned}$$

March 5, 2012:
$$\begin{aligned} R &= \log_{10} I \\ 4.0 &= \log_{10} I \\ I &= 10^{4.0} \end{aligned}$$

$$\text{Ratio: } \frac{\text{April 18}}{\text{March 5}} = \frac{10^{8.3}}{10^{4.0}} = 10^{4.3} \approx 19{,}953$$

The earthquake of 1906 was about 19,953 times as intense.

Chapter Test for Chapter 9

1. (a) $f(-1) = 54\left(\frac{2}{3}\right)^{-1}$
$= 54\left(\frac{3}{2}\right)$
$= 81$

(b) $f(0) = 54\left(\frac{2}{3}\right)^{0}$
$= 54$

(c) $f\left(\frac{1}{2}\right) = 54\left(\frac{2}{3}\right)^{1/2}$
≈ 44.09

(d) $f(2) = 54\left(\frac{2}{3}\right)^{2}$
$= 54\left(\frac{4}{9}\right)$
$= 24$

2. $f(x) = 2^{x-5}$

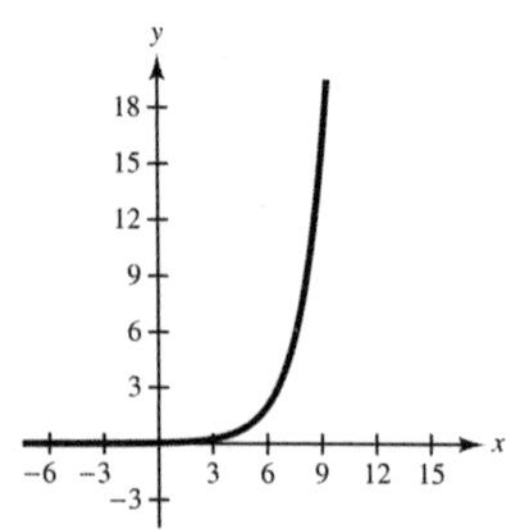

Horizontal asymptote: $y = 0$

3. (a) $f \circ g = f(g(x)) = f(5 - 3x)$
$= 2(5 - 3x)^2 + (5 - 3x)$
$= 2(25 - 30x + 9x^2) + 5 - 3x$
$= 50 - 60x + 18x^2 + 5 - 3x$
$= 18x^2 - 63x + 55$

Domain: $(-\infty, \infty)$

(b) $g \circ f = g(f(x)) = g(2x^2 + x)$
$= 5 - 3(2x^2 + x)$
$= 5 - 6x^2 - 3x$
$= -6x^2 - 3x + 5$

Domain: $(-\infty, \infty)$

4.
$$f(x) = 9x - 4$$
$$y = 9x - 4$$
$$x = 9y - 4$$
$$x + 4 = 9y$$
$$\frac{x + 4}{9} = y$$
$$\frac{1}{9}(x + 4) = f^{-1}(x)$$

5. $f(g(x)) = f(-2x + 6)$
$= -\frac{1}{2}(-2x + 6) + 3$
$= x - 3 + 3$
$= x$

$g(f(x)) = g\left(-\frac{1}{2}x + 3\right)$
$= -2\left(-\frac{1}{2}x + 3\right) + 6$
$= x - 6 + 6$
$= x$

6. $\log_4 \frac{1}{256} = -4$ because $4^{-4} = \frac{1}{256}$.

7. The graph of g is a reflection in the line $y = x$ of the graph of f.

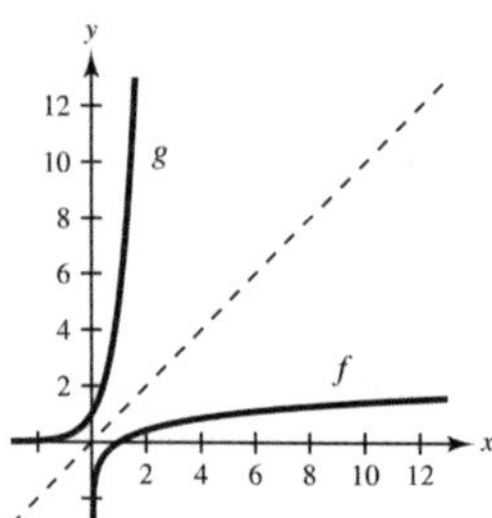

8. $\log_8\left(\frac{4\sqrt{x}}{y^4}\right) = \log_8 4\sqrt{x} - \log_8 y^4$
$= \log_8 4 + \log_8 x^{1/2} - \log_8 y^4$
$= \frac{2}{3} + \frac{1}{2}\log_8 x - 4\log_8 y$

9. $\ln x - 4\ln y = \ln \frac{x}{y^4},\ y > 0$

10. $\log_2 x = 5$
$2^{\log_2 x} = 2^5$
$x = 32$

11. $$\begin{aligned} 9^{2x} &= 182 \\ \log_9 9^{2x} &= \log_9 182 \\ 2x &= \frac{\log 182}{\log 9} \\ x &= \frac{\log 182}{2 \log 9} \\ x &\approx 1.18 \end{aligned}$$

12. $$\begin{aligned} 400e^{0.08t} &= 1200 \\ e^{0.08t} &= 3 \\ \ln e^{0.08t} &= \ln 3 \\ 0.08t &= \ln 3 \\ t &= \frac{\ln 3}{0.08} \\ t &\approx 13.73 \end{aligned}$$

13. $$\begin{aligned} 3 \ln(2x - 3) &= 10 \\ \ln(2x - 3) &= \frac{10}{3} \\ e^{\ln(2x-3)} &= e^{10/3} \\ 2x - 3 &= e^{10/3} \\ x &= \frac{e^{10/3} + 3}{2} \\ x &\approx 15.52 \end{aligned}$$

14. $$\begin{aligned} 12(7 - 2^x) &= -300 \\ 7 - 2^x &= -\tfrac{300}{12} \\ 7 - 2^x &= -25 \\ -2^x &= -32 \\ 2^x &= 32 \\ 2^x &= 2^5 \\ x &= 5 \end{aligned}$$

15. $$\begin{aligned} \log_2 x + \log_2 4 &= 5 \\ \log_2 x(4) &= 5 \\ 2^{\log_2 4x} &= 2^5 \\ 4x &= 32 \\ x &= 8 \end{aligned}$$

16. $$\begin{aligned} \ln x - \ln 2 &= 4 \\ \ln \frac{x}{2} &= 4 \\ e^{\ln(x/2)} &= e^4 \\ \frac{x}{2} &= e^4 \\ x &= 2e^4 \\ x &\approx 109.20 \end{aligned}$$

17. $$\begin{aligned} 30(e^x + 9) &= 300 \\ e^x + 9 &= 10 \\ e^x &= 1 \\ e^x &= e^0 \\ \text{so } x &= 0. \end{aligned}$$

18. (a) $A = 2000\left(1 + \dfrac{0.07}{4}\right)^{4(20)} \approx \$8012.78$

(b) $A = 2000e^{0.07(20)} \approx \$8110.40$

19. $$\begin{aligned} 100{,}000 &= P\left(1 + \frac{0.09}{4}\right)^{4(25)} \\ \frac{100{,}000}{(1.0225)^{100}} &= P \\ \$10{,}806.08 &\approx P \end{aligned}$$

20. $$\begin{aligned} 1006.88 &= 500e^{r(10)} \\ 2.01376 &= e^{10r} \\ \ln 2.01376 &= \ln e^{10r} \\ \ln 2.01376 &= 10r \\ \frac{\ln 2.01376}{10} &= r \\ 0.07 &\approx r \\ 7\% &\approx r \end{aligned}$$

21. $$\begin{aligned} y &= Ce^{kt} \\ 15{,}000 &= 20{,}000e^{k(1)} \\ \tfrac{3}{4} &= e^k \\ \ln \tfrac{3}{4} &= k \end{aligned}$$

$$\begin{aligned} y &= 20{,}000e^{[\ln(3/4)](5)} \\ y &\approx \$4746.09 \end{aligned}$$

22. $p(0) = \dfrac{2400}{1 + 3e^{-0/4}} = 600$ foxes

23. $p(4) = \dfrac{2400}{1 + 3e^{-4/4}} \approx 1141$ foxes

24. $$\begin{aligned} 1200 &= \frac{2400}{1 + 3e^{-t/4}} \\ 1 + 3e^{-t/4} &= \frac{2400}{1200} \\ 3e^{-t/4} &= 1 \\ e^{-t/4} &= \frac{1}{3} \\ \ln e^{-t/4} &= \ln \frac{1}{3} \\ -\frac{t}{4} &= \ln \frac{1}{3} \\ t &= -4 \ln \frac{1}{3} \approx 4.4 \text{ years} \end{aligned}$$

CHAPTER 10
Conics

Section 10.1 Circles and Parabolas 319

Section 10.2 Ellipses 322

Mid-Chapter Quiz 325

Section 10.3 Hyperbolas 327

Section 10.4 Solving Nonlinear Systems of Equations 331

Review Exercises 336

Chapter Test 342

Cumulative Test for Chapters 8–10 345

CHAPTER 10
Conics

Section 10.1 Circles and Parabolas

1. The graph is a parabola.

3. The graph is a circle.

5. Center: $(0, 0)$; radius: 5

$$x^2 + y^2 = r^2$$
$$x^2 + y^2 = 5^2$$
$$x^2 + y^2 = 25$$

7. Center: $(0, 0)$; radius: $\frac{2}{3}$

$$x^2 + y^2 = r^2$$
$$x^2 + y^2 = \left(\frac{2}{3}\right)^2$$
$$x^2 + y^2 = \frac{4}{9} \text{ or } 9x^2 + 9y^2 = 4$$

9. $x^2 + y^2 = 16$

Center: $(0, 0)$; radius: 4

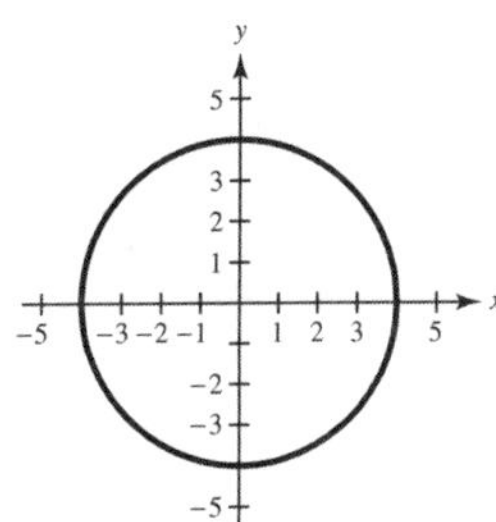

11. $x^2 + y^2 = 36$

Center: $(0, 0)$; radius: 6

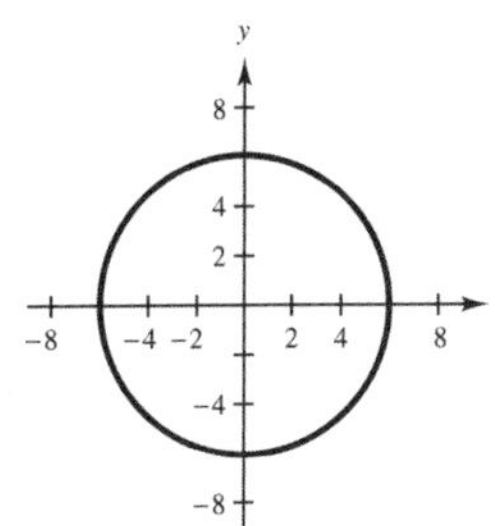

13. Center: $(4, 3)$; radius: 10

$$(x - h)^2 + (y - k)^2 = r^2$$
$$(x - 4)^2 + (y - 3)^2 = 10^2$$
$$(x - 4)^2 + (y - 3)^2 = 100$$

15. Center: $(6, -5)$; radius: 3

$$(x - 6)^2 + (y + 5)^2 = 9$$

17. Center: $(-2, 1)$; point: $(0, 1)$

$$r = \sqrt{\left[0 - (-2)^2\right] + (1 - 1)^2}$$
$$r = \sqrt{4 + 0}$$
$$r = 2$$
$$(x - h)^2 + (y - k)^2 = r^2$$
$$\left[x - (-2)\right]^2 + (y - 1)^2 = 2^2$$
$$(x + 2)^2 + (y - 1)^2 = 4$$

19.
$$x^2 + y^2 + 2x + 6y + 6 = 0$$
$$x^2 + 2x + y^2 + 6y = -6$$
$$\left(x^2 + 2x + 1\right) + \left(y^2 + 6y + 9\right) = -6 + 1 + 9$$
$$(x + 1)^2 + (y + 3)^2 = 4$$

Center: $(-1, -3)$; radius: 2

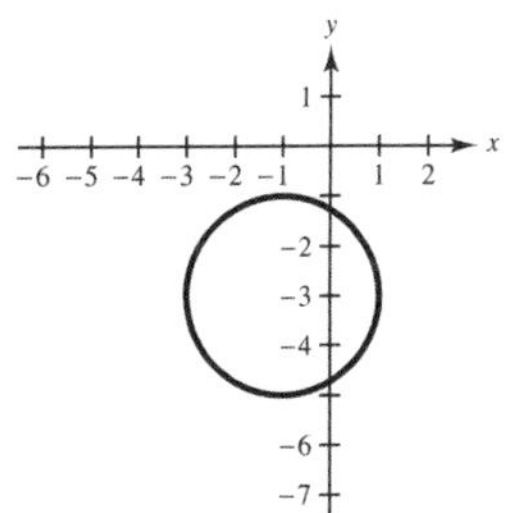

21. The center of the circle is $(0, 0)$ and the radius of the circle $r = 40$ feet. This implies

$$(x - 0)^2 + (y - 0)^2 = 40^2$$
$$x^2 + y^2 = 1600$$
$$y = \pm\sqrt{1600 - x^2}.$$

Finally, take the positive square root to obtain the equation for the semicircle, $y = \sqrt{1600 - x^2}$.

23. Vertex: $(0, 0)$, focus: $\left(0, -\frac{3}{2}\right)$, vertical parabola

$$x^2 = 4py$$
$$p = -\frac{3}{2}$$
$$x^2 = 4\left(-\frac{3}{2}\right)y$$
$$x^2 = -6y$$

25. Vertex: $(0, 0)$, focus: $(-2, 0)$, horizontal parabola

$$y^2 = 4px$$
$$p = -2$$
$$y^2 = 4(-2)x$$
$$y^2 = -8x$$

27. Vertex: $(3, 2)$, focus: $(1, 2)$, horizontal parabola

$$p = -2$$
$$(y - k)^2 = 4p(x - h)$$
$$(y - 2)^2 = 4(-2)(x - 3)$$
$$(y - 2)^2 = -8(x - 3)$$

29. $y = \frac{1}{2}x^2$

$2y = x^2$

Vertical parabola

$4p = 2$

$p = \frac{2}{4} - \frac{1}{2}$

Vertex: $(0, 0)$

Focus: $\left(0, \frac{1}{2}\right)$

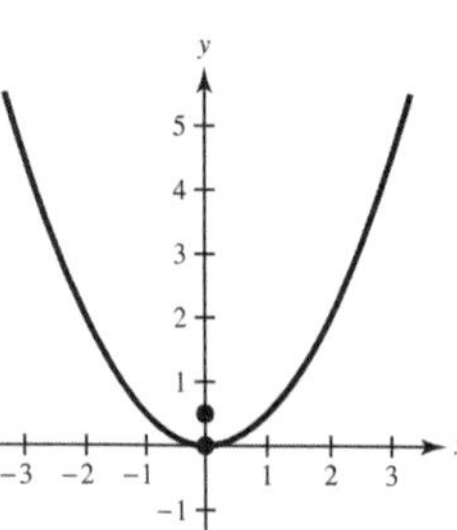

31. $y^2 = -10x$

Horizontal parabola

$4p = -10$

$p = -\frac{5}{2}$

Vertex: $(0, 0)$

Focus: $\left(-\frac{5}{2}, 0\right)$

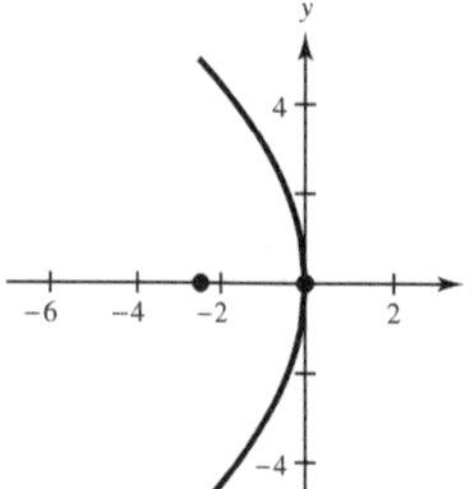

33. $x^2 + 8y = 0$

$x^2 = -8y$

Vertical parabola

$4p = -8$

$p = -\frac{8}{4} = -2$

Vertex: $(0, 0)$

Focus: $(0, -2)$

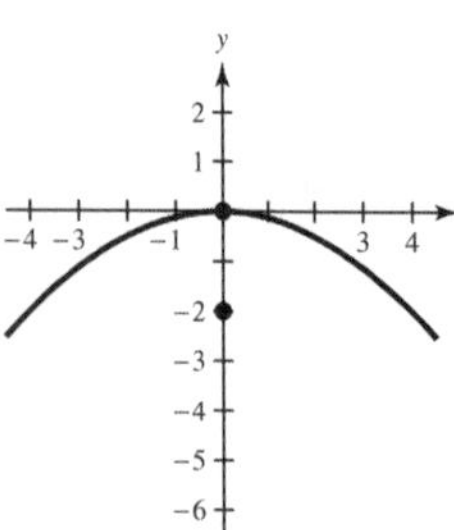
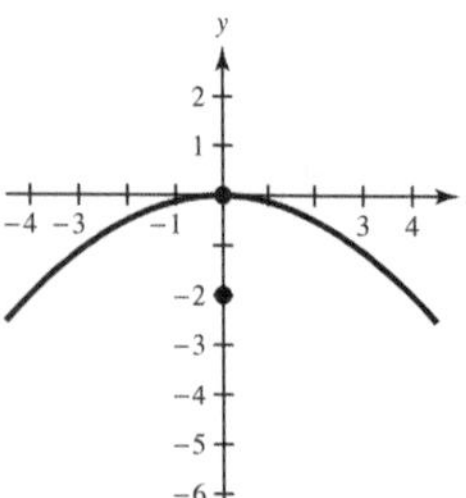

35. $(x - 1)^2 + 8(y + 2) = 0$

$(x - 1)^2 = -8(y + 2)$

Vertical parabola

$4p = -8$

$p = -2$

Vertex: $(1, -2)$

Focus: $(1, -4)$

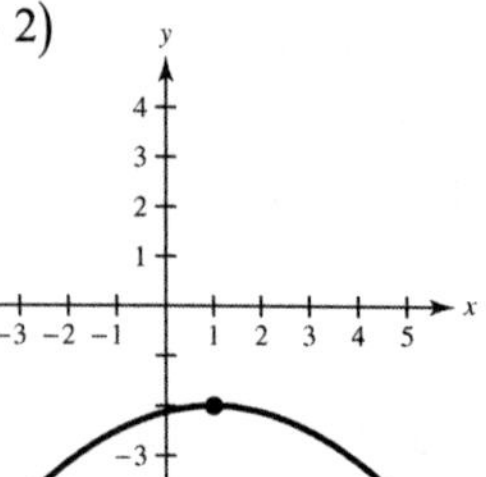

37. $\left(y + \frac{1}{2}\right)^2 = 2(x - 5)$

Horizontal parabola

$4p = 2$

$p = \frac{2}{4} = \frac{1}{2}$

Vertex: $\left(5, -\frac{1}{2}\right)$

Focus: $\left(\frac{11}{2}, -\frac{1}{2}\right)$

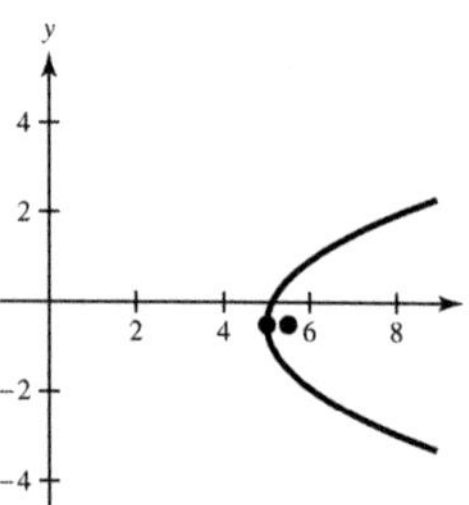

39.
$$y = \frac{1}{3}(x^2 - 2x + 10)$$
$$x^2 - 2x + 10 = 3y$$
$$x^2 - 2x + 1 = 3y - 10 + 1$$
$$(x - 1)^2 = 3y - 9$$
$$(x - 1)^2 = 3(y - 3)$$

Vertical parabola

$4p = 3$

$p = \frac{3}{4}$

Vertex: $(1, 3)$

Focus: $\left(1, \frac{15}{4}\right)$

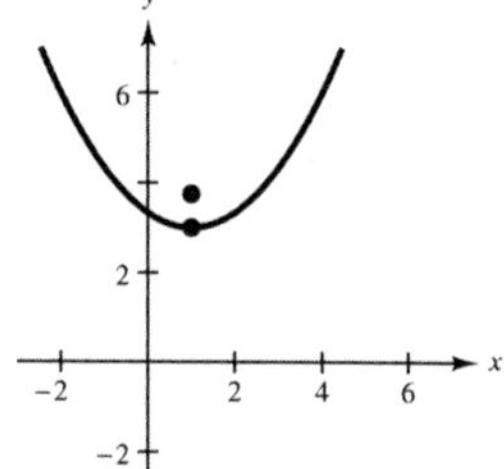

41. $y^2 + 6y + 8x + 25 = 0$

$$(y^2 + 6y + 9) = -8x - 25 + 9$$
$$(y + 3)^2 = -8x - 16$$
$$(y + 3)^2 = -8(x + 2)$$

Horizontal parabola

$4p = -8$

$p = -2$

Vertex: $(-2, -3)$

Focus: $(-4, -3)$

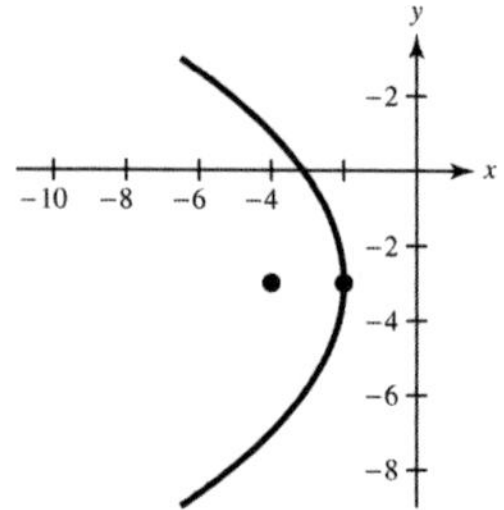

43. First write the equation in standard form: $x^2 + y^2 = 6^2$. Because $h = 0$ and $k = 0$, the center is the origin. Because $r^2 = 6^2$, the radius is six units.

45. The equation $(x - h)^2 = 4p(y - k),\ p \neq 0,$ should be used because the vertex and the focus lie on the same vertical axis.

47. $\left(x + \frac{9}{4}\right)^2 + (y - 4)^2 = 16$

Center: $\left(-\frac{9}{4}, 4\right)$; radius: 4

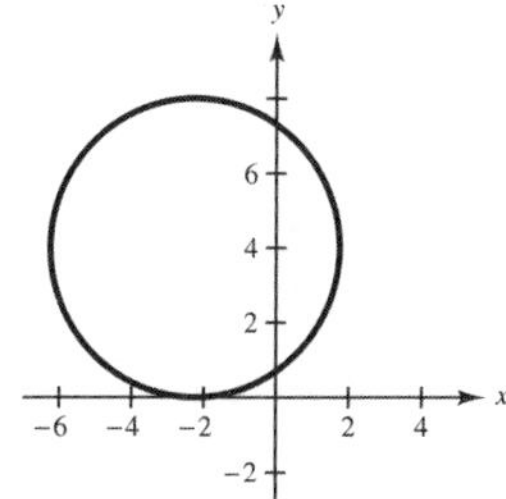

49.
$$x^2 + y^2 + 10x - 4y - 7 = 0$$
$$x^2 + y^2 + 10x - 4y = 7$$
$$\left(x^2 + 10x + 25\right) + \left(y^2 - 4y + 4\right) = 7 + 25 + 4$$
$$(x + 5)^2 + (y - 2)^2 = 36$$

Center: $(-5, 2)$; radius: 6

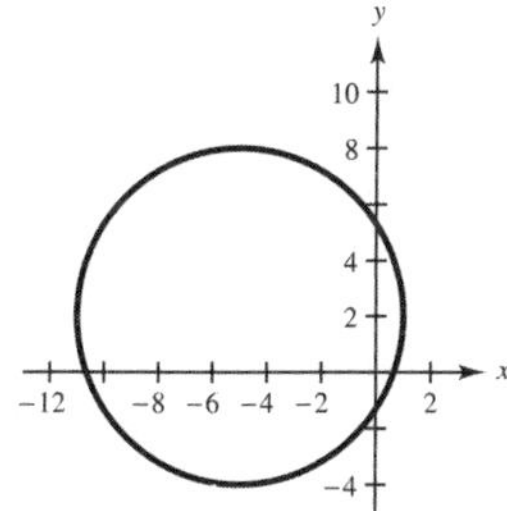

51. $x^2 + y^2 = r^2$

$x^2 + y^2 = 75^2$

53. (a) Vertex: $(0, 0)$, point: $(60, 20)$

Vertical parabola

$$x^2 = 4py$$
$$60^2 = 4p(20)$$
$$3600 = 4p(20)$$
$$180 = 4p$$
$$x^2 = 180y$$
$$\frac{x^2}{180} = y$$

(b) Let $x = 0$.

$$y = \frac{0^2}{180} = 0$$

Let $x = 20$.

$$y = \frac{20^2}{180} = \frac{400}{180} = \frac{20}{9} = 2\frac{2}{9}$$

Let $x = 40$.

$$y = \frac{40^2}{180} = \frac{1600}{180} = \frac{80}{9} = 8\frac{8}{9}$$

Let $x = 60$.

$$y = \frac{60^2}{180} = \frac{3600}{180} = 20$$

x	0	20	40	60
y	0	$2\frac{2}{9}$	$8\frac{8}{9}$	20

55. No. The equation of the circle is $(x + 1)^2 + (y - 1)^2 = 13,$ and the point $(3, 2)$ does not satisfy the equation.

57. No. For each $x > 0$, there correspond two values of y.

59. Yes. The directrix of a parabola is perpendicular to the line through the vertex and focus.

61.
$$x^2 + 6x = -4$$
$$x^2 + 6x + 9 = -4 + 9$$
$$(x + 3)^2 = 5$$
$$x + 3 = \pm\sqrt{5}$$
$$x = -3 \pm \sqrt{5}$$

63. $$\begin{aligned} 4x^2 - 12x - 10 &= 0 \\ 4x^2 - 12x &= 10 \\ x^2 - 3x &= \frac{5}{2} \\ x^2 - 3x + \frac{9}{4} &= \frac{5}{2} + \frac{9}{4} \\ \left(x - \frac{3}{2}\right)^2 &= \frac{19}{4} \\ x - \frac{3}{2} &= \pm\sqrt{\frac{19}{4}} \\ x &= \frac{3}{2} \pm \frac{\sqrt{19}}{2} \end{aligned}$$

65. $$\begin{aligned} 9x^2 - 12x &= 14 \\ x^2 - \tfrac{4}{3}x &= \tfrac{14}{9} \\ x^2 - \tfrac{4}{3}x + \tfrac{4}{9} &= \tfrac{14}{9} + \tfrac{4}{9} \\ \left(x - \tfrac{2}{3}\right)^2 &= \tfrac{18}{9} \\ x - \tfrac{2}{3} &= \pm\sqrt{2} \\ x &= \tfrac{2}{3} \pm \sqrt{2} \end{aligned}$$

67. $$\begin{aligned} \log_{10}\sqrt{xy^3} &= \tfrac{1}{2}\log_{10}\left(xy^3\right) \\ &= \tfrac{1}{2}\left[\log_{10}x + \log_{10}y^3\right] \\ &= \tfrac{1}{2}\left(\log_{10}x + 3\log_{10}y\right) \end{aligned}$$

69. $\ln\frac{x}{y^4} = \ln x - \ln y^4 = \ln x - 4\ln y$

71. $2\log_3 x - \log_3 y = \log_3 x^2 - \log_3 y = \log_3\frac{x^2}{y}$

73. $$\begin{aligned} 4(\ln x + \ln y) - \ln\left(x^4 + y^4\right) &= 4\ln x + 4\ln y - \ln\left(x^4 + y^4\right) \\ &= \ln x^4 + \ln y^4 - \ln\left(x^4 + y^4\right) \\ &= \ln\frac{x^4y^4}{x^4 + y^4} \end{aligned}$$

Section 10.2 Ellipses

1. Because the major axis is vertical, the standard form of the equation of the ellipse is $\frac{x^2}{b^2} + \frac{y^2}{a^2} = 1$.

3. Because the major axis is horizontal, the standard form of the equation of the ellipse is $\frac{x^2}{a^2} + \frac{y^2}{b^2} = 1$.

5. Center: $(0, 0)$

Vertices: $(-4, 0), (4, 0)$

Co-vertices: $(0, -3), (0, 3)$

$$\frac{x^2}{a^2} + \frac{y^2}{b^2} = 1$$

Major axis is x-axis so $a = 4$.

Minor axis is y-axis so $b = 3$.

$$\frac{x^2}{4^2} + \frac{y^2}{3^2} = 1$$

$$\frac{x^2}{16} + \frac{y^2}{9} = 1$$

7. Center: $(0, 0)$

Vertices: $(0, -6), (0, 6)$

Co-vertices: $(-3, 0), (3, 0)$

$$\frac{x^2}{b^2} + \frac{y^2}{a^2} = 1$$

Major axis is y-axis so $b = 3$.

Minor axis is x-axis so $a = 6$.

$$\frac{x^2}{3^2} + \frac{y^2}{6^2} = 1$$

$$\frac{x^2}{9} + \frac{y^2}{36} = 1$$

9. Vertices: $(-4, 0), (4, 0)$

Co-vertices: $(0, 2), (0, -2)$

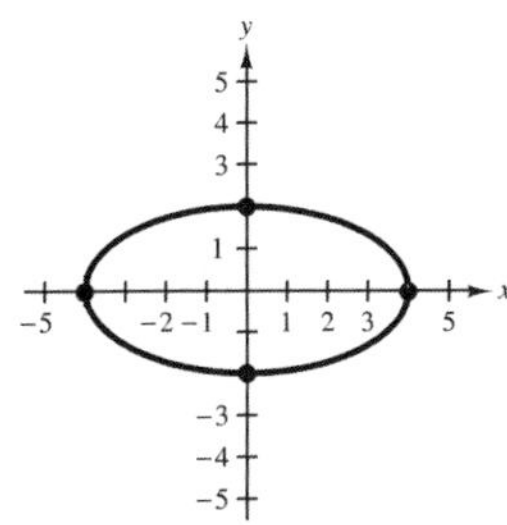

11. $4x^2 + y^2 - 4 = 0$

$$\frac{x^2}{1} + \frac{y^2}{4} = 1$$

Vertices: $(0, 2), (0, -2)$

Co-vertices: $(1, 0), (-1, 0)$

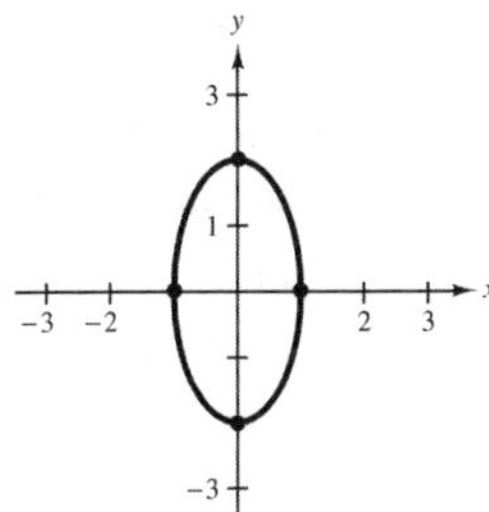

13. Because the major axis is vertical, the standard form of the equation of the ellipse is $\frac{(x - b)^2}{b^2} + \frac{(y - k)^2}{a^2} = 1.$

15. Because the major axis is horizontal, the standard form of the equation of the ellipse is

$$\frac{(x - b)^2}{a^2} + \frac{(y - k)^2}{b^2} = 1.$$

17. Center: $(2, 2)$

Vertices: $(-1, 2), (5, 2)$

Co-vertices: $(2, 0), (2, 4)$

Major axis is horizontal so $a = 3$.

Minor axis is vertical so $b = 2$.

$$\frac{(x - 2)^2}{3^2} + \frac{(y - 2)^2}{2^2} = 1$$

$$\frac{(x - 2)^2}{9} + \frac{(y - 2)^2}{4} = 1$$

19. Center: $(4, 0)$

Vertices: $(4, 4), (4, -4)$

Co-vertices: $(1, 0), (7, 0)$

Major axis is vertical so $a = 4$.

Minor axis is horizontal so $b = 3$.

$$\frac{(x - 4)^2}{3^2} + \frac{(y - 0)^2}{4^2} = 1$$

$$\frac{(x - 4)^2}{9} + \frac{y^2}{16} = 1$$

21. $4(x - 2)^2 + 9(y + 2)^2 = 36$

$$\frac{(x - 2)^2}{9} + \frac{(y + 2)^2}{4} = 1$$

Center: $(2, -2)$

Vertices: $(-1, -2), (5, -2)$

$a^2 = 9$ $\quad$ $b^2 = 4$

$a = 3$ $\quad$ $b = 2$

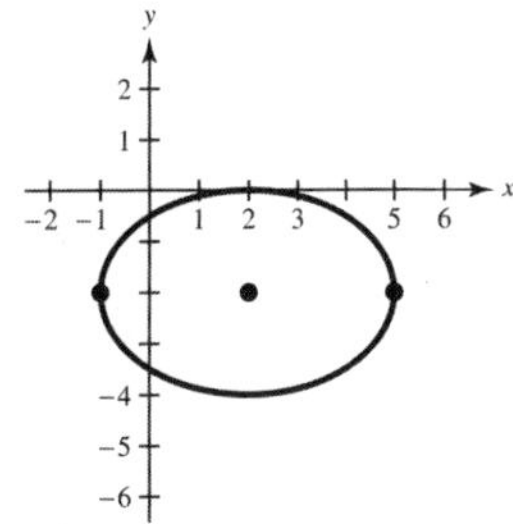

23.

$$9x^2 + 4y^2 + 36x - 24y + 36 = 0$$

$$(9x^2 + 36x) + (4y^2 - 24y) = -36$$

$$9(x^2 + 4x + 4) + 4(y^2 - 6y + 9) = -36 + 36 + 36$$

$$9(x + 2)^2 + 4(y - 3)^2 = 36$$

$$\frac{(x + 2)^2}{4} + \frac{(y - 3)^2}{9} = 1$$

Center: $(-2, 3)$

Vertices: $(-2, 0), (-2, 6)$

$a^2 = 9$ $\quad$ $b^2 = 4$

$a = 3$ $\quad$ $b = 2$

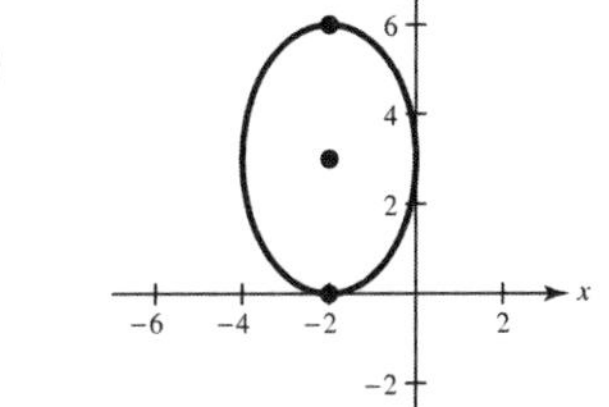

25.
$$25x^2 + 9y^2 - 200x + 54y + 256 = 0$$
$$\left(25x^2 - 200x\right) + \left(9y^2 + 54y\right) = -256$$
$$25\left(x^2 - 8x + 16\right) + 9\left(y^2 + 6y + 9\right) = -256 + 400 + 81$$
$$25(x-4)^2 + 9(y+3)^2 = 225$$
$$\frac{(x-4)^2}{9} + \frac{(y+3)^2}{25} = 1$$

Center: $(4, -3)$

Vertices: $(4, 2), (4, -8)$

$a^2 = 25 \qquad b^2 = 9$

$a = 5 \qquad b = 3$

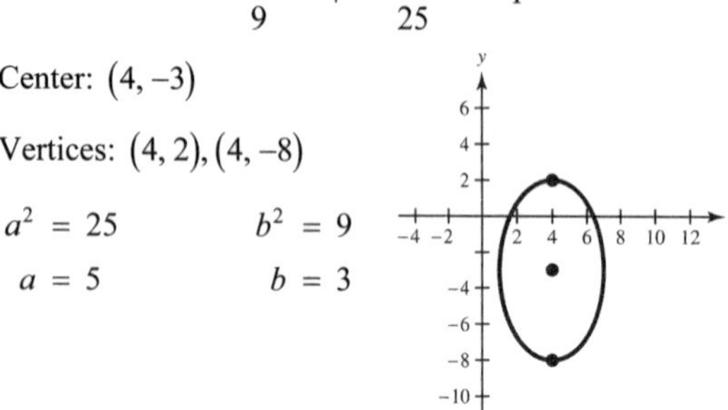

27. Center: $(0, 0)$

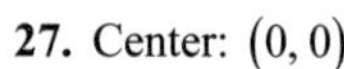

Length of major axis: 1230

Length of minor axis: 580

$2a = 1230 \qquad 2b = 580$

$a = 615 \qquad b = 290$

$$\frac{x^2}{615^2} + \frac{y^2}{290^2} = 1$$

or

$$\frac{x^2}{290^2} + \frac{y^2}{615^2} = 1$$

29. An ellipse is the set of all points (x, y) such that the sum of the distances between (x, y) and two distinct fixed points is a constant.

$$\frac{x^2}{a^2} + \frac{y^2}{b^2} = 1 \text{ or } \frac{x^2}{b^2} + \frac{y^2}{a^2} = 1$$

31. The length of the major axis is $2a$ and the length of the minor axis is $2b$.

33. Center: $(0, 0)$

Major axis (vertical) 10 units

Minor axis 6 units

$$\frac{x^2}{b^2} + \frac{y^2}{a^2} = 1$$

$b = 3, \quad a = 5$

$$\frac{x^2}{3^2} + \frac{y^2}{5^2} = 1$$

$$\frac{x^2}{9} + \frac{y^2}{25} = 1$$

35.
$$16x^2 + 25y^2 - 9 = 0$$
$$\frac{16x^2}{9} + \frac{25y^2}{9} = 1$$
$$\frac{x^2}{9/16} + \frac{y^2}{9/25} = 1$$

Vertices: $\left(\pm\frac{3}{4}, 0\right)$

Co-vertices: $\left(0, \pm\frac{3}{5}\right)$

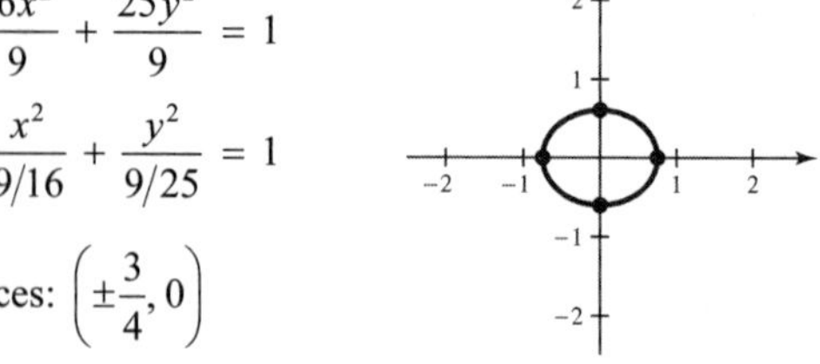

37.
$$\frac{x^2}{324} + \frac{y^2}{196} = 1$$

$a^2 = 324 \qquad b^2 = 196$

$a = 18 \qquad b = 14$

$2a = 36 \qquad 2b = 28$

36 feet = longest distance

24 feet = shortest distance

39. (a) Every point on the ellipse represents the maximum distance (800 miles) that the plane can safely fly with enough fuel to get from airport A to airport B.

(b) Center: $(0, 0)$ Airport A: $(-250, 0)$

$2c = 500$ Airport B: $(250, 0)$

$c = 250$

(c) Airplane flies maximum of 800 miles without refueling, so $800 - 500 = 300$.

300 = twice the distance from the airport to the vertex.

150 = the distance from the airport to the vertex.

Vertices: $(\pm 400, 0)$

$-250 - 150 = -400$

$250 + 150 = 400$

(d)
$$c^2 = a^2 - b^2$$
$$250^2 = 400^2 - b^2$$
$$400^2 - 250^2 = b^2$$
$$97{,}500 = b^2$$
$$\sqrt{97{,}500} = b$$
$$50\sqrt{39} = 6$$

$$\frac{x^2}{a^2} + \frac{y^2}{b^2} = 1$$
$$\frac{x^2}{400^2} + \frac{y^2}{\left(50\sqrt{39}\right)^2} = 1$$

(e)
$$A = \pi ab$$
$$A = \pi(400)\left(50\sqrt{39}\right)$$
$$= 20{,}000\sqrt{39}\pi \approx 392{,}385 \text{ square miles}$$

41. A circle is an ellipse in which the major axis and the minor axis have the same length. Both circles and ellipses have foci; however, in a circle the foci are both at the same point, whereas in an ellipse they are not.

43. The sum of the distances between each point on the ellipse and the two foci is a constant.

45. The graph of an ellipse written in the standard form $\frac{(x-h)^2}{a^2} + \frac{(y-k)^2}{b^2} = 1$ intersects the y-axis if $|h| > a$ and intersects the x-axis if $|k| > b$. Similarly, the graph of $\frac{(x-h)^2}{b^2} + \frac{(y-k)^2}{a^2} = 1$ intersects the y-axis if $|h| > b$ and intersects the x-axis if $|k| > a$.

47. $f(x) = 3^{-x}$

(a) $f(-2) = 3^{-(-2)} = 3^2 = 9$

(b) $f(2) = 3^{-2} = \frac{1}{3^2} = \frac{1}{9}$

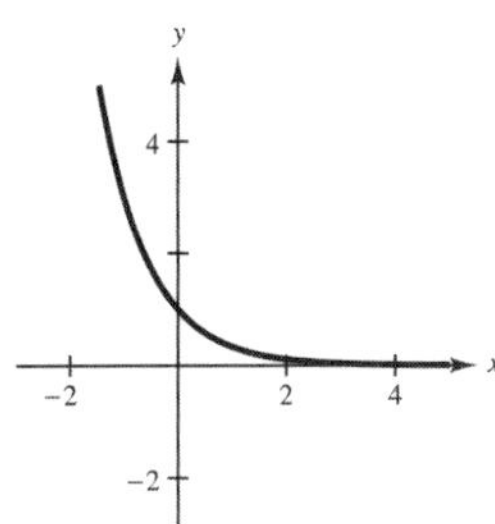

49. $g(x) = 6e^{0.5x}$

(a) $g(-1) = 6e^{0.5(-1)}$

(b) $g(2) = 6e^{0.5(2)} \approx 16.310$

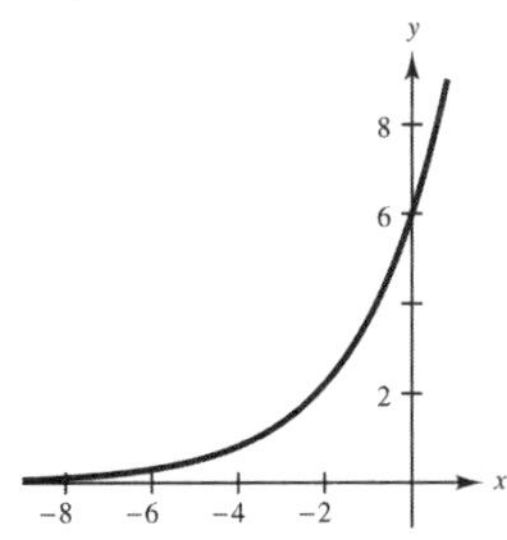

51. $h(x) = \log_{16} 4x$

(a) $h(4) = \log_{16} 4(4) = \log_{16} 16 = 1$

(b) $h(64) = \log_{16} 4(64) = \log_{16} 256 = \log_{16} 16^2 = 2$

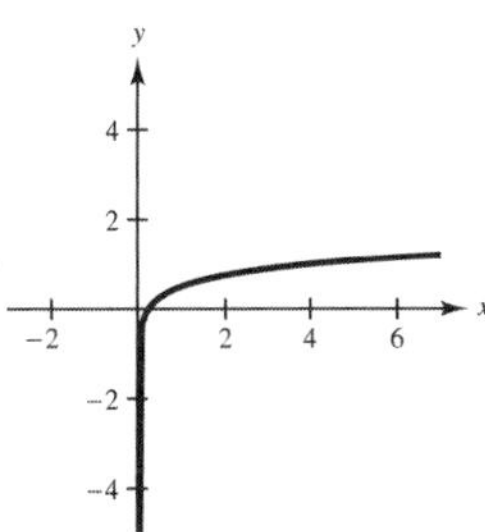

53. $f(x) = \log_4(x-3)$

(a) $f(3) = \log_4(3-3) = \log_4 0 =$ does not exist

(b) $f(35) = \log4(35-3)$

$= \log_4 32$

$= \log_4 2^5$

$= 5 \log_4 2 = 5 \cdot \frac{1}{2} = \frac{5}{2}$

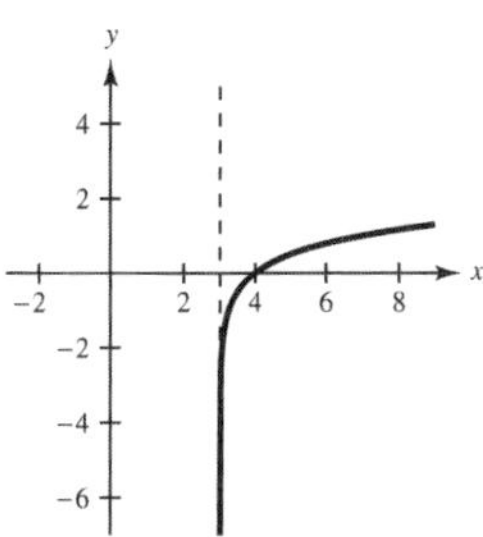

Mid-Chapter Quiz for Chapter 10

1. Center: $(0, 0)$; radius: 5

$$(x-h)^2 + (y-k)^2 = r^2$$
$$(x-0)^2 + (y-0)^2 = 5^2$$
$$x^2 + y^2 = 25$$

2. Center: $(3, -5)$; passes through the point: $(0, -1)$

$$\begin{aligned} r &= \sqrt{(3-0)^2 + (-5-(-1))^2} \\ &= \sqrt{3^2 + (-4)^2} \\ &= \sqrt{9+16} \\ &= \sqrt{25} \\ &= 5 \end{aligned}$$

$$\begin{aligned} (x-h)^2 + (y-k)^2 &= r^2 \\ (x-3)^2 + (y-(-5))^2 &= 5^2 \\ (x-3)^2 + (y+5)^2 &= 25 \end{aligned}$$

3. Vertex: $(-2, 1)$; focus: $(0, 1)$

$$\begin{aligned} (y-k)^2 &= 4p(x-h) \Rightarrow p = 2 \\ (y-1)^2 &= 4(2)(x-(-2))^2 \\ (y-1)^2 &= 8(x+2) \end{aligned}$$

4. Vertex: $(2, 3)$; focus: $(2, 1)$

$$\begin{aligned} (x-h)^2 &= 4p(y-k) \Rightarrow p = -2 \\ (x-2)^2 &= 4(-2)(y-3) \\ (x-2)^2 &= -8(y-3) \end{aligned}$$

5. Center: $(-2, -1)$

Vertices: $(-6, -1), (2, -1)$

Co-vertices: $(-2, 1), (-2, -3)$

Major axis is horizontal so $a = 4$.

Minor axis is vertical so $b = 2$.

$$\frac{[x-(-2)]^2}{4^2} + \frac{[y-(-1)]^2}{2^2} = 1$$

$$\frac{(x+2)^2}{16} + \frac{(y+1)^2}{4} = 1$$

6. Vertices: $(0, -10), (0, 10)$

Co-vertices: $(-6, 0), (6, 0)$

Vertical ellipse

$$\frac{x^2}{b^2} + \frac{y^2}{a^2} = 1$$

$$\frac{x^2}{6^2} + \frac{y^2}{10^2} = 1$$

$$\frac{x^2}{36} + \frac{y^2}{100} = 1$$

7. $x^2 + y^2 + 6y - 7 = 0$

$$\begin{aligned} x^2 + y^2 + 6y + 9 &= 7 + 9 \\ x^2 + (y+3)^2 &= 16 \end{aligned}$$

Center: $(0, -3)$

Radius: 4

8.

$$\begin{aligned} x^2 + y^2 + 2x - 4y + 4 &= 0 \\ (x^2 + 2x + 1) + (y^2 - 4y + 4) &= -4 + 1 + 4 \\ (x+1)^2 + (y-2)^2 &= 1 \end{aligned}$$

Center: $(-1, 2)$

Radius: 1

9.

$$\begin{aligned} x &= y^2 - 6y - 7 \\ x + 7 &= (y^2 - 6y + 9) - 9 \\ x + 16 &= (y-3)^2 \end{aligned}$$

Vertex: $(-16, 3)$

$$\begin{aligned} 4p &= 1 \\ p &= \tfrac{1}{4} \end{aligned}$$

Focus: $\left(-16 + \frac{1}{4}, 3\right) = \left(-\frac{64}{4} + \frac{1}{4}, 3\right) = \left(-\frac{63}{4}, 3\right)$

10. $x^2 - 8x + y + 12 = 0$

$$\begin{aligned} x^2 - 8x + 16 &= -y - 12 + 16 \\ (x-4)^2 &= -y + 4 \\ (x-4)^2 &= -1(y-4) \end{aligned}$$

Vertex: $(4, 4)$

$$\begin{aligned} 4p &= -1 \\ p &= -\tfrac{1}{4} \end{aligned}$$

Focus: $\left(4, 4 - \frac{1}{4}\right) = \left(4, \frac{16}{4} - \frac{1}{4}\right) = \left(4, \frac{15}{4}\right)$

11. $4x^2 + y^2 - 16x - 20 = 0$

$$\begin{aligned} 4x^2 - 16x + y^2 &= 20 \\ 4(x^2 - 4x) + y^2 &= 20 \\ 4(x^2 - 4x + 4) + y^2 &= 20 + 16 \\ 4(x-2)^2 + y^2 &= 36 \\ \frac{(x-2)^2}{9} + \frac{y^2}{36} &= 1 \end{aligned}$$

Center: $(2, 0)$

Vertices: $(2, -6), (2, 6)$

$$\begin{aligned} a^2 &= 36 \\ a &= 6 \end{aligned}$$

12.
$$4x^2 + 9y^2 - 48x + 36y + 144 = 0$$
$$(4x^2 - 48x) + (9y^2 + 36y) = -144$$
$$4(x^2 - 12x + 36) + 9(y^2 + 4y + 4) = -144 + 144 + 36$$
$$4(x - 6)^2 + 9(y + 2)^2 = 36$$
$$\frac{(x - 6)^2}{9} + \frac{(y + 2)^2}{4} = 1$$

Center: $(6, -2)$

Vertices: $(3, -2), (9, -2)$

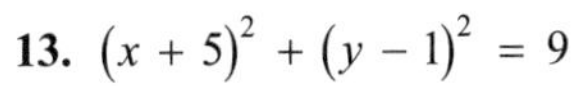

13. $(x + 5)^2 + (y - 1)^2 = 9$

Circle:

Center: $(-5, 1)$

Radius: 3

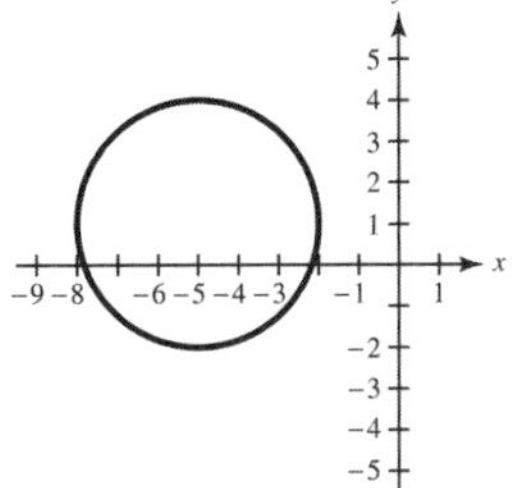

14. $9x^2 + y^2 = 81$

$$\frac{x^2}{9} + \frac{y^2}{81} = 1$$

Vertical ellipse

$b^2 = 9$ $\qquad a^2 = 81$

$b = \pm 3$ $\qquad a = \pm 9$

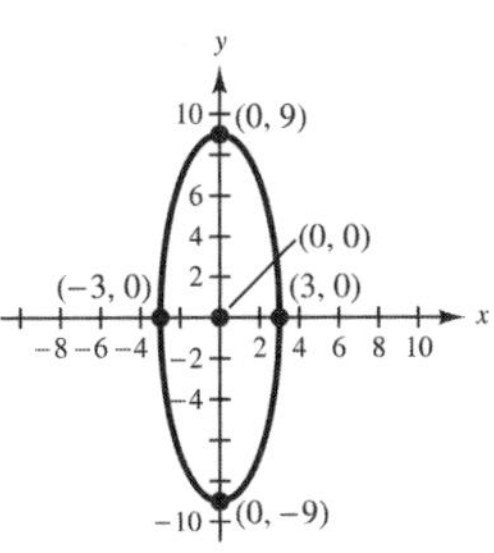

15.
$$x = -y^2 - 4y$$
$$x = -1(y^2 + 4y + 4) + 4$$
$$x - 4 = -1(y + 2)^2$$
$$-1(x - 4) = (y + 2)^2$$

Parabola

Vertex: $(4, -2)$

$4p = -1$

$p = -\frac{1}{4}$

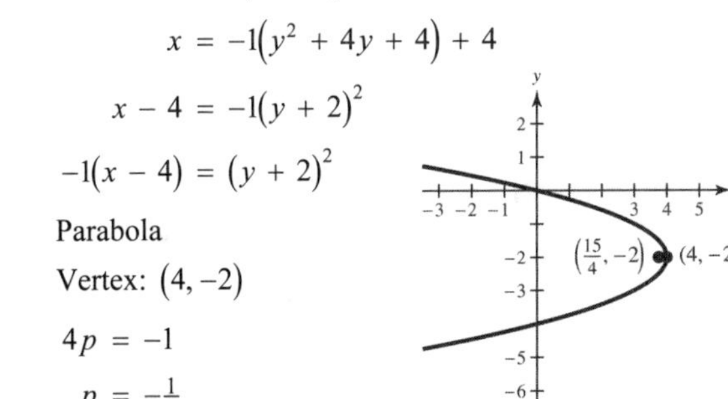

Focus: $\left(4 - \frac{1}{4}, -2\right) = \left(\frac{16}{4} - \frac{1}{4}, -2\right) = \left(\frac{15}{4}, -2\right)$

16. $x^2 + (y + 4)^2 = 1$

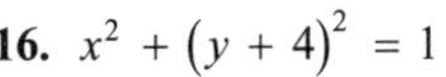

Circle

Center: $(0, -4)$

Radius: 1

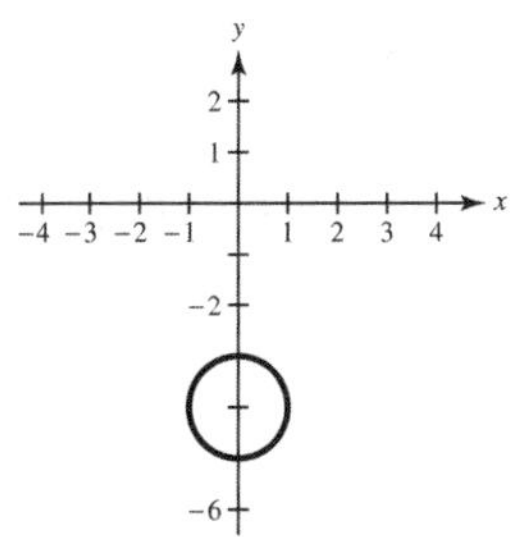

17. $y = x^2 - 2x + 1$

$y = (x - 1)^2$

Parabola

Vertex: $(1, 0)$

$4p = 1$

$p = \frac{1}{4}$

Focus: $\left(1, 0 + \frac{1}{4}\right) = (1, 0.25)$

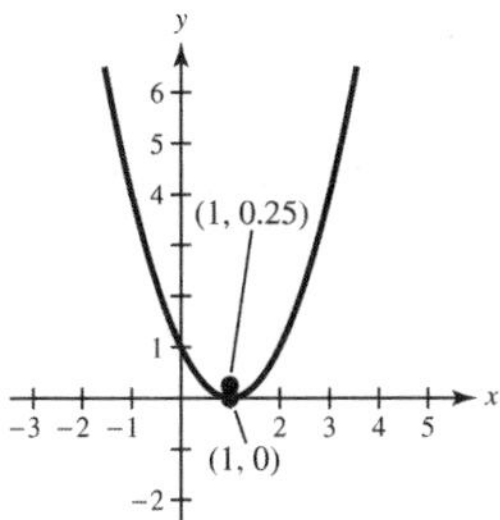

18. $4(x + 3)^2 + (y - 2)^2 = 16$

$$\frac{4(x + 3)^2}{16} + \frac{(y - 2)^2}{16} = 1$$
$$\frac{(x + 3)^2}{4} + \frac{(y - 2)^2}{16} = 1$$

Ellipse

$a^2 = 16$ $\qquad b^2 = 4$

$a = 4$ $\qquad b = 2$

Center: $(-3, 2)$

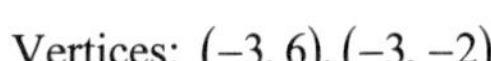

Vertices: $(-3, 6), (-3, -2)$

Co-vertices: $(-5, 2), (-1, 2)$

Section 10.3 Hyperbolas

1. Because the transverse axis is vertical, the standard form of the equation of the hyperbola is $\dfrac{y^2}{a^2} - \dfrac{x^2}{b^2} = 1$.

3. Because the transverse axis is horizontal, the standard form of the equation of the hyperbola is $\dfrac{x^2}{a^2} - \dfrac{y^2}{b^2} = 1$.

5. $\frac{x^2}{9} - \frac{y^2}{25} = 1$

Vertices: $(3, 0), (-3, 0)$

Asymptotes: $y = \frac{5}{3}x$

$y = -\frac{5}{3}x$

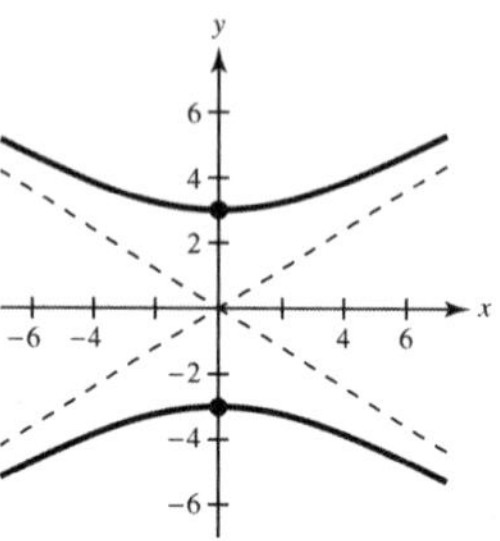

7. $\frac{y^2}{9} - \frac{x^2}{25} = 1$

Vertices: $(0, \pm 3)$

Asymptotes: $y = \pm\frac{3}{5}x$

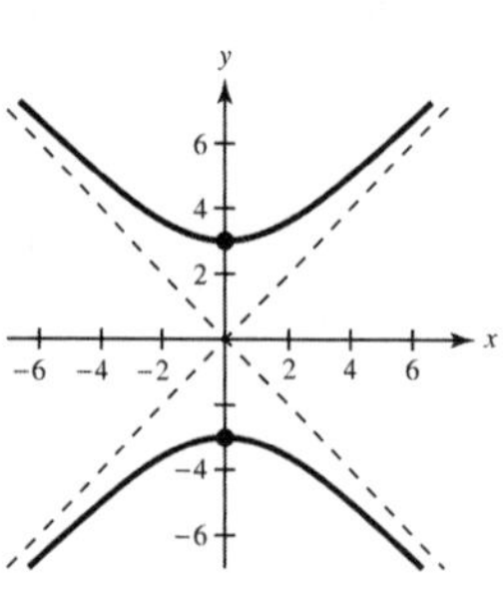

9. $y^2 - x^2 = 9$

$\frac{y^2}{9} - \frac{x^2}{9} = 1$

Vertices: $(0, \pm 3)$

Asymptotes: $y = \pm x$

11. Vertices: $(\pm 4, 0)$

Asymptotes: $y = \pm 2x$

$a = 4 \qquad \frac{b}{a} = \pm 2$

$\frac{b}{4} = \pm 2$

$b = \pm 8$

$\frac{x^2}{a^2} - \frac{y^2}{b^2} = 1$

$\frac{x^2}{4^2} - \frac{y^2}{8^2} = 1$

$\frac{x^2}{16} - \frac{y^2}{64} = 1$

13. Vertices: $(0, \pm 4)$

Asymptotes: $y = \pm\frac{1}{2}x$

$a = 4 \qquad \frac{a}{b} = \pm\frac{1}{2}$

$\frac{4}{b} = \pm\frac{1}{2}$

$b = 8$

$\frac{y^2}{a^2} - \frac{x^2}{b^2} = 1$

$\frac{y^2}{4^2} - \frac{x^2}{8^2} = 1$

$\frac{y^2}{16} - \frac{x^2}{64} = 1$

15. Vertices: $(\pm 9, 0)$

Asymptotes: $y = \pm\frac{2}{3}x$

$a = 9 \qquad \frac{b}{a} = \pm\frac{2}{3}$

$\frac{b}{9} = \pm\frac{2}{3}$

$b = 6$

$\frac{x^2}{a^2} - \frac{y^2}{b^2} = 1$

$\frac{x^2}{9^2} - \frac{y^2}{6^2} = 1$

$\frac{x^2}{81} - \frac{y^2}{36} = 1$

17. $\frac{(x-1)^2}{4} - \frac{(y+2)^2}{1} = 1$

Center: $(1, -2)$

Vertices: $(-1, -2), (3, -2)$

$a^2 = 4 \qquad b^2 = 1$

$a = 2 \qquad b = 1$

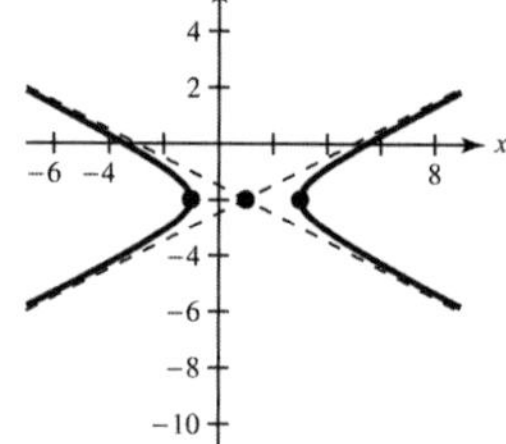

19. $(y+4)^2 - (x-3)^2 = 25$

$\frac{(y+4)^2}{25} - \frac{(x-3)^2}{25} = 1$

Center: $(3, -4)$

Vertices: $(3, 1), (3, -9)$

$a = 5 \qquad b = 5$

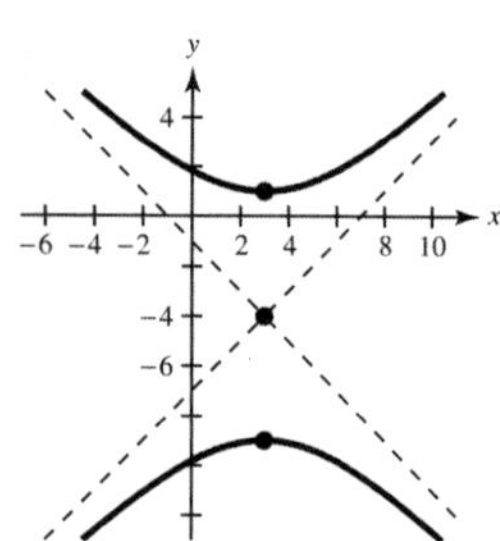

21.
$$9x^2 - y^2 - 36x - 6y + 18 = 0$$
$$(9x^2 - 36x) - (y^2 + 6y) = -18$$
$$9(x^2 - 4x + 4) - (y^2 + 6y + 9) = -18 + 36 - 9$$
$$9(x-2)^2 - (y+3)^2 = 9$$
$$\frac{(x-2)^2}{1} - \frac{(y+3)^2}{9} = 1$$

Center: $(2, -3)$

Vertices: $(3, -3), (1, -3)$

$a^2 = 1 \qquad b^2 = 9$

$a = 1 \qquad b = 3$

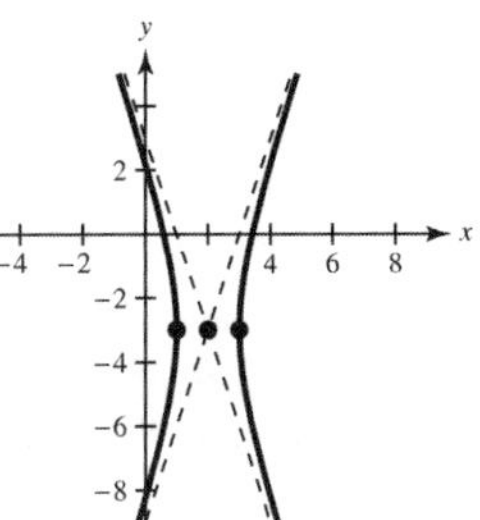

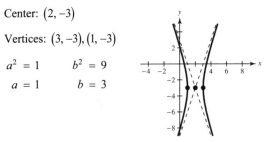

23.
$$4x^2 - y^2 + 24x + 4y + 28 = 0$$
$$(4x^2 + 24x) - (y^2 - 4y) = -28$$
$$4(x^2 + 6x + 9) - (y^2 - 4y + 4) = -28 + 36 - 4$$
$$4(x+3)^2 - (y-2)^2 = 4$$
$$\frac{(x+3)^2}{1} - \frac{(y-2)^2}{4} = 1$$

Center: $(-3, 2)$

Vertices: $(-4, 2), (-2, 2)$

$a^2 = 1 \qquad b^2 = 4$

$a = 1 \qquad b = 2$

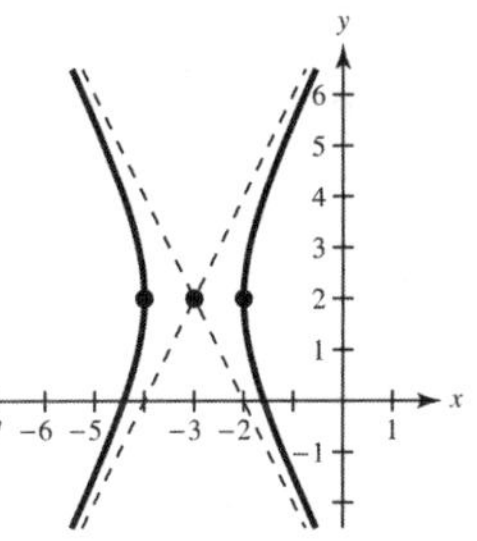

25.
$$89x^2 - 55y^2 - 4272x + 31{,}684 = 0$$
$$89(x^2 - 48x + 576) - 55y^2 = -31{,}684 + 51{,}264$$
$$89(x-24)^2 - 55y^2 = 19{,}580$$
$$\frac{(x-24)^2}{220} - \frac{y^2}{356} = 1$$

$a^2 = 220$

$a = \sqrt{220}$

Center: $(24, 0)$

Vertex: $(24 + \sqrt{220}, 0) \approx (38.8, 0)$

27. The sides of the central rectangle will pass through the points $(\pm a, 0)$ and $(0, \pm b)$. To sketch the asymptotes, draw and extend the rectangles of the central rectangle.

29. Central rectangle dimensions: $2a \times 2b$; Center: (h, k)

31. Vertices: $(0, \pm 3)$

Vertical axis

Point: $(-2, 5)$

Center: $(0, 0)$

$$\frac{y^2}{a^2} - \frac{x^2}{b^2} = 1$$
$$\frac{(5)^2}{3^2} - \frac{(-2)^2}{b^2} = 1$$
$$\frac{25}{9} - \frac{4}{b^2} = 1$$
$$\frac{25}{9} - \frac{9}{9} = \frac{4}{b^2}$$
$$\frac{16}{9} = \frac{4}{b^2}$$
$$b^2 = \frac{36}{16} = \frac{9}{4}$$
$$\frac{y^2}{9} - \frac{x^2}{9/4} = 1$$

33. Vertices: $(1, 2), (5, 2)$

Horizontal axis

Point: $(0, 0)$

Center: $(3, 2)$

$$\frac{(x-h)^2}{a^2} - \frac{(y-k)^2}{b^2} = 1$$
$$\frac{(0-3)^2}{2^2} - \frac{(0-2)^2}{b^2} = 1$$
$$\frac{9}{4} - \frac{4}{b^2} = 1$$
$$\frac{9}{4} - \frac{4}{4} = \frac{4}{b^2}$$
$$\frac{5}{4} = \frac{4}{b^2}$$
$$b^2 = \frac{16}{5}$$
$$\frac{(x-3)^2}{4} - \frac{(y-2)^2}{16/5} = 1$$

35. $\dfrac{x^2}{1} - \dfrac{y^2}{9/4} = 1$

Vertices: $(\pm 1, 0)$

Asymptotes: $y = \pm\dfrac{3/2}{1}x$

$y = \pm\dfrac{3}{2}x$

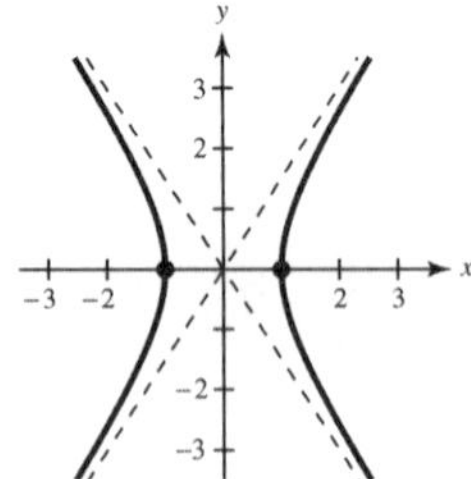

37. $4y^2 - x^2 + 16 = 0$

$$\frac{4y^2}{-16} - \frac{x^2}{-16} = \frac{-16}{-16}$$
$$\frac{-y^2}{4} + \frac{x^2}{16} = 1$$
$$\frac{x^2}{16} - \frac{y^2}{4} = 1$$

Vertices: $(4, 0), (-4, 0)$

Asymptotes: $y = \dfrac{2}{4}x = \dfrac{1}{2}x$

$y = -\dfrac{2}{4}x = -\dfrac{1}{2}x$

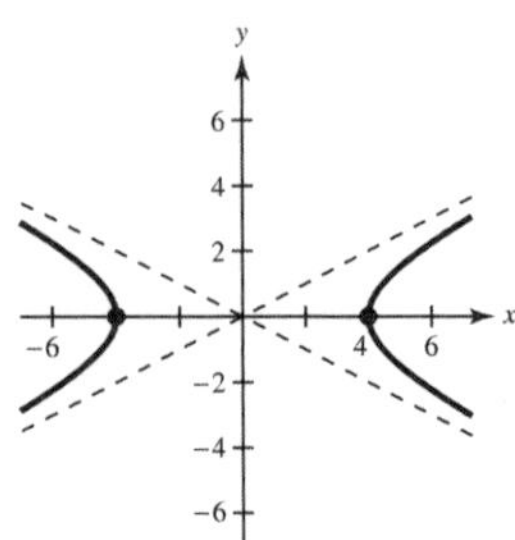

39. $\dfrac{(x-3)^2}{4^2} + \dfrac{(y-4)^2}{6^2} = 1$

Ellipse of the form $\dfrac{(x-h)^2}{b^2} + \dfrac{(y-k)^2}{a^2} = 1$

where $(3, 4)$ is the center.

$a^2 = 6^2$ and $b^2 = 4^2$

41. $x^2 - y^2 = 1$

Hyperbola of the form $\dfrac{x^2}{a^2} - \dfrac{y^2}{b^2} = 1$ where $(0, 0)$ is the center and $a^2 = b^2$.

43.
$$y^2 - x^2 - 2y + 8x - 19 = 0$$
$$(y^2 - 2y) - (x^2 - 8x) = 19$$
$$(y^2 - 2y + 1) - (x^2 - 8x + 16) = 19 + 1 - 16$$
$$(y-1)^2 - (x-4)^2 = 4$$
$$\frac{(y-1)^2}{4} - \frac{(x-4)^2}{4} = 1$$

Hyperbola of the form $\dfrac{(y-k)^2}{a^2} - \dfrac{(x-h)^2}{b^2} = 1$ where $(1, 4)$ is the center and $a^2 = b^2$.

45. $\dfrac{x^2}{100} - \dfrac{y^2}{4} = 1$

The shortest horizontal distance would be the distance between the center and a vertex of the hyperbola, that is, 10 miles.

47. Infinitely many. The constant difference of the distances can be different for an infinite number of hyperbolas that have the same set of foci.

49. Answers will vary.

51. $\begin{cases} x - 3y = 5 \\ 2x - 6y = -5 \end{cases}$

$-3y = -x + 5$ $\qquad$ $-6y = -2x - 5$

$y = \frac{1}{3}x - \frac{5}{3}$ $\qquad$ $y = \frac{1}{3}x + \frac{5}{6}$

$m_1 = m_2$ and $b_1 \neq b_2$; inconsistent

53. $\begin{cases} 4x + 3y = 3 \\ x - 2y = 9 \end{cases}$

$$\begin{aligned} 4x + 3y &= 3 \\ -4x + 8y &= -36 \\ \hline 11y &= -33 \\ y &= -3 \end{aligned}$$

$$\begin{aligned} x - 2(-3) &= 9 \\ x + 6 &= 9 \\ x &= 3 \end{aligned}$$

$(3, -3)$

Section 10.4 Solving Non-linear Systems of Equations

1. $\begin{cases} x + y = 2 \\ x^2 - y = 0 \end{cases}$

Solve for y.

$y = -x + 2$

$y = x^2$

Solutions: $(-2, 4), (1, 1)$

Check:

$-2 + 4 \stackrel{?}{=} 2$

$2 = 2$

$(-2)^2 - 4 \stackrel{?}{=} 0$

$0 = 0$

Check:

$1 + 1 \stackrel{?}{=} 2$

$2 = 2$

$1^2 - 1 \stackrel{?}{=} 0$

$0 = 0$

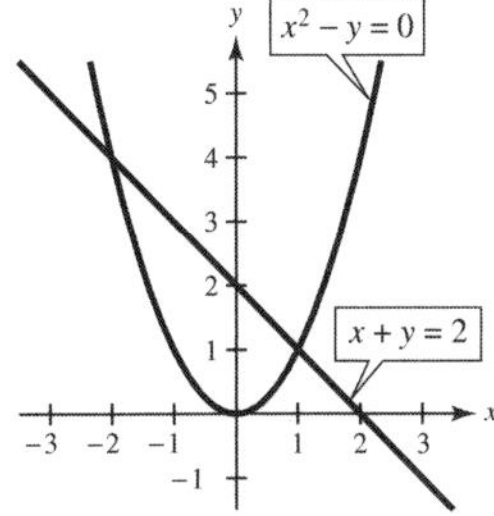

3. $\begin{cases} x^2 + y = 9 \\ x - y = -3 \end{cases}$

Solve for y.

$y = -x^2 + 9$

$y = x + 3$

Solutions: $(2, 5), (-3, 0)$

Check:

$2^2 + 5 \stackrel{?}{=} 9$

$9 = 9$

$2 - 5 \stackrel{?}{=} -3$

$-3 = -3$

Check:

$(-3)^2 + 0 \stackrel{?}{=} 9$

$9 = 9$

$-3 - 0 \stackrel{?}{=} -3$

$-3 = -3$

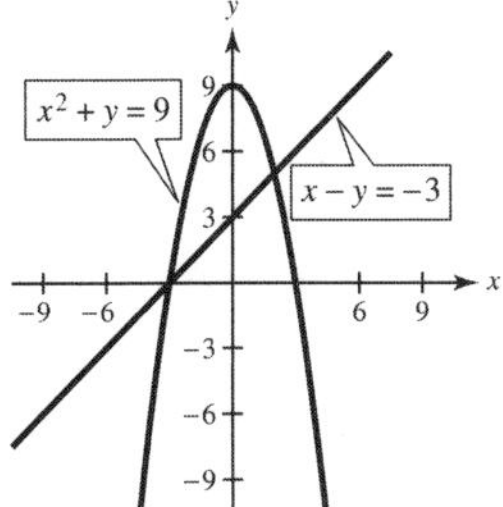

5. $\begin{cases} x^2 + y^2 = 100 \\ x + y = 2 \end{cases}$

Solve for y.

$y = \pm\sqrt{100 - x^2}$

$y = 2 - x$

Solutions: $(-6, 8), (8, -6)$

Check:

$(-6)^2 + 8^2 \stackrel{?}{=} 100$

$100 = 100$

$-6 + 8 \stackrel{?}{=} 2$

$2 = 2$

Check:

$8^2 + (-6)^2 \stackrel{?}{=} 100$

$100 = 100$

$8 + (-6) \stackrel{?}{=} 2$

$2 = 2$

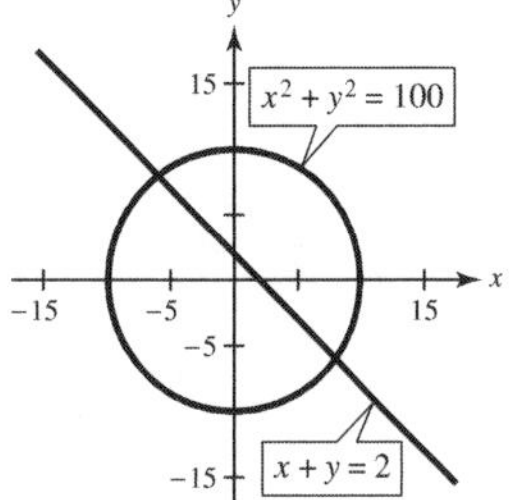

7. $\begin{cases} x^2 + y^2 = 25 \\ 2x - y = -5 \end{cases}$

Solve for y.

$y = \pm\sqrt{25 - x^2}$

$y = 2x + 5$

Solutions: $(0, 5), (-4, -3)$

Check:

$0^2 + 5^2 \stackrel{?}{=} 25$

$25 = 25$

$2(0) - 5 \stackrel{?}{=} -5$

$-5 = -5$

Check:

$(-4)^2 + (-3)^2 \stackrel{?}{=} 25$

$25 = 25$

$2(-4) - (-3) \stackrel{?}{=} -5$

$-5 = -5$

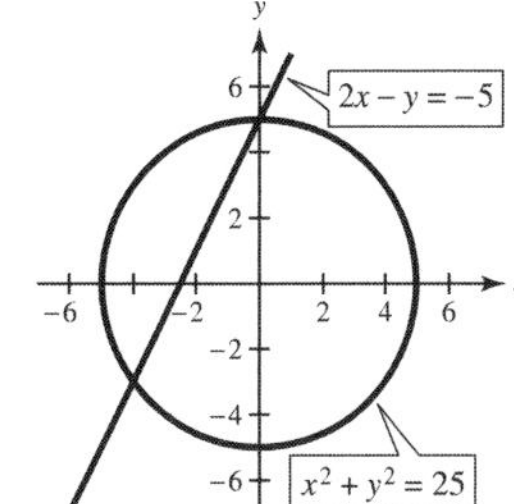

9. $\begin{cases} x - 2y = 4 \\ x^2 - y = 0 \end{cases}$

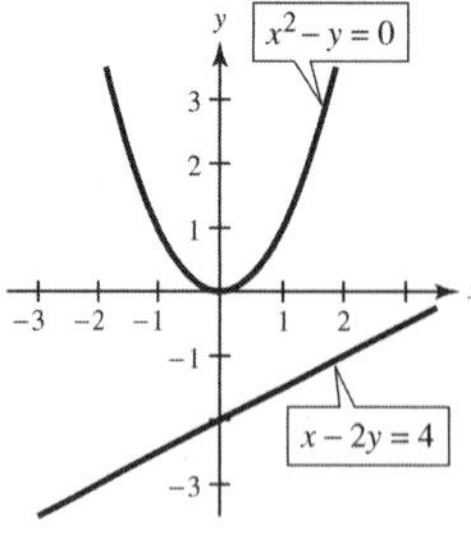

Solve for y.

$$x - 4 = 2y$$
$$\tfrac{1}{2}x - 2 = y$$
$$x^2 = y$$

Solutions: none

11. $\begin{cases} x^2 - y^2 = 16 \\ 3x - y = 12 \end{cases}$

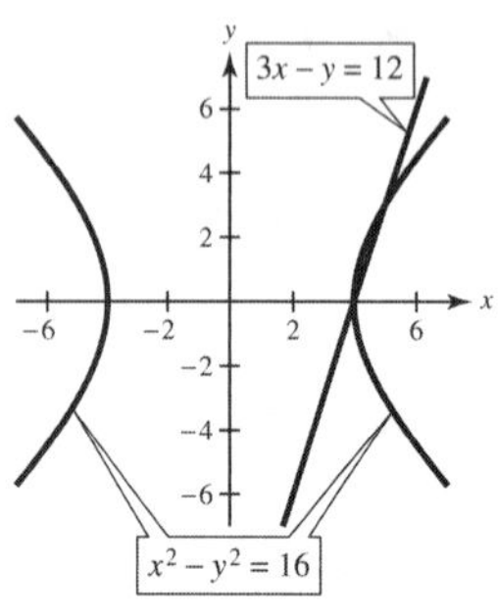

Rewrite.

$$\frac{x^2}{16} - \frac{y^2}{16} = 1$$

Solve for y.

$$y = 3x - 12$$

Solutions: $(5, 3), (4, 0)$

Check:

$$5^2 - 3^2 \stackrel{?}{=} 16$$
$$16 = 16$$
$$3(5) - 3 \stackrel{?}{=} 12$$
$$12 = 12$$

Check:

$$4^2 - 0^2 \stackrel{?}{=} 16$$
$$16 = 16$$
$$3(4) - 0 \stackrel{?}{=} 12$$
$$12 = 12$$

13. $\begin{cases} y = 2x^2 \\ y = 6x - 4 \end{cases}$

$$2x^2 = 6x - 4$$
$$2x^2 - 6x + 4 = 0$$
$$2(x^2 - 3x + 2) = 0$$
$$2(x - 2)(x - 1) = 0$$

$x = 2$ $\qquad$ $x = 1$

$y = 2(2)^2 = 8$ $\qquad$ $y = 2(1)^2 = 1$

$(2, 8)$ $\qquad$ $(1, 2)$

15. $\begin{cases} x^2 + y^2 = 4 \\ x + y = 2 \end{cases}$

$$y = 2 - x$$
$$x^2 + (2 - x)^2 = 4$$
$$x^2 + 4 - 4x + x^2 = 4$$
$$2x^2 - 4x = 0$$
$$2x(x - 2) = 0$$

$2x = 0$ $\qquad$ $x - 2 = 0$

$x = 0$ $\qquad$ $x = 2$

$y = 2 - 0$ $\qquad$ $y = 2 - 2$

$y = 2$ $\qquad$ $y = 0$

$(0, 2)$ $\qquad$ $(2, 0)$

17. $\begin{cases} x^2 - 4y^2 = 16 \\ x^2 + y^2 = 1 \end{cases}$

$$y^2 = 1 - x^2$$
$$x^2 - 4(1 - x^2) = 16$$
$$x^2 - 4 + 4x^2 = 16$$
$$5x^2 = 20$$
$$x^2 = 4$$
$$x = \pm 2$$

$y^2 = 1 - 4 = -3$

No real solution

19. $\begin{cases} y = x^2 - 3 \\ x^2 + y^2 = 9 \end{cases}$

$$x^2 = y + 3$$
$$y + 3 + y^2 = 9$$
$$y^2 + y - 6 = 0$$
$$(y + 3)(y - 2) = 0$$

$y + 3 = 0$ $\qquad$ $y - 2 = 0$

$y = -3$ $\qquad$ $y = 2$

$x^2 = -3 + 3$ $\qquad$ $x^2 = 2 + 3$

$x^2 = 0$ $\qquad$ $x^2 = 5$

$x = 0$ $\qquad$ $x = \pm\sqrt{5}$

$(0, -3)$ $\qquad$ $(\pm\sqrt{5}, 2)$

21. $\begin{cases} 16x^2 + 9y^2 = 144 \\ 4x + 3y = 12 \end{cases}$

$$4x = 12 - 3y$$
$$x = \frac{12 - 3y}{4}$$
$$16\left(\frac{12 - 3y}{4}\right)^2 + 9y^2 = 144$$
$$16\left(\frac{144 - 72y + 9y^2}{16}\right) + 9y^2 = 144$$
$$144 - 72y + 9y^2 + 9y^2 = 144$$
$$18y^2 - 72y = 0$$
$$18y(y - 4) = 0$$

$18y = 0$ | $y - 4 = 0$

$y = 0$ | $y = 4$

$x = \frac{12 - 3(0)}{4}$ | $x = \frac{12 - 3(4)}{4}$

$x = 3$ | $x = 0$

$(3, 0)$ | $(0, 4)$

23. $\begin{cases} x^2 - y^2 = 9 \\ x^2 + y^2 = 1 \end{cases}$

$$x^2 = 1 - y^2$$
$$1 - y^2 - y^2 = 9$$
$$-2y^2 = 8$$
$$y^2 = -4$$

No real solution

25. $\begin{cases} x^2 + 2y = 1 \\ x^2 + y^2 = 4 \end{cases}$

$$\begin{array}{rcrcr} x^2 & + & 2y & = & 1 \\ -x^2 & - & y^2 & = & -4 \\ \hline -y^2 & + & 2y & = & -3 \end{array}$$

$$y^2 - 2y - 3 = 0$$
$$(y - 3)(y + 1) = 0$$

$y - 3 = 0$ | $y + 1 = 0$

$y = 3$ | $y = -1$

$x^2 + 3^2 = 4$ | $x^2 + (-1)^2 = 4$

$x^2 = -5$ | $x^2 = 3$

$x = \pm\sqrt{3}$

No real solution | $\left(\pm\sqrt{3}, -1\right)$

27. $\begin{cases} -x + y^2 = 10 \\ x^2 - y^2 = -8 \end{cases}$

$$-x + x^2 = 2$$
$$x^2 - x - 2 = 0$$
$$(x - 2)(x + 1) = 0$$

$x - 2 = 0$ | $x + 1 = 0$

$x = 2$ | $x = -1$

$-2 + y^2 = 10$ | $-1(-1) + y^2 = 10$

$y^2 = 12$ | $y^2 = 9$

$y = \pm\sqrt{12}$ | $y = \pm 3$

$y = \pm 2\sqrt{3}$

$\left(2, \pm 2\sqrt{3}\right)$ | $(-1, \pm 3)$

29. Equation of circle: $(x-0)^2 + (y-0)^2 = 1^2$

$$x^2 + y^2 = 1$$

Equation of Clarke Street (line): points: $(-2,-1)$ and $(5,0)$; slope: $m = \dfrac{0-(-1)}{5-(-2)} = \dfrac{1}{7}$;

Equation: $y - 0 = \dfrac{1}{7}(x-5)$

$$y = \frac{1}{7}x - \frac{5}{7}$$

$$7y = x - 5$$

Find the intersection of the circle and the line by solving the system:

$$\begin{aligned} x^2 + y^2 &= 1 \\ 7y &= x - 5 \\ x &= 7y + 5 \\ (7y+5)^2 + y^2 &= 1 \\ 49y^2 + 70y + 25 + y^2 &= 1 \\ 50y^2 + 70y + 24 &= 0 \\ 25y^2 + 35y + 12 &= 0 \\ (5y+4)(5y+3) &= 0 \end{aligned}$$

$y = -\dfrac{4}{5}$ | $y = -\dfrac{3}{5}$

$x = 7\left(-\dfrac{4}{5}\right) + 5 = -\dfrac{28}{5} + \dfrac{25}{5}$ | $x = 7\left(-\dfrac{3}{5}\right) + 5 = -\dfrac{21}{5} + \dfrac{25}{5}$

$x = -\dfrac{3}{5}$ | $x = \dfrac{4}{5}$

$\left(-\dfrac{3}{5}, -\dfrac{4}{5}\right)$ | $\left(\dfrac{4}{5}, -\dfrac{3}{5}\right)$

31. In addition to zero, one, or infinitely many solutions, a system of nonlinear equations can have two or more solutions.

33. Substitution

35. $\begin{cases} y = x^2 - 5 \\ 3x + 2y = 10 \end{cases}$

$$\begin{aligned} 3x + 2(x^2 - 5) &= 10 \\ 3x + 2x^2 - 10 &= 10 \\ 2x^2 + 3x - 20 &= 0 \\ (2x-5)(x+4) &= 0 \end{aligned}$$

$2x - 5 = 0$ | $x + 4 = 0$

$x = \frac{5}{2}$ | $x = -4$

$y = \left(\frac{5}{2}\right)^2 - 5 = \frac{25}{4} - \frac{20}{4} = \frac{5}{4}$ | $y = (-4)^2 - 5 = 16 - 5 = 11$

$\left(\frac{5}{2}, \frac{5}{4}\right)$ | $(-4, 11)$

37. $\begin{cases} y = \sqrt{4-x} \\ x + 3y = 6 \end{cases}$

$$\begin{aligned} x + 3\sqrt{4-x} &= 6 \\ 3\sqrt{4-x} &= 6 - x \\ 9(4-x) &= 36 - 12x + x^2 \\ 36 - 9x &= 36 - 12x + x^2 \\ 0 &= x^2 - 3x \\ 0 &= x(x-3) \end{aligned}$$

$0 = x$ | $x - 3 = 0$

$y = \sqrt{4-0} = 2$ | $x = 3$

| $y = \sqrt{4-3} = 1$

$(0, 2)$ | $(3, 1)$

39. $\begin{cases} \dfrac{x^2}{4} + y^2 = 1 \\ x^2 + \dfrac{y^2}{4} = 1 \end{cases}$

$$x^2 + 4y^2 = 4$$
$$4x^2 + y^2 = 4$$
$$x^2 + 4y^2 = 4$$
$$-16x^2 - 4y^2 = -16$$
$$-15x^2 = -12$$
$$x^2 = \frac{12}{15}$$
$$x^2 = \frac{4}{5}$$
$$x = \pm\sqrt{\frac{4}{5}}$$
$$x = \pm\frac{2}{\sqrt{5}} \cdot \frac{\sqrt{5}}{\sqrt{5}}$$
$$x = \pm\frac{2\sqrt{5}}{5}$$

$$\left(\pm\frac{2\sqrt{5}}{5}\right)^2 + 4y^2 = 4$$
$$\frac{4}{5} + 4y^2 = 4$$
$$4y^2 = \frac{16}{5}$$
$$y^2 = \frac{4}{5}$$
$$y = \pm\frac{2\sqrt{5}}{5}$$

$$\left(\pm\frac{2\sqrt{5}}{5}, \pm\frac{2\sqrt{5}}{5}\right)$$

41. $\begin{cases} y^2 - x^2 = 10 \\ x^2 + y^2 = 16 \end{cases}$

$$-x^2 + y^2 = 10$$
$$x^2 + y^2 = 16$$
$$2y^2 = 26$$
$$y^2 = 13$$
$$y = \pm\sqrt{13}$$
$$13 - x^2 = 10$$
$$-x^2 = -3$$
$$x^2 = 3$$
$$x = \pm\sqrt{3}$$
$$\left(\pm\sqrt{3}, \pm\sqrt{13}\right)$$

43. *Verbal Model:* 15 + [Base] + [Height] = [Perimeter]

[Base]^2 + [Height]^2 = [Hypotenuse]^2

Labels: Base $= x$

Height $= y$

System: $\begin{matrix} 15 + x + y = 36 \\ x^2 + y^2 = 15^2 \end{matrix} \Rightarrow \begin{matrix} x + y = 21 \\ x^2 + y^2 = 225 \end{matrix}$

$$y = 21 - x$$
$$x^2 + (21 - x)^2 = 225$$
$$x^2 + 441 - 42x + x^2 = 225$$
$$2x^2 - 42x + 216 = 0$$
$$x^2 - 21x + 108 = 0$$
$$(x - 9)(x - 12) = 0$$

$x = 9$ $\quad x = 12$

$y = 12$ $\quad y = 9$

9 meters × 12 meters

45. Find the equation of the line with points $(0, 10)$ and $(5, 0)$.

$$m = \frac{0 - 10}{5 - 0} = -2, y = -2x + 10$$

Find the point of intersection of the line and the hyperbola.

$\begin{cases} y = -2x + 10 \\ \dfrac{x^2}{9} - \dfrac{y^2}{16} = 1 \end{cases} \Rightarrow \begin{matrix} y = -2x + 10 \\ 16x^2 - 9y^2 = 144 \end{matrix}$

$$16x^2 - 9(-2x + 10)^2 = 144$$
$$16x^2 - 9(4x^2 - 40x + 100) = 144$$
$$16x^2 - 36x^2 + 360x - 900 - 144 = 0$$
$$-20x^2 + 360x - 1044 = 0$$
$$5x^2 - 90x + 261 = 0$$
$$x = \frac{90 \pm \sqrt{90^2 - 4(5)(261)}}{10} \approx 3.633$$
$$y = -2(3.633) + 10$$
$$y \approx 2.733$$
$$(3.633, 2.733)$$

47. Solve one of the equations for one variable in terms of the other. Substitute that expression into the other equation. Solve the equation. Back-substitute the solution into the first equation to find the value of the other variable. Check the solution to see that it satisfies both of the original equations.

49. Two. The line can intersect a branch of the hyperbola at most twice, and it can intersect only one point on each branch at the same time.

51. $\sqrt{6 - 2x} = 4$

$\left(\sqrt{6 - 2x}\right)^2 = 4^2$

$6 - 2x = 16$

$-2x = 10$

$x = -5$

Check: $\sqrt{6 - 2(-5)} \stackrel{?}{=} 4$

$\sqrt{6 + 10} \stackrel{?}{=} 4$

$\sqrt{16} \stackrel{?}{=} 4$

$4 = 4$

53. $\sqrt{x} = x - 6$

$\left(\sqrt{x}\right)^2 = (x - 6)^2$

$x = x^2 - 12x + 36$

$0 = x^2 - 13x + 36$

$0 = (x - 9)(x - 4)$

$9 = x \qquad x = 4$

Check: $\sqrt{9} \stackrel{?}{=} 9 - 6$

$3 = 3$

Check: $\sqrt{4} \stackrel{?}{=} 4 - 6$

$2 \neq -2$

Solution: $x = 9$

55. $3^x = 243$

$3^x = 3^5$

$x = 5$

Check: $3^5 \stackrel{?}{=} 243$

$243 = 243$

57. $5^{x-1} = 310$

$\log 5^{x-1} = \log 310$

$(x - 1)\log 5 = \log 310$

$x \log 5 - \log 5 = \log 310$

$x \log 5 = \log 310 + \log 5$

$x = \dfrac{\log 310 + \log 5}{\log 5}$

$x \approx 4.564$

Check: $5^{4.564-1} \stackrel{?}{=} 310$

$5^{3.564} \stackrel{?}{=} 310$

$310 \stackrel{?}{=} 310$

59. $\log_{10} x = 0.01$

$10^{0.01} = x$

$1.023 \approx x$

Check: $\log_{10} 1.023 \stackrel{?}{=} 0.01$

$0.01 = 0.01$

61. $2 \ln(x + 1) = -2$

$\ln(x + 1) = -1$

$e^{-1} = x + 1$

$e^{-1} - 1 = x$

$-0.632 \approx x$

Check: $2 \ln(-0.632 + 1) = -2$

$2 \ln(0.368) \stackrel{?}{=} -2$

$-2 = -2$

Review Exercises for Chapter 10

1. Ellipse

3. Circle

5. Hyperbola

7. Circle

9. Parabola

11. Center: $(0, 0)$

Radius: 6

$x^2 + y^2 = 6^2$

$x^2 + y^2 = 36$

13. $x^2 + y^2 = 64$

Center: $(0, 0)$

Radius: 8

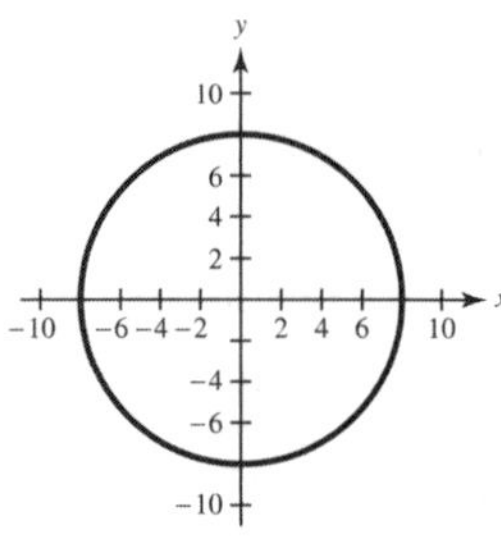

15. Center: $(2, 6)$; Radius: 3

$(x - h)^2 + (y - k)^2 = r^2$

$(x - 2)^2 + (y - 6)^2 = 9$

17. $x^2 + y^2 + 6x + 8y + 21 = 0$

$$(x^2 + 6x + 9) + (y^2 + 8y + 16) = -21 + 9 + 16$$

$$(x + 3)^2 + (y + 4)^2 = 4$$

Center: $(-3, -4)$

Radius: 2

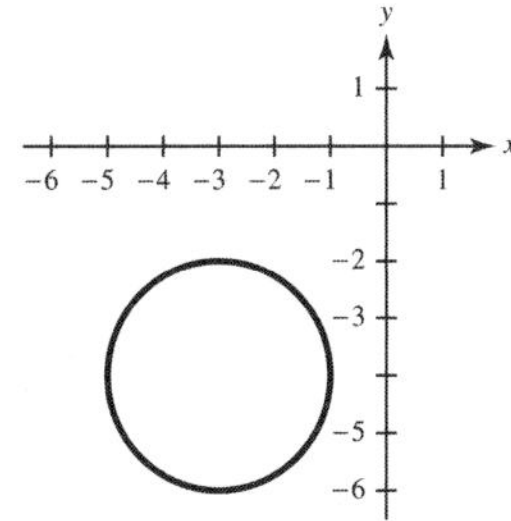

19. Vertex: $(0, 0)$; focus: $(6, 0)$

$$(y - k)^2 = 4p(x - h)$$

$$p = 6$$

$$(y - 0)^2 = 4(6)(x - 0)$$

$$y^2 = 24x$$

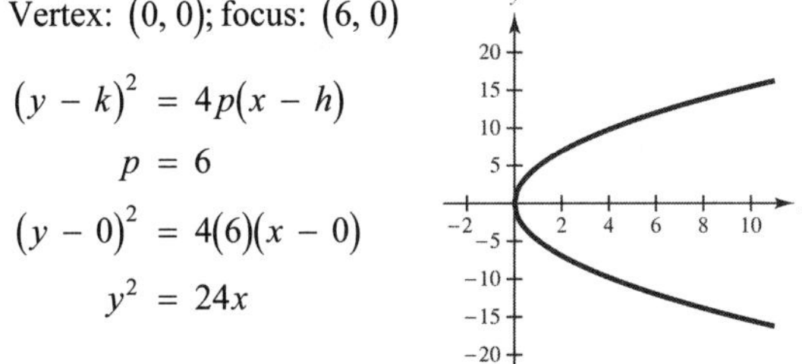

21. Vertex: $(0, 5)$; focus: $(2, 5)$

$$(y - k)^2 = 4p(x - h)$$

$$p = 2$$

$$(y - 5)^2 = 4(2)(x - 0)$$

$$(y - 5)^2 = 8x$$

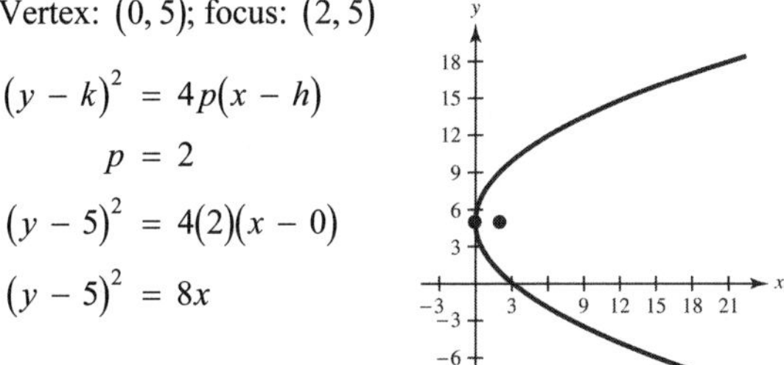

23. $y = \frac{1}{2}x^2 - 8x + 7$

$$2y = x^2 - 16x + 14$$

$$2y - 14 = x^2 - 16x$$

$$2y - 14 + 64 = x^2 - 16x + 64$$

$$2y + 50 = (x - 8)^2$$

$$2(y + 25) = (x - 8)^2$$

Vertex: $(8, -25)$

$$4p = 2$$

$$p = \frac{1}{2}$$

Focus: $\left(8, -25 + \frac{1}{2}\right)$

$$\left(8, -\frac{50}{2} + \frac{1}{2}\right)$$

$$\left(8, -\frac{49}{2}\right)$$

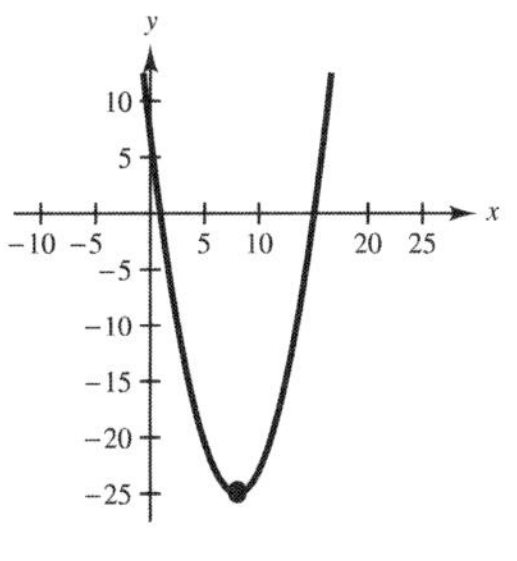

25. Vertices: $(0, -5), (0, 5)$

Co-vertices: $(-2, 0), (2, 0)$

$$\frac{x^2}{a^2} + \frac{y^2}{b^2} = 1$$

$$\frac{x^2}{2^2} + \frac{y^2}{5^2} = 1$$

$$\frac{x^2}{4} + \frac{y^2}{25} = 1$$

27. Major axis (vertical) 6 units

Minor axis 4 units

$$\frac{x^2}{a^2} + \frac{y^2}{b^2} = 1$$

$2b = 6$ $\quad$ $2a = 4$

$b = 3$ $\quad$ $a = 2$

$$\frac{x^2}{2^2} + \frac{y^2}{3^2} = 1$$

$$\frac{x^2}{4} + \frac{y^2}{9} = 1$$

29. $\frac{x^2}{64} + \frac{y^2}{16} = 1$

$a^2 = 64$ $\quad$ $b^2 = 16$

$a = 8$ $\quad$ $b = 4$

Vertices: $(\pm 8, 0)$

Co-vertices: $(0, \pm 4)$

Horizontal axis

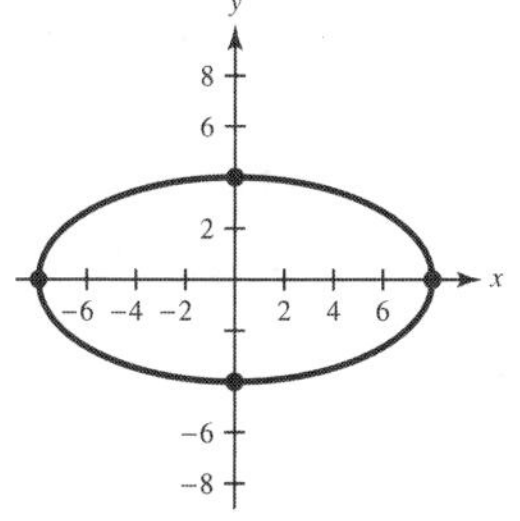

31. $36x^2 + 9y^2 - 36 = 0$

$$x^2 + \frac{y^2}{4} = 1$$

$a^2 = 4$ $\quad$ $b^2 = 1$

$a = \pm 2$ $\quad$ $b = \pm 1$

Vertices: $(0, \pm 2)$

Co-vertices: $(\pm 1, 0)$

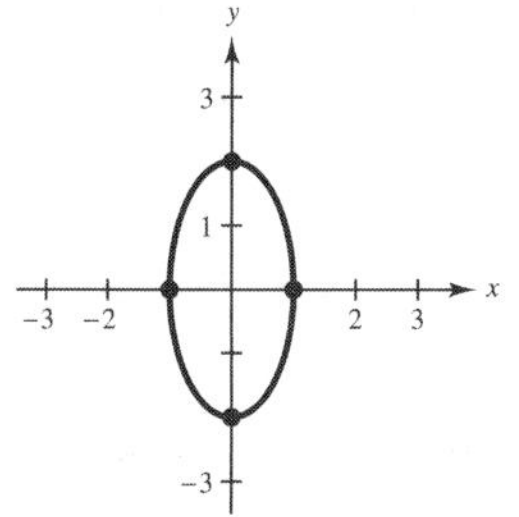

33. Vertices: $(-2, 4), (8, 4)$

Co-vertices: $(3, 0), (3, 8)$

Center: $(3, 4)$

Major axis is horizontal so $a = 5$.

Minor axis is vertical so $b = 4$.

$$\frac{(x-h)^2}{a^2} + \frac{(y-k)^2}{b^2} = 1$$

$$\frac{(x-3)^2}{5^2} + \frac{(y-4)^2}{4^2} = 1$$

$$\frac{(x-3)^2}{25} + \frac{(y-4)^2}{16} = 1$$

35. Vertices: $(0, 0), (0, 8)$

Co-vertices: $(-3, 4), (3, 4)$

Center: $(0, 4)$

Major axis is vertical so $b = 4$.

Minor axis is horizontal so $a = 3$.

$$\frac{(x-h)^2}{a^2} + \frac{(y-k)^2}{b^2} = 1$$

$$\frac{(x-0)^2}{3^2} + \frac{(y-4)^2}{4^2} = 1$$

$$\frac{x^2}{9} + \frac{(y-4)^2}{16} = 1$$

37. $9(x+1)^2 + 4(y-2)^2 = 144$

$$\frac{(x+1)^2}{16} + \frac{(y-2)^2}{36} = 1$$

Center: $(-1, 2)$

Vertices: $(-1, -4), (-1, 8)$

$a^2 = 36 \qquad b^2 = 16$

$a = 6 \qquad b = 4$

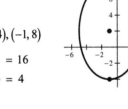

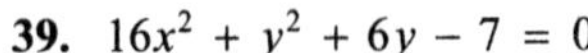

39. $16x^2 + y^2 + 6y - 7 = 0$

$$16x^2 + (y^2 + 6y + 9) = 7 + 9$$

$$16x^2 + (y+3)^2 = 16$$

$$x^2 + \frac{(y+3)^2}{16} = 1$$

Center: $(0, -3)$

Vertices: $(0, -7), (0, 1)$

$a^2 = 16 \qquad b^2 = 1$

$a = 4 \qquad b = 1$

Vertical axis

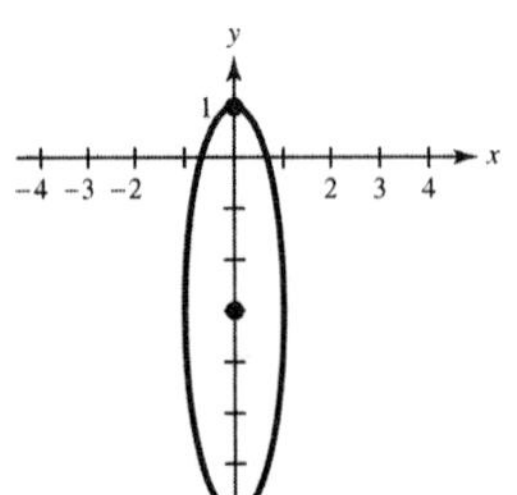

41. $x^2 - y^2 = 25$

$$\frac{x^2}{25} - \frac{y^2}{25} = 1$$

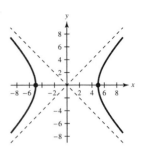

Center: $(0, 0)$

Vertices: $(\pm 5, 0)$

Asymptotes: $y = \pm x$

$a^2 = 25 \qquad b^2 = 25$

$a = 5 \qquad b = 5$

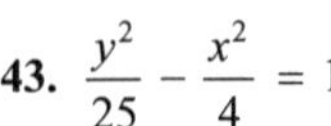

43. $\dfrac{y^2}{25} - \dfrac{x^2}{4} = 1$

Center: $(0, 0)$

Vertices: $(0, \pm 5)$

Asymptotes: $y = \pm\frac{5}{2}x$

$a^2 = 25 \qquad b^2 = 4$

$a = 5 \qquad b = 2$

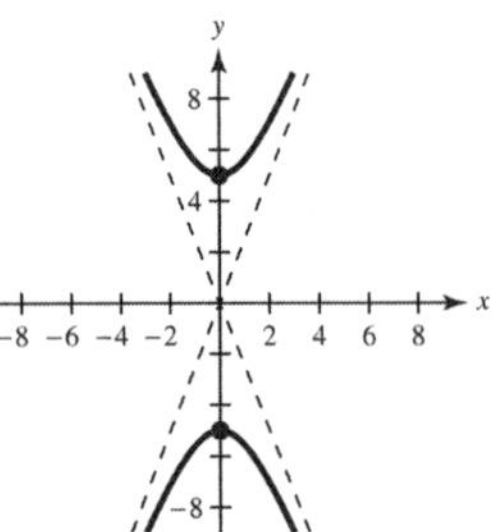

45. Vertices: $(\pm 2, 0)$

Asymptotes: $y = \pm\frac{3}{2}x$

Center: $(0, 0)$

$a = 2 \qquad \frac{b}{a} = \pm\frac{3}{2}$

$\frac{b}{2} = \pm\frac{3}{2}$

$b = 3$

$$\frac{x^2}{a^2} - \frac{y^2}{b^2} = 1$$

$$\frac{x^2}{2^2} - \frac{y^2}{3^2} = 1$$

$$\frac{x^2}{4} - \frac{y^2}{9} = 1$$

47. Center: $(0, 0)$

Vertices: $(0, -8)$, $(0, 8)$

Asymptotes: $y = \frac{4}{5}x$, $y = -\frac{4}{5}x$

Vertical transverse axis

$$\frac{y^2}{a^2} - \frac{x^2}{b^2} = 1$$

$$a = 8 \qquad \frac{a}{b} = \frac{4}{5}$$

$$\frac{8}{b} = \frac{4}{5}$$

$$b = 10$$

$$\frac{y^2}{8^2} - \frac{x^2}{10^2} = 1$$

$$\frac{y^2}{64} - \frac{x^2}{100} = 1$$

49. $\frac{(x-3)^2}{9} - \frac{(y+1)^2}{4} = 1$

Center: $(3, -1)$

Vertices: $(0, -1), (6, -1)$

$$a^2 = 9 \qquad b^2 = 4$$

$$a = 3 \qquad b = 2$$

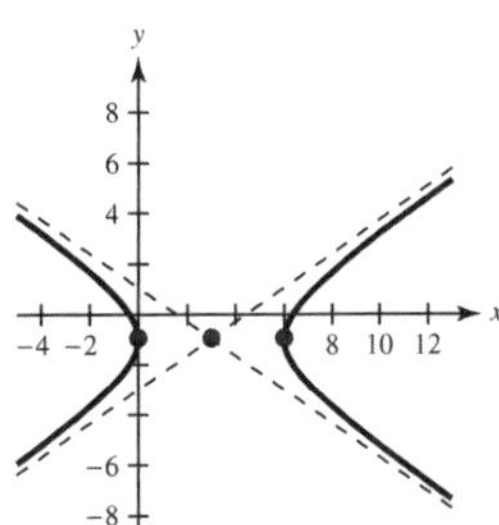

51.

$$8y^2 - 2x^2 + 48y + 16x + 8 = 0$$

$$(8y^2 + 48y) - (2x^2 - 16x) = -8$$

$$8(y^2 + 6y + 9) - 2(x^2 - 8x + 16) = -8 + 72 - 32$$

$$8(y+3)^2 - 2(x-4)^2 = 32$$

$$\frac{(y+3)^2}{4} - \frac{(x-4)^2}{16} = 1$$

Center: $(4, -3)$

Vertices: $(4, -1), (4, -5)$

$$a^2 = 4 \qquad b^2 = 16$$

$$a = 2 \qquad b = 4$$

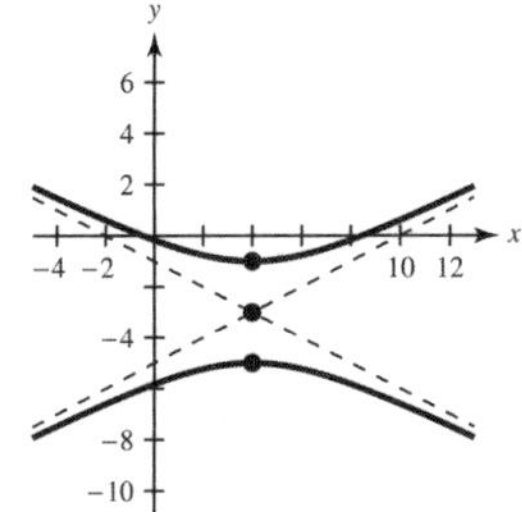

53. Center: $(-4, 6)$

Vertices: $(-6, 6), (-2, 6)$

Point: $(0, 12)$

Horizontal hyperbola

$$\frac{(x-h)^2}{a^2} - \frac{(y-k)^2}{b^2} = 1$$

$$a = 2$$

$$a^2 = 4$$

$$\frac{[0-(-4)]^2}{4} - \frac{(12-6)^2}{b^2} = 1$$

$$\frac{16}{4} - \frac{36}{b^2} = 1$$

$$4 - 1 = \frac{36}{b^2}$$

$$3 = \frac{36}{b^2}$$

$$b^2 = 12$$

$$\frac{(x+4)^2}{4} - \frac{(y-6)^2}{12} = 1$$

55. $\begin{cases} y = x^2 \\ y = 3x \end{cases}$

Solutions: $(0, 0), (3, 9)$

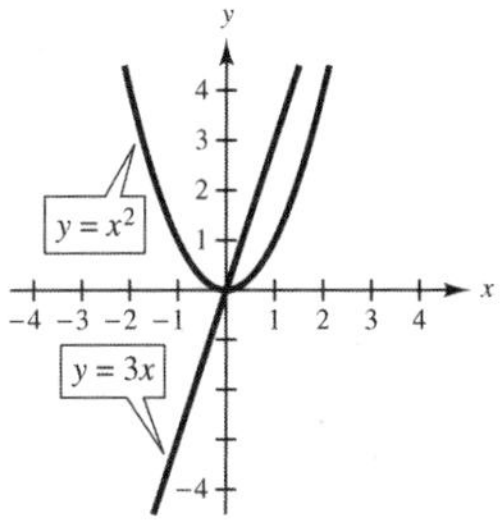

Check:

$0 = 0^2$

$0 = 3(0)$

Check:

$9 \stackrel{?}{=} 3^2$

$9 = 9$

$9 \stackrel{?}{=} 3(3)$

$9 = 9$

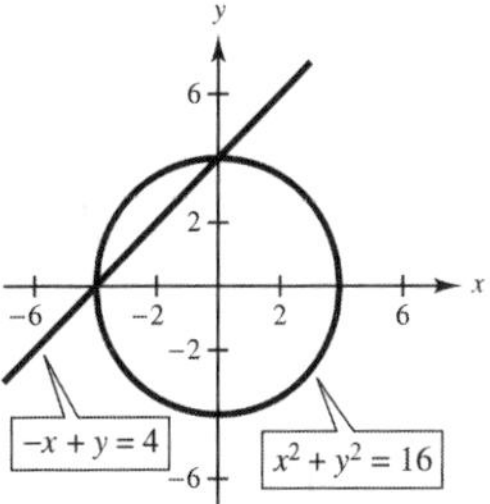

57. $\begin{cases} x^2 + y^2 = 16 \\ -x + y = 4 \end{cases} \Rightarrow \begin{matrix} y = \pm\sqrt{16 - x^2} \\ y = x + 4 \end{matrix}$

Solutions: $(-4, 0), (0, 4)$

Check:

$(-4)^2 + 0^2 \stackrel{?}{=} 16$

$16 = 16$

$-(-4) + 0 \stackrel{?}{=} 4$

$4 = 4$

Check:

$0^2 + 4^2 \stackrel{?}{=} 16$

$16 = 16$

$-0 + 4 \stackrel{?}{=} 4$

$4 = 4$

59. $\begin{cases} y = 5x^2 \\ y = -15x - 10 \end{cases}$

$$5x^2 = -15x - 10$$
$$5x^2 + 15x + 10 = 0$$
$$x^2 + 3x + 2 = 0$$
$$(x + 2)(x + 1) = 0$$

$x + 2 = 0$ $\quad$ $x + 1 = 0$

$x = -2$ $\quad$ $x = -1$

$y = 5(-2)^2 = 5(4) = 20$ $\quad$ $y = 5(-1)^2 = 5(1) = 5$

$(-2, 20)$ $\quad$ $(-1, 5)$

61. $\begin{cases} x^2 + y^2 = 1 \\ x + y = -1 \end{cases}$

$$y = -1 - x$$
$$x^2 + (-1 - x)^2 = 1$$
$$x^2 + 1 + 2x + x^2 - 1 = 0$$
$$2x^2 + 2x = 0$$
$$2x(x + 1) = 0$$

$2x = 0$ $\quad$ $x + 1 = 0$

$x = 0$ $\quad$ $x = -1$

$y = -1 - 0 = -1$ $\quad$ $y = -1 - (-1) = 0$

$(0, -1)$ $\quad$ $(-1, 0)$

63. $\begin{cases} 4x + y^2 = 2 \\ 2x - y = -11 \end{cases}$

$$y = 2x + 11$$
$$4x + (2x + 11)^2 = 2$$
$$4x + 4x^2 + 44x + 121 = 2$$
$$4x^2 + 48x + 119 = 0$$
$$(2x + 17)(2x + 7) = 0$$

$2x + 17 = 0$ $\quad$ $2x + 7 = 0$

$x = -\frac{17}{2}$ $\quad$ $x = -\frac{7}{2}$

$y = 2\left(-\frac{17}{2}\right) + 11 = -17 + 11 = -6$ $\quad$ $y = 2\left(-\frac{7}{2}\right) + 11 = -7 + 11 = 4$

$\left(-\frac{17}{2}, -6\right)$ $\quad$ $\left(-\frac{7}{2}, 4\right)$

65. $\begin{cases} x^2 + y^2 = 9 \\ x + 2y = 3 \end{cases}$

$$x = 3 - 2y$$
$$(3 - 2y)^2 + y^2 = 9$$
$$9 - 12y + 4y^2 + y^2 = 9$$
$$5y^2 - 12y = 0$$
$$y(5y - 12) = 0$$
$$5y - 12 = 0$$

$y = 0$ $\quad$ $y = \frac{12}{5}$

$x = 3 - 2(0) = 3$ $\quad$ $x = 3 - 2\left(\frac{12}{5}\right) = \frac{15}{5} - \frac{24}{5} = -\frac{9}{5}$

$(3, 0)$ $\quad$ $\left(-\frac{9}{5}, \frac{12}{5}\right)$

67. $\begin{cases} 6x^2 - y^2 = 15 \\ x^2 + y^2 = 13 \end{cases}$

$$7x^2 = 28$$
$$x^2 = 4$$
$$x = \pm 2$$
$$(\pm 2)^2 + y^2 = 13$$
$$y^2 = 9$$
$$y = \pm 3$$

$(\pm 2, \pm 3)$

69. $\begin{cases} x^2 + y^2 = 7 \\ x^2 - y^2 = 1 \end{cases}$

$$2x^2 = 8$$
$$x^2 = 4$$
$$x = \pm 2$$
$$(\pm 2)^2 + y^2 = 7$$
$$y^2 = 3$$
$$y = \pm\sqrt{3}$$

$\left(\pm 2, \pm\sqrt{3}\right)$

71. $\begin{cases} x^2 - y^2 = 4 \\ x^2 + y^2 = 4 \end{cases}$

$$2x^2 = 8$$
$$x^2 = 4$$
$$x = \pm 2$$
$$(\pm 2)^2 - y^2 = 4$$
$$-y^2 = 0$$
$$y = 0$$

$(\pm 2, 0)$

73. $\begin{cases} x^2 + y^2 = 13 \\ 2x^2 + 3y^2 = 30 \end{cases}$

$$-2x^2 - 2y^2 = -26$$
$$2x^2 + 3y^2 = 30$$
$$y^2 = 4$$
$$y = \pm 2$$

$$x^2 + (\pm 2)^2 = 13$$
$$x^2 = 9$$
$$x = \pm 3$$

$(\pm 3, \pm 2)$

75. $\begin{cases} 4x^2 + 9y^2 = 36 \\ 2x^2 - 9y^2 = 18 \end{cases}$

$$6x^2 = 54$$
$$x^2 = 9$$
$$x = \pm 3$$

$$4(\pm 3)^2 + 9y^2 = 36$$
$$36 + 9y^2 = 36$$
$$9y^2 = 0$$
$$y^2 = 0$$
$$y = 0$$

$(\pm 3, 0)$

77. *Verbal Model:* 2 · [Length] + 2 · [Width] = [Perimeter]

[Length]^2 + [Width]^2 = [Diagonal]^2

Labels: Length = x

Width = y

System:

$$2x + 2y = 28$$
$$x^2 + y^2 = 10^2$$
$$x + y = 14$$
$$y = 14 - x$$

$$x^2 + (14 - x)^2 = 100$$
$$x^2 + 196 - 28x + x^2 - 100 = 0$$
$$2x^2 - 28x + 96 = 0$$
$$x^2 - 14x + 48 = 0$$
$$(x - 8)(x - 6) = 0$$

$x = 8$ $x = 6$

$y = 6$ $y = 8$

6 cm × 8 cm

79. *Verbal Model:* [Length] × [Width] = [Area]

[Length]^2 + [Width]^2 = [Diagonal]^2

Labels: Length = x

Width = y

System:

$$xy = 3000$$
$$x^2 + y^2 = 85^2$$
$$y = \frac{3000}{x}$$
$$x^2 + \left(\frac{3000}{x}\right)^2 = 7225$$
$$x^2 + \frac{9{,}000{,}000}{x^2} = 7225$$
$$x^4 + 9{,}000{,}000 = 7225x^2$$
$$x^4 - 7225x^2 + 9{,}000{,}000 = 0$$
$$(x^2 - 5625)(x^2 - 1600) = 0$$

$x^2 = 5625$ $x^2 = 1600$

$x = 75$ $x = 40$

$y = \frac{3000}{75}$ $y = \frac{3000}{40}$

$= 40$ $= 75$

40 feet × 75 feet

81. *Verbal Model:* [Length of first piece] + [Length of second piece] = 100

[Area of Square 1] = [Area of Square 2] + 144

Labels: Length of first piece = x

Length of second piece = y

System:

$$x + y = 100$$
$$\left(\frac{x}{4}\right)^2 = \left(\frac{y}{4}\right)^2 + 144$$
$$x = 100 - y$$
$$\left(\frac{100 - y}{4}\right)^2 = \left(\frac{y}{4}\right)^2 + 144$$
$$\frac{10{,}000 - 200y + y^2}{16} = \frac{y^2}{16} + 144$$
$$10{,}000 - 200y + y^2 = y^2 + 2304$$
$$-200y = -7696$$
$$y \approx 38.48$$
$$x \approx 100 - 38.48$$
$$\approx 61.52$$

38.48 inches; 61.52 inches

Chapter Test for Chapter 10

1. Center: $(-2, -3)$

Radius: 4

$(x - h)^2 + (y - k)^2 = r^2$

$(x + 2)^2 + (y + 3)^2 = 16$

2.
$$\begin{aligned} x^2 + y^2 - 2x - 6y + 1 &= 0 \\ (x^2 - 2x) + (y^2 - 6y) &= -1 \\ (x^2 - 2x + 1) + (y^2 - 6y + 9) &= -1 + 1 + 9 \\ (x - 1)^2 + (y - 3)^2 &= 9 \end{aligned}$$

Center: $(1, 3)$

Radius: 3

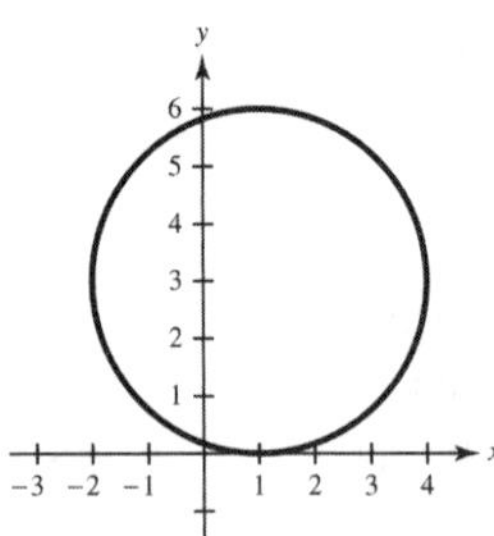

3.
$$\begin{aligned} x^2 + y^2 + 4x - 6y + 4 &= 0 \\ (x^2 + 4x) + (y^2 - 6y) &= -4 \\ (x^2 + 4x + 4) + (y^2 - 6y + 9) &= -4 + 4 + 9 \\ (x + 2)^2 + (y - 3)^2 &= 9 \end{aligned}$$

Center: $(-2, 3)$

Radius: 3

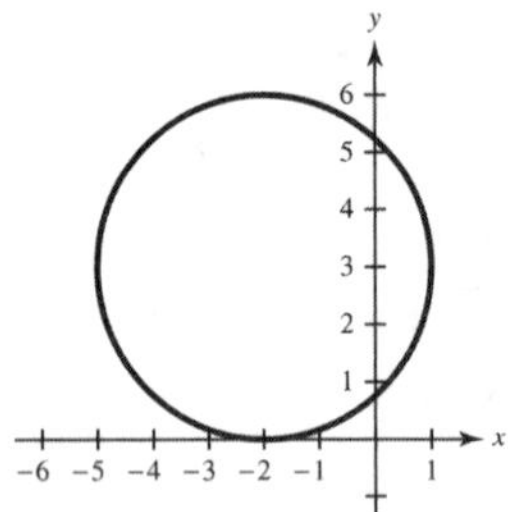

4.
$$\begin{aligned} x &= -3y^2 + 12y - 8 \\ x + 8 &= -3(y^2 - 4y) \\ x + 8 - 12 &= -3(y^2 - 4y + 4) \\ x - 4 &= -3(y - 2)^2 \end{aligned}$$

Vertex: $(4, 2)$

$4p = -\frac{1}{3}$

$p = -\frac{1}{12}$

Focus: $\left(4 - \frac{1}{12}, 2\right) = \left(\frac{48}{12} - \frac{1}{12}, 2\right) = \left(\frac{47}{12}, 2\right)$

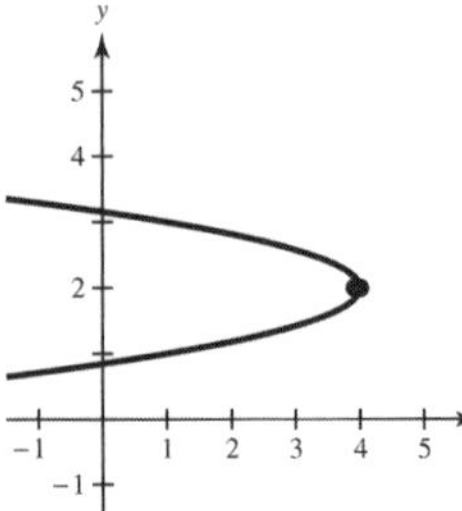

5. Vertex: $(7, -2)$

Focus: $(7, 0)$

Vertical parabola

$$\begin{aligned} (x - h)^2 &= 4p(y - k) \\ (x - 7)^2 &= 4p[y - (-2)] \\ p &= 2 \\ (x - 7)^2 &= 4(2)(y + 2) \\ (x - 7)^2 &= 8(y + 2) \end{aligned}$$

6. Vertices: $(-3, 0), (7, 0)$

Co-vertices: $(2, 3), (2, -3)$

Horizontal axis

Center: $(2, 0)$

$a = 5 \qquad b = 3$

$a^2 = 25 \qquad b^2 = 9$

$$\frac{(x - h)^2}{a^2} + \frac{(y - k)^2}{b^2} = 1$$

$$\frac{(x - 2)^2}{25} + \frac{y^2}{9} = 1$$

7. $16x^2 + 4y^2 = 64$

$$\frac{x^2}{4} + \frac{y^2}{16} = 1$$

Center: $(0, 0)$

Vertices: $(0, \pm 4)$

$a^2 = 16 \qquad b^2 = 4$

$a = 4 \qquad b = 2$

8.

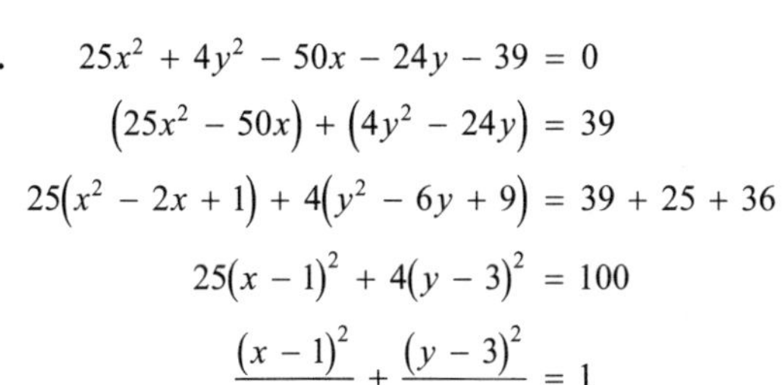

$$25x^2 + 4y^2 - 50x - 24y - 39 = 0$$
$$(25x^2 - 50x) + (4y^2 - 24y) = 39$$
$$25(x^2 - 2x + 1) + 4(y^2 - 6y + 9) = 39 + 25 + 36$$
$$25(x - 1)^2 + 4(y - 3)^2 = 100$$
$$\frac{(x - 1)^2}{4} + \frac{(y - 3)^2}{25} = 1$$

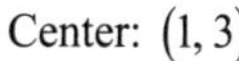

Center: $(1, 3)$

$a^2 = 25$

$a = 5$

Vertices: $(1, -2), (1, 8)$

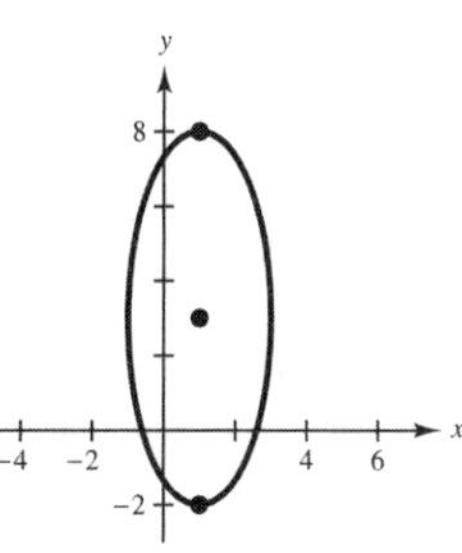

9. Vertices: $(\pm 3, 0)$

Asymptotes: $y = \pm\frac{2}{3}x$

$a = 3 \qquad \frac{b}{a} = \pm\frac{2}{3}$

$\frac{b}{3} = \pm\frac{2}{3}$

$b = 2$

$$\frac{x^2}{a^2} - \frac{y^2}{b^2} = 1$$
$$\frac{x^2}{3^2} - \frac{y^2}{2^2} = 1$$
$$\frac{x^2}{9} - \frac{y^2}{4} = 1$$

10. Vertices: $(0, -5), (0, 5)$

Asymptotes: $y = \pm\frac{5}{2}x$

Vertical transverse axis

$$\frac{y^2}{a^2} - \frac{x^2}{b^2} = 1$$

$a = 5 \qquad \frac{a}{b} = \frac{5}{2}$

$\frac{5}{b} = \frac{5}{2}$

$b = 2$

$$\frac{y^2}{25} - \frac{x^2}{4} = 1$$

11. $4x^2 - 2y^2 - 24x + 20 = 0$

$$(4x^2 - 24x) - 2y^2 = -20$$
$$4(x^2 - 6x + 9) - 2y^2 = -20 + 36$$
$$4(x - 3)^2 - 2y^2 = 16$$
$$\frac{(x - 3)^2}{4} - \frac{y^2}{8} = 1$$

Center: $(3, 0)$

Vertices: $(1, 0), (5, 0)$

$a^2 = 4$

$a = 2$

12. $16y^2 - 25x^2 + 64y + 200x - 736 = 0$

$$(16y^2 + 64y) - (25x^2 - 200x) = 736$$
$$16(y^2 + 4y + 4) - 25(x^2 - 8x + 16) = 736 + 64 - 400$$
$$16(y + 2)^2 - 25(x - 4)^2 = 400$$
$$\frac{(y + 2)^2}{25} - \frac{(x - 4)^2}{16} = 1$$

Center: $(4, -2)$

Vertices: $(4, -7), (4, 3)$

$a^2 = 25 \qquad b^2 = 16$

$a = 5 \qquad b = 4$

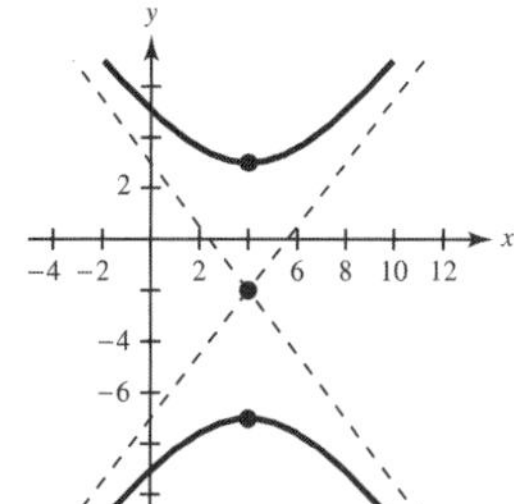

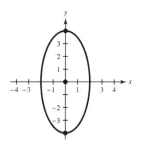

13. $\begin{cases} \dfrac{x^2}{16} + \dfrac{y^2}{9} = 1 \\ 3x + 4y = 12 \end{cases}$

Solve for y in second equation.

$4y = 3x + 12$

$y = -\dfrac{3}{4}x + 3$

Substitute into first equation after multiplying by 144.

$$9x^2 + 16y^2 = 144$$

$$9x^2 + 16\left(-\frac{3}{4}x + 3\right)^2 = 144$$

$$9x^2 + 16\left(\frac{9}{16}x^2 - \frac{9}{2}x + 9\right) = 144$$

$$9x^2 + 9x^2 - 72x + 144 = 144$$

$$18x^2 - 72x = 0$$

$$18x(x - 4) = 0$$

$18x = 0$ $\qquad$ $x - 4 = 0$

$x = 0$ $\qquad$ $x = 4$

$y = -\dfrac{3}{4}(0) + 3$ $\qquad$ $y = -\dfrac{3}{4}(4) + 3$

$= 3$ $\qquad$ $= 0$

$(0, 3)$ $\qquad$ $(4, 0)$

14. $\begin{cases} x^2 + y^2 = 16 \\ \dfrac{x^2}{16} - \dfrac{y^2}{9} = 1 \end{cases}$

Multiply second equation by 144.

$x^2 + y^2 = 16$

$9x^2 + 16y^2 = 144$

Multiply first equation by 16.

$$\begin{aligned} 16x^2 + 16y^2 &= 256 \\ 9x^2 - 16y^2 &= 144 \\ \hline 25x^2 &= 400 \\ x^2 &= 16 \\ x &= \pm 4 \end{aligned}$$

$(\pm 4)^2 + y^2 = 16$

$16 + y^2 = 16$

$y^2 = 0$

$y = 0$

$(4, 0), (-4, 0)$

15. $\begin{cases} x^2 + y^2 = 10 \\ x^2 = y^2 + 2 \end{cases}$

$$\begin{aligned} x^2 + y^2 &= 10 \\ x^2 - y^2 &= 2 \\ \hline 2x^2 &= 12 \\ x^2 &= 6 \\ x &= \pm\sqrt{6} \end{aligned}$$

$(\pm\sqrt{6})^2 + y^2 = 10$

$6 + y^2 = 10$

$y^2 = 4$

$y = \pm 2$

$(\sqrt{6}, 2), (-\sqrt{6}, 2), (\sqrt{6}, -2), (-\sqrt{6}, -2)$

16. $r = 5000$

$x^2 + y^2 = r^2$

$x^2 + y^2 = 5000^2$

$x^2 + y^2 = 25{,}000{,}000$

17. *Verbal Model:* 2 · [Length] + 2 · [Width] = [Perimeter]

[Length]^2 + [Width]^2 = [Diagonal]^2

Labels: Length $= x$

Width $= y$

System:

$$2x + 2y = 56$$

$$x^2 + y^2 = 20^2$$

$$x + y = 28$$

$$y = 28 - x$$

$$x^2 + (28 - x)^2 = 400$$

$$x^2 + 784 - 56x + x^2 - 400 = 0$$

$$2x^2 - 56x + 384 = 0$$

$$x^2 - 28x + 192 = 0$$

$$(x - 16)(x - 12) = 0$$

$x = 16$ $\qquad$ $x = 12$

$y = 28 - 16$ $\qquad$ $y = 28 - 12$

$= 12$ $\qquad$ $= 16$

12 inches × 16 inches

Cumulative Test for Chapters 8–10

1. $4x^2 - 9x - 9 = 0$

$(4x + 3)(x - 3) = 0$

$4x + 3 = 0 \qquad x - 3 = 0$

$x = -\frac{3}{4} \qquad x = 3$

2. $(x - 5)^2 - 64 = 0$

$(x - 5)^2 = 64$

$x - 5 = \pm 8$

$x = 5 \pm 8$

$x = 13, -3$

3. $x^2 - 10x - 25 = 0$

$x^2 - 10x = 25$

$x^2 - 10x + 25 = 25 + 25$

$(x - 5)^2 = 50$

$x - 5 = \pm\sqrt{50}$

$x = 5 \pm 5\sqrt{2}$

4. $3x^2 + 6x + 2 = 0$

$a = 3, b = 6, c = 2$

$$x = \frac{-6 \pm \sqrt{6^2 - 4(3)(2)}}{2(3)}$$

$$x = \frac{-6 \pm \sqrt{36 - 24}}{6}$$

$$x = \frac{-6 \pm \sqrt{12}}{6}$$

$$x = \frac{-6}{6} \pm \frac{2\sqrt{3}}{6}$$

$$x = -1 \pm \frac{\sqrt{3}}{3}$$

5. $x^4 - 8x^2 + 15 = 0$

$(x^2 - 5)(x^2 - 3) = 0$

$x^2 = 5 \qquad x^2 = 3$

$x = \pm\sqrt{5} \qquad x = \pm\sqrt{3}$

6. $3x^2 + 8x \le 3$

$3x^2 + 8x - 3 \le 0$

$(3x - 1)(x + 3) \le 0$

Critical numbers: $-3, \frac{1}{3}$

Test intervals:

Positive: $(-\infty, -3)$

Negative: $\left(-3, \frac{1}{3}\right)$

Positive: $\left(\frac{1}{3}, \infty\right)$

Solution: $\left[-3, \frac{1}{3}\right]$

$\frac{1}{3}$

$-3 \quad -2 \quad -1 \quad 0 \quad 1 \quad x$

7. $\dfrac{3x + 4}{2x - 1} < 0$

Critical numbers: $x = -\frac{4}{3}, \frac{1}{2}$

Test intervals:

Positive: $\left(-\infty, -\frac{4}{3}\right)$

Negative: $\left(-\frac{4}{3}, \frac{1}{2}\right)$

Positive: $\left(\frac{1}{2}, \infty\right)$

Solution: $\left(-\frac{4}{3}, \frac{1}{2}\right)$

$-\frac{4}{3} \quad \frac{1}{2}$

$-2 \quad -1 \quad 0 \quad 1 \quad x$

8. $x = -2$ and $x = 6$

$x + 2 = 0$ and $x - 6 = 0$

$(x + 2)(x - 6) = 0$

$x^2 - 4x - 12 = 0$

9. $f(x) = 2x^2 - 3, g(x) = 5x - 1$

(a) $(f \circ g)(x) = f[g(x)]$
$= f[5x - 1]$
$= 2(5x - 1)^2 - 3$
$= 2(25x^2 - 10x + 1) - 3$
$= 50x^2 - 20x + 2 - 3$
$= 50x^2 - 20x - 1$

Domain: $(-\infty, \infty)$

(b) $(g \circ f)(x) = g[f(x)]$
$= g[2x^2 - 3]$
$= 5(2x^2 - 3) - 1$
$= 10x^2 - 15 - 1$
$= 10x^2 - 16$

Domain: $(-\infty, \infty)$

10. $f(x) = \dfrac{5 - 3x}{4}$

$$y = \frac{5 - 3x}{4}$$
$$x = \frac{5 - 3y}{4}$$
$$4x = 5 - 3y$$
$$4x - 5 = -3y$$
$$-\frac{4}{3}x + \frac{5}{3} = y$$
$$-\frac{4}{3}x + \frac{5}{3} = f^{-1}(x)$$

11. $f(x) = 7 + 2^{-x}$

(a) $f(1) = 7 + 2^{-1} = 7 + \dfrac{1}{2} = \dfrac{15}{2}$

(b) $f(0.5) = 7 + 2^{-0.5}$
$= 7 + \dfrac{1}{\sqrt{2}} = \dfrac{14}{2} + \dfrac{\sqrt{2}}{2} = \dfrac{14 + \sqrt{2}}{2}$

(c) $f(3) = 7 + 2^{-3} = 7 + \dfrac{1}{8} = \dfrac{56}{8} + \dfrac{1}{8} = \dfrac{57}{8}$

12. $f(x) = 4^{x-1}$

Horizontal asymptote: $y = 0$

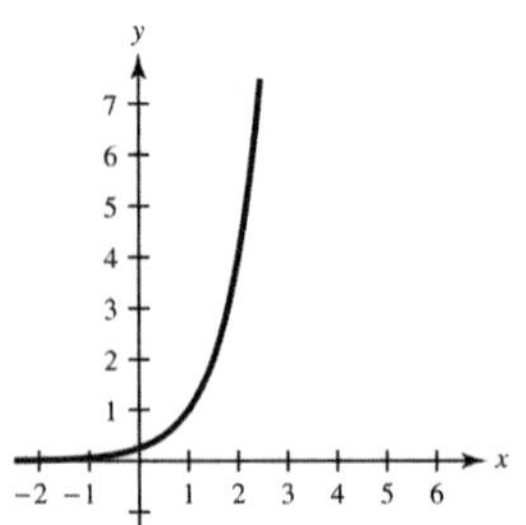

13.

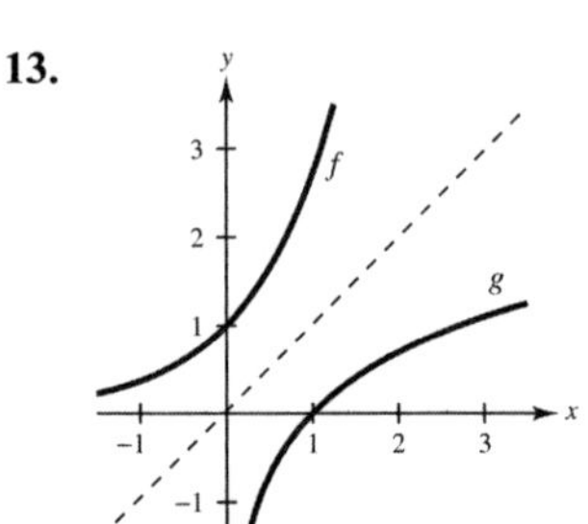

f and g are inverse functions, so the graphs are reflections in the line $y = x$.

14. $g(x) = \log_3(x - 1)$

Vertical asymptote: $x = 1$

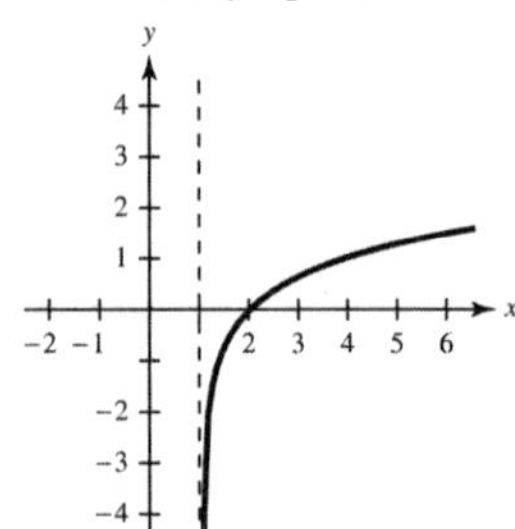

15. $\log_4 \frac{1}{16} = -2$ because $4^{-2} = \frac{1}{16}$.

16. $3(\log_2 x + \log_2 y) - \log_2 z = \log_2(xy)^3 - \log_2 z$
$= \log_2 \dfrac{(xy)^3}{z} = \log_2 \dfrac{x^3y^3}{z}$

17. $\log_{10} \dfrac{\sqrt{x + 1}}{x^4} = \log_{10}\sqrt{x + 1} - \log_{10} x^4$
$= \log_{10}(x + 1)^{1/2} - 4 \log_{10} x$
$= \dfrac{1}{2}\log_{10}(x + 1) - 4 \log_{10} x$

18.
$$\log_x\left(\frac{1}{9}\right) = -2$$
$$\frac{1}{9} = x^{-2}$$
$$\frac{1}{9} = \frac{1}{x^2}$$
$$9 = x^2$$
$$3 = x$$

19.
$$4 \ln x = 10$$
$$\ln x = \tfrac{10}{4}$$
$$\ln x = \tfrac{5}{2}$$
$$e^{\ln x} = e^{5/2}$$
$$x = e^{5/2}$$
$$x \approx 12.182$$

20.
$$500(1.08)^t = 2000$$
$$1.08^t = \frac{2000}{500}$$
$$1.08^t = 4$$
$$\log_{1.08} 1.08^t = \log_{1.08} 4$$
$$t = \frac{\log 4}{\log 1.08}$$
$$t \approx 18.013$$

21.
$$3\left(1 + e^{2x}\right) = 20$$
$$1 + e^{2x} = \frac{20}{3}$$
$$e^{2x} = \frac{17}{3}$$
$$\ln e^{2x} = \ln \frac{17}{3}$$
$$2x = \ln \frac{17}{3}$$
$$x = \frac{\ln(17/3)}{2}$$
$$x \approx 0.867$$

22.
$$C(t) = P(1.028)^t$$
$$C(5) = 29.95(1.028)^5 \approx \$34.38$$

23.
$$A = Pe^{rt}$$
$$A = 1000e^{0.08(1)}$$
$$A = \$1083.29$$
$$\text{Effective yield} = \frac{83.29}{1000} = 0.08329 \approx 8.33\%$$

24.
$$A = Pe^{rt}$$
$$6000 = 1500e^{0.07t}$$
$$4 = e^{0.07t}$$
$$\ln 4 = \ln e^{0.07t}$$
$$\ln 4 = 0.07t$$
$$\frac{\ln 4}{0.07} = t$$
$$19.8 \approx t \text{ years}$$

25.
$$x^2 + y^2 - 6x + 14y - 6 = 0$$
$$\left(x^2 - 6x\right) + \left(y^2 + 14y\right) = 6$$
$$\left(x^2 - 6x + 9\right) + \left(y^2 + 14y + 49\right) = 6 + 9 + 49$$
$$(x - 3)^2 + (y + 7)^2 = 64$$

Center: $(3, -7)$

Radius: 8

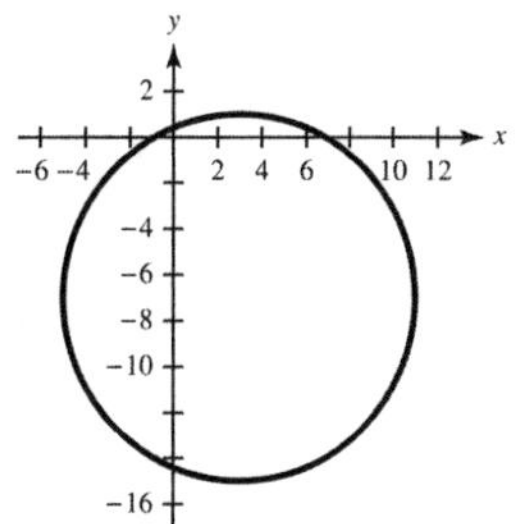

26.
$$y = 2x^2 - 20x + 5$$
$$y - 5 = 2\left(x^2 - 10x\right)$$
$$y - 5 + 50 = 2\left(x^2 - 10x + 25\right)$$
$$y + 45 = 2(x - 5)^2$$
$$\tfrac{1}{2}(y + 45) = (x - 5)^2$$

Vertex: $(5, -45)$

$$4p = \tfrac{1}{2}$$
$$p = \tfrac{1}{8}$$

Focus: $\left(5, 45 + \frac{1}{8}\right) = \left(5, -\frac{360}{8} + \frac{1}{8}\right) = \left(5, -\frac{359}{8}\right)$

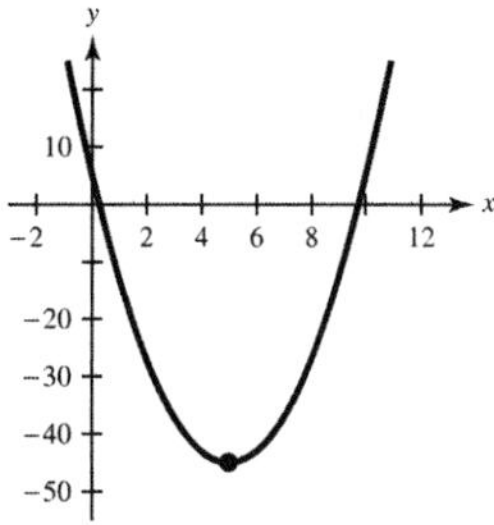

27. $$\frac{(x-h)^2}{b^2} + \frac{(y-k)^2}{a^2} = 1$$

$$\frac{(x+3)^2}{4} + \frac{(y-2)^2}{25} = 1$$

$b = 2 \qquad a = 5$

$b^2 = 4 \qquad a^2 = 25$

28. 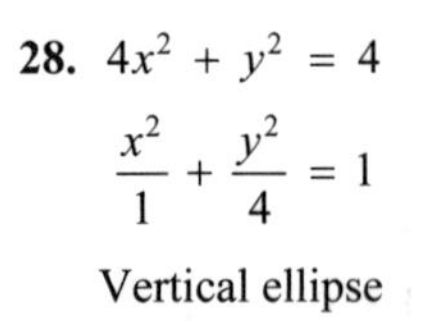

$4x^2 + y^2 = 4$

$$\frac{x^2}{1} + \frac{y^2}{4} = 1$$

Vertical ellipse

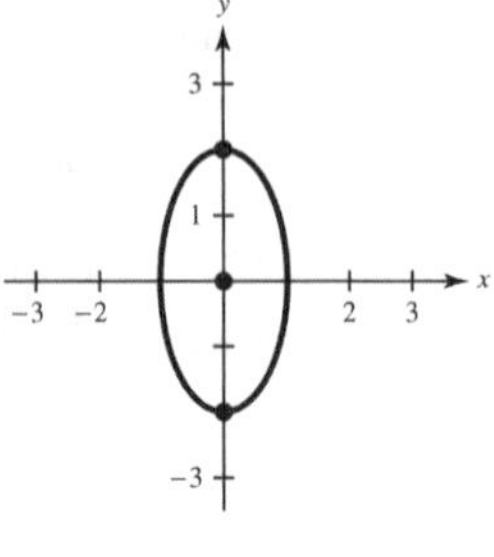

Center: $(0, 0)$

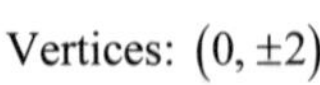

Vertices: $(0, \pm 2)$

$a^2 = 4 \qquad b^2 = 1$

$a = 2 \qquad b = 1$

29. Hyperbola (Vertical)

Vertices: $(0, \pm 3)$

Asymptotes: $y = \pm 3x$

Center: $(0, 0)$

$a = 3 \qquad \frac{a}{b} = \pm 3$

$a^2 = 9 \qquad \frac{3}{b} = \pm 3$

$b = 1$

$$\frac{y^2}{a^2} - \frac{x^2}{b^2} = 1$$

$$\frac{y^2}{9} - \frac{x^2}{1} = 1$$

30. $$\begin{aligned} x^2 - 9y^2 + 18y &= 153 \\ x^2 - (9y^2 - 18y) &= 153 \\ x^2 - 9(y^2 - 2y + 1) &= 153 - 9 \\ x^2 - 9(y-1)^2 &= 144 \\ \frac{x^2}{144} - \frac{(y-1)^2}{16} &= 1 \end{aligned}$$

Center: $(0, 1)$

Vertices: $(\pm 12, 1)$

$a^2 = 144 \qquad b^2 = 16$

$a = 12 \qquad b = 4$

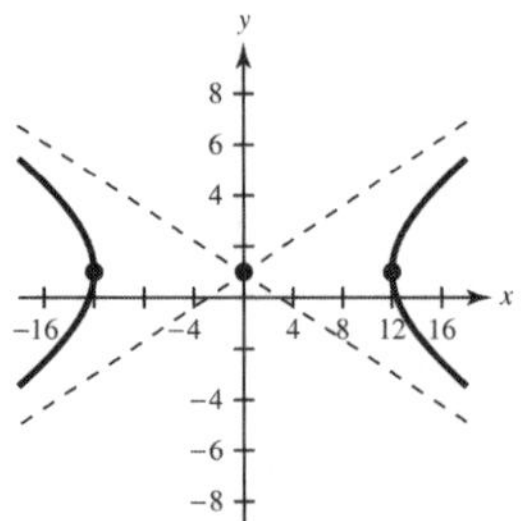

31. $$\begin{cases} y = x^2 - x - 1 \\ 3x - y = 4 \end{cases}$$

Substitute for y in second equation.

$$\begin{aligned} 3x - (x^2 - x - 1) &= 4 \\ 3x - x^2 + x + 1 - 4 &= 0 \\ -x^2 + 4x - 3 &= 0 \\ x^2 - 4x + 3 &= 0 \\ (x-3)(x-1) &= 0 \end{aligned}$$

$x = 3 \qquad\qquad x = 1$

$3(3) - y = 4 \qquad 3(1) - y = 4$

$-y = -5 \qquad\qquad -y = 1$

$y = 5 \qquad\qquad y = -1$

$(3, 5) \qquad\qquad (1, -1)$

32. $\begin{cases} x^2 + 5y^2 = 21 \\ -x + y^2 = 5 \end{cases}$

$$y^2 = 5 + x$$
$$x^2 + 5(5 + x) = 21$$
$$x^2 + 25 + 5x - 21 = 0$$
$$x^2 + 5x + 4 = 0$$
$$(x + 4)(x + 1) = 0$$

$x = -4$	$x = -1$
$y^2 = 5 + (-4)$	$y^2 = 5 + (-1)$
$y^2 = 1$	$y^2 = 4$
$y = \pm 1$	$y = \pm 2$
$(-4, \pm 1)$	$(-1, \pm 2)$

33. *Verbal Model:* [Length] · [Width] = [Area]

2 · [Length] + 2 · [Width] = [Perimeter]

Labels: Length $= x$

Width $= y$

System: $xy = 32$

$$2x + 2y = 24$$
$$x + y = 12$$
$$y = 12 - x$$
$$x(12 - x) = 32$$
$$12x - x^2 = 32$$
$$0 = x^2 - 12x + 32$$
$$0 = (x - 8)(x - 4)$$

$x = 8$	$x = 4$
$y = 4$	$y = 8$

4 feet × 8 feet

34. *Verbal Model:* [Length] · [Width] = [Area]

2 · [Length] + 2 · [Width] = [Perimeter]

Labels: Length $= x$

Width $= y$

System: $\begin{cases} xy = 21 \\ 2x + 2y = 20 \end{cases}$

$$2x + 2y = 20$$
$$x + y = 10$$
$$y = 10 - x$$
$$x(10 - x) = 21$$
$$10x - x^2 = 21$$
$$0 = x^2 - 10x + 21$$
$$0 = (x - 7)(x - 3)$$

$x = 7$	$x = 3$
$y = 3$	$y = 7$

Dimensions: 7 feet × 3 feet

CHAPTER 11
Sequences, Series, and the Binomial Theorem

Section 11.1 Sequences and Series....................351

Section 11.2 Arithmetic Sequences....................355

Mid-Chapter Quiz....................358

Section 11.3 Geometric Sequences and Series....................359

Section 11.4 The Binomial Theorem....................362

Review Exercises....................365

Chapter Test....................368

CHAPTER 11
Sequences, Series, and the Binomial Theorem

Section 11.1 Sequences and Series

1. $a_1 = 2(1) = 2$

$a_2 = 2(2) = 4$

$a_3 = 2(3) = 6$

$a_4 = 2(4) = 8$

$a_5 = 2(5) = 10$

$2, 4, 6, 8, 10, \ldots, 2n, \ldots$

3. $a_1 = \left(\frac{1}{4}\right)^1 = \frac{1}{4}$

$a_2 = \left(\frac{1}{4}\right)^2 = \frac{1}{16}$

$a_3 = \left(\frac{1}{4}\right)^3 = \frac{1}{64}$

$a_4 = \left(\frac{1}{4}\right)^4 = \frac{1}{256}$

$a_5 = \left(\frac{1}{4}\right)^5 = \frac{1}{1024}$

$\frac{1}{4}, \frac{1}{16}, \frac{1}{64}, \frac{1}{256}, \frac{1}{1024}, \ldots, \left(\frac{1}{4}\right)^n$

5. $a_1 = 5(1) - 2 = 3$

$a_2 = 5(2) - 2 = 8$

$a_3 = 5(3) - 2 = 13$

$a_4 = 5(4) - 2 = 18$

$a_5 = 5(5) - 2 = 23$

$3, 8, 13, 18, 23, \ldots, 5h - 2$

7. $a_1 = \dfrac{4}{1 + 3} = 1$

$a_2 = \dfrac{4}{2 + 3} = \dfrac{4}{5}$

$a_3 = \dfrac{4}{3 + 3} = \dfrac{4}{6} = \dfrac{2}{3}$

$a_4 = \dfrac{4}{4 + 3} = \dfrac{4}{7}$

$a_5 = \dfrac{4}{5 + 3} = \dfrac{4}{8} = \dfrac{1}{2}$

$1, \dfrac{4}{5}, \dfrac{2}{3}, \dfrac{4}{7}, \dfrac{1}{2}, \ldots, \dfrac{4}{n + 3}$

9. $a_1 = \dfrac{3(1)}{5(1) - 1} = \dfrac{3}{4}$

$a_2 = \dfrac{3(2)}{5(2) - 1} = \dfrac{6}{9} = \dfrac{2}{3}$

$a_3 = \dfrac{3(3)}{5(3) - 1} = \dfrac{9}{14}$

$a_4 = \dfrac{3(4)}{5(4) - 1} = \dfrac{12}{19}$

$a_5 = \dfrac{3(5)}{5(5) - 1} = \dfrac{15}{24} = \dfrac{5}{8}$

$\dfrac{3}{4}, \dfrac{2}{3}, \dfrac{9}{14}, \dfrac{12}{19}, \dfrac{15}{24}, \ldots, \dfrac{3n}{5n - 1}$

11. $a_1 = (-1)^1 2(1) = -2$

$a_2 = (-1)^2 2(2) = 4$

$a_3 = (-1)^3 2(3) = -6$

$a_4 = (-1)^4 2(4) = 8$

$a_5 = (-1)^5 2(5) = -10$

$-2, 4, -6, 8, -10, \ldots$

13. $a_1 = \left(-\frac{1}{2}\right)^2 = \frac{1}{4}$

$a_2 = \left(-\frac{1}{2}\right)^3 = -\frac{1}{8}$

$a_3 = \left(-\frac{1}{2}\right)^4 = \frac{1}{16}$

$a_4 = \left(-\frac{1}{2}\right)^5 = -\frac{1}{32}$

$a_5 = \left(-\frac{1}{2}\right)^6 = \frac{1}{64}$

$\frac{1}{4}, -\frac{1}{8}, \frac{1}{16}, -\frac{1}{32}, \frac{1}{64}, \ldots, \left(-\frac{1}{2}\right)^{n+1}, \ldots$

15. $a_1 = \frac{(1)^1}{1^2} = -1$

$a_2 = \frac{(-1)^2}{2^2} = \frac{1}{4}$

$a_3 = \frac{(-1)^3}{3^2} = -\frac{1}{9}$

$a_4 = \frac{(-1)^4}{4^2} = \frac{1}{16}$

$a_5 = \frac{(-1)^5}{5^2} = -\frac{1}{25}$

$-1, \frac{1}{4}, -\frac{1}{9}, \frac{1}{16}, -\frac{1}{25}, \ldots, \frac{(-1)^n}{n^2}, \ldots$

17. $6! = 1 \cdot 2 \cdot 3 \cdot 4 \cdot 5 \cdot 6 = 720$

19. $9! = 1 \cdot 2 \cdot 3 \cdot 4 \cdot 5 \cdot 6 \cdot 7 \cdot 8 \cdot 9 = 362{,}880$

21. $a_1 = \frac{1!}{1} = 1$

$a_2 = \frac{2!}{2} = 1$

$a_3 = \frac{3!}{3} = 2$

$a_4 = \frac{4!}{4} = 6$

$a_5 = \frac{5!}{5} = 24$

$1, 1, 2, 6, 24, \ldots, \frac{n!}{n}, \ldots$

23. $a_1 = \frac{(1+1)!}{1!} = \frac{2!}{1!} = \frac{2 \cdot 1}{1} = 2$

$a_2 = \frac{(2+1)!}{2!} = \frac{3!}{2!} = \frac{3 \cdot 2!}{2!} = 3$

$a_3 = \frac{(3+1)!}{3!} = \frac{4!}{3!} = \frac{4 \cdot 3!}{3!} = 4$

$a_4 = \frac{(4+1)!}{4!} = \frac{5!}{4!} = \frac{5 \cdot 4!}{4!} = 5$

$a_5 = \frac{(5+1)!}{5!} = \frac{6!}{5!} = \frac{6 \cdot 5!}{5!} = 6$

$2, 3, 4, 5, 6 \ldots, \frac{(n+1)!}{n!}$

25. $\frac{5!}{4!} = \frac{5 \cdot 4 \cdot 3 \cdot 2 \cdot 1}{4 \cdot 3 \cdot 2 \cdot 1} = 5$

27. $\frac{25!}{20!5!} = \frac{25 \cdot 24 \cdot 23 \cdot 22 \cdot 21 \cdot 20!}{20!5!}$

$= \frac{25 \cdot 24 \cdot 23 \cdot 22 \cdot 21}{5 \cdot 4 \cdot 3 \cdot 2 \cdot 1}$

$= 5 \cdot 6 \cdot 23 \cdot 11 \cdot 7$

$= 53{,}130$

29. $\frac{n!}{(n+1)!} = \frac{n \cdot 1}{(n+1)n \cdot 1} = \frac{1}{n+1}$

31. $\frac{(n+1)!}{(n-1)!} = \frac{(n+1)n(n-1)!}{(n-1)!} = (n+1)n$

33. (b); $a_n = \frac{6}{n+1}$

$a_1 = \frac{6}{2} = 3$

$a_2 = \frac{6}{3} = 2$

$a_3 = \frac{6}{4} = \frac{3}{2}$

$a_4 = \frac{6}{5}$

$a_5 = \frac{6}{6} = 1$

$a_6 = \frac{6}{7}$

$a_7 = \frac{6}{8} = \frac{3}{4}$

$a_8 = \frac{6}{9} = \frac{2}{3}$

$a_9 = \frac{6}{10} = \frac{3}{5}$

$a_{10} = \frac{6}{11}$

35. (c); $a_n = (0.6)^{n-1}$

$a_1 = (0.6)^0 = 1$

$a_2 = (0.6)^1 = 0.6$

$a_3 = (0.6)^2 = 0.36$

$a_4 = (0.6)^3 \approx 0.22$

$a_5 = (0.6)^4 \approx 0.13$

$a_6 = (0.6)^5 \approx 0.08$

$a_7 = (0.6)^6 \approx 0.05$

$a_8 = (0.6)^7 \approx 0.03$

$a_9 = (0.6)^8 \approx 0.02$

$a_{10} = (0.8)^9 \approx 0.01$

37. n: 1 2 3 4 5

Terms: 1 3 5 7 9

Apparent pattern: Each term is twice n minus 1.

$a_n = 2n - 1$

39. n: 1 2 3 4 5

Terms: 0 3 8 15 24

Apparent pattern: Each term is the square of n minus 1.

$a_n = n^2 - 1$

41. n: 1 2 3 4 5

Terms: $\frac{-1}{5}$ $\frac{1}{25}$ $\frac{-1}{125}$ $\frac{1}{625}$ $\frac{-1}{3125}$

Apparent pattern: The fraction $-\frac{1}{5}$ is raised to the n power.

$a_n = \left(-\frac{1}{5}\right)^n$

43.
$$\begin{aligned}
a_1 &= 2(1) + 5 = 7\\
a_2 &= 2(2) + 5 = 9\\
a_3 &= 2(3) + 5 = 11\\
a_4 &= 2(4) + 5 = 13\\
a_5 &= 2(5) + 5 = 15\\
a_6 &= 2(6) + 5 = 17
\end{aligned}$$

$$\begin{aligned}
S_1 &= a_1 = 7\\
S_2 &= a_1 + a_2 = 7 + 9 = 16\\
S_6 &= a_1 + a_2 + a_3 + a_4 + a_5 + a_6\\
&= 7 + 9 + 11 + 13 + 15 + 17\\
&= 72
\end{aligned}$$

45.
$$\begin{aligned}
a_1 &= \frac{1}{1} = 1\\
a_2 &= \frac{1}{2}\\
a_3 &= \frac{1}{3}\\
a_4 &= \frac{1}{4}\\
a_5 &= \frac{1}{5}\\
a_6 &= \frac{1}{6}\\
a_7 &= \frac{1}{7}\\
a_8 &= \frac{1}{8}\\
a_9 &= \frac{1}{9}
\end{aligned}$$

$$\begin{aligned}
S_2 &= a_1 + a_2 = 1 + \frac{1}{2} = \frac{3}{2}\\
S_3 &= a_1 + a_2 + a_3 = 1 + \frac{1}{2} + \frac{1}{3} = \frac{6 + 3 + 2}{6} = \frac{11}{6}\\
S_9 &= a_1 + a_2 + a_3 + a_4 + a_5 + a_6 + a_7 + a_8 + a_9\\
&= 1 + \frac{1}{2} + \frac{1}{3} + \frac{1}{4} + \frac{1}{5} + \frac{1}{6} + \frac{1}{7} + \frac{1}{8} + \frac{1}{9}\\
&= \frac{2520 + 1260 + 840 + 630 + 504 + 420 + 360 + 315 + 280}{2520} = \frac{7129}{2520}
\end{aligned}$$

47. $\sum_{k=1}^{5} 6 = 6 + 6 + 6 + 6 + 6 = 30$

49.
$$\begin{aligned}
\sum_{i=0}^{6} (2i + 5) &= [2(0) + 5] + [2(1) + 5] + [2(2) + 5] + [2(3) + 5] + [2(4) + 5] + [2(5) + 5] + [2(6) + 5]\\
&= 5 + 7 + 9 + 11 + 13 + 15 + 17 = 77
\end{aligned}$$

51.
$$\begin{aligned}
\sum_{j=0}^{3} \frac{1}{j^2 + 1} &= \frac{1}{0^2 + 1} + \frac{1}{1^2 + 1} + \frac{1}{2^2 + 1} + \frac{1}{3^2 + 1}\\
&= \frac{1}{0 + 1} + \frac{1}{1 + 1} + \frac{1}{4 + 1} + \frac{1}{9 + 1} = \frac{1}{1} + \frac{1}{2} + \frac{1}{5} + \frac{1}{10} = \frac{10}{10} + \frac{5}{10} + \frac{2}{10} + \frac{1}{10} = \frac{18}{10} = \frac{9}{5}
\end{aligned}$$

53. $\displaystyle\sum_{n=0}^{5}\left(-\frac{1}{3}\right)^n = \left(-\frac{1}{3}\right)^0 + \left(-\frac{1}{3}\right)^1 + \left(-\frac{1}{3}\right)^2 + \left(-\frac{1}{3}\right)^3 + \left(-\frac{1}{3}\right)^4 + \left(-\frac{1}{3}\right)^5$

$$= 1 + \left(-\frac{1}{3}\right) + \frac{1}{9} + \left(-\frac{1}{27}\right) + \frac{1}{81} + \left(-\frac{1}{243}\right) = \frac{243 - 81 + 27 - 9 + 3 - 1}{243} = \frac{182}{243}$$

55. $\displaystyle\sum_{k=1}^{5} k$

57. $\displaystyle\sum_{k=1}^{11} \frac{k}{k+1}$

59. The third term in the sequence $a_n = \dfrac{2}{3n}$ is $\dfrac{2}{9}$.

61. To find the 6th term of the sequence $a_n = \dfrac{2}{3n}$, substitute 6 for n in a_n: $a_6 = \dfrac{2}{3(6)} = \dfrac{1}{9}$.

63. (a) $A_1 = 500(1 + 0.07)^1 = \$535.00$

$A_2 = 500(1 + 0.07)^2 = \$572.45$

$A_3 = 500(1 + 0.07)^3 = \$612.52$

$A_4 = 500(1 + 0.07)^4 = \$655.40$

$A_5 = 500(1 + 0.07)^5 = \$701.28$

$A_6 = 500(1 + 0.07)^6 = \$750.37$

$A_7 = 500(1 + 0.07)^7 = \$802.89$

$A_8 = 500(1 + 0.07)^8 = \$859.09$

(b) $A_{40} = 500(1 + 0.07)^{40} = \$7487.23$

(c) *Keystrokes* (calculator in sequence and dot mode):

[Y=] 500 [(] 1 [+] 0.07 [)] [^] [n] [TRACE]

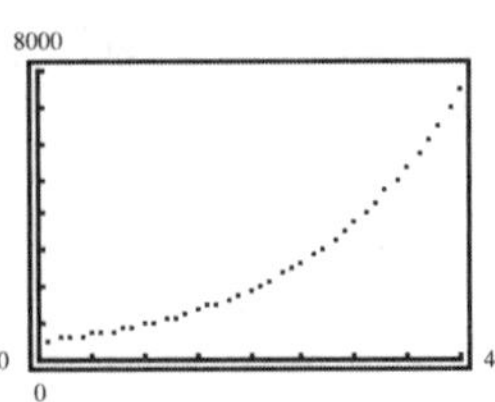

(d) No. Investment earning compound interest increases at an increasing rate.

65. $d_5 = \dfrac{180(5-4)}{5} = \dfrac{180(1)}{5} = \dfrac{180}{5} = 36°$

$d_6 = \dfrac{180(6-4)}{6} = \dfrac{180(2)}{6} = \dfrac{360}{6} = 60°$

$d_7 = \dfrac{180(7-4)}{7} = \dfrac{180(3)}{7} = \dfrac{540}{7} \approx 77.1°$

$d_8 = \dfrac{180(8-4)}{8} = \dfrac{180(4)}{8} = \dfrac{720}{8} = 90°$

$d_9 = \dfrac{180(9-4)}{9} = \dfrac{180(5)}{9} = \dfrac{900}{9} = 100°$

$d_{10} = \dfrac{180(10-4)}{10} = \dfrac{180(6)}{10} = \dfrac{1080}{10} = 108°$

67. A sequence is a function because there is only one value for each term of the sequence.

69. $a_n = 4n! = 4(1 \cdot 2 \cdot 3 \cdot 3 \cdot \ldots \cdot (n-1) \cdot n)$

$a_n = (4n)! = (4 \cdot 1) \cdot (4 \cdot 2) \cdot (4 \cdot 3) \cdot (4 \cdot 4) \cdot \ldots \cdot (4n - 1) \cdot (4n)$

71. True.

$\displaystyle\sum_{k=1}^{4} 3k = 3 + 6 + 9 + 12 = 3(1 + 2 + 3 + 4) = 3\sum_{k=1}^{4} k$

73. $-2n + 15$ for $n = 3$: $-2(3) + 15 = 9$

75. $25 - 3(n + 4)$ for $n = 8$:

$$\begin{aligned} 25 - 3(8 + 4) &= 25 - 3(12) \\ &= 25 - 36 \\ &= -11 \end{aligned}$$

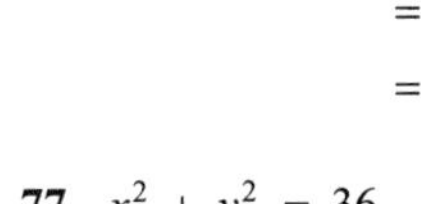

77. $x^2 + y^2 = 36$

Center: $(0, 0)$

Radius: 6

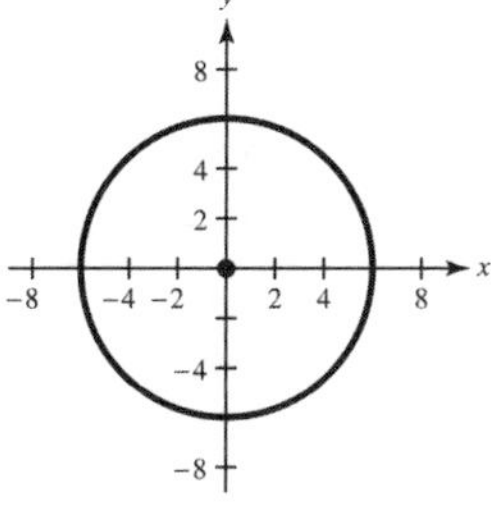

79. $x^2 + y^2 + 4x - 12 = 0$

$$\left(x^2 + 4x + 4\right) + y^2 = 12 + 4$$

$$(x + 2)^2 + y^2 = 16$$

Center: $(-2, 0)$

Radius: 4

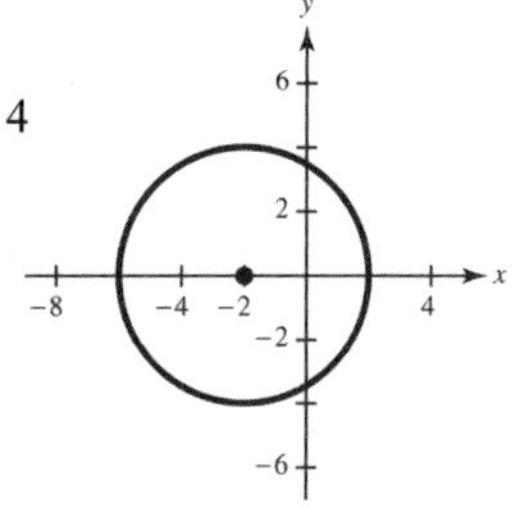

81. $x^2 = 6y$

Vertex: $(0, 0)$

$4p = 6$

$p = \frac{6}{4} = \frac{3}{2}$

Focus: $\left(0, \frac{3}{2}\right)$

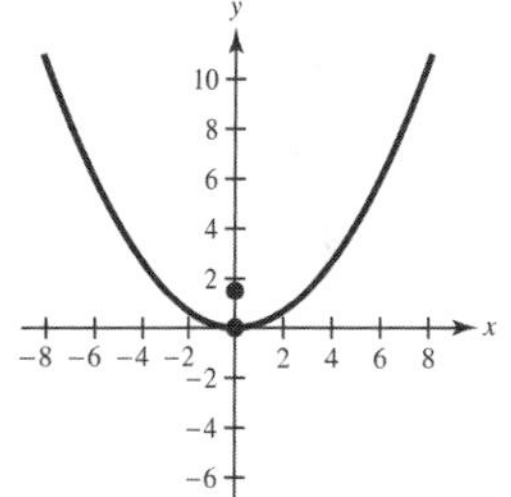

83. $x^2 + 8y + 32 = 0$

$$x^2 + 8(y + 4) = 0$$

$$x^2 = -8(y + 4)$$

Vertex: $(0, -4)$

$4p = -8$

$p = -\frac{8}{4} = -2$

Focus: $(0, -6)$

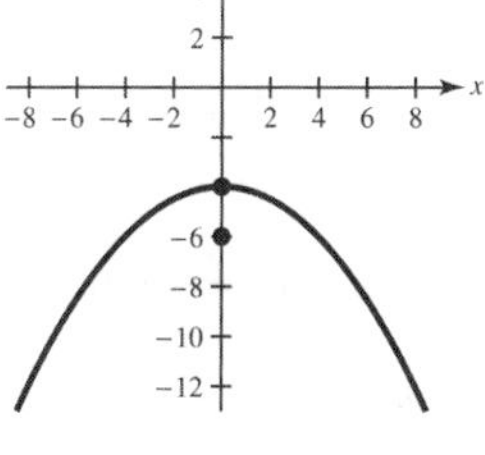

Section 11.2 Arithmetic Sequences

1. $2, 5, 8, 11, \ldots$

$d = 3$

$5 - 2 = 3, 8 - 5 = 3, 11 - 8 = 3$

3. $100, 94, 88, 82, \ldots$

$d = -6$

$94 - 100 = -6, 88 - 94 = -6, 82 - 88 = -6$

5. $4, \frac{9}{2}, 5, \frac{11}{2}, 6, \ldots$

$\frac{9}{2} - 4 = \frac{1}{2}$

$5 - \frac{9}{2} = \frac{1}{2}$

$\frac{11}{2} - 5 = \frac{1}{2}$

$6 - \frac{11}{2} = \frac{1}{2}$

$d = \frac{1}{2}$

7. $a_n = 4n + 5$

$a_1 = 9$

$a_2 = 13$

$a_3 = 17$

$a_4 = 21$

$d = 4$

9. $a_n = 8 - 3n$

$a_1 = 5$

$a_2 = 2$

$a_3 = -1$

$a_4 = -4$

$d = -3$

11. $a_n = \frac{1}{2}(n + 1)$

$a_1 = 1$

$a_2 = \frac{3}{2}$

$a_3 = 2$

$a_4 = \frac{5}{2}$

$d = \frac{1}{2}$

13. $a_1 = 4, d = 3$

$a_n = a_1 + (n - 1)d$

$a_n = 4 + (n - 1)3$

$a_n = 4 + 3n - 3$

$a_n = 3n + 1$

15. $a_1 = \frac{1}{2}, d = \frac{3}{2}$

$$a_n = a_1 + (n-1)d$$
$$a_n = \frac{1}{2} + (n-1)\frac{3}{2}$$
$$a_n = \frac{1}{2} + \frac{3}{2}n - \frac{3}{2}$$
$$a_n = \frac{3}{2}n - \frac{2}{2}$$
$$a_n = \frac{3}{2}n - 1$$

17. $a_1 = 100, d = -5$

$$a_n = a_1 + (n-1)d$$
$$a_n = 100 + (n-1)(-5)$$
$$a_n = 100 - 5n + 5$$
$$a_n = -5n + 105$$

19. $a_3 = 6, d = \frac{3}{2}$

$$a_n = a_1 + (n-1)d$$
$$6 = a_1 + (3-1)\frac{3}{2}$$
$$6 = a_1 + 3$$
$$3 = a_1$$

$$a_n = 3 + (n-1)\frac{3}{2}$$
$$= 3 + \frac{3}{2}n - \frac{3}{2}$$
$$a_n = \frac{3}{2}n + \frac{3}{2}$$

21. $a_n = a_1 + (n-1)d$

$$15 = 5 + (5-1)d$$
$$15 = 5 + 4d$$
$$10 = 4d$$
$$\frac{5}{2} = \frac{10}{4} = d$$

So,

$$a_n = 5 + (n-1)\frac{5}{2}$$
$$a_n = 5 + \frac{5}{2}n - \frac{5}{2}$$
$$a_n = \frac{5}{2}n + \frac{5}{2}.$$

23. $d = a_2 - a_1$

$$= 11 - 5$$
$$= 6$$

$$a_n = a_1 + (n-1)d$$
$$a_{10} = 5 + (10-1)(6)$$
$$a_{10} = 59$$

25. $a_1 = 14, a_{k+1} = a_k + 6$ so, $d = 6$.

$$a_2 = a_{1+1} = a_1 + 6 = 14 + 6 = 20$$
$$a_3 = a_{2+1} = a_2 + 6 = 20 + 6 = 26$$
$$a_4 = a_{3+1} = a_3 + 6 = 26 + 6 = 32$$
$$a_5 = a_{4+1} = a_4 + 6 = 32 + 6 = 38$$

27. $a_1 = 23, a_{k+1} = a_k - 5$ so, $d = -5$.

$$a_2 = a_{1+1} = a_1 - 5 = 23 - 5 = 18$$
$$a_3 = a_{2+1} = a_2 - 5 = 18 - 5 = 13$$
$$a_4 = a_{3+1} = a_3 - 5 = 13 - 5 = 8$$
$$a_5 = a_{4+1} = a_4 - 5 = 8 - 5 = 3$$

29. (a) $a_4 = a_1 = (n-1)d$

$$23 = a_1 + (4-1)\cdot 6$$
$$23 = a_1 + 18$$
$$a_1 = 5$$

(b) $a_n = a_1 + (n-1)d$

$$a_5 = 5 + (5-1)\cdot 6$$
$$a_5 = 29$$

31. $\sum_{k=1}^{20} k = \frac{20}{2}(1 + 20) = 210$

33. $\sum_{k=1}^{50} (k + 3) = \frac{50}{2}(4 + 53) = 1425$

35. $\sum_{k=1}^{10} (5k - 2) = \frac{10}{2}(3 + 48) = 255$

37. $\sum_{n=1}^{500} \frac{n}{2} = \frac{500}{2}\left(\frac{1}{2} + 250\right) = 62{,}625$

39. $\sum_{n=1}^{30} \left(\frac{1}{3}n - 4\right) = \frac{30}{2}\left(-\frac{11}{3} + 6\right) = 35$

41. $\sum_{n=1}^{12} (7n - 2) = \frac{12}{2}(5 + 82) = 522$

43. $\sum_{n=1}^{50} (12n - 62) = \frac{50}{2}(-50 + 538) = 12{,}200$

45. $\sum_{n=1}^{12} (3.5n - 2.5) = \frac{12}{2}(1 + 39.5) = 243$

47. $\sum_{n=1}^{10} (0.4n + 0.1) = \frac{10}{2}(0.5 + 4.1) = 23$

49. $\sum_{n=1}^{75} n = \frac{75}{2}(1 + 75) = 2850$

51. $a_1 = \$54{,}000$

$a_2 = \$57{,}000$

$a_3 = \$60{,}000$

$a_4 = \$63{,}000$

$a_5 = \$66{,}000$

$a_6 = \$69{,}000$

$$\text{Total salary} = \tfrac{6}{2}(54{,}000 + 69{,}000) = 3(123{,}000) = \$369{,}000$$

53. Sequence $= 93, 89, 85, 81, \ldots$

$$\sum_{n=1}^{8}(97 - 4n) = \tfrac{8}{2}(93 + 65) = 632 \text{ bales}$$

55. To find the next term in the sequence, add 3 to the value of the term.

57. The nth partial sum can be found by multiplying the number of terms by the average of the first term and the nth term.

59. $a_1 = 7, d = 5$

$a_1 = 7$

$a_2 = 7 + 5 = 12$

$a_3 = 12 + 5 = 17$

$a_4 = 17 + 5 = 22$

$a_5 = 22 + 5 = 27$

61. $a_1 = 11, d = 4$

$a_1 = 11$

$a_2 = 11 + 4 = 15$

$a_3 = 15 + 4 = 19$

$a_4 = 19 + 4 = 23$

$a_5 = 23 + 4 = 27$

63. $a_1 = 3(1) + 4 = 7$

$a_2 = 3(2) + 4 = 10$

$a_3 = 3(3) + 4 = 13$

$a_4 = 3(4) + 4 = 16$

$a_5 = 3(5) + 4 = 19$

65. $a_1 = \tfrac{5}{2}(1) - 1 = \tfrac{3}{2}$

$a_2 = \tfrac{5}{2}(2) - 1 = 4$

$a_3 = \tfrac{5}{2}(3) - 1 = \tfrac{13}{2}$

$a_4 = \tfrac{5}{2}(4) - 1 = 9$

$a_5 = \tfrac{5}{2}(5) - 1 = \tfrac{23}{2}$

67. (d)

68. (c)

69. (a)

70. (b)

71. Sequence $= 16, 48, 80, \ldots, n = 8, d = 32$

$a_n = a_1 + (n - 1)d$

$a_n = 16 + (n - 1)32$

$= 16 + 32n - 32$

$a_n = 32n - 16$

$a_n = 16 + (8 - 1)32$

$a_n = 16 + 224$

$a_n = 240$

$$\sum_{n=1}^{8}(32n - 16) = \tfrac{8}{2}(16 + 240) = 1024 \text{ feet}$$

73. Sequence $= 1, 2, 3, 4, \ldots$

$a_n = a_1 + (n - 1)d$

$a_n = 1 + (n - 1)(1)$

$a_n = 1 + n - 1$

$a_n = n$

$$\sum_{n=1}^{12} n = \tfrac{12}{2}(1 + 12) = 78 \text{ chimes}$$

3 chimes each hour $\times$ 12 hours $=$ 36 chimes

Total chimes $= 78 + 36 = 114$ chimes

75. A recursion formula gives the relationship between the terms a_{n+1} and a_n.

77. Yes. Because a_{2n} is n terms away from a_n, add n times the difference d to a_n.

$a_{2n} = a_n + nd$

79. (a) $1 + 3 = 4$

$1 + 3 + 5 = 9$

$1 + 3 + 5 + 7 = 16$

$1 + 3 + 5 + 7 + 9 = 25$

$1 + 3 + 5 + 7 + 9 + 11 = 36$

(b) No. There is no common difference between consecutive terms of the sequence.

(c) $1 + 3 + 5 + 7 + 9 + 11 + 13 = 49$

$$\sum_{k=1}^{n} (2k - 1) = n^2$$

(d) Looking at the sums in part (a), each sum is the square of a consecutive positive integer.

$1 + 3 = 2^2$

$1 + 3 + 5 = 3^2$

$1 + 3 + 5 + 7 = 4^2$

$1 + 3 + 5 + 7 + 9 = 5^2$

$1 + 3 + 5 + 7 + 9 + 11 = 6^2$

So the sum of n odd integers is n^2. An odd integer is represented by $2k - 1$ which gives the formula

$$\sum_{k=1}^{n} (2k - 1) = n^2.$$

81. $\dfrac{(x + 2)^2}{4} + (y - 8)^2 = 1$

Center: $(-2, 8)$

$a^2 = 4$

$a = 2$, horizontal axis

Vertices: $(-2 + 2, 8), (-2 - 2, 8)$

$(0, 8), (-4, 8)$

83. $x^2 + 4y^2 - 8x + 12 = 0$

$(x^2 - 8x + 16) + 4y^2 = -12 + 16$

$(x - 4)^2 + 4y^2 = 4$

$\dfrac{(x - 4)^2}{4} + y^2 = 1$

Center: $(4, 0)$

$a^2 = 4$

$a = 2$, horizontal axis

Vertices: $(4 + 2, 0), (4 - 2, 0)$

$(6, 0), (2, 0)$

85. $3 + 6 + 9 + 12 + 15 = \sum_{k=1}^{5} 3k$

87. $2 + 2^2 + 2^3 + 2^4 + 2^5 = \sum_{k=1}^{5} 2^k$

Mid-Chapter Quiz for Chapter 11

1. $a_n = 4n$

$a_1 = 4(1) = 4$

$a_2 = 4(2) = 8$

$a_3 = 4(3) = 12$

$a_4 = 4(4) = 16$

$a_5 = 4(5) = 20$

2. $a_n = 2n + 5$

$a_1 = 2(1) + 5 = 7$

$a_2 = 2(2) + 5 = 9$

$a_3 = 2(3) + 5 = 11$

$a_4 = 2(4) + 5 = 13$

$a_5 = 2(5) + 5 = 15$

3. $a_n = 32\left(\frac{1}{4}\right)^{n-1}$

$a_1 = 32\left(\frac{1}{4}\right)^{1-1} = 32$

$a_2 = 32\left(\frac{1}{4}\right)^{2-1} = 8$

$a_3 = 32\left(\frac{1}{4}\right)^{3-1} = 2$

$a_4 = 32\left(\frac{1}{4}\right)^{4-1} = \frac{1}{2}$

$a_5 = 32\left(\frac{1}{4}\right)^{5-1} = \frac{1}{8}$

4. $a_n = \dfrac{(-3)^n n}{n + 4}$

$a_1 = \dfrac{(-3)^1 \cdot 1}{1 + 4} = -\dfrac{3}{5}$

$a_2 = \dfrac{(-3)^2 \cdot 2}{2 + 4} = 3$

$a_3 = \dfrac{(-3)^3 \cdot 3}{3 + 4} = -\dfrac{81}{7}$

$a_4 = \dfrac{(-3)^4 \cdot 4}{4 + 4} = \dfrac{81}{2}$

$a_5 = \dfrac{(-3)^5 \cdot 5}{5 + 4} = -135$

5. $\sum_{k=1}^{4} 10k = \frac{4}{2}(10 + 40) = 100$

6. $\sum_{i=1}^{10} 4 = \frac{10}{2}(4 + 4) = 40$

7. $\sum_{j=1}^{5} \dfrac{60}{j + 1} = \dfrac{60}{2} + \dfrac{60}{3} + \dfrac{60}{4} + \dfrac{60}{5} + \dfrac{60}{6}$

$= 30 + 20 + 15 + 12 + 10$

$= 87$

8. $\sum_{n=1}^{4} \dfrac{12}{n} = 12 + 6 + 4 + 3 = 25$

9. $\sum_{n=1}^{5} (3n - 1) = \frac{5}{2}(2 + 14) = 40$

10. $\sum_{k=1}^{4} \left(k^2 - 1\right) = 0 + 3 + 8 + 15 = 26$

11. $\sum_{k=1}^{20} \dfrac{2}{3k}$

12. $\sum_{k=1}^{25} \dfrac{(-1)^{k+1}}{k^3}$

13. $\sum_{k=1}^{20} \dfrac{k - 1}{k}$

14. $\sum_{k=1}^{10} \dfrac{k^2}{2}$

15. $d = \frac{1}{2}$

16. $d = -6$

17. $a_n = a_1 + (n - 1)d$

$11 = 20 + (4 - 1)d$

$-9 = 3d$

$-3 = d$

$a_n = 20 + (n - 1)(-3)$

$a_n = 20 - 3n + 3$

$a_n = -3n + 23$

18. $a_1 = 32, d = -4$

$a_n = a_1 + (n - 1)d$

$a_n = 32 + (n - 1)(-4)$

$a_n = 32 - 4n + 4$

$a_n = -4n + 36$

19. $\sum_{n=1}^{200} 2n = \frac{200}{2}(2 + 400) = 100(402) = 40{,}200$

20. $a_n = a_1 + (n - 1)d$

$= 0.50 + (n - 1)(0.50)$

$= 0.50 + 0.50n - 0.50$

$a_n = 0.50n$

$\sum_{n=1}^{365} 0.50n = \frac{365}{2}(0.50 + 182.5) = \$33{,}397.50$

Section 11.3 Geometric Sequences and Series

1. 3, 6, 12, 24, …

$r = 2$ since

$\frac{6}{3} = 2, \frac{12}{6} = 2, \frac{24}{12} = 2.$

3. $1, \pi, \pi^2, \pi^3, \ldots$

$r = \pi$ since

$\dfrac{\pi}{1} = \pi, \dfrac{\pi^2}{\pi} = \pi, \dfrac{\pi^3}{\pi} = \pi.$

5. The sequence is not geometric, because

$\frac{15}{10} = \frac{3}{2}$ and $\frac{20}{15} = \frac{4}{3}$.

7. The sequence is geometric.

$r = \frac{1}{2}$ since

$\frac{32}{64} = \frac{1}{2}, \frac{16}{32} = \frac{1}{2}, \frac{8}{16} = \frac{1}{2}.$

9. The sequence is geometric: $r = -\frac{1}{2}$ since

$$\frac{a_{n+1}}{a_n} = \frac{4\left(-\frac{1}{2}\right)^{n+1}}{4\left(-\frac{1}{2}\right)^n} = -\frac{1}{2}.$$

11. $a_n = a_1 r^{n-1}$

$a_n = 1(2)^{n-1}$

13. $a_n = a_1 r^{n-1}$

$a_n = 9\left(\frac{2}{3}\right)^{n-1}$

15. $a_n = a_1 r^{n-1} \Rightarrow r = \frac{4}{2} = 2$

$a_n = 2(2)^{n-1}$

$a_7 = 2(2)^{7-1} = 128$

17. $a_n = a_1 r^{n-1} \Rightarrow r = \frac{2}{8} = \frac{1}{4}$

$a_n = 8\left(\frac{1}{4}\right)^{n-1}$

19. $\sum_{i=1}^{8} 4(3)^{i-1} = 4\left(\frac{3^8 - 1}{3 - 1}\right) = 4\left(\frac{6560}{2}\right) = 13{,}120$

21. $\sum_{i=1}^{10} 1(-3)^{i-1} = \left(\frac{(-3)^{10} - 1}{-3 - 1}\right) = -14{,}762$

23. $\sum_{i=1}^{15} 8\left(\frac{1}{2}\right)^{i-1} = 8\left(\frac{\left(\frac{1}{2}\right)^{15} - 1}{\frac{1}{2} - 1}\right) \approx 16.00$

25. $\sum_{i=1}^{6}\left(\frac{3}{4}\right)^n = \frac{3}{4}\left(\frac{\left(\frac{3}{4}\right)^6 - 1}{\left(\frac{3}{4}\right) - 1}\right) = \frac{3}{4}\left(\frac{-\frac{3367}{4096}}{-\frac{1}{4}}\right) = \frac{3}{4}\left(\frac{3367}{1024}\right) \approx 2.47$

27. $\sum_{i=1}^{10} 2^{i-1} = 1\left(\frac{2^{10} - 1}{2 - 1}\right) = \frac{1024 - 1}{1} = 1023$

29. $\sum_{i=1}^{12} 3\left(\frac{3}{2}\right)^{i-1} = 3\left(\frac{\left(\frac{3}{2}\right)^{12} - 1}{\frac{3}{2} - 1}\right) = 3\left(\frac{128.74634}{0.5}\right) \approx 772.48$

31. $\sum_{n=1}^{\infty} 8\left(\frac{3}{4}\right)^{n-1} = 8\left(\frac{1}{1 - \frac{3}{4}}\right) = 8\left(\frac{1}{\frac{1}{4}}\right) = 8(4) = 32$

33. $\sum_{n=0}^{\infty} 2\left(\frac{2}{3}\right)^n = \frac{2}{1 - \frac{2}{3}} = \frac{2}{\frac{1}{3}} = 6$

35. $\sum_{n=0}^{\infty}\left(-\frac{3}{7}\right)^n = \frac{1}{1 - \left(-\frac{3}{7}\right)} = \frac{1}{\frac{10}{7}} = \frac{7}{10}$

37. $\sum_{n=1}^{\infty}\left(\frac{1}{2}\right)^{n-1} = \frac{1}{1 - \frac{1}{2}} = \frac{1}{\frac{1}{2}} = 2$

39. $\sum_{n=1}^{\infty}\left(-\frac{1}{2}\right)^{n-1} = \frac{1}{1 - \left(\frac{1}{2}\right)} = \frac{1}{\frac{3}{2}} = \frac{2}{3}$

41. $\sum_{n=0}^{\infty}\left(\frac{1}{10}\right)^n = \frac{1}{1 - \frac{1}{10}} = \frac{1}{\frac{9}{10}} = \frac{10}{9}$

43. Total salary $= \sum_{n=1}^{40} 30{,}000(1.05)^n$

$= 30{,}000\left(\frac{1.05^{40} - 1}{1.05 - 1}\right) \approx \$3{,}623{,}993.23$

45. $a_n = 0.01(2)^{n-1}$

(a) Total income $= \sum_{n=1}^{29} 0.01(2)^{n-1}$

$= 0.01\left[\frac{2^{29} - 1}{2 - 1}\right]$

$\approx \$5{,}368{,}709.11$

(b) Total income $= \sum_{n=1}^{30} 0.01(2)^{n-1}$

$= 0.01\left[\frac{2^{30} - 1}{2 - 1}\right]$

$\approx \$10{,}737{,}418.23$

47. $A = P\left(1 + \frac{r}{n}\right)^{nt}$

$a_{120} = 50\left(1 + \frac{0.09}{12}\right)^{12(10)} = 50(1.0075)^{120}$

$a_1 = 50(1.0075)^1$

$A = \left[50(1.0075)\right]\left[\frac{1.0075^{120} - 1}{1.0075 - 1}\right]$

$\approx \$9748.28$

49. $A = P\left(1 + \frac{r}{n}\right)^{nt}$

$a_{480} = 30\left(1 + \frac{0.08}{12}\right)^{12(40)} = 30\left(\frac{151}{150}\right)^{480}$

$a_1 = 30\left(\frac{151}{150}\right)^1$

$\text{Balance} = \left[30\left(\frac{151}{150}\right)\right]\left[\frac{\left(\frac{151}{150}\right)^{480} - 1}{\left(\frac{151}{150}\right) - 1}\right] \approx \$105{,}428.44$

51. $A = P\left(1 + \frac{r}{n}\right)^{nt}$

$a_{360} = 100\left(1 + \frac{0.06}{12}\right)^{12(30)} = 100(1.005)^{360}$

$a_1 = 100(1.005)^1$

$A = \left[100(1.005)\right]\left[\frac{1.005^{360} - 1}{1.005 - 1}\right] \approx \$100{,}953.76$

53. The partial sum $\sum_{i=1}^{6} 3(2)^{i-1}$ is the sum of the first 6 terms of the geometric sequence whose nth term is $a_n = 3(2)^{i-1}$.

55. The common ratio of the geometric series with the sum $\sum_{n=0}^{\infty} 5\left(\frac{4}{5}\right)^n$ is $\frac{4}{5}$.

57. $a_n = a_1r^{n-1}$

$a_n = 5(-2)^{n-1}$

$a_1 = 5(-2)^{1-1} = 5(-2)^0 = 5(1) = 5$

$a_2 = 5(-2)^{2-1} = 5(-2)^1 = 5(-2) = -10$

$a_3 = 5(-2)^{3-1} = 5(-2)^2 = 5(4) = 20$

$a_4 = 5(-2)^{4-1} = 5(-2)^3 = 5(-8) = -40$

$a_5 = 5(-2)^{5-1} = 5(-2)^4 = 5(16) = 80$

59. $a_n = a_1r^{n-1}$

$a_n = (-4)\left(-\frac{1}{2}\right)^{n-1}$

$a_1 = (-4)\left(-\frac{1}{2}\right)^{1-1} = -4$

$a_2 = (-4)\left(-\frac{1}{2}\right)^{2-1} = (-4)\left(-\frac{1}{2}\right) = 2$

$a_3 = (-4)\left(-\frac{1}{2}\right)^{3-1} = (-4)\left(\frac{1}{4}\right) = -1$

$a_4 = (-4)\left(-\frac{1}{2}\right)^{4-1} = (-4)\left(-\frac{1}{8}\right) = \frac{1}{2}$

$a_5 = (-4)\left(-\frac{1}{2}\right)^{5-1} = (-4)\left(\frac{1}{16}\right) = -\frac{1}{4}$

61. $\sum_{n=0}^{\infty} 8\left(\frac{3}{4}\right)^n = \frac{8}{1 - \frac{3}{4}} = \frac{8}{\frac{1}{4}} = 32$

63. Total area of shaded region

$= \frac{1}{4} + \frac{3}{4} \cdot \frac{1}{4} + \frac{9}{16} \cdot \frac{1}{4} = \frac{16}{64} + \frac{12}{64} + \frac{9}{64} = \frac{37}{64}$

or

Total area of shaded region

$= \sum_{i=1}^{3} \frac{1}{4}\left(\frac{3}{4}\right)^{n-1} = \frac{1}{4}\left[\frac{\left(\frac{3}{4}\right)^3 - 1}{\frac{3}{4} - 1}\right] = \frac{37}{64} \approx 0.578 \text{ square unit}$

65. $a_n = a_1r^n$

$T_n = 70(0.8)^n$

$T_6 = 70(0.8)^6 \approx 18.4°$

67. The general formula for the nth term of a geometric sequence is $a_n = a_1r^{n-1}$.

69. An arithmetic sequence has a common difference between consecutive terms and a geometric sequence has a common ratio between terms.

71. The terms of a geometric sequence decrease when $a_1 > 0$ and $a < r < 1$ because raising a real number between 0 and 1 to higher powers yields smaller numbers.

73. $\begin{cases} y = 2x^2 \\ y = 2x + 4 \end{cases}$

$$\begin{aligned} 2x^2 &= 2x + 4 \\ 2x^2 - 2x - 4 &= 0 \\ x^2 - x - 2 &= 0 \\ (x - 2)(x + 1) &= 0 \end{aligned}$$

$x = 2 \qquad x = -1$

$y = 2(2)^2 = 8 \qquad y = 2(-1)^2 = 2$

$(2, 8), (-1, 2)$

75. $A = P\left(1 + \dfrac{r}{n}\right)^{nt}$

$$\begin{aligned} 2219.64 &= 1000\left(1 + \frac{r}{12}\right)^{12(10)} \\ 2.21964 &= \left(1 + \frac{r}{12}\right)^{120} \\ 2.21964^{1/120} &= 1 + \frac{r}{12} \\ 12\left(2.21964^{1/120} - 1\right) &= r \\ 0.08 &\approx r \\ 8\% &\approx r \end{aligned}$$

77. $A = P\left(1 + \dfrac{r}{n}\right)^{nt}$

$$\begin{aligned} 10{,}619.63 &= 2500\left(1 + \frac{r}{1}\right)^{1(20)} \\ 4.247852 &= (1 + r)^{20} \\ 4.247852^{1/20} &= 1 + r \\ 4.247852^{1/20} - 1 &= r \\ 0.075 &\approx r \\ 7.5\% &\approx r \end{aligned}$$

79. $\dfrac{x^2}{16} - \dfrac{y^2}{9} = 1$

Horizontal axis

Vertices: $(\pm 4, 0)$

Asymptotes: $y = \pm\dfrac{3}{4}x$

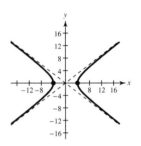

Section 11.4 The Binomial Theorem

1. ${}_6C_4 = {}_6C_2 = \dfrac{6 \cdot 5}{2 \cdot 1} = 15$

3. ${}_{10}C_5 = \dfrac{10 \cdot 9 \cdot 8 \cdot 7 \cdot 6}{5 \cdot 4 \cdot 3 \cdot 2 \cdot 1} = 252$

5. ${}_{12}C_{12} = \dfrac{12!}{0!12!} = 1$

7. ${}_{20}C_6 = \dfrac{20!}{14!6!} = \dfrac{20 \cdot 19 \cdot 18 \cdot 17 \cdot 16 \cdot 15}{6 \cdot 5 \cdot 4 \cdot 3 \cdot 2 \cdot 1} = 38{,}760$

9. ${}_{20}C_{14} = \dfrac{20!}{6!14!} = \dfrac{20 \cdot 19 \cdot 18 \cdot 17 \cdot 16 \cdot 15}{6 \cdot 5 \cdot 4 \cdot 3 \cdot 2 \cdot 1} = 38{,}760$

11.

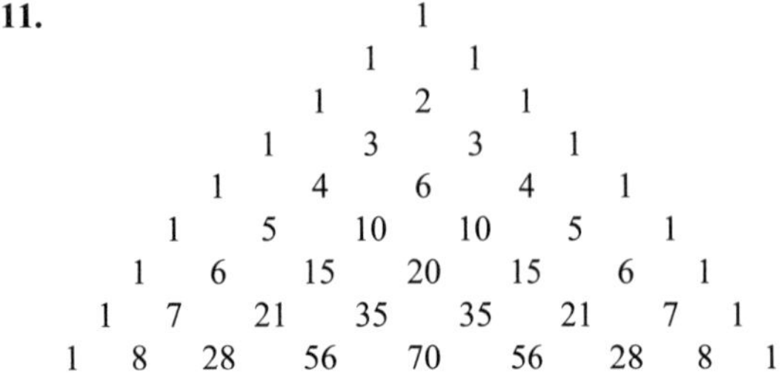

1
1 1
1 2 1
1 3 3 1
1 4 6 4 1
1 5 10 10 5 1
1 6 15 20 15 6 1
1 7 21 35 35 21 7 1
1 8 28 56 70 56 28 8 1

13. ${}_6C_2 = 15$

Row 6: 1 6 15 20 15 6 1

↑ entry 2 (15)

15. ${}_7C_3 = 35$

Row 7: 1 7 21 35 35 21 7 1

↑ entry 3 (35)

17. ${}_8C_4 = 70$

Row 8: 1 8 28 56 70 56 28 8 1

↑ entry 4 (70)

19. ${}_5C_3$

Row 5: 1 5 10 10 5 1

↑ entry 3 (10)

${}_5C_3 = 10$

21. ${}_7C_4$

Row 7: 1 7 21 35 35 21 7 1

↑ entry 4

${}_7C_4 = 35$

23. $(t+5)^3 = (1)t^3 + (3)t^2(5) + (3)t(5)^2 + (1)5^3$

$= t^3 + 15t^2 + 75t + 125$

25. $(m-n)^5 = (1)m^5 + (5)m^4(-n) + (10)m^3(-n)^2 + 10m^2(-n)^3 + 5m(-n)^4 + 1(-n)^5$

$= m^5 - 5m^4n + 10m^3n^2 - 10m^2n^3 + 5mn^4 - n^5$

27. $(x+3)^6 = 1x^6 + 6x^5(3) + 15x^4(3)^2 + 20x^3(3)^3 + 15x^2(3)^4 + 6x(3)^5 + 1(3)^6$

$= x^6 + 18x^5 + 135x^4 + 540x^3 + 1215x^2 + 1458x + 729$

29. $(u-v)^3 = 1(u^3) + 3(u^2)(-v) + 3(u)(-v)^2 + 1(-v)^3$

$= u^3 - 3u^2v + 3uv^2 - v^3$

31. $(3a-1)^5 = (1)(3a)^5 + (5)(3a)^4(-1) + (10)(3a)^3(-1)^2 + (10)(3a)^2(-1)^3 + (5)(3a)(-1)^4 + (1)(-1)^5$

$= 243a^5 - 405a^4 + 270a^3 - 90a^2 + 15a - 1$

33. $(2y+z)^6 = (1)(2y)^6 + 6(2y)^5z + 15(y)^4z^2 + 20(2y)^3z^3 + 15(2y)^2z^4 + 6(2y)z^5 + 1z^6$

$= 64y^6 + 192y^5z + 240y^4z^2 + 160y^3z^3 + 60y^2z^4 + 12yz^5 + z^6$

35. $(x^2+2)^4 = 1(x^2)^4 + 4(x^2)^3(2) + 6(x^2)^2(2)^2 + 4(x^2)(2)^3 + 1(2)^4$

$= x^8 + 8x^6 + 24x^4 + 32x^2 + 16$

37. $(3a+2b)^4 = 1(3a)^4 + 4(3a)^3(2b) + 6(3a)^2(2b)^2 + 4(3a)(2b)^3 + 1(2b)^4$

$= 81a^4 + 216a^3b + 216a^2b^2 + 96ab^3 + 16b^4$

39. $\left(x + \frac{2}{y}\right)^4 = 1(x)^4 + 4(x)^3\left(\frac{2}{y}\right) + 6(x)^2\left(\frac{2}{y}\right)^2 + 4(x)\left(\frac{2}{y}\right)^3 + 1\left(\frac{2}{y}\right)^4$

$= x^4 + \frac{8x^3}{y} + \frac{24x^2}{y^2} + \frac{32x}{y^3} + \frac{16}{y^4}$

41. $(2x^2-y)^5 = 1(2x^2)^5 + 5(2x^2)^4(-y) + 10(2x^2)^3(-y)^2 + 10(2x^2)^2(-y)^3 + 5(2x^2)(-y)^4 + 1(-y)^5$

$= 32x^{10} - 80x^8y + 80x^6y^2 - 40x^4y^3 + 10x^2y^4 - y^5$

43. $(x+y)^{10}$, 4th term

$(r+1)$th term $= {}_nC_rx^{n-r}y^r$

$r+1 = 4 \quad n = 10 \quad x = x \quad y = y$

$r = 3$

4th term $= {}_{10}C_3x^{10-3}y^3 = 120x^7y^3$

45. $(a+6b)^9$, 5th term

$(r+1)$th term $= {}_nC_rx^{n-r}y^r$

$r+1 = 5 \quad n = 9 \quad x = a \quad y = 6b$

$r = 4$

5th term $= {}_9C_4a^{9-4}(6b)^4$

$= 126a^5(1296b^4)$

$= 163{,}296a^5b^4$

47. $(4x + 3y)^9$, 8th terms

$(r + 1)\text{th term} = {}_nC_r x^{n-r} y^r$

$r + 1 = 8 \quad n = 9 \quad x = 4x \quad y = 3y$

$r = 7$

$$\begin{aligned}\text{8th term} &= {}_9C_7(4x)^2(3y)^7\\ &= 36(16x^2)(2187y^7)\\ &= 1{,}259{,}712x^2y^7\end{aligned}$$

49. ${}_nC_r x^{n-r} y^r$

$n = 10 \quad n - r = 7 \quad r = 3 \quad x = x \quad y = 1$

${}_{10}C_3 x^7 1^3$

$${}_{10}C_3 = \frac{10 \cdot 9 \cdot 8}{3 \cdot 2 \cdot 1} = 120$$

51. $${}_{12}C_3 = \frac{12 \cdot 11 \cdot 10}{3 \cdot 2 \cdot 1} = 220$$

$\text{Coefficient} = (3)^3\,{}_{12}C_3 = 27(220) = 5940$

53. Pascal's Triangle is formed by making the first and last numbers in each row 1. Every other number in the row is formed by adding the two numbers immediately above the number.

55. The signs in the expansion of $(x + y)^n$ are all positive.

The signs in the expansion of $(x - y)^n$ alternate.

57. $$\begin{aligned}\left(\sqrt{x} + 5\right)^3 &= \left(\sqrt{x}\right)^3 + 3\left(\sqrt{x}\right)^2(5) + 3\left(\sqrt{x}\right)(5^2) + 5^3\\ &= x^{3/2} + 15x + 75x^{1/2} + 125\end{aligned}$$

59. $$\left(x^{2/3} - y^{1/3}\right)^3 = \left(x^{2/3}\right)^3 - 3\left(x^{2/3}\right)^2\left(y^{1/3}\right) + 3\left(x^{2/3}\right)\left(y^{1/3}\right)^2 - \left(y^{1/3}\right)^3 = x^2 - 3x^{4/3}y^{1/3} + 3x^{2/3}y^{2/3} - y$$

61. $$\begin{aligned}\left(3\sqrt{t} + \sqrt[4]{t}\right)^4 &= \left(3\sqrt{t}\right)^4 + 4\left(3\sqrt{t}\right)^3\left(\sqrt[4]{t}\right) + 6\left(3\sqrt{t}\right)^2\left(\sqrt[4]{t}\right)^2 + 4\left(3\sqrt{t}\right)\left(\sqrt[4]{t}\right)^3 + \left(\sqrt[4]{t}\right)^4\\ &= 81t^2 + 108t^{7/4} + 54t^{3/2} + 12t^{5/4} + t\end{aligned}$$

63. Keystrokes:

30 [MATH] [PRB 3] 6 [ENTER] ${}_{30}C_6 = 593{,}775$

65. Keystrokes:

52 [MATH] [PRB 3] 5 [ENTER] ${}_{52}C_5 = 2{,}598{,}960$

67. *Keystrokes:*

800 [MATH] [PRB 3] 797 [ENTER]

${}_{800}C_{797} = 85{,}013{,}600$

69. $$\begin{aligned}(1 + i)^4 &= 1^4 + 4 \cdot 1^3 i + 6 \cdot 1^2 \cdot i^2 + 4 \cdot 1 \cdot i^3 + i^4\\ &= 1 + 4i - 6 - 4i + 1\\ &= -4\end{aligned}$$

71. $$\begin{aligned}(1.02)^8 &= (1 + 0.02)^8\\ &= (1)^8 + 8(1)^7(0.02) + 28(1)^6(0.02)^2 + 56(1)^5(0.02)^3 + \dots\\ &\approx 1 + 0.16 + 0.0112 + 0.000448\\ &\approx 1.172\end{aligned}$$

73. $$\begin{aligned}(2.99)^{12} &= (3 - 0.01)^{12}\\ &= 1(3)^{12} - 12(3)^{11}(0.01) + 66(3)^{10}(0.01)^2 - 220(3)^9(0.01)^3 + 495(3)^8(0.01)^4 - 792(3)^7(0.01)^5 + \dots\\ &\approx 531{,}441 - 21{,}257.64 + 389.7234 - 4.33026 + 0.03247695 - 0.0001732104\\ &\approx 510{,}568.785\end{aligned}$$

75. $$\begin{aligned}\left(\tfrac{1}{2} + \tfrac{1}{2}\right)^5 &= 1\left(\tfrac{1}{2}\right)^5 + 5\left(\tfrac{1}{2}\right)^4\left(\tfrac{1}{2}\right) + 10\left(\tfrac{1}{2}\right)^3\left(\tfrac{1}{2}\right)^2 + 10\left(\tfrac{1}{2}\right)^2\left(\tfrac{1}{2}\right)^3 + 5\left(\tfrac{1}{2}\right)\left(\tfrac{1}{2}\right)^4 + 1\left(\tfrac{1}{2}\right)^5\\ &= \tfrac{1}{32} + \tfrac{5}{32} + \tfrac{10}{32} + \tfrac{10}{32} + \tfrac{5}{32} + \tfrac{1}{32}\end{aligned}$$

77. $$\left(\tfrac{1}{4} + \tfrac{3}{4}\right)^4 = 1\left(\tfrac{1}{4}\right)^4 + 4\left(\tfrac{1}{4}\right)^3\left(\tfrac{3}{4}\right) + 6\left(\tfrac{1}{4}\right)^2\left(\tfrac{3}{4}\right)^2 + 4\left(\tfrac{1}{4}\right)\left(\tfrac{3}{4}\right)^3 + 1\left(\tfrac{3}{4}\right)^4 = \tfrac{1}{256} + \tfrac{12}{256} + \tfrac{54}{256} + \tfrac{108}{256} + \tfrac{81}{256}$$

79. The sum of the numbers in each row is a power of 2. Because Row 2 is $1 + 2 + 1 = 4 = 2^2$, Row n is 2^n.

81. No. The coefficient of x^4y^6 is ${}_{10}C_6$ because $n = 10$ and $r = 6$.

83. The value for r is 6 because $r + 1 = 7$.

85. $\sum_{i=1}^{15} (2 + 3i) = \frac{15}{2}(5 + 47) = \frac{15}{2}(52) = 390$

$a_1 = 2 + 3(1) = 5, \quad a_n = 2 + 3(15) = 47$

87. $\sum_{k=1}^{8} 5^{k-1} = 1\left[\frac{5^8 - 1}{5 - 1}\right] = 97{,}656$

Review Exercises for Chapter 11

1. $a_n = 3n + 5$

$a_1 = 3(1) + 5 = 8$

$a_2 = 3(2) + 5 = 11$

$a_3 = 3(3) + 5 = 14$

$a_4 = 3(4) + 5 = 17$

$a_5 = 3(5) + 5 = 20$

3. $a_n = \frac{n}{3n - 1}$

$a_1 = \frac{1}{3(1) - 1} = \frac{1}{2}$

$a_2 = \frac{2}{3(2) - 1} = \frac{2}{5}$

$a_3 = \frac{3}{3(3) - 1} = \frac{3}{8}$

$a_4 = \frac{4}{3(4) - 1} = \frac{4}{11}$

$a_5 = \frac{5}{3(5) - 1} = \frac{5}{14}$

5. $a_n = (n + 1)!$

$a_1 = (1 + 1)! = 2! = 2 \cdot 1 = 2$

$a_2 = (2 + 1)! = 3! = 3 \cdot 2 \cdot 1 = 6$

$a_3 = (3 + 1)! = 4! = 4 \cdot 3 \cdot 2 \cdot 1 = 24$

$a_4 = (4 + 1)! = 5! = 5 \cdot 4 \cdot 3 \cdot 2 \cdot 1 = 120$

$a_5 = (5 + 1)! = 6! = 6 \cdot 5 \cdot 4 \cdot 3 \cdot 2 \cdot 1 = 720$

7. $a_n = \frac{n!}{2n}$

$a_1 = \frac{1!}{2 \cdot 1} = \frac{1}{2}$

$a_2 = \frac{2!}{2 \cdot 2} = \frac{2}{4} = \frac{1}{2}$

$a_3 = \frac{3!}{2 \cdot 3} = \frac{3 \cdot 2 \cdot 1}{2 \cdot 3} = 1$

$a_4 = \frac{4!}{2 \cdot 4} = \frac{4 \cdot 3 \cdot 2 \cdot 1}{2 \cdot 4} = 3$

$a_5 = \frac{5!}{2 \cdot 5} = \frac{5 \cdot 4 \cdot 3 \cdot 2 \cdot 1}{2 \cdot 5} = 12$

9. $a_n = 3n + 1$

11. $a_n = \frac{1}{n^2 + 1}$

13. $a_n = -2n + 5$

15. $a_n = \frac{3n^2}{n^2 + 1}$

17. $\sum_{k=1}^{4} 7 = 7 + 7 + 7 + 7 = 28$

19. $\sum_{i=1}^{5} \frac{i - 2}{i + 1} = -\frac{1}{2} + 0 + \frac{1}{4} + \frac{2}{5} + \frac{1}{2}$

$= \frac{1}{4} + \frac{2}{5}$

$= \frac{5}{20} + \frac{8}{20}$

$= \frac{13}{20}$

21. $\sum_{k=1}^{4} (5k - 3)$

23. $\sum_{k=1}^{6} \frac{1}{3k}$

25. 50, 44.5, 39, 33.5, 28, …

$44.5 - 50 = -5.5$

$39 - 44.5 = -5.5$

$33.5 - 39 = -5.5$

$28 - 33.5 = -5.5$

$d = -5.5$

27. $a_1 = 132 - 5(1) = 127$

$a_2 = 132 - 5(2) = 122$

$a_3 = 132 - 5(3) = 117$

$a_4 = 132 - 5(4) = 112$

$a_5 = 132 - 5(5) = 107$

29. $a_1 = \frac{1}{3}(1) + \frac{5}{3} = \frac{6}{3} = 2$

$a_2 = \frac{1}{3}(2) + \frac{5}{3} = \frac{7}{3}$

$a_3 = \frac{1}{3}(3) + \frac{5}{3} = \frac{8}{3}$

$a_4 = \frac{1}{3}(4) + \frac{5}{3} = \frac{9}{3} = 3$

$a_5 = \frac{1}{3}(5) + \frac{5}{3} = \frac{10}{3}$

31. $a_1 = 80$

$a_2 = 80 - \frac{5}{2} = \frac{160}{2} - \frac{5}{2} = \frac{155}{2}$

$a_3 = \frac{155}{2} - \frac{5}{2} = \frac{150}{2} = 75$

$a_4 = \frac{150}{2} - \frac{5}{2} = \frac{145}{2}$

$a_5 = \frac{145}{2} - \frac{5}{2} = \frac{140}{2} = 70$

33. $a_n = dn + c$

$10 = 4(1) + c$

$6 = c$

$a_n = 4n + 6$

35. $a_n = dn + c$

$1000 = -50(1) + c$

$1050 = c$

$a_n = -50n + 1050$

37. $\sum_{k=1}^{12} (7k - 5) = \frac{12}{2}(2 + 79) = 486$

39. $\sum_{j=1}^{120} \left(\frac{1}{4}j + 1\right) = \frac{120}{2}\left(\frac{5}{4} + 31\right) = 60\left(\frac{129}{4}\right) = 1935$

41. $a_n = a_1 + (n - 1)d$

$6.5 = 5.25 + (2 - 1)d$

$6.5 = 5.25 + d$

$1.25 = d$

$a_n = a_1 + (n - 1)d$

$a_{60} = 5.25 + (60 - 1)(1.25)$

$a_{60} = 79$

$\sum_{i=1}^{n} a_i = \frac{n}{2}(a_1 + a_n)$

$\sum_{i=1}^{60} a_i = \frac{60}{2}(5.25 + 79) = 2527.5$

43. $\sum_{n=1}^{50} 4n = \frac{50}{2}(4 + 200) = 5100$

45. $\sum_{n=1}^{12} (3n + 19) = \frac{12}{2}(22 + 55) = 462$ seats

47. $8, 20, 50, 125, \frac{625}{2}, \ldots$

$r = \frac{5}{2}$ since

$\frac{20}{8} = \frac{5}{2}, \frac{50}{20} = \frac{5}{2}, \frac{125}{50} = \frac{5}{2}, \frac{\frac{625}{2}}{125} = \frac{5}{2}.$

49. $a_n = a_1r^{n-1}$

$a_n = 10(3)^{n-1}$

$a_1 = 10(3)^{1-1} = 10$

$a_2 = 10(3)^{2-1} = 30$

$a_3 = 10(3)^{3-1} = 90$

$a_4 = 10(3)^{4-1} = 270$

$a_5 = 10(3)^{5-1} = 810$

51. $a_n = a_1r^{n-1}$

$a_n = 100\left(-\frac{1}{2}\right)^{n-1}$

$a_1 = 100\left(-\frac{1}{2}\right)^{1-1} = 100$

$a_2 = 100\left(-\frac{1}{2}\right)^{2-1} = -50$

$a_3 = 100\left(-\frac{1}{2}\right)^{3-1} = 25$

$a_4 = 100\left(-\frac{1}{2}\right)^{4-1} = -12.5$

$a_5 = 100\left(-\frac{1}{2}\right)^{5-1} = 6.25$

53. $a_n = a_1r^{n-1}$

$a_n = 4\left(\frac{3}{2}\right)^{n-1}$

$a_1 = 4\left(\frac{3}{2}\right)^{1-1} = 4\left(\frac{3}{2}\right)^0 = 4(1) = 4$

$a_2 = 4\left(\frac{3}{2}\right)^{2-1} = 4\left(\frac{3}{2}\right)^1 = 4\left(\frac{3}{2}\right) = 6$

$a_3 = 4\left(\frac{3}{2}\right)^{3-1} = 4\left(\frac{3}{2}\right)^2 = 4\left(\frac{9}{4}\right) = 9$

$a_4 = 4\left(\frac{3}{2}\right)^{4-1} = 4\left(\frac{3}{2}\right)^3 = 4\left(\frac{27}{8}\right) = \frac{27}{2}$

$a_5 = 5\left(\frac{3}{2}\right)^{5-1} = 4\left(\frac{3}{2}\right)^4 = 4\left(\frac{81}{16}\right) = \frac{81}{4}$

55. $a_n = a_1r^{n-1}$

$a_n = 1\left(-\frac{2}{3}\right)^{n-1}$

57. $a_n = a_1r^{n-1}$

$a_n = 24(3)^{n-1}$

$r = \frac{a_2}{a_1} = \frac{72}{24} = 3$

59. $a_n = a_1 r^{n-1}$

$a_n = 12\left(-\frac{1}{2}\right)^{n-1}$

61. $a_n = a_1 r^{n-1} \Rightarrow r = \dfrac{280}{200} = \dfrac{7}{5}$

$a_n = 200\left(\dfrac{7}{5}\right)^{n-1}$

$\displaystyle\sum_{i=1}^{n} a_1 r^{i-1} = a_1\left(\frac{r^{n-1}}{r-1}\right)$

$\displaystyle\sum_{i=1}^{12} 200\left(\frac{7}{5}\right)^{i-1} = 200\left(\frac{\left(\frac{7}{5}\right)^{12} - 1}{\frac{7}{5} - 1}\right) \approx 27{,}846.96$

63. $a_n = a_1 r^{n-1} \Rightarrow r = -\dfrac{36}{27} = -\dfrac{4}{3}$

$a_n = 27\left(-\dfrac{4}{3}\right)^{n-1}$

$\displaystyle\sum_{i=1}^{n} a_1 r^{i-1} = a_1\left(\frac{r^{n-1}}{r-1}\right)$

$\displaystyle\sum_{i=1}^{14} 27\left(-\frac{4}{3}\right)^{i-1} = 27\left(\frac{\left(-\frac{4}{3}\right)^{14} - 1}{\left(-\frac{4}{3}\right) - 1}\right) \approx -637.85$

65. $\displaystyle\sum_{n=1}^{12} 2^n = 2\left(\frac{2^{12} - 1}{2 - 1}\right) = 8190$

67. $\displaystyle\sum_{k=1}^{8} 5\left(-\frac{3}{4}\right)^k = -\frac{15}{4}\left(\frac{\left(-\frac{3}{4}\right)^8 - 1}{-\frac{3}{4} - 1}\right) \approx -1.928$

69. $\displaystyle\sum_{i=1}^{\infty} \left(\frac{7}{8}\right)^{i-1} = \frac{1}{1 - \frac{7}{8}} = \frac{1}{\frac{1}{8}} = 8$

71. $\displaystyle\sum_{k=0}^{\infty} 4\left(\frac{2}{3}\right)^k = \frac{4}{1 - \frac{2}{3}} = \frac{4}{\frac{1}{3}} = 12$

73. (a) $a_n = a_1 r^n$

$a_n = 120{,}000(0.7)^n$

(b) $a_5 = 120{,}000(0.7)^5 = \$20{,}168.40$

75. $\displaystyle\sum_{i=1}^{90} 1000(1.125)^i = 1000\left[\frac{1.125^{90} - 1}{1.125 - 1}\right]$

$\approx 321{,}222{,}672$ visitors

77. ${}_8C_3 = \dfrac{8!}{3!5!} = \dfrac{8 \cdot 7 \cdot 6 \cdot 5!}{3 \cdot 2 \cdot 5!} = 56$

79. ${}_{15}C_4 = \dfrac{15!}{11!4!} = \dfrac{15 \cdot 14 \cdot 13 \cdot 12}{4 \cdot 3 \cdot 2 \cdot 1} = 1365$

81. ${}_{40}C_4 = \dfrac{40!}{36!4!} = 91{,}390$

83. ${}_{25}C_6 = \dfrac{25!}{19!6!} = 177{,}100$

85. ${}_4C_2$

Row 4: 1 4 6 4 1

↑ entry 2 (under 6)

${}_4C_2 = 6$

87. ${}_{10}C_3$

Row 10: 1 10 45 120 210 252 210 120 45 10 1

↑ entry 3 (under 120)

${}_{10}C_3 = 120$

89. $(x-5)^4 = 1(x)^4 + 4(x)^3(-5) + 6(x)^2(-5)^2 + 4(x)(-5)^3 + 1(-5)^4 = x^4 - 20x^3 + 150x^2 - 500x + 625$

91. $(5x+2)^3 = (1)(5x)^3 + (3)(5x)^2(2) + (3)(5x)(2)^2 + 1(2)^3 = 125x^3 + 150x^2 + 60x + 8$

93. $(x+1)^{10} = 1x^{10} + 10x^9(1) + 45x^8(1)^2 + 120x^7(1)^3 + 210x^6(1)^4 + 252x^5(1)^5 + 210x^4(1)^6 + 120x^3(1)^7 + 45x^2(1)^8$
$+ 10x(1)^9 + 1(1)^{10}$
$= x^{10} + 10x^9 + 45x^8 + 120x^7 + 210x^6 + 252x^5 + 210x^4 + 120x^3 + 45x^2 + 10x + 1$

95. $(3x - 2y)^4 = 1(3x)^4 + 4(3x)^3(-2y) + 6(3x)^2(-2y)^2 + 4(3x)(-2y)^3 + (-2y)^4$
$= 81x^4 - 216x^3y + 216x^2y^2 - 96xy^3 + 16y^4$

97. $(u^2 + v^3)^5 = (1)(u^2)^5 + (5)(u^2)^4(v^3) + (10)(u^2)^3(v^3)^2 + (10)(u^2)^2(v^3)^3 + (5)(u^2)(v^3)^4 + (1)(v^3)^5$
$= u^{10} + 5u^8v^3 + 10u^6v^6 + 10u^4v^9 + 5u^2v^{12} + v^{15}$

99. $(x + 2)^{10}$, 7th term

$(r + 1)$th term $= {}_nC_rx^{n-r}y^r$

$r + 1 = 7,\ n = 10,\ x = x,\ y = 2$

$r = 6$

7th term $= {}_{10}C_6x^{10-6}(2)^6 = 210x^4(64) = 13{,}440x^4$

101. ${}_nC_rx^{n-r}y^r$

$n = 10,\ n - r = 5,\ r = 5,\ x = x,\ y = (-3)$

${}_{10}C_5 = 252 \cdot (-3)^5 = -61{,}236$

Chapter Test for Chapter 11

1. $a_n = \left(-\frac{3}{5}\right)^{n-1}$

$a_1 = \left(-\frac{3}{5}\right)^{1-1} = 1$

$a_2 = \left(-\frac{3}{5}\right)^{2-1} = -\frac{3}{5}$

$a_3 = \left(-\frac{3}{5}\right)^{3-1} = \frac{9}{25}$

$a_4 = \left(-\frac{3}{5}\right)^{4-1} = -\frac{27}{125}$

$a_5 = \left(-\frac{3}{5}\right)^{5-1} = \frac{81}{625}$

2. $a_n = 3n^2 - n$

$a_1 = 3(1)^2 - 1 = 2$

$a_2 = 3(2)^2 - 2 = 12 - 2 = 10$

$a_3 = 3(3)^2 - 3 = 27 - 3 = 24$

$a_4 = 3(4)^2 - 4 = 48 - 4 = 44$

$a_5 = 3(5)^2 - 5 = 75 - 5 = 70$

3. $\sum_{n=1}^{12} 5 = 12(5) = 60$

4. $\sum_{k=0}^{8} (2k - 3) = -3 + (-1) + 1 + 3 + 5 + 7 + 9 + 11 + 13$
$= 45$

5. $\sum_{n=1}^{5} (3 - 4n) = \frac{5}{2}[-1 + (-17)] = -45$

6. $\sum_{n=1}^{12} \frac{2}{3n + 1}$

7. $\sum_{k=1}^{6} \left(\frac{1}{2}\right)^{2k-2}$

8. $a_n = a_1 + (n - 1)d$

$a_n = 12 + (n - 1)4 = 12 + 4n - 4 = 4n + 8$

$a_1 = 4(1) + 8 = 12$

$a_2 = 4(2) + 8 = 16$

$a_3 = 4(3) + 8 = 20$

$a_4 = 4(4) + 8 = 24$

$a_5 = 4(5) + 8 = 28$

9. $a_n = a_1 + (n - 1)d$

$a_n = 5000 + (n - 1)(-100)$

$a_n = 50000 - 100n + 100$

$a_n = -100n + 5100$

10. $\sum_{n=1}^{50} 3n = \frac{50}{2}(3 + 150) = 3825$

11. $-4, 3, -\frac{9}{4}, \frac{27}{16}, \ldots$

$r = -\frac{3}{4}$ since

$\frac{3}{-4} = -\frac{3}{4}, \frac{-\frac{9}{4}}{3} = -\frac{3}{4}, \frac{\frac{27}{16}}{-\frac{9}{4}} = -\frac{3}{4}.$

12. $a_n = a_1r^{n-1}$

$a_n = 4\left(\frac{1}{2}\right)^{n-1}$

13. $\sum_{n=1}^{8} 2(2^n) = 4\left(\frac{2^8 - 1}{2 - 1}\right) = 1020$

14. $\sum_{n=1}^{10} 3\left(\frac{1}{2}\right)^n = \frac{3}{2}\left(\frac{\frac{1^{10}}{2} - 1}{\frac{1}{2} - 1}\right) = \frac{3069}{1024}$

15. $\sum_{i=1}^{\infty} \left(\frac{1}{2}\right)^i = \frac{\frac{1}{2}}{1 - \frac{1}{2}} = \frac{\frac{1}{2}}{\frac{1}{2}} = 1$

16. $\sum_{i=1}^{\infty} 10(0.4)^{i-1} = \frac{10}{1 - 0.4} = \frac{10}{0.6} = \frac{100}{6} = \frac{50}{3}$

17. ${}_{20}C_3 = \frac{20 \cdot 19 \cdot 18}{3 \cdot 2 \cdot 1} = 1140$

18. $(x - 2)^5 = 1(x^5) - 5x^4(2) + 10x^3(2)^2 - 10x^2(2)^3 + 5x(2)^4 - 1(2)^5$

$= x^5 - 10x^4 + 40x^3 - 80x^2 + 80x - 32$

19. The coefficient of x^3y^5 in expansion of $(x + y)^8$ is 56, since ${}_8C_3 = 56$.

20. $14.7 - 4.9 = 9.8$

$24.5 - 14.7 = 9.8$

So, $d = 9.8$.

$a_n = a_1 + (n - 1)d$

$a_n = 4.9 + (n - 1)9.8$

$a_n = 4.9 + 9.8n - 9.8$

$a_n = 9.8n - 4.9$

$\sum_{n=1}^{10} 9.8n - 4.9 = \frac{10}{2}(4.9 + 93.1) = 490$ meters

21. Balance $= 80\left(1 + \frac{0.048}{12}\right)^1 + \cdots + 80\left(1 + \frac{0.048}{12}\right)^{540}$

$= 80(1.004)\left(\frac{1.004^{540} - 1}{1.004 - 1}\right)$

$= \$153{,}287.87$